Communications and Control Engineering

AF768281

Springer-Verlag London Ltd.

Published titles include:

Loop Transfer Recovery: Analysis and Design
A. Saberi, B.M. Chen and P. Sannuti

Markov Chains and Stochastic Stability
S.P. Meyn and R.L. Tweedie

Robust Control: Systems with Uncertain Physical Parameters
J. Ackermann in co-operation with A. Bartlett, D. Kaesbauer, W. Sienel and R. Steinhauser

Optimal Sampled-Data Control Systems
Tongwen Chen and Bruce Francis

Nonlinear Control Systems (3rd edition)
Alberto Isidori

Theory of Robot Control
C. Canudas de Wit, B. Siciliano and G. Bastin (Eds)

Fundamental Limitations in Filtering and Control
María M. Seron, Julio Braslavsky and Graham C. Goodwin

Constructive Nonlinear Control
R. Sepulchre, M. Janković and P.V. Kokotović

A Theory of Learning and Generalization
M. Vidyasagar

Adaptive Control
I.D. Landau, R. Lozano and M. M'Saad

Stabilization of Nonlinear Uncertain Systems
Miroslav Krstić and Hua Deng

Passivity-based Control of Euler-Lagrange Systems
Romeo Ortega, Antonio Loría, Per Johan Nicklasson and Hebertt Sira-Ramírez

Stability and Stabilization of Infinite Dimensional Systems with Applications
Zheng-Hua Luo, Bao-Zhu Guo and Omer Morgul

Nonsmooth Mechanics (2nd edition)
Bernard Brogliato

Nonlinear Control Systems II
Alberto Isidori

L_2-Gain and Passivity Techniques in Nonlinear Control
Arjan van der Schaft

Control of Linear Systems with Regulation and Input Constraints
Ali Saberi, Anton A. Stoorvogel and Peddapullaiah Sannuti

Ben M. Chen

Robust and H_∞ Control

With 67 Figures

Springer

Ben M. Chen, PhD
Department of Electrical Engineering, National University of Singapore,
Singapore 117576

Series Editors
E.D. Sontag • M. Thoma

ISBN 978-1-84996-858-4

British Library Cataloguing in Publication Data
Chen, Ben M., 1963-
Robust and H[infinity symbol] control. - (Communications
and control engineering)
1.H[infinity symbol] control
I.Title
629.8'312
ISBN 978-1-84996-858-4   ISBN 978-1-4471-3653-8 (eBook)
DOI 10.1007/978-1-4471-3653-8
Library of Congress Cataloging-in-Publication Data
A catalog record for this book is available from the Library of Congress

Apart from any fair dealing for the purposes of research or private study, or criticism or review, as permitted under the Copyright, Designs and Patents Act 1988, this publication may only be reproduced, stored or transmitted, in any form or by any means, with the prior permission in writing of the publishers, or in the case of reprographic reproduction in accordance with the terms of licences issued by the Copyright Licensing Agency. Enquiries concerning reproduction outside those terms should be sent to the publishers.

© Springer-Verlag London 2000

Originally published by Springer-Verlag London Limited in 2000

Softcover reprint of the hardcover 1st edition 2000

The use of registered names, trademarks, etc. in this publication does not imply, even in the absence of a specific statement, that such names are exempt from the relevant laws and regulations and therefore free for general use.

The publisher makes no representation, express or implied, with regard to the accuracy of the information contained in this book and cannot accept any legal responsibility or liability for any errors or omissions that may be made.

Typesetting: Camera ready by author

69/3830-543210 Printed on acid-free paper SPIN 10745597

This work is dedicated to

My Grand Uncle, Very Reverend Paul Chan,

My Parents, and

Feng, Andy, Jamie and Wen

Preface

THE FORMULATION OF the optimization theory has certainly become one of the mile stones of modern control theory. In a typical analytical design of control systems, the given specifications are first transformed into a performance criterion, and then control laws which would minimize the performance criterion are sought. Two important and well-known criteria are the H_2 norm and the H_∞ norm of the transfer matrix from an exogenous disturbance to a pertinent controlled output of a given linear time invariant plant. This book aims to study the H_∞ control wherein the control design problem is modeled as a problem of minimizing the H_∞ norm of a certain closed-loop transfer matrix under appropriate feedback control laws. Our aim is to examine both the theoretical and practical aspects of H_∞ control from the angle of the structural properties of linear systems. Our objectives are to provide constructive algorithms for finding solutions to general singular H_∞ control problems, to general H_∞ almost disturbance decoupling problems, and to newly formulated robust and perfect tracking problems, as well as to apply these techniques to solve several practical problems.

The preliminary edition of this work was published earlier by the publisher under the title, *H_∞ Control and Its Applications*, Volume 235 of the Series of *Lecture Notes in Control and Information Sciences*. I am thankful to Nicholas Pinfield, the Engineering Editor, for urging me to upgrade it to the current series and for his kindly assistance. I have taken this opportunity to enhance the overall presentation of the work, and to include several newly developed theoretical and practical results, namely, Chapters 9, 13 and 14, which deal with the theory of robust and perfect tracking and its application to a hard disk drive servo system design.

The intended audience of this manuscript includes practicing control engineers and researchers in areas related to control engineering. An appropriate background for this monograph would be some first year graduate courses in linear systems and multivariable control. A little bit of knowledge of the geometrical theory of linear systems would certainly be helpful.

I have been fortunate to have the benefit of the cooperation of several coworkers. Foremost, I am indebted to Zongli Lin of University of Virginia. Many parts of this book were born as the result of our continual collaboration and our numerous discussions over the past few years. In general, I would like to thank Professors Ali Saberi of Washington State University, Yacov Shamash of State University of New York at Stony Brook, Chang C. Hang and Tong H. Lee of National University of Singapore, Uy-Loi Ly of University of Washington, Yaling Chen of Xiamen University, Anton Stoorvogel of Eindhoven University of Technology, and Steve Weller of University of Newcastle, for their various contributions to certain results presented in this book. Also, I am thankful to Professors Pedda Sannuti of Rutgers University, Dazhong Zheng of Tsinghua University, Cishen Zhang of University of Melbourne, and Shuzhi S. Ge of National University of Singapore for many beneficial discussions over the past few years, and to Drs. Teck-Seng Low, Tow-Chong Chong and Guoxiao Guo of the Data Storage Institute of Singapore for their generous support to my research project on the dual actuator systems of hard disk drives.

I am particularly thankful to my current and former graduate students, especially Boon-Choy Siew, Yi Guo, Jun He, Kexiu Liu, Zhongming Li and Teck-Beng Goh, for their contributions and for applying and testing parts of the results of this book to real life problems such as gyro-stabilized mirror platform, piezoelectric actuator, and dual actuator systems of hard disk drives. I am also indebted to Andra Leo, my good friend and English teacher at the National University of Singapore, for her kindest help in correcting English errors throughout the preliminary edition of this manuscript.

This work was completed mainly using my 'spare' time, i.e., evenings, weekends and holidays. I owe a debt of deepest gratitude to my parents, my wife Feng, and my children Andy, Jamie and Wen, for their sacrifice, understanding and encouragement. Last but certainly not the least, I would like to give my hearty thanks to my grand uncle, Very Reverend Paul Chan, and to his Sino-American Amity Fund and Chinese Catholic Information Center, New York. It would not have been possible for me to build my academic career without the spiritual and financial support that I received from them during my course of studies at Gonzaga University and Washington State University. It is natural that I dedicate this work to all of them.

Ben M. Chen
Singapore, 2000

Contents

Chapter 1

Introduction

1.1. Introduction

THE ULTIMATE GOAL of a control system designer is to build a system that will work in a real environment. Since the real environment may change and operating conditions may vary from time to time, the control system must be able to withstand these variations. Even if the environment does not change, other factors of life are the model uncertainties as well as noises. Any mathematical representation of a system often involves simplifying assumptions. Nonlinearities are either unknown and hence unmodeled, or are modeled and later ignored in order to simplify analysis. High frequency dynamics are often ignored at the design stage as well. In consequence, control systems designed based on simplified models may not work on real plants in real environments. The particular property that a control system must possess for it to operate properly in realistic situations is commonly called *robustness*. Mathematically, this means that the controller must perform satisfactorily not just for one plant, but for a family of plants. If a controller can be designed such that the whole system to be controlled remains stable when its parameters vary within certain expected limits, the system is said to possess robust stability. In addition, if it can satisfy performance specifications such as steady state tracking, disturbance rejection and speed of response requirements, it is said to possess robust performance. The problem of designing controllers that satisfy both robust stability and performance requirements is called robust control. Optimization theory is one of the cornerstones of modern control theory and was developed in an attempt to solve such a problem. In a typical control system design, the given specifications are at first transformed into a performance index, and then control laws which would minimize certain norm, say H_2 or H_∞ norm, of the performance index are sought. This book focuses on the H_∞ optimal control theory, and its related problems such as the H_∞ almost disturbance decoupling problem, and the robust and perfect tracking problem.

Over the past decades we have witnessed a proliferation of literature on H_∞ optimal control since it was first introduced by Zames [139]. The main focus of the work has been and continues to be on the formulation of the problem for robust multivariable control and its solution. Since the original formulation of the H_∞ problem in Zames [139], a great deal of work has been done on finding the solution to this problem. Practically all the research results of the early years involved a mixture of time-domain and frequency-domain techniques including the following: 1) *Interpolation approach* (see e.g., Limbeer and Anderson [77]); 2) *Frequency domain approach* (see e.g., Doyle [47], Francis [54] and Glover [57]); 3) *Polynomial approach* (see e.g., Kwakernaak [69]); and 4) *J-spectral factorization approach* (see e.g., Kimura [67]). Recently, considerable attention has been focussed on purely *time-domain methods* based on algebraic Riccati equations (ARE) (see e.g., Chen, Guo and Lin [21], Chen, Saberi and Ly [31], Doyle and Glover [48], Doyle, Glover, Khargonekar and Francis [49], Khargonekar, Petersen and Rotea [65], Petersen [103], Saberi, Chen and Lin [108], Sampei, Mita and Nakamichi [115], Scherer [117–119], Stoorvogel [124], Stoorvogel, Saberi and Chen [125], Tadmor [129], Zhou, Doyle and Glover [140], and Zhou and Khargonekar [141]). Along this line of research, connections are also made between H_∞ optimal control and differential games (see e.g., Başar and Bernhard [4], and Papavassilopoulos and Safonov [100]).

Most of the results in the literature are restricted to the so-called regular H_∞ control problem (see Definition 1.3.13). Unfortunately, many real life problems do not satisfy these conditions and must be formulated in terms of the regular case by adding some dummy controlled outputs and/or disturbances in order to apply the theory that deals with only the regular problem. The problem we treat in this book is general, i.e., it does not necessarily satisfy the regularity assumptions. The existence conditions for H_∞ suboptimal controllers for this type of problem are well studied in Stoorvogel [124] and Scherer [119]. The main focus of this book is, however, very different. We concentrate on 1) the computation of infimum of H_∞ optimization problems, which must be known before one can carry out any meaningful design; 2) solutions to general H_∞ optimization problems; 3) solutions to general H_∞ disturbance decoupling problems, which themselves are a very important subject; 4) solutions to robust and perfect tracking problems; and 5) the practical applications of these theories.

Most of the results presented in this book are from research carried out by the author and his co-workers over the last decade. The purpose of this book is to discuss various aspects of the subject under a single cover.

1.2. Notations

Throughout this book, we shall adopt the following notations:

$\mathbb{R}$:= the set of real numbers;

$\mathbb{C}$:= the entire complex plane;

$\mathbb{C}^{\odot}$:= the set of complex numbers inside the unit circle;

$\mathbb{C}^{\otimes}$:= the set of complex numbers outside the unit circle;

$\mathbb{C}^{\circ}$:= the unit circle in the complex plane;

$\mathbb{C}^{-}$:= the open left-half complex plane;

$\mathbb{C}^{+}$:= the open right-half complex plane;

$\mathbb{C}^{0}$:= the imaginary axis in the complex plane;

I := an identity matrix;

I_k := an identity matrix of dimension $k \times k$;

X' := the transpose of X;

X^{H} := the complex conjugate transpose of X;

$\det(X)$:= the determinant of X;

$\mathrm{rank}(X)$:= the rank of X;

$\mathrm{Im}\,(X)$:= the range space of X;

$\mathrm{Ker}\,(X)$:= the null space of X;

$X^{\dagger}$:= the Moore-Penrose (pseudo) inverse of X;

$\lambda(X)$:= the set of eigenvalues of X;

$\lambda_{\max}(X)$:= the maximum eigenvalues of X where $\lambda(X) \subset \mathbb{R}$;

$\sigma_{\max}(X)$:= the maximum singular value of X;

$\rho(X)$:= the spectral radius of X which is equal to $\max_i |\lambda_i(X)|$;

$|X|$:= the usual 2-norm of a matrix X;

$|x|$:= the Euclidean norm of a vector x;

$\|G\|_2$:= the H_2-norm of a stable system $G(s)$ or $G(z)$;

$\|g\|_2$:= the l_2-norm of a signal $g(t)$ or $g(k)$;

L_2 := the set of all functions whose l_2 norms are finite;

$$\begin{aligned}
\|g\|_p &:= \text{the } l_p\text{-norm of a signal } g(t) \text{ or } g(k);\\
L_p &:= \text{the set of all functions whose } l_p \text{ norms are finite;}\\
\|G\|_{\mathcal{L}_\infty} &:= \text{the } \mathcal{L}_\infty\text{-norm of a system } G(s) \text{ or } G(z);\\
\|G\|_\infty &:= \text{the } H_\infty\text{-norm of a stable system } G(s) \text{ or } G(z);\\
\dim(\mathcal{X}) &:= \text{the dimension of a subspace } \mathcal{X};\\
\mathcal{X}^\perp &:= \text{the orthogonal complement of a subspace } \mathcal{X} \text{ of } \mathbb{R}^n;\\
C^{-1}\{\mathcal{X}\} &:= \{x \,|\, Cx \in \mathcal{X}\}, \text{ where } \mathcal{X} \text{ is a subspace and } C \text{ is a matrix;}\\
\Sigma_* &:= \text{a linear system characterized by } (A_*, B_*, C_*, D_*);\\
\Sigma_*^\star &:= \text{a dual system of } \Sigma_* \text{ \& is characterized by } (A_*', C_*', B_*', D_*');
\end{aligned}$$

[A] := the end of an algorithm or assumption;

[C] := the end of a corollary;

[D] := the end of a definition;

[E] := the end of an example;

[L] := the end of a lemma;

[O] := the end of an observation;

[P] := the end of a property or proposition;

[R] := the end of a remark;

[T] := the end of a theorem;

[X] := the end of the proof of an interim result;

[⊠] := the end of a proof.

Finally, we denote normrank $\{X(\varsigma)\}$ the rank of $X(\varsigma)$ with entries in the field of rational functions of ς.

1.3. The Standard H_∞ Optimization Problem

We consider a generalized system Σ with a state-space description,

$$\Sigma : \left\{ \begin{array}{rl} \delta(x) &= A\,x + B\,u + E\,w, \\ y &= C_1\,x + D_{11}\,u + D_1\,w, \\ h &= C_2\,x + D_2\,u + D_{22}\,w, \end{array} \right. \tag{1.3.1}$$

where $\delta(x) = \dot{x}(t)$ if Σ is a continuous-time system, or $\delta(x) = x(k+1)$ if Σ is a discrete-time system. As usual, $x \in \mathbb{R}^n$ is the state, $u \in \mathbb{R}^m$ is the control input, $w \in \mathbb{R}^q$ is the external disturbance input, $y \in \mathbb{R}^p$ is the measurement output, and $h \in \mathbb{R}^\ell$ is the controlled output of Σ. They represent $x(t)$, $u(t)$,

$w(t)$, $y(t)$ and $h(t)$, respectively, if Σ is of continuous-time, or represent $x(k)$, $u(k)$, $w(k)$, $y(k)$ and $h(k)$, respectively, if Σ is of discrete-time. For the sake of simplicity in future development, throughout this book, we let $\Sigma_{\rm P}$ be the subsystem characterized by the matrix quadruple (A, B, C_2, D_2) and $\Sigma_{\rm Q}$ be the subsystem characterized by (A, E, C_1, D_1).

Generally, we can assume that matrix D_{11} in (1.3.1) is zero. This can be justified as follows: If $D_{11} \neq 0$, we define a new measurement output,

$$y_{\rm new} = y - D_{11}u = C_1 x + D_1 w, \tag{1.3.2}$$

which does not have a direct feedthrough term from u. Suppose we carry on our control system design using this new measurement output to obtain a proper control law, say,

$$u = \mathcal{K}(\varsigma) y_{\rm new}, \tag{1.3.3}$$

where $\varsigma = s$, the Laplace transform operator, if Σ is a continuous-time system, or $\varsigma = z$, the z-transform operator, if Σ is a discrete-time one. Then, it is straightforward to verify that the control law (1.3.3) is equivalent to the following one,

$$u = [I + \mathcal{K}(\varsigma) D_{11}]^{-1} \mathcal{K}(\varsigma) y, \tag{1.3.4}$$

provided that $[I + \mathcal{K}(\varsigma) D_{11}]^{-1}$ is well posed, i.e., the inverse exists for almost all $\varsigma \in \mathbb{C}$. Thus, for simplicity, we will assume throughout the book that $D_{11} = 0$.

The standard H_∞ optimal control problem is to find an internally stabilizing proper measurement feedback control law,

$$\Sigma_{\rm cmp} \; : \; \begin{cases} \delta(v) = A_{\rm cmp}\, v + B_{\rm cmp}\, y, \\ \quad u \;\; = C_{\rm cmp}\, v + D_{\rm cmp}\, y, \end{cases} \tag{1.3.5}$$

such that the H_∞-norm of the overall closed-loop transfer matrix function from w to h is minimized (see also Figure 1.3.1). To be more specific, we will say that the control law $\Sigma_{\rm cmp}$ of (1.3.5) is internally stabilizing when applied to the system Σ of (1.3.1), if the following matrix is asymptotically stable:

$$A_{\rm cl} := \begin{bmatrix} A + BD_{\rm cmp}C_1 & BC_{\rm cmp} \\ B_{\rm cmp}C_1 & A_{\rm cmp} \end{bmatrix}, \tag{1.3.6}$$

i.e., all its eigenvalues lie in the open left-half complex plane for a continuous-time system or in the open unit disc for a discrete-time system. It is straightforward to verify that the closed-loop transfer matrix from the disturbance w to the controlled output h is given by

$$T_{hw}(\varsigma) = C_e(\varsigma I - A_e)^{-1} B_e + D_e, \tag{1.3.7}$$

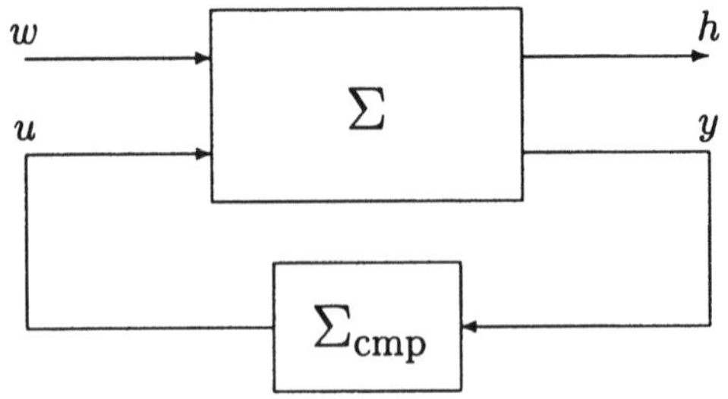

Figure 1.3.1: The standard H_∞-optimization problem.

where

$$A_e := \begin{bmatrix} A + BD_{\rm cmp}C_1 & BC_{\rm cmp} \\ B_{\rm cmp}C_1 & A_{\rm cmp} \end{bmatrix}, \quad B_e := \begin{bmatrix} E + BD_{\rm cmp}D_1 \\ B_{\rm cmp}D_1 \end{bmatrix}, \tag{1.3.8}$$

and

$$C_e := [\, C_2 + D_2D_{\rm cmp}C_1 \quad D_2C_{\rm cmp} \,], \quad D_e := D_2D_{\rm cmp}D_1 + D_{22}. \tag{1.3.9}$$

It is simple to note that if $\Sigma_{\rm cmp}$ is a static state feedback law, i.e., $u = Fx$, then the closed-loop transfer matrix from w to h is given by

$$T_{hw}(\varsigma) = (C_2 + D_2F)(\varsigma I - A - BF)^{-1}E + D_{22}. \tag{1.3.10}$$

Similarly, if $\Sigma_{\rm cmp}$ is given by $u = F_1x + F_2w$, i.e., a static full information feedback control law, then we have

$$T_{hw}(\varsigma) = (C_2 + D_2F_1)(\varsigma I - A - BF_1)^{-1}(E + BF_2) + (D_{22} + D_2F_2). \tag{1.3.11}$$

The following definitions will be convenient in our future development.

Definition 1.3.1. (Euclidean norm and 2-norm). Given a vector $x \in \mathbb{C}^n$ with entries $x_1, x_2, \cdots, x_n$, its Euclidean norm is defined as,

$$|x| = \sqrt{|x_1|^2 + |x_2|^2 + \cdots + |x_n|^2}. \tag{1.3.12}$$

Given a matrix $A \in \mathbb{C}^{n\times m}$, its 2-norm is defined as,

$$|A| = \sup_{x\neq 0} \frac{|Ax|}{|x|} = \max_i \sqrt{\lambda_i(A^{\rm H} A)}. \tag{1.3.13}$$

The 2-norm of the matrix A is also called its spectral norm. ◻

Definition 1.3.2. (l_2-norm). The l_2-norm of a continuous-time signal $g(t)$, which is either a column vector or a row vector, is defined as

$$\|g\|_2 := \left(\text{trace}\left[\int_0^\infty g(t)g(t)'dt\right]\right)^{\frac{1}{2}}. \tag{1.3.14}$$

Similarly, for a discrete-time signal vector $g(k)$, we have

$$\|g\|_2 := \left(\text{trace}\left[\sum_{k=0}^{\infty} g(k)g(k)'\right]\right)^{\frac{1}{2}}. \tag{1.3.15}$$

The square of the l_2-norm of $g(t)$ or $g(k)$ is commonly termed the total energy in the signal $g(t)$ or $g(k)$. In many areas of engineering, energy or l_2-norm is used as a measure of the size of a transient signal $g(t)$ or $g(k)$ which decays to zero as time t or shift k progresses towards infinity. By Parseval's theorem, $\|g\|_2$ can also be computed in the frequency domain as follows: for the continuous-time case,

$$\|g\|_2 = \left(\frac{1}{2\pi}\text{trace}\left[\int_{-\infty}^{\infty} G(j\omega)G(j\omega)^{\text{H}}d\omega\right]\right)^{\frac{1}{2}}, \tag{1.3.16}$$

where $G(j\omega)$ is the Fourier transform of $g(t)$; similarly, for the discrete-time case,

$$\|g\|_2 = \left(\frac{1}{2\pi}\text{trace}\left[\int_{-\pi}^{\pi} G(e^{j\omega})G(e^{j\omega})^{\text{H}}d\omega\right]\right)^{\frac{1}{2}}, \tag{1.3.17}$$

where $G(z)$ is the z-transform of $g(k)$. ◻D

Definition 1.3.3. (l_p-norm and L_p). Let $p \in [1, \infty)$. The l_p-norm of a continuous-time vector signal $g(t)$ is defined as

$$\|g\|_p := \left(\int_0^\infty |g(t)|^p dt\right)^{1/p}. \tag{1.3.18}$$

Similarly, for a discrete-time signal vector $g(k)$, we define,

$$\|g\|_p := \left(\sum_{k=0}^{\infty} |g(k)|^p\right)^{1/p}. \tag{1.3.19}$$

For $p = 2$, the above definitions coincide with those in Definition 1.3.2. Also, L_p denotes the set of vector functions, whose l_p-norms are bounded. ◻D

Definition 1.3.4. (l_∞-norm and L_∞). The l_∞-norm of a continuous-time signal vector $g(t)$ is defined as

$$\|g\|_\infty := \operatorname*{ess\ sup}_{t \geq 0} |g(t)|. \tag{1.3.20}$$

Similarly, for a discrete-time signal vector $g(k)$, we define,

$$\|g\|_\infty := \sup_{k\geq 0} |g(k)|. \tag{1.3.21}$$

Also, L_∞ denotes the set of vector functions, whose l_∞-norms are bounded. ◻

Definition 1.3.5. (H_2-norm). The H_2-norm of an asymptotically stable and proper continuous-time transfer matrix $G(s)$ is defined as

$$\|G\|_2 := \left(\frac{1}{2\pi}\text{trace}\left[\int_{-\infty}^{\infty} G(j\omega)G(j\omega)^{\mathrm{H}} d\omega\right]\right)^{\frac{1}{2}}. \tag{1.3.22}$$

By Parseval's theorem, $\|G\|_2$ can equivalently be defined as

$$\|G\|_2 = \left(\text{trace}\left[\int_{0}^{\infty} g(t)g(t)'dt\right]\right)^{\frac{1}{2}}, \tag{1.3.23}$$

where $g(t)$ is the unit impulse response matrix of $G(s)$. Thus, $\|G\|_2 = \|g\|_2$. Similarly, the H_2-norm of an asymptotically stable and proper discrete-time transfer matrix $G(z)$ is defined as

$$\|G\|_2 := \left(\frac{1}{2\pi}\text{trace}\left[\int_{-\pi}^{\pi} G(e^{j\omega})G(e^{j\omega})^{\mathrm{H}} d\omega\right]\right)^{\frac{1}{2}}. \tag{1.3.24}$$

Again, by Parseval's theorem, $\|G\|_2$ can equivalently be defined as

$$\|G\|_2 = \left(\text{trace}\left[\sum_{k=0}^{\infty} g(k)g(k)'\right]\right)^{\frac{1}{2}}, \tag{1.3.25}$$

where $g(k)$ is the unit impulse response matrix of $G(k)$. Thus, as in the continuous-time case, $\|G\|_2 = \|g\|_2$. ◻

Definition 1.3.6. (H_∞-norm). The H_∞-norm of an asymptotically stable and proper continuous-time transfer matrix $G(s)$ is defined as

$$\|G\|_\infty := \sup_{\omega\in[0,\infty)} \sigma_{\max}[G(j\omega)] = \sup_{\|w\|_2=1} \frac{\|h\|_2}{\|w\|_2}, \tag{1.3.26}$$

where w and h are respectively the input and output of $G(s)$. Similarly, the H_∞-norm of an asymptotically stable and proper discrete-time transfer matrix $G(z)$ is defined as

$$\|G\|_\infty := \sup_{\omega\in[0,2\pi]} \sigma_{\max}[G(e^{j\omega})] = \sup_{\|w\|_2=1} \frac{\|h\|_2}{\|w\|_2}, \tag{1.3.27}$$

where w and h are respectively the input and output of $G(z)$. ◻

Definition 1.3.7. ($\mathcal{L}_\infty$-norm). The $\mathcal{L}_\infty$-norm of a continuous-time transfer matrix $G(s)$, which is not necessarily stable, is defined as

$$\|G\|_{\mathcal{L}_\infty} := \sup_{\omega\in[0,\infty)} \sigma_{\max}[G(j\omega)]. \tag{1.3.28}$$

Similarly, the $\mathcal{L}_\infty$-norm of a discrete-time transfer matrix $G(z)$, which is not necessarily stable, is defined as

$$\|G\|_{\mathcal{L}_\infty} := \sup_{\omega\in[0,2\pi]} \sigma_{\max}[G(e^{j\omega})]. \tag{1.3.29}$$

Obviously, if G is stable, its $\mathcal{L}_\infty$-norm coincides with its H_∞-norm. ◻

Definition 1.3.8. (γ-Suboptimal Controller). Consider the given system Σ of (1.3.1) and the controller $\Sigma_{\rm cmp}$ of (1.3.5). $\Sigma_{\rm cmp}$ is said to be an H_∞ γ-suboptimal controller, or in short a γ-suboptimal controller, for Σ if, when $\Sigma_{\rm cmp}$ is applied to Σ, the resulting closed-loop is internally stable and the H_∞-norm of the closed-loop transfer matrix is less than γ. ◻

Definition 1.3.9. (Infimum γ^*). Consider the given system Σ of (1.3.1) and the controller $\Sigma_{\rm cmp}$ of (1.3.5). The infimum of the H_∞-norm of the closed-loop transfer matrix $T_{hw}(\Sigma\times\Sigma_{\rm cmp})$ over all stabilizing controllers $\Sigma_{\rm cmp}$ is denoted by γ^*, namely,

$$\gamma^* := \inf\left\{ \|T_{hw}(\Sigma\times\Sigma_{\rm cmp})\|_\infty \mid \Sigma_{\rm cmp} \text{ internally stabilizes } \Sigma \right\}. \tag{1.3.30}$$

Obviously, $\gamma^* \geq 0$. Occasionally, when it is clear in the context, we may also say that γ^* is the infimum of the given system Σ. ◻

Definition 1.3.10. (H_∞ Optimal Controller). Consider the given system Σ of (1.3.1) and the controller $\Sigma_{\rm cmp}$ of (1.3.5). $\Sigma_{\rm cmp}$ is said to be an H_∞ optimal controller for Σ if, when $\Sigma_{\rm cmp}$ is applied to Σ, the resulting closed-loop is internally stable and the H_∞-norm of the closed-loop transfer matrix is equal to γ^*. ◻

Definition 1.3.11. (Full Information Feedback Case). Consider the given system Σ of (1.3.1). The H_∞ optimization problem for Σ is called a full information feedback case if

$$y = \begin{pmatrix} x \\ w \end{pmatrix}, \text{ i.e., } C_1 = \begin{pmatrix} I \\ 0 \end{pmatrix}, \ D_1 = \begin{pmatrix} 0 \\ I \end{pmatrix}. \tag{1.3.31}$$

We will also call such a system Σ a full information feedback system. ◻

Definition 1.3.12. (Full State Feedback Case). Consider the given system Σ of (1.3.1). The H_∞ optimization problem for Σ is called a full state feedback case if $y = x$, i.e., $C_1 = I$ and $D_1 = 0$. We will also call such a system Σ a full state feedback system. ◻

Definition 1.3.13. (Regular Case). Consider the given system Σ of (1.3.1). The H_∞ optimization problem for Σ is said to be a regular case or a regular problem provided that:

1. The following conditions are satisfied if Σ is a continuous-time system,
   (a) D_2 is of full column rank and Σ_P is free of imaginary invariant zeros;
   (b) D_1 is of full row rank and Σ_Q is free of imaginary invariant zeros.
2. The following conditions are satisfied if Σ is a discrete-time system,
   (a) Σ_P is left invertible and is free of unit circle invariant zeros;
   (b) Σ_Q is right invertible and is free of unit circle invariant zeros.

Also, we will call such a system Σ a regular system. We note that the characterizations of the regular case for discrete-time systems precisely correspond to those for continuous-time systems under a bilinear mapping. This will be seen clearly in Chapters 3 and 5. ◻

Definition 1.3.14. (Singular Case). Consider the given system Σ of (1.3.1). The H_∞ optimization problem for Σ is said to be a singular case or a singular problem if it is not a regular one. We will occasionally call such a system Σ a singular system. ◻

1.4. Some Common Robust Control Problems

There are many common robust control problems that can be converted into the standard H_∞ optimization problem. Once a problem is translated into the standard one, it can be readily solved using the results of the coming chapters. For example, Figure 1.4.1 illustrates the robust stability problem for an unstructurally perturbed system: Suppose that Σ_cmp is an H_∞ γ-suboptimal controller for the nominal system Σ of the uncertain plant, i.e., $\|T_{hw}(\Sigma \times \Sigma_\text{cmp})\|_\infty < \gamma$. Then, by the well-known *Small-Gain Theorem*, the overall closed-loop system in Figure 1.4.1 will remain stable for all possible perturbations of the plant, i.e., Δ, with $\|\Delta\|_\infty < 1/\gamma$. Thus, by pushing γ closer to γ^* of the nominal system, the overall system will be more robust with respect to the perturbation

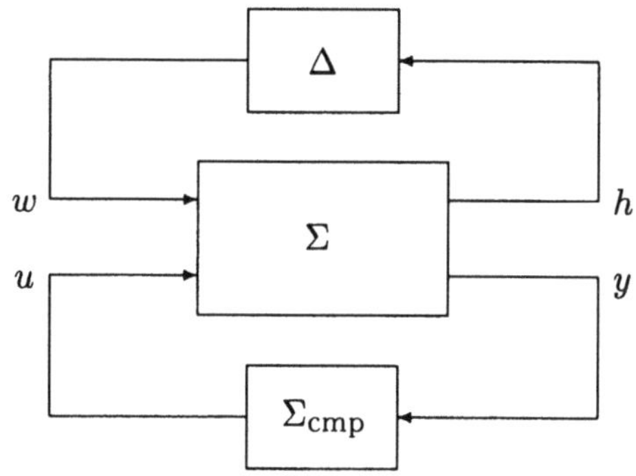

Figure 1.4.1: Robust stability problem of unstructurally perturbed systems.

Δ. Similarly, robust stabilization problems for systems with structural perturbations such as additive perturbation and multiplicative perturbation can also be easily converted into a standard H_∞ control problem. In what follows, we will show how to cast the following four common problems into the standard H_∞ optimization problem:

- The mixed-sensitivity problem;
- Maximization of complex stability radius;
- Robust stabilization with additive perturbations;
- Robust stabilization with multiplicative perturbations.

1.4.1. The Mixed-sensitivity Problem

The mix-sensitivity problem is associated with a widely used control system configuration as depicted in Figure 1.4.2. In the author's opinion, such a configuration is definitely not a good structure to design a tracking control system, because it feeds only the error signal e into the controller Σ_{F}. As it will be seen later in Chapters 9 and 13, a controller structure, which utilizes both the reference r and the measurement y independently, will in general yield a much better performance. Anyhow, it is up to readers to make their own judgment.

We note that Σ_{m} in Figure 1.4.2 represents the nominal model of a given plant. Let the transfer matrix of Σ_{m} be given as

$$G_{\mathrm{m}}(s) = C_{\mathrm{m}}(sI - A_{\mathrm{m}})^{-1}B_{\mathrm{m}} + D_{\mathrm{m}}, \tag{1.4.1}$$

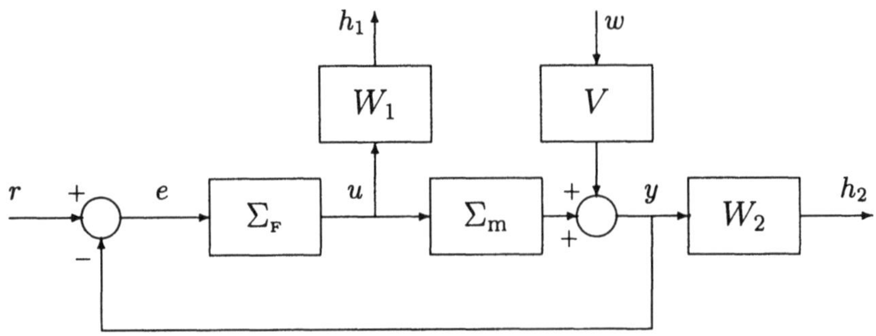

Figure 1.4.2: The configuration associated with the mix-sensitivity problem.

with (A_m, B_m) being stabilizable and (A_m, C_m) being detectable. Let the transfer matrix of Σ_F, which is the controller to be designed, be given as

$$G_F(s) = C_F(sI - A_F)^{-1}B_F + D_F, \tag{1.4.2}$$

and the minimal realizations of the weighting functions V, W_1 and W_2 be respectively given as

$$V(s) = C_V(sI - A_V)^{-1}B_V + D_V, \tag{1.4.3}$$

$$W_1(s) = C_{W1}(sI - A_{W1})^{-1}B_{W1} + D_{W1}, \tag{1.4.4}$$

and

$$W_2(s) = C_{W2}(sI - A_{W2})^{-1}B_{W2} + D_{W2}. \tag{1.4.5}$$

We note that the choices of these weighting functions V, W_1 and W_2 are subject to the design specifications of the overall system. Then, the sensitivity function S and the complementary sensitivity function T are respectively defined as

$$S := (I + G_m G_F)^{-1} \quad \text{and} \quad T := G_F(I + G_m G_F)^{-1}. \tag{1.4.6}$$

It is simple to see that the sensitivity function S and the complementary sensitivity function T are respectively the transfer matrices from r to e, and from r to u, if the disturbance $w = 0$. Clearly, a small S would yield a small e, i.e., a good tracking performance, while a small T would yield a small control u. Unfortunately, it is clear from the definitions of S and T that we can never make them both small simultaneously. In general, some kinds of trade-offs (by properly choosing weighting functions V, W_1 and W_2) are always needed in practical situations. Next, we note that if $r = 0$, then the transfer matrices from w to h_1 and from w to h_2 are respectively given by $-W_1TV$ and W_2SV.

The mixed-sensitivity problem is to find an internally stabilizing control law Σ_{F} such that the H_∞-norm of the closed-loop system from the disturbance w to the controlled output $h = (h_1', h_2')'$, i.e.,

$$\left\| \begin{matrix} -W_1 TV \\ W_2 SV \end{matrix} \right\|_\infty = \left\| \begin{matrix} W_1 TV \\ W_2 SV \end{matrix} \right\|_\infty, \tag{1.4.7}$$

is minimized. Such a problem can be translated into a state-space setting as a special case of the standard H_∞ optimization problem. The problem is equivalent to design an internally stabilizing control law

$$\Sigma_{\mathrm{F}} : \begin{cases} \dot{v} = A_{\mathrm{F}}\, v - B_{\mathrm{F}}\, y \\ u = C_{\mathrm{F}}\, v - D_{\mathrm{F}}\, y \end{cases} \tag{1.4.8}$$

for the following auxiliary system,

$$\Sigma_{\mathrm{mix}} : \begin{cases} \dot{x} = A\, x + B\, u + E\, w, \\ y = C_1\, x + D_{11}\, u + D_1\, w, \\ h = C_2\, x + D_2\, u + D_{22}\, w, \end{cases} \tag{1.4.9}$$

where

$$A = \begin{bmatrix} A_{\mathrm{m}} & 0 & 0 & 0 \\ 0 & A_{\mathrm{v}} & 0 & 0 \\ 0 & 0 & A_{\mathrm{w1}} & 0 \\ B_{\mathrm{w2}} C_{\mathrm{m}} & B_{\mathrm{w2}} C_{\mathrm{v}} & 0 & A_{\mathrm{w2}} \end{bmatrix}, \quad B = \begin{bmatrix} B_{\mathrm{m}} \\ 0 \\ B_{\mathrm{w1}} \\ B_{\mathrm{w2}} D_{\mathrm{m}} \end{bmatrix}, \tag{1.4.10}$$

$$E = \begin{bmatrix} 0 \\ B_{\mathrm{v}} \\ 0 \\ B_{\mathrm{w2}} D_{\mathrm{v}} \end{bmatrix}, \quad D_{22} = \begin{bmatrix} 0 \\ D_{\mathrm{w2}} D_{\mathrm{v}} \end{bmatrix}, \tag{1.4.11}$$

$$C_1 = [\, C_{\mathrm{m}} \quad C_{\mathrm{v}} \quad 0 \quad 0 \,], \quad D_{11} = D_{\mathrm{m}}, \quad D_1 = D_{\mathrm{v}}, \tag{1.4.12}$$

and

$$C_2 = \begin{bmatrix} 0 & 0 & C_{\mathrm{w1}} & 0 \\ D_{\mathrm{w2}} C_{\mathrm{m}} & D_{\mathrm{w2}} C_{\mathrm{v}} & 0 & C_{\mathrm{w2}} \end{bmatrix}, \quad D_2 = \begin{bmatrix} D_{\mathrm{w1}} \\ D_{\mathrm{w2}} D_{\mathrm{m}} \end{bmatrix}. \tag{1.4.13}$$

such that when Σ_{F} is applied to Σ_{mix}, the H_∞-norm of the resulting closed-loop system from w to h is minimized. Obviously, depending on the choices of the weighting functions, the above problem can be either a regular or a singular one. All results for the standard H_∞ optimization problem in the coming chapters can be utilized to solve this mixed-sensitivity problem.

1.4.2. Maximization of Complex Stability Radius

We introduce in this subsection the problem of the maximization of complex stability radius for uncertain systems. Let us consider an uncertain linear time-invariant system Σ_Δ characterized by

$$\dot{x} = Ax + Bu + E\Delta Cx, \tag{1.4.14}$$

where $A \in \mathbb{R}^{n\times n}$, $B \in \mathbb{R}^{n\times m}$, $E \in \mathbb{R}^{n\times \ell}$ and $C \in \mathbb{R}^{p\times n}$ are given constant matrices while Δ expresses the uncertainty which is structured by the matrices E and C. Moreover, we assume that (A, B) is stabilizable. For any stabilizing state feedback law

$$u = Fx, \tag{1.4.15}$$

with $F \in \mathbb{R}^{m\times n}$, the complex stability radius for Σ_Δ is defined as (see e.g., Hinrichsen and Pritchard [62]),

$$\gamma_c(\Sigma_\Delta, F) := \inf\Big\{|\Delta| \ : \ \Delta \in \mathbb{C}^{\ell\times p} \text{ such that } A+BF+E\Delta C \text{ is unstable}\Big\}. \tag{1.4.16}$$

The supremum of the complex stability radii that can be achieved over the stabilizing linear state feedback law of the form (1.4.15) is defined as

$$\gamma_c^*(\Sigma_\Delta) = \sup\Big\{\gamma_c(\Sigma_\Delta, F) \ : \ F \in \mathbb{R}^{m\times n} \text{ and } A + BF \text{ is stable}\Big\}. \tag{1.4.17}$$

At a first glance, it seems that the complex perturbation is not natural and should not play a role in robustness analysis. However, it turned out that the complex stability radius is important for two good reasons: First of all, it provides a lower bound for the real stability radius (defined as the complex stability radius but with the restriction that Δ be a real matrix), and there are important special cases where the real and complex stability radii coincide. Moreover, there are elegant results for the complex stability radii while that is not the case for the real stability radii. Secondly, it turned out that the complex stability radii are equivalent to the real dynamic stability radii, i.e., Δ is a real dynamic perturbation (see Hinrichsen and Pritchard [62] for further details and a survey of the literature). Following the result of [62], i.e.,

$$\gamma_c(\Sigma_\Delta, F) = \|G_{\mathrm{F}}\|_\infty^{-1}, \tag{1.4.18}$$

where $G_{\mathrm{F}}(s) = C(sI - A - BF)^{-1}E$, we can show that the supremum of the complex stability radii is given by

$$\gamma_c^*(\Sigma_\Delta) = \frac{1}{\gamma^*}, \tag{1.4.19}$$

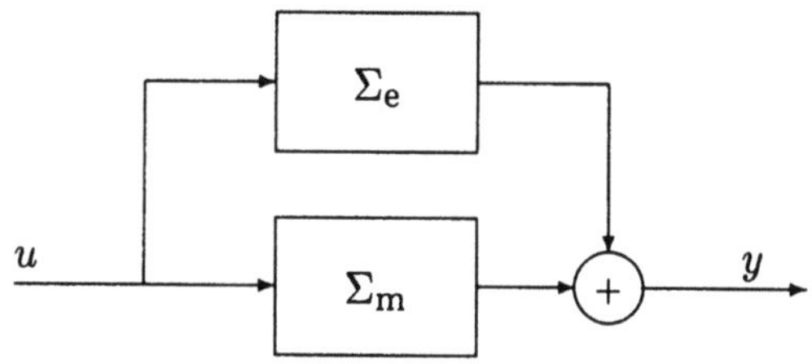

Figure 1.4.3: Plant with additive perturbations.

where γ^* is the infimum of the following standard state feedback H_∞ optimization problem,

$$\begin{cases} \dot{x} = A\,x + B\,u + E\,w, \\ y = \quad x \\ h = C\,x + 0\,u + 0\,w. \end{cases} \tag{1.4.20}$$

Clearly, the above problem belongs to the singular case. The computation of γ^* and thus $\gamma_c^*(\Sigma_\Delta)$ can be done using algorithms given later in Chapter 6.

1.4.3. Robust Stabilization with Additive Perturbations

We consider the problem of robust stabilization for plants with additive perturbations. To be more specific, we consider a stabilization problem for an uncertain system described in Figure 1.4.3, in which Σ_m is the nominal model of the given plant and Σ_e is the unknown perturbation. We assume that Σ_m and $\Sigma_m + \Sigma_e$ have the same number of unstable poles. Let the transfer matrix of Σ_m be characterized by

$$G_m(s) = C_m(sI - A_m)^{-1}B_m + D_m, \tag{1.4.21}$$

with (A_m, B_m) being stabilizable and (A_m, C_m) being detectable. Given a scalar $\gamma_a > 0$, the problem of robust stabilization for the plant with additive perturbations is to find a controller of the form (1.3.5) such that when it is applied to the uncertain system of Figure 1.4.3, the resulting closed-loop system is internally stable for all possible perturbations Σ_e with their $\mathcal{L}_\infty$-norm less than or equal to γ_a.

Following the result of Vidyasagar [132] (see also [124]), we can show that the above problem is equivalent to find an H_∞ γ-suboptimal controller (with $\gamma = 1/\gamma_a$) for the following auxiliary system,

$$\Sigma_{add} : \begin{cases} \dot{x} = A_m\,x + B_m\,u + 0\,w, \\ y = C_m\,x + D_m\,u + I\,w, \\ h = \;\;0\;\;x + \;\;I\;\;u + 0\,w. \end{cases} \tag{1.4.22}$$

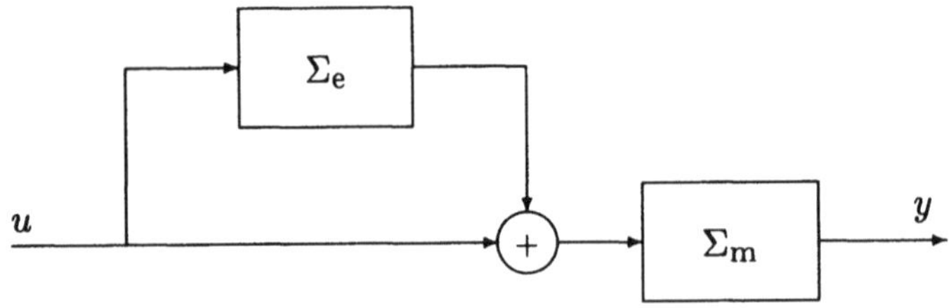

Figure 1.4.4: Plant with multiplicative perturbations.

It is straightforward to see that both subsystems characterized respectively by $(A_m, B_m, 0, I)$ and $(A_m, 0, C_m, I)$ are invertible. In fact, both subsystems share common invariant zeros, which coincide with the eigenvalues of A_m. The results of Chapter 6 can be utilized to exactly compute the infimum γ^* for Σ_{add} without imposing any additional condition. The result of [124] requires A_m to be free of eigenvalues on the imaginary axis.

We note that $1/\gamma^*$ is corresponding to the largest possible $\mathcal{L}_\infty$-norm bound on Σ_e for which the uncertain system of Figure 1.4.3 can still be made asymptotically stable.

1.4.4. Robust Stabilization with Multiplicative Perturbations

In this subsection, we consider the problem of robust stabilization for plants with multiplicative perturbations, i.e., we consider a stabilization problem for an uncertain system described in Figure 1.4.4, in which Σ_m is the nominal model of the given plant and Σ_e is the unknown perturbation. We assume that Σ_m, and the uncertain system comprising Σ_m and Σ_e as in Figure 1.4.4, have the same number of unstable poles. Let the transfer matrix of Σ_m be characterized by

$$G_m(s) = C_m(sI - A_m)^{-1}B_m + D_m, \tag{1.4.23}$$

with (A_m, B_m) being stabilizable and (A_m, C_m) being detectable. Given a scalar $\gamma_m > 0$, the problem of robust stabilization for the plant with multiplicative perturbations is to find a controller of the form (1.3.5) such that when it is applied to the uncertain system of Figure 1.4.4, the resulting closed-loop system is internally stable for all possible perturbations Σ_e with their $\mathcal{L}_\infty$-norm less than or equal to γ_m.

Similarly, following the result of Vidyasagar [132] (see also [124]), we can show that the above problem is equivalent to find an H_∞ γ-suboptimal controller (with $\gamma = 1/\gamma_m$) for the following auxiliary system,

$$\Sigma_{\rm multi} : \begin{cases} \dot{x} = A_{\rm m}\, x + B_{\rm m}\, u + B_{\rm m}\, w, \\ y = C_{\rm m}\, x + D_{\rm m}\, u + D_{\rm m}\, w, \\ h = \ 0\ x + \ I\ u + \ 0\ w. \end{cases} \tag{1.4.24}$$

Here we note that the above system always satisfies those conditions posed in Chapter 6. An exact computation of the infimum γ^* for $\Sigma_{\rm multi}$ is always feasible without imposing any additional condition. Again, $1/\gamma^*$ is corresponding to the largest possible $\mathcal{L}_\infty$-norm bound on $\Sigma_{\rm e}$ for which the uncertain system of Figure 1.4.4 can still be made asymptotically stable.

1.5. Preview of Each Chapter

A preview of each chapter is given next. The book can naturally be divided into three parts. The first part consists of Chapters 1 to 5 and contains some preliminary results and background materials. Chapter 2 recalls some linear system tools such as the Jordan and real Jordan canonical forms and several structural decompositions of linear systems such as the controllability structural decomposition and the special coordinate basis. The latter has the distinct feature of explicitly displaying the finite and infinite zero structures of a given system. It plays a dominant role in the development of the whole book. Chapter 3 presents a comprehensive study on the structural mapping of bilinear and inverse bilinear transformations. Chapter 4 recalls results on the existence conditions of H_∞ suboptimal controllers for both continuous- and discrete-time systems, which are to be used in the proofs of results developed in the second part of the book. Finally, Chapter 5 provides solutions to several types of discrete-time Riccati equations. Results in Chapters 3 and 5 are instrumental in the development of main results in discrete-time H_∞ optimization problems.

The second part of the book consists of Chapters 6 to 13 and is also the heart of the book. Chapter 6 deals with the computation of infimum in continuous-time H_∞ optimization problems. For a fairly large class of singular problem in which the given system satisfies certain geometric conditions, we present a non-iterative procedure that computes its infimum exactly. For the case when the geometric conditions are not satisfied, we modify our algorithm to yield an iterative scheme for approximating this infimum based on an auxiliary reduced order regular system, which generally has a much smaller dynamical order than that of the original system. Chapter 7 deals with finding H_∞ γ-suboptimal controllers for the state feedback case, and the full order and reduced order measurement feedback cases. We provide closed-form solutions to the H_∞ suboptimal control problem for the class of singular systems which satisfy the above mentioned geometric conditions. Here by closed-form solutions we mean solutions which are explicitly parameterized in terms of γ and are obtained without explicitly requiring a value of γ. Hence, one can easily tune the parameter γ in order to obtain the desired level of disturbance attenuation. This method will be adapted to find γ-suboptimal control laws for general systems when the geometric conditions are not satisfied. Chapter 8 gives solutions to the general H_∞ almost disturbance decoupling problem with either state feedback or measurement feedback and with internal stability for plants whose subsystems have invariant zeros on the imaginary axis of the complex plane. Some newly developed results on a so-called robust and perfect tracking problem are presented in Chapter 9. The problem is to design a family of appropriate control laws for a given plant and a given reference such that the resulting closed-loop system is asymptotically stable and the controlled output of the given plant is capable of tracking the reference arbitrarily fast from any initial condition in face of external disturbances. We will first derive a set of necessary and sufficient conditions under which robust and perfect tracking performance for a given continuous-time system can be achieved. Constructive algorithms will then be given to realize the required controllers. Similarly, Chapters 10 to 13 focus on the discrete-time counterparts of Chapters 6 to 9, respectively.

The last part of the book concerns with some real-life applications. Chapter 14 studies a servo system design for a voice-coil-motor actuator of computer hard disk drives. The purpose of this chapter is to challenge the widely used PID structure with our newly developed robust and perfect tracking approach. It turns out that the robust and perfect tracking controller, which has a dynamical order of 1, beats the conventional PID one in every category examined. Chapter 15 deals with a case study on a piezoelectric actuator control system design using the H_∞ almost disturbance decoupling approach. Such a piezoelectric actuator system has a potential application in forming a dual actuator system for the hard disk drives of the next generation. Chapter 16 presents another case study on a gyro-stabilized mirror targeting system design using the robust and perfect tracking approach. The gyro-stabilized system has some crucial military applications. Finally, we note that all these designs are carried out with a clear understanding of the theories and the properties of the given systems. Simulation and real implementation results show that these applications turn out to be very satisfactory.

Chapter 2

Linear System Tools

2.1. Introduction

AS WILL BE evident in the coming chapters, the finite and infinite zero structures as well as the invertibility structures of the given system play dominant roles in the computation of the infima and the solutions to both continuous-time and discrete-time H_∞ optimization problems. Thus a good non-ambiguous understanding of linear system structures is essential for our study. In our opinion, the best way to display all the structural properties of linear systems is to transform them into a so-called special coordinate basis (SCB) developed by Sannuti and Saberi [116] and Saberi and Sannuti [111]. However, quite often it happens that the original special coordinate basis of Sannuti and Saberi is not fine enough to characterize all the details of the properties of linear systems. In order to see all the fine points of a given system, we would have to further decompose certain subsystems of its SCB using some well-known canonical forms such as the Jordan canonical form and controllability structural decomposition. Keeping this in mind, we recall in this chapter the following results: 1) the Jordan and real Jordan canonical forms for a square constant matrix; 2) the controllability structural decomposition and block diagonal control canonical form for a constant matrix pair; and 3) the special coordinate basis of a linear time invariant system characterized by either a matrix triple or a matrix quadruple. These canonical forms and the special coordinate basis will form a transformer for linear systems. Once a system is touched by this transformer, all its structural properties become clear and transparent.

2.2. Jordan and Real Jordan Canonical Forms

We recall in this section the Jordan canonical form and the real Jordan canonical form of a square constant matrix. We first have the following theorem.

Theorem 2.2.1. Consider a constant matrix $A \in \mathbb{R}^{n\times n}$. There exists a nonsingular transformation $T \in \mathbb{C}^{n\times n}$ and an integer k such that

$$T^{-1}AT = \text{blkdiag}\Big\{J_1, J_2, \cdots, J_k\Big\}, \tag{2.2.1}$$

where J_i, $i = 1, 2, \cdots, k$, are some $n_i \times n_i$ Jordan blocks, i.e.,

$$J_i = \begin{bmatrix} \lambda_i & 1 & & \\ & \ddots & \ddots & \\ & & \lambda_i & 1 \\ & & & \lambda_i \end{bmatrix}. \tag{2.2.2}$$

Obviously, $\lambda_i \in \lambda(A)$, $i = 1, 2, \cdots, k$, and $\sum_{i=1}^{k} n_i = n$. ▣

The result of the above theorem is very well-known. The realization of this Jordan canonical form in MATLAB can be found in Chen [14]. The following theorem is to find a real Jordan canonical form.

Theorem 2.2.2. Consider a constant matrix $A \in \mathbb{R}^{n\times n}$. There exists a nonsingular transformation $P \in \mathbb{R}^{n\times n}$ and an integer k such that

$$P^{-1}AP = \text{blkdiag}\Big\{J_1, J_2, \cdots, J_k\Big\}, \tag{2.2.3}$$

where each block J_i, $i = 1, 2, \cdots, k$, has the following form: if $\lambda_i \in \lambda(A)$ is real,

$$J_i = \begin{bmatrix} \lambda_i & 1 & & \\ & \ddots & \ddots & \\ & & \lambda_i & 1 \\ & & & \lambda_i \end{bmatrix}, \tag{2.2.4}$$

or if $\lambda_i = \mu_i + j\omega_i \in \lambda(A)$ and $\bar{\lambda}_i = \mu_i - j\omega_i \in \lambda(A)$ with $\omega_i \neq 0$,

$$J_i = \begin{bmatrix} \Lambda_i & I_2 & & \\ & \ddots & \ddots & \\ & & \Lambda_i & I_2 \\ & & & \Lambda_i \end{bmatrix}, \quad \Lambda_i = \begin{bmatrix} \mu_i & \omega_i \\ -\omega_i & \mu_i \end{bmatrix}. \tag{2.2.5}$$

The above structure of $P^{-1}AP$ is called *the real Jordan canonical form*. ▣

The proof of the above theorem can be found in many texts (see e.g., Wonham [138]). The following is a constructive algorithm for obtaining the transformation P that will transform the given matrix A into the real Jordan canonical form. First, we compute a non-singular transformation $T \in \mathbb{R}^{n\times n}$ such that

$$T^{-1}AT = \text{blkdiag}\Big\{A_1,\, A_2,\, \cdots,\, A_\ell\Big\}, \tag{2.2.6}$$

where sub-matrices $A_i \in \mathbb{R}^{n_i\times n_i}$, $i = 1, 2, \cdots, \ell$, have either a single or one repeated (if $n_i > 1$) eigenvalue λ_i, if λ_i is real, or two or two repeated (if $n_i > 2$) eigenvalues λ_i and $\bar{\lambda}_i$, if λ_i is not real. Also, we have $\lambda_i \neq \lambda_j$, if $i \neq j$. Note that such a transformation T can easily be obtained using some numerically very stable algorithms such as the real Schur decomposition.

For each A_i with its corresponding λ_i being a real number, we use the result of Theorem 2.2.1 to obtain a non-singular transformation $\tilde{S}_i = S_i \in \mathbb{R}^{n_i\times n_i}$ such that A_i can be transformed into the Jordan canonical form. For each A_i which has eigenvalues $\lambda_i = \mu_i + j\omega_i$ and $\bar{\lambda}_i = \mu_i - j\omega_i$ with $\omega_i > 0$, we follow the result of Fama and Matthews [51] to define a new $(2n_i) \times (2n_i)$ matrix,

$$Z_i := \begin{bmatrix} A_i - \mu_i I_{n_i} & \omega_i I_{n_i} \\ -\omega_i I_{n_i} & A_i - \mu_i I_{n_i} \end{bmatrix}. \tag{2.2.7}$$

It is simple to show that Z_i has n_i real eigenvalues at 0 and n_i purely imaginary eigenvalues at $\pm j2\omega_i$. Then, we use the real Schur decomposition technique to find a non-singular transformation $S_i^0 \in \mathbb{R}^{(2n_i)\times(2n_i)}$ such that

$$(S_i^0)^{-1} Z_i S_i^0 = \begin{bmatrix} Z_{i0} & 0 \\ 0 & Z_{ix} \end{bmatrix}, \tag{2.2.8}$$

where Z_{i0} has all its eigenvalues at 0 while Z_{ix} has no eigenvalue at 0. Next, we utilize the result of Theorem 2.2.1 to obtain a non-singular transformation $S_i^1 \in \mathbb{R}^{n_i\times n_i}$ such that

$$(S_i^1)^{-1} Z_{i0} S_i^1 = \text{blkdiag}\Big\{J_0^1,\, J_0^1,\, J_0^2,\, J_0^2,\, \cdots,\, J_0^{\sigma_i}, J_0^{\sigma_i}\Big\}, \tag{2.2.9}$$

where J_0^m, $m = 1, 2, \cdots, \sigma_i$, have the form,

$$J_0^m = \begin{bmatrix} 0 & I_{n_{im}-1} \\ 0 & 0 \end{bmatrix}. \tag{2.2.10}$$

Let us partition

$$S_i := S_i^0 \begin{bmatrix} S_i^1 & 0 \\ 0 & I_{n_i} \end{bmatrix} = \begin{bmatrix} S_{i,1}^{1,1} & \cdots & S_{i,1}^{1,n_{i1}} & X_{i,1}^{1,1} & \cdots & X_{i,1}^{1,n_{i1}} & \cdots\cdots \\ S_{i,1}^{2,1} & \cdots & S_{i,1}^{2,n_{i1}} & X_{i,1}^{2,1} & \cdots & X_{i,1}^{2,n_{i1}} & \cdots\cdots \end{bmatrix}$$

$$\begin{matrix} S_{i,\sigma_i}^{1,1} & \cdots & S_{i,\sigma_i}^{1,n_{i\sigma_i}} & X_{i,\sigma_i}^{1,1} & \cdots & X_{i,\sigma_i}^{1,n_{i\sigma_i}} & \star \\ S_{i,\sigma_i}^{2,1} & \cdots & S_{i,\sigma_i}^{2,n_{i\sigma_i}} & X_{i,\sigma_i}^{2,1} & \cdots & X_{i,\sigma_i}^{2,n_{i\sigma_i}} & \star \end{matrix}\Bigg], \tag{2.2.11}$$

where $S^{1,k}_{i,m}$, $S^{2,k}_{i,m}$, $X^{1,k}_{i,m}$ and $X^{2,k}_{i,m}$, $m = 1, 2, \cdots, \sigma_i$ and $k = 1, 2, \cdots, n_{im}$, are $n_i \times 1$ column vectors. In fact, they are all real-valued. Next, define an $n_i \times n_i$ real-valued matrix,

$$\tilde{S}_i = \begin{bmatrix} S^{1,1}_{i,1} & S^{2,1}_{i,1} & \cdots & S^{1,n_{i1}}_{i,1} & S^{2,n_{i1}}_{i,1} & \cdots & S^{1,1}_{i,\sigma_i} & S^{2,1}_{i,\sigma_i} & \cdots & S^{1,n_{i\sigma_i}}_{i,\sigma_i} & S^{2,n_{i\sigma_i}}_{i,\sigma_i} \end{bmatrix}.$$

Finally, let

$$S = \text{blkdiag}\left\{ \tilde{S}_1, \cdots, \tilde{S}_\ell \right\}, \tag{2.2.12}$$

and $P = TS \in \mathbb{R}^{n \times n}$. It is now straightforward to show that $P^{-1}AP$ is in the real Jordan canonical form as described in Theorem 2.2.2. The algorithm has been implemented in Chen [14].

2.3. Structural Decompositions of Matrix Pairs

In this section, we will first recall the controllability structural decomposition (CSD) for a linear system characterized by a matrix pair (A, B), which was called a Brunovsky canonical form by many researchers in the literature (see e.g., [64]), as well as in the preliminary edition of this book [18]. However, it is noted that such a decomposition was actually first discovered by Luenberger [89] in 1967, which was three years earlier than the publication of Brunovsky's results [7] in 1970. As such, we rename such a canonical form the controllability structural decomposition, since it has a direct connection with the controllability structure of (A, B). We will next introduce a so-called block diagonal control canonical form (BDCCF) for a controllable matrix pair (A, B). Both the CSD and BDCCF will be the keys in the derivations of some important results later in the book. The derivation of the former is well-known in the literature and its software realization can be found in Chen [14]. We will give an explicit constructing algorithm for the latter to find non-singular transformations, say T_s and T_i, such that $T_s^{-1}AT_s$ has a special block diagonal form and $T_s^{-1}BT_i$ has an upper block triangular form. Such special forms of A and B will play an important role in constructing solutions to the general H_∞ almost disturbance decoupling problems later in this book. The existence of this block diagonal control canonical form was proved by Wonham [138].

We have the following theorems regarding the controllability structural decomposition and the block diagonal control canonical form for a given matrix pair.

Theorem 2.3.1. Consider a constant matrix pair (A, B) with $A \in \mathbb{R}^{n \times n}$ and $B \in \mathbb{R}^{n \times m}$ with B being of full rank. There exist nonsingular state and input transformations T_s and T_i such that $(\tilde{A}, \tilde{B}) := (T_s^{-1}AT_s, T_s^{-1}BT_i)$ has the following form,

$$\left(\begin{bmatrix} A_o & 0 & 0 & \cdots & 0 & 0 \\ 0 & 0 & I_{k_1-1} & \cdots & 0 & 0 \\ \star & \star & \star & \cdots & \star & \star \\ \vdots & \vdots & \vdots & \ddots & \vdots & \vdots \\ 0 & 0 & 0 & \cdots & 0 & I_{k_m-1} \\ \star & \star & \star & \cdots & \star & \star \end{bmatrix}, \begin{bmatrix} 0 & \cdots & 0 \\ 0 & \cdots & 0 \\ 1 & \cdots & 0 \\ \vdots & \ddots & \vdots \\ 0 & \cdots & 0 \\ 0 & \cdots & 1 \end{bmatrix} \right), \tag{2.3.1}$$

where $k_i > 0$, $i = 1, \cdots, m$, A_o is of dimension $n_o := n - \sum_{i=1}^{m} k_i$ and its eigenvalues are the uncontrollable modes of (A, B). Moreover, the set of integers, $\mathcal{C} := \{ n_o, k_1, \cdots, k_m \}$, is called the *controllability index* of (A, B). ◻

Proof. See Luenberger [89]. The software realization of such a canonical form can be found in Chen [14]. ◼

Theorem 2.3.2. Consider a constant matrix pair (A, B) with $A \in \mathbb{R}^{n\times n}$ and $B \in \mathbb{R}^{n\times m}$ and with (A, B) being completely controllable. Then there exist an integer $k \le m$, a set of κ integers $k_1, k_2, \cdots, k_\kappa$, and nonsingular transformations T_s and T_i such that

$$T_s^{-1}AT_s = \begin{bmatrix} A_1 & 0 & 0 & \cdots & 0 \\ 0 & A_2 & 0 & \cdots & 0 \\ 0 & 0 & A_3 & \cdots & 0 \\ \vdots & \vdots & \vdots & \ddots & \vdots \\ 0 & 0 & 0 & \cdots & A_\kappa \end{bmatrix}, \tag{2.3.2}$$

and

$$T_s^{-1}BT_i = \begin{bmatrix} B_1 & \star & \star & \cdots & \star & \star \\ 0 & B_2 & \star & \cdots & \star & \star \\ 0 & 0 & B_3 & \cdots & \star & \star \\ \vdots & \vdots & \vdots & \ddots & \vdots & \vdots \\ 0 & 0 & 0 & \cdots & B_\kappa & \star \end{bmatrix}, \tag{2.3.3}$$

where $\star$s represent some matrices of less interest, and A_i and B_i, $i = 1, 2, \cdots, \kappa$, have the following control canonical form,

$$A_i = \begin{bmatrix} 0 & 1 & 0 & \cdots & 0 \\ 0 & 0 & 1 & \cdots & 0 \\ \vdots & \vdots & \vdots & \ddots & \vdots \\ 0 & 0 & 0 & \cdots & 1 \\ -a_{k_i}^i & -a_{k_i-1}^i & -a_{k_i-2}^i & \cdots & -a_1^i \end{bmatrix}, \quad B_i = \begin{bmatrix} 0 \\ 0 \\ \vdots \\ 0 \\ 1 \end{bmatrix}, \tag{2.3.4}$$

for some scalars $a_1^i, a_2^i, \cdots, a_{k_i}^i$. Obviously, $\sum_{i=1}^{\kappa} k_i = n$. We call the above structure of A and B a *block diagonal control canonical form*. ◻

Proof. The existence of the block diagonal control canonical form was shown in [138]. In what follows, we will give an explicit constructing algorithm for realizing such a canonical form. First, we follow Theorem 2.2.2 to find a nonsingular transformation $Q \in \mathbb{R}^{n\times n}$ such that matrix A is transformed into a real Jordan canonical form, i.e.,

$$\tilde{A} = Q^{-1}AQ = \text{blkdiag}\Big\{J_{\lambda_1}^1, \cdots, J_{\lambda_1}^{\sigma_1}, J_{\lambda_2}^1, \cdots, J_{\lambda_2}^{\sigma_2}, \cdots\cdots, J_{\lambda_\ell}^1, \cdots, J_{\lambda_\ell}^{\sigma_\ell}\Big\}, \tag{2.3.5}$$

where $\lambda_i = \mu_i + j\omega_i \in \lambda(A)$ with $\omega_i \geq 0$, and also $\lambda_{i_1} \neq \lambda_{i_2}$, if $i_1 \neq i_2$. Moreover, for each $i \in \{1, 2, \cdots, \ell\}$ and $s = 1, 2, \cdots, \sigma_i$, $J_{\lambda_i}^s \in \mathbb{R}^{n_{is}\times n_{is}}$ has the following real Jordan form,

$$J_{\lambda_i}^s = \begin{bmatrix} \mu_i & 1 & & \\ & \ddots & \ddots & \\ & & \mu_i & 1 \\ & & & \mu_i \end{bmatrix}, \tag{2.3.6}$$

if $\omega_i = 0$, or

$$J_{\lambda_i}^s = \begin{bmatrix} \Lambda_i & I_2 & & \\ & \ddots & \ddots & \\ & & \Lambda_i & I_2 \\ & & & \Lambda_i \end{bmatrix}, \quad \Lambda_i = \begin{bmatrix} \mu_i & \omega_i \\ -\omega_i & \mu_i \end{bmatrix}, \tag{2.3.7}$$

if $\omega_i > 0$. For the sake of easy presentation later, we arrange the Jordan blocks in the way that $n_{i1} \geq n_{i2} \geq \cdots \geq n_{i\sigma_i}$. Next, compute

$$\tilde{B} = Q^{-1}B = \begin{bmatrix} B_{11}^1 & B_{11}^2 & \cdots & B_{11}^m \\ \vdots & \vdots & \ddots & \vdots \\ B_{1\sigma_1}^1 & B_{1\sigma_1}^2 & \cdots & B_{1\sigma_1}^m \\ B_{21}^1 & B_{21}^2 & \cdots & B_{21}^m \\ \vdots & \vdots & \ddots & \vdots \\ B_{2\sigma_2}^1 & B_{2\sigma_2}^2 & \cdots & B_{2\sigma_2}^m \\ \vdots & \vdots & \ddots & \vdots \\ \vdots & \vdots & \ddots & \vdots \\ B_{\ell 1}^1 & B_{\ell 1}^2 & \cdots & B_{\ell 1}^m \\ \vdots & \vdots & \ddots & \vdots \\ B_{\ell\sigma_\ell}^1 & B_{\ell\sigma_\ell}^2 & \cdots & B_{\ell\sigma_\ell}^m \end{bmatrix}. \tag{2.3.8}$$

It is straightforward to verify that the controllability of (A, B) implies: there exists a B_{i1}^ν with $\nu \in \{1, 2, \cdots, m\}$ such that $(J_{\lambda_i}^1, B_{i1}^\nu)$ is completely controllable, which is equivalent to the last row of B_{i1}^ν being nonzero if λ_i is real, or at least one of the last two rows of B_{i1}^{ν} being nonzero if λ_i is not real. Thus, it is simple to find a vector

$$T_1 = \begin{bmatrix} t_{11} \\ t_{21} \\ \vdots \\ t_{m1} \end{bmatrix}, \quad t_{11} \neq 0, \tag{2.3.9}$$

and partition

$$\tilde{B}_1 = \tilde{B}T_1 = \begin{bmatrix} \tilde{B}_{11}^1 \\ \vdots \\ \tilde{B}_{1\sigma_1}^1 \\ \tilde{B}_{21}^1 \\ \vdots \\ \tilde{B}_{2\sigma_2}^1 \\ \vdots \\ \vdots \\ \tilde{B}_{\ell 1}^1 \\ \vdots \\ \tilde{B}_{\ell\sigma_\ell}^1 \end{bmatrix}, \tag{2.3.10}$$

such that $(J_{\lambda_i}^1, \tilde{B}_{i1}^1)$ is completely controllable. Because of the special structure of the real Jordan form and the fact that $n_{i1} \geq n_{i2} \geq \cdots \geq n_{i\sigma_i}$, the eigenstructures associated with $J_{\lambda_i}^s$ with $s > 1$ are totally uncontrollable by $\tilde{B}_1$. Thus, it is straightforward to show that there exist nonsingular transformations T_{s1}^i, $i = 1, 2, \cdots, \ell$, such that

$$(T_{s1}^i)^{-1} \begin{bmatrix} J_{\lambda_i}^1 & & & \\ & J_{\lambda_i}^2 & & \\ & & \ddots & \\ & & & J_{\lambda_i}^{\sigma_i} \end{bmatrix} T_{s1}^i = \begin{bmatrix} J_{\lambda_i}^1 & & & \\ & J_{\lambda_i}^2 & & \\ & & \ddots & \\ & & & J_{\lambda_i}^{\sigma_i} \end{bmatrix}, \tag{2.3.11}$$

and

$$(T_{s1}^i)^{-1} \begin{bmatrix} \tilde{B}_{i1}^1 \\ \tilde{B}_{i2}^1 \\ \vdots \\ \tilde{B}_{i\sigma_i}^1 \end{bmatrix} = \begin{bmatrix} \breve{B}_{i1}^1 \\ 0 \\ \vdots \\ 0 \end{bmatrix}, \tag{2.3.12}$$

with $(J_{\lambda_i}^1, \breve{B}_{i1}^1)$ being completely controllable. This can be done by utilizing the special structure of the controllability structural decomposition (see Theorem 2.3.1). Next, perform a permutation transformation P_{s1} such that

$$
\begin{aligned}
(P_{s1})^{-1}\begin{bmatrix} T_{s1}^1 & & & \\ & T_{s1}^2 & & \\ & & \ddots & \\ & & & T_{s1}^\ell \end{bmatrix}^{-1} \tilde{A} \begin{bmatrix} T_{s1}^1 & & & \\ & T_{s1}^2 & & \\ & & \ddots & \\ & & & T_{s1}^\ell \end{bmatrix} P_{s1} \\
= \operatorname{blkdiag}\left\{ J_{\lambda_1}^1, \cdots, J_{\lambda_\ell}^1, J_{\lambda_1}^2, \cdots, J_{\lambda_1}^{\sigma_1}, \cdots\cdots, J_{\lambda_\ell}^2, \cdots, J_{\lambda_\ell}^{\sigma_\ell} \right\},
\end{aligned}
\tag{2.3.13}
$$

and

$$
(P_{s1})^{-1}\begin{bmatrix} T_{s1}^1 & & & \\ & T_{s1}^2 & & \\ & & \ddots & \\ & & & T_{s1}^\ell \end{bmatrix}^{-1} \tilde{B} \begin{bmatrix} t_{11} & 0 & \cdots & 0 \\ t_{21} & 1 & \cdots & 0 \\ \vdots & 0 & \ddots & 0 \\ t_{m1} & 0 & \cdots & 1 \end{bmatrix}
= \begin{bmatrix}
\breve{B}_{11}^1 & \breve{B}_{11}^2 & \cdots & \breve{B}_{11}^m \\
\vdots & \vdots & \ddots & \vdots \\
\breve{B}_{\ell 1}^1 & \breve{B}_{\ell 1}^2 & \cdots & \breve{B}_{\ell 1}^m \\
0 & \breve{B}_{12}^2 & \cdots & \breve{B}_{12}^m \\
\vdots & \vdots & \ddots & \vdots \\
0 & \breve{B}_{1\sigma_1}^2 & \cdots & \breve{B}_{1\sigma_1}^m \\
\vdots & \vdots & \ddots & \vdots \\
\vdots & \vdots & \ddots & \vdots \\
0 & \breve{B}_{\ell 2}^2 & \cdots & \breve{B}_{\ell 2}^m \\
\vdots & \vdots & \ddots & \vdots \\
0 & \breve{B}_{\ell\sigma_\ell}^2 & \cdots & \breve{B}_{\ell\sigma_\ell}^m
\end{bmatrix}.
\tag{2.3.14}
$$

Because λ_i, $i = 1, 2, \cdots, \ell$, are distinct, the controllability of $(J_{\lambda_i}^1, \breve{B}_{i1}^1)$ implies that the pair

$$
(\breve{A}_1, \breve{B}_1) := \left(\begin{bmatrix} J_{\lambda_1}^1 & & & \\ & J_{\lambda_2}^1 & & \\ & & \ddots & \\ & & & J_{\lambda_\ell}^1 \end{bmatrix}, \begin{bmatrix} \breve{B}_{11}^1 \\ \breve{B}_{21}^1 \\ \vdots \\ \breve{B}_{i1}^1 \end{bmatrix} \right),
\tag{2.3.15}
$$

is completely controllable. Hence, there exists a nonsingular transformation $X_1 \in \mathbb{R}^{k_1 \times k_1}$, where $k_1 = \sum_{i=1}^{\ell} n_{i1}$, such that

$$X_1^{-1} \check{A}_1 X_1 = A_1 = \begin{bmatrix} 0 & 1 & 0 & \cdots & 0 \\ 0 & 0 & 1 & \cdots & 0 \\ \vdots & \vdots & \vdots & \ddots & \vdots \\ 0 & 0 & 0 & \cdots & 1 \\ -a_{k_1}^1 & -a_{k_1-1}^1 & -a_{k_1-2}^1 & \cdots & -a_1^1 \end{bmatrix}, \tag{2.3.16}$$

and

$$X_1^{-1} \check{B}_1 = B_1 = \begin{bmatrix} 0 \\ 0 \\ \vdots \\ 0 \\ 1 \end{bmatrix}. \tag{2.3.17}$$

Next, repeating the above procedure for the following pair

$$\left(\operatorname{blkdiag}\left\{ J_{\lambda_1}^2, \cdots, J_{\lambda_1}^{\sigma_1}, \cdots\cdots, J_{\lambda_\ell}^2, \cdots, J_{\lambda_\ell}^{\sigma_\ell} \right\}, \begin{bmatrix} \check{B}_{12}^2 & \cdots & \check{B}_{12}^m \\ \vdots & \ddots & \vdots \\ \check{B}_{1\sigma_1}^2 & \cdots & \check{B}_{1\sigma_1}^m \\ \vdots & \ddots & \vdots \\ \vdots & \ddots & \vdots \\ \check{B}_{\ell 2}^2 & \cdots & \check{B}_{\ell 2}^m \\ \vdots & \ddots & \vdots \\ \check{B}_{\ell\sigma_\ell}^2 & \cdots & \check{B}_{\ell\sigma_\ell}^m \end{bmatrix} \right), \tag{2.3.18}$$

one is able to separate (A_2, B_2). Keep repeating the same procedure for $\kappa - 2$ more steps, where $\kappa = \max\{\sigma_1, \sigma_2, \cdots, \sigma_\ell\}$, one is able to obtain the block diagonal control canonical form as in Theorem 2.3.2. This completes the proof of the theorem. The result has been implemented in Chen [14]. ⊠

We illustrate the above results in the following example.

Example 2.3.1. Consider a matrix pair (A, B) characterized by

$$A = \begin{bmatrix} 1 & 1 & 0 & 0 & 0 & 0 & 0 & 0 \\ 0 & 1 & 0 & 0 & 0 & 0 & 0 & 0 \\ 0 & 0 & 1 & 1 & 0 & 0 & 0 & 0 \\ 0 & 0 & 0 & 1 & 0 & 0 & 0 & 0 \\ 0 & 0 & 0 & 0 & 0 & 1 & 1 & 0 \\ 0 & 0 & 0 & 0 & -1 & 0 & 0 & 1 \\ 0 & 0 & 0 & 0 & 0 & 0 & 0 & 1 \\ 0 & 0 & 0 & 0 & 0 & 0 & -1 & 0 \end{bmatrix}, \quad B = \begin{bmatrix} 1 & 8 \\ 2 & 7 \\ 3 & 6 \\ 4 & 5 \\ 5 & 4 \\ 6 & 3 \\ 7 & 2 \\ 8 & 1 \end{bmatrix}, \tag{2.3.19}$$

where matrix A is already in the form of the real Jordan canonical form with $\lambda_1 = 1$, $\sigma_1 = 2$ and $\lambda_2 = j$, $\sigma_2 = 1$. Following the proof of Theorem 2.3.2, we obtain

$$T_s = \begin{bmatrix} 0.1508 & 0.1508 & 0.3015 & 0.3015 & 0.1508 & 0.1508 & -0.4002 & 1.8189 \\ -0.3015 & 0.3015 & -0.6030 & 0.6030 & -0.3015 & 0.3015 & -1.4188 & 1.4188 \\ 0.1508 & 0.4523 & 0.3015 & 0.9045 & 0.1508 & 0.4523 & 0.3274 & -1.1641 \\ -0.6030 & 0.6030 & -1.2061 & 1.2061 & -0.6030 & 0.6030 & 0.8367 & -0.8367 \\ -0.1508 & 3.4674 & -4.5227 & 0 & 0.4523 & 0.7538 & 0 & 0 \\ -1.9598 & 2.7136 & 0.9045 & -1.2061 & -1.3568 & 0.9045 & 0 & 0 \\ 1.2061 & -1.3568 & 0.3015 & -0.3015 & -0.9045 & 1.0553 & 0 & 0 \\ -1.0553 & 3.3166 & -4.5227 & 4.5227 & -3.4674 & 1.2061 & 0 & 0 \end{bmatrix},$$

$$T_i = \begin{bmatrix} 0.1508 & 0 \\ 0 & 0.4083 \end{bmatrix},$$

and

$$T_s^{-1} A T_s = \begin{bmatrix} 0 & 1 & 0 & 0 & 0 & 0 & 0 & 0 \\ 0 & 0 & 1 & 0 & 0 & 0 & 0 & 0 \\ 0 & 0 & 0 & 1 & 0 & 0 & 0 & 0 \\ 0 & 0 & 0 & 0 & 1 & 0 & 0 & 0 \\ 0 & 0 & 0 & 0 & 0 & 1 & 0 & 0 \\ -1 & 2 & -3 & 4 & -3 & 2 & 0 & 0 \\ 0 & 0 & 0 & 0 & 0 & 0 & 0 & 1 \\ 0 & 0 & 0 & 0 & 0 & 0 & -1 & 2 \end{bmatrix}, \quad T_s^{-1} B T_i = \begin{bmatrix} 0 & -1.4368 \\ 0 & -0.2982 \\ 0 & 0.5207 \\ 0 & 1.3969 \\ 0 & 2.8085 \\ 1 & 4.6900 \\ 0 & 0 \\ 0 & 1 \end{bmatrix}.$$

This verifies the results of Theorem 2.3.2. ⓔ

2.4. Special Coordinate Basis

Let us consider a general proper linear time-invariant (LTI) system Σ_*, which could be of either continuous-time or discrete-time, characterized by a matrix quadruple (A_*, B_*, C_*, D_*) or in the state space form,

$$\Sigma_* : \begin{cases} \delta(x) = A_* \, x + B_* \, u, \\ \quad y \;\; = C_* \, x + D_* \, u, \end{cases} \tag{2.4.1}$$

where $\delta(x) = \dot{x}(t)$, if Σ_* is a continuous-time system, or $\delta(x) = x(k+1)$, if Σ_* is a discrete-time system. Similarly, $x \in \mathbb{R}^n$, $u \in \mathbb{R}^m$ and $y \in \mathbb{R}^p$ are the state, the input and the output of Σ_*. They represent $x(t)$, $u(t)$ and $y(t)$, respectively, if the given system is of continuous-time, or represent $x(k)$, $u(k)$ and $y(k)$, respectively, if Σ_* is of discrete-time. Without loss of any generality, we assume that both $[\,B_*' \;\; D_*'\,]$ and $[\,C_* \;\; D_*\,]$ are of full rank. The transfer function of Σ_* is then given by

$$H_*(\varsigma) = C_*(\varsigma I - A_*)^{-1} B_* + D_*, \tag{2.4.2}$$

where $\varsigma = s$, the Laplace transform operator, if Σ_* is of continuous-time, or $\varsigma = z$, the z-transform operator, if Σ_* is of discrete-time. It is simple to verify that there exist nonsingular transformations U and V such that

$$UD_*V = \begin{bmatrix} I_{m_0} & 0 \\ 0 & 0 \end{bmatrix}, \tag{2.4.3}$$

where m_0 is the rank of matrix D_*. In fact, U can be chosen as an orthogonal matrix. This fact will be used later in the computation of γ^* throughout this book. Hence hereafter, without loss of generality, it is assumed that the matrix D_* has the form given on the right hand side of (2.4.3). One can now rewrite system Σ_* of (2.4.1) as,

$$\begin{cases} \delta(x) = A_* \, x + [B_{*,0} \;\; B_{*,1}] \begin{pmatrix} u_0 \\ u_1 \end{pmatrix}, \\ \begin{pmatrix} y_0 \\ y_1 \end{pmatrix} = \begin{bmatrix} C_{*,0} \\ C_{*,1} \end{bmatrix} x + \begin{bmatrix} I_{m_0} & 0 \\ 0 & 0 \end{bmatrix} \begin{pmatrix} u_0 \\ u_1 \end{pmatrix}, \end{cases} \tag{2.4.4}$$

where the matrices $B_{*,0}$, $B_{*,1}$, $C_{*,0}$ and $C_{*,1}$ have appropriate dimensions. We have the following theorem.

Theorem 2.4.1 (SCB). Given the linear system Σ_* of (2.4.1), there exist

1. Coordinate free nonnegative integers n_a^-, n_a^0, n_a^+, n_b, n_c, n_d, $m_d \le m - m_0$ and q_i, $i = 1, \cdots, m_d$, and
2. Non-singular state, output and input transformations Γ_s, Γ_o and Γ_i which take the given Σ_* into a special coordinate basis that displays explicitly both the finite and infinite zero structures of Σ_*.

The special coordinate basis is described by the following set of equations:

$$x = \Gamma_s \tilde{x}, \quad y = \Gamma_o \tilde{y}, \quad u = \Gamma_i \tilde{u}, \tag{2.4.5}$$

$$\tilde{x} = \begin{pmatrix} x_a \\ x_b \\ x_c \\ x_d \end{pmatrix}, \quad x_a = \begin{pmatrix} x_a^- \\ x_a^0 \\ x_a^+ \end{pmatrix}, \quad x_d = \begin{pmatrix} x_1 \\ x_2 \\ \vdots \\ x_{m_d} \end{pmatrix}, \tag{2.4.6}$$

$$\tilde{y} = \begin{pmatrix} y_0 \\ y_d \\ y_b \end{pmatrix}, \; y_d = \begin{pmatrix} y_1 \\ y_2 \\ \vdots \\ y_{m_d} \end{pmatrix}, \; \tilde{u} = \begin{pmatrix} u_0 \\ u_d \\ u_c \end{pmatrix}, \; u_d = \begin{pmatrix} u_1 \\ u_2 \\ \vdots \\ u_{m_d} \end{pmatrix}, \tag{2.4.7}$$

and

$$\delta(x_a^-) = A_{aa}^- x_a^- + B_{0a}^- y_0 + L_{ad}^- y_d + L_{ab}^- y_b, \tag{2.4.8}$$

$$\delta(x_a^0) = A_{aa}^0 x_a^0 + B_{0a}^0 y_0 + L_{ad}^0 y_d + L_{ab}^0 y_b, \tag{2.4.9}$$

$$\delta(x_a^+) = A_{aa}^+ x_a^+ + B_{0a}^+ y_0 + L_{ad}^+ y_d + L_{ab}^+ y_b, \tag{2.4.10}$$

$$\delta(x_b) = A_{bb} x_b + B_{0b} y_0 + L_{bd} y_d, \quad y_b = C_b x_b, \tag{2.4.11}$$

$$\delta(x_c) = A_{cc} x_c + B_{0c} y_0 + L_{cb} y_b + L_{cd} y_d + B_c \left[E_{ca}^- x_a^- + E_{ca}^0 x_a^0 + E_{ca}^+ x_a^+ \right] + B_c u_c, \tag{2.4.12}$$

$$y_0 = C_{0c} x_c + C_{0a}^- x_a^- + C_{0a}^+ x_a^0 + C_{0a}^+ x_a^+ + C_{0d} x_d + C_{0b} x_b + u_0, \tag{2.4.13}$$

and for each $i = 1, \cdots, m_d$,

$$\delta(x_i) = A_{q_i} x_i + L_{i0} y_0 + L_{id} y_d + B_{q_i} \left[u_i + E_{ia} x_a + E_{ib} x_b + E_{ic} x_c + \sum_{j=1}^{m_d} E_{ij} x_j \right], \tag{2.4.14}$$

$$y_i = C_{q_i} x_i, \quad y_d = C_d x_d. \tag{2.4.15}$$

Here the states x_a^-, x_a^0, x_a^+, x_b, x_c and x_d are respectively of dimensions n_a^-, n_a^0, n_a^+, n_b, n_c and $n_d = \sum_{i=1}^{m_d} q_i$, while x_i is of dimension q_i for each $i = 1, \cdots, m_d$. The control vectors u_0, u_d and u_c are respectively of dimensions m_0, m_d and $m_c = m - m_0 - m_d$ while the output vectors y_0, y_d and y_b are respectively of dimensions $p_0 = m_0$, $p_d = m_d$ and $p_b = p - p_0 - p_d$. The matrices A_{q_i}, B_{q_i} and C_{q_i} have the following form:

$$A_{q_i} = \begin{bmatrix} 0 & I_{q_i-1} \\ 0 & 0 \end{bmatrix}, \quad B_{q_i} = \begin{bmatrix} 0 \\ 1 \end{bmatrix}, \quad C_{q_i} = [1, \ 0, \ \cdots, \ 0]. \tag{2.4.16}$$

Assuming that x_i, $i = 1, 2, \cdots, m_d$, are arranged such that $q_i \leq q_{i+1}$, the matrix L_{id} has the particular form

$$L_{id} = [\, L_{i1} \quad L_{i2} \quad \cdots \quad L_{ii-1} \quad 0 \quad \cdots \quad 0 \,]. \tag{2.4.17}$$

The last row of each L_{id} is identically zero. Moreover,

1. If Σ_* is a continuous-time system, then

$$\lambda(A_{aa}^-) \subset \mathbb{C}^-, \quad \lambda(A_{aa}^0) \subset \mathbb{C}^0, \quad \lambda(A_{aa}^+) \subset \mathbb{C}^+. \tag{2.4.18}$$

2. If Σ_* is a discrete-time system, then

$$\lambda(A_{aa}^-) \subset \mathbb{C}^{\odot}, \quad \lambda(A_{aa}^0) \subset \mathbb{C}^{\circ}, \quad \lambda(A_{aa}^+) \subset \mathbb{C}^{\otimes}. \tag{2.4.19}$$

Also, the pair (A_{cc}, B_c) is controllable and the pair (A_{bb}, C_b) is observable. ▣

Proof. For strictly proper systems, using a modified structural algorithm of Silverman [122], an explicit procedure of constructing the above special coordinate basis is given in [116]. The required modifications for non-strictly proper systems are given in [111].

Here in Theorem 2.4.1 by another change of basis, the variable x_a is further decomposed into x_a^-, x_a^0 and x_a^+. For continuous-time systems, one can use the real Schur algorithm to obtain such a decomposition. For discrete-time systems, the algorithm of Chen [13] can be used.

The software toolboxes that realize the continuous-time SCB can be found in LAS by Chen [11] or in MATLAB by Lin [79]. The realization of this unified SCB can be found in Chen [14]. A numerical example will be given at the end of this section to illustrate the procedure of constructing the SCB and all its associated properties. ⊠

We can rewrite the special coordinate basis of the quadruple (A_*, B_*, C_*, D_*) given by Theorem 2.4.1 in a more compact form,

$$\begin{aligned}\tilde{A}_* &= \Gamma_s^{-1}(A_* - B_{*,0}C_{*,0})\Gamma_s \\ &= \begin{bmatrix} A_{aa}^- & 0 & 0 & L_{ab}^-C_b & 0 & L_{ad}^-C_d \\ 0 & A_{aa}^0 & 0 & L_{ab}^0C_b & 0 & L_{ad}^0C_d \\ 0 & 0 & A_{aa}^+ & L_{ab}^+C_b & 0 & L_{ad}^+C_d \\ 0 & 0 & 0 & A_{bb} & 0 & L_{bd}C_d \\ B_cE_{ca}^- & B_cE_{ca}^0 & B_cE_{ca}^+ & L_{cb}C_b & A_{cc} & L_{cd}C_d \\ B_dE_{da}^- & B_dE_{da}^0 & B_dE_{da}^+ & B_dE_{db} & B_dE_{dc} & A_{dd} \end{bmatrix},\end{aligned} \tag{2.4.20}$$

$$\tilde{B}_* = \Gamma_s^{-1}\,[\,B_{*,0} \quad B_{*,1}\,]\,\Gamma_i = \begin{bmatrix} B_{0a}^- & 0 & 0 \\ B_{0a}^0 & 0 & 0 \\ B_{0a}^+ & 0 & 0 \\ B_{0b} & 0 & 0 \\ B_{0c} & 0 & B_c \\ B_{0d} & B_d & 0 \end{bmatrix}, \tag{2.4.21}$$

$$\tilde{C}_* = \Gamma_o^{-1}\begin{bmatrix} C_{*,0} \\ C_{*,1} \end{bmatrix}\Gamma_s = \begin{bmatrix} C_{0a}^- & C_{0a}^0 & C_{0a}^+ & C_{0b} & C_{0c} & C_{0d} \\ 0 & 0 & 0 & 0 & 0 & C_d \\ 0 & 0 & 0 & C_b & 0 & 0 \end{bmatrix}, \tag{2.4.22}$$

$$\tilde{D}_* = \Gamma_o^{-1}D_*\Gamma_i = \begin{bmatrix} I_{m_0} & 0 & 0 \\ 0 & 0 & 0 \\ 0 & 0 & 0 \end{bmatrix}. \tag{2.4.23}$$

A block diagram of the special coordinate basis of Theorem 2.4.1 is given in Figure 2.4.1. In this figure, a signal given by a double-edged arrow is some

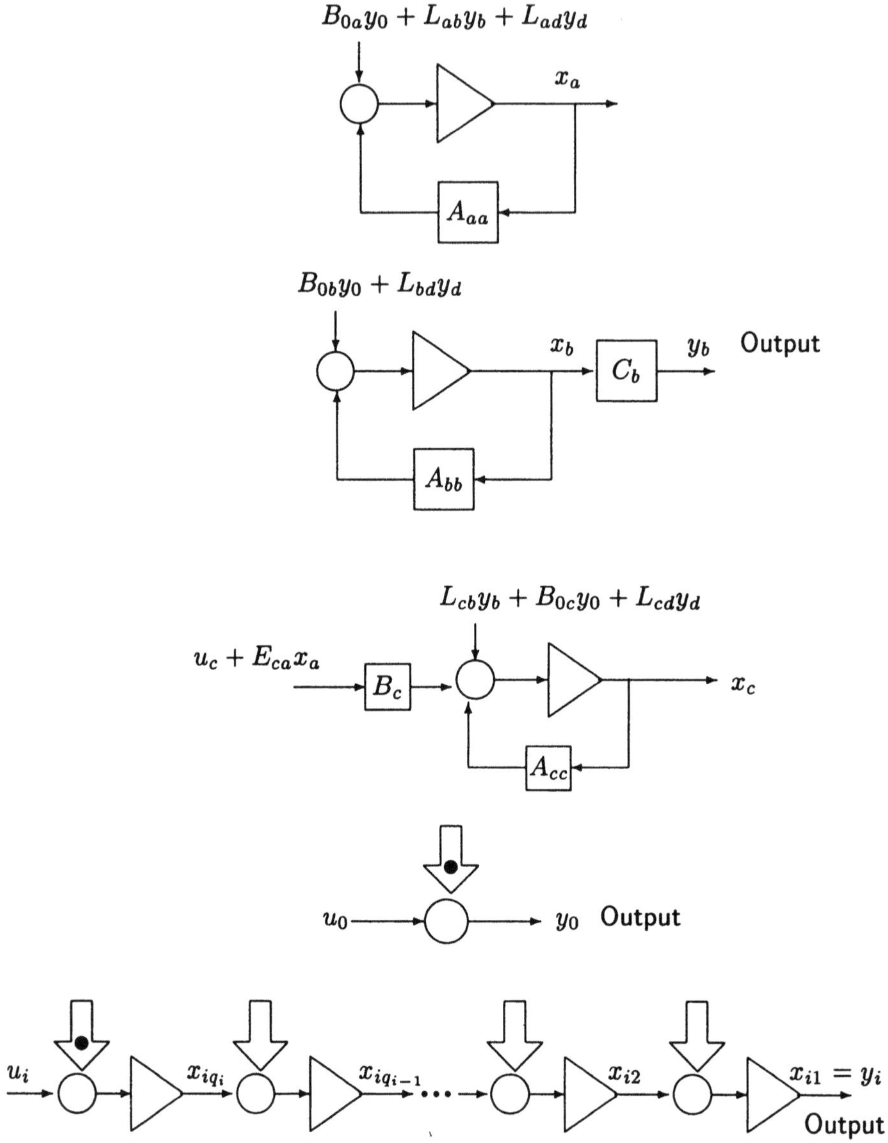

Note that a signal given by a double-edged arrow with a solid dot is some linear combination of all the states, whereas a signal given by a simple double-edged arrow is some linear combination of only output y_d. Also, matrices B_{0a}, L_{ab}, L_{ad} and E_{ca} are to be defined in Property 2.4.1.

Figure 2.4.1: A block diagram representation of the special coordinate basis.

linear combination of outputs y_i, $i = 0$ to m_d, where as a signal given by the double-edged arrow with a solid dot is some linear combination of all the states. Also, the block $\triangleright$ is either an integrator if Σ_* is of continuous-time or a backward shifting operator if Σ_* is of discrete-time.

We note the following intuitive points regarding the special coordinate basis.

1. The variable u_i controls the output y_i through a stack of q_i integrators (or backward shifting operators), while x_i is the state associated with those integrators (or backward shifting operators) between u_i and y_i. Moreover, (A_{q_i}, B_{q_i}) and (A_{q_i}, C_{q_i}) respectively form controllable and observable pairs. This implies that all the states x_i are both controllable and observable.

2. The output y_b and the state x_b are not directly influenced by any inputs, however they could be indirectly controlled through the output y_d. Moreover, (A_{bb}, C_b) forms an observable pair. This implies that the state x_b is observable.

3. The state x_c is directly controlled by the input u_c, but it does not directly affect any output. Moreover, (A_{cc}, B_c) forms a controllable pair. This implies that the state x_c is controllable.

4. The state x_a is neither directly controlled by any input nor does it directly affect any output.

In what follows, we state some important properties of the above special coordinate basis which are pertinent to our present work and will be used throughout this book. The proofs of these properties will be given in the next section.

Property 2.4.1. The given system Σ_* is observable (detectable) if and only if the pair $(A_{\rm obs}, C_{\rm obs})$ is observable (detectable), where

$$A_{\rm obs} := \begin{bmatrix} A_{aa} & 0 \\ B_c E_{ca} & A_{cc} \end{bmatrix}, \quad C_{\rm obs} := \begin{bmatrix} C_{0a} & C_{0c} \\ E_{da} & E_{dc} \end{bmatrix}, \tag{2.4.24}$$

and where

$$A_{aa} := \begin{bmatrix} A_{aa}^- & 0 & 0 \\ 0 & A_{aa}^0 & 0 \\ 0 & 0 & A_{aa}^+ \end{bmatrix}, \quad C_{0a} := [\, C_{0a}^- \quad C_{0a}^0 \quad C_{0a}^+ \,], \tag{2.4.25}$$

$$E_{da} := [\, E_{da}^- \quad E_{da}^0 \quad E_{da}^+ \,], \quad E_{ca} := [\, E_{ca}^- \quad E_{ca}^0 \quad E_{ca}^+ \,]. \tag{2.4.26}$$

Also, define

$$A_{\rm con} := \begin{bmatrix} A_{aa} & L_{ab}C_b \\ 0 & A_{bb} \end{bmatrix}, \quad B_{\rm con} := \begin{bmatrix} B_{0a} & L_{ad} \\ B_{0b} & L_{bd} \end{bmatrix}, \tag{2.4.27}$$

$$B_{0a} := \begin{bmatrix} B_{0a}^- \\ B_{0a}^0 \\ B_{0a}^+ \end{bmatrix}, \quad L_{ab} := \begin{bmatrix} L_{ab}^- \\ L_{ab}^0 \\ L_{ab}^+ \end{bmatrix}, \quad L_{ad} := \begin{bmatrix} L_{ad}^- \\ L_{ad}^0 \\ L_{ad}^+ \end{bmatrix}. \tag{2.4.28}$$

Similarly, Σ_* is controllable (stabilizable) if and only if the pair $(A_{\rm con}, B_{\rm con})$ is controllable (stabilizable). Ⓟ

The invariant zeros of a system Σ_* characterized by (A_*, B_*, C_*, D_*) can be defined via the Smith canonical form of the (Rosenbrock) system matrix [107] of Σ_*,

$$P_{\Sigma_*}(\varsigma) := \begin{bmatrix} \varsigma I - A_* & -B_* \\ C_* & D_* \end{bmatrix}. \tag{2.4.29}$$

We have the following definition for the invariant zeros (see also [91]).

Definition 2.4.1. (Invariant Zeros). A complex scalar $\alpha \in \mathbb{C}$ is said to be an invariant zero of Σ_* if

$$\text{rank}\,\{P_{\Sigma_*}(\alpha)\} < n + \text{normrank}\,\{H_*(\varsigma)\}, \tag{2.4.30}$$

where normrank $\{H_*(\varsigma)\}$ denotes the normal rank of $H_*(\varsigma)$, which is defined as its rank over the field of rational functions of ς with real coefficients. Ⓓ

The special coordinate basis of Theorem 2.4.1 shows explicitly the invariant zeros and the normal rank of Σ_*. To be more specific, we have the following properties.

Property 2.4.2.

1. The normal rank of $H_*(\varsigma)$ is equal to $m_0 + m_d$.

2. Invariant zeros of Σ_* are the eigenvalues of A_{aa}, which are the unions of the eigenvalues of A_{aa}^-, A_{aa}^0 and A_{aa}^+. Moreover, the given system Σ_* is of minimum phase if and only if A_{aa} has only stable eigenvalues, marginal minimum phase if and only if A_{aa} has no unstable eigenvalue but has at least one marginally stable eigenvalue, and nonminimum phase if and only if A_{aa} has at least one unstable eigenvalue. Ⓟ

In order to display various multiplicities of invariant zeros, let X_a be a nonsingular transformation matrix such that A_{aa} can be transformed into a Jordan canonical form (see Theorem 2.2.1), i.e.,

$$X_a^{-1} A_{aa} X_a = J = \text{blkdiag}\left\{J_1, J_2, \cdots, J_k\right\}, \tag{2.4.31}$$

where J_i, $i = 1, 2, \cdots, k$, are some $n_i \times n_i$ Jordan blocks:

$$J_i = \text{diag}\left\{\alpha_i, \alpha_i, \cdots, \alpha_i\right\} + \begin{bmatrix} 0 & I_{n_i-1} \\ 0 & 0 \end{bmatrix}. \tag{2.4.32}$$

For any given $\alpha \in \lambda(A_{aa})$, let there be τ_α Jordan blocks of A_{aa} associated with α. Let $n_{\alpha,1}$, $n_{\alpha,2}$, $\cdots$, n_{α,τ_α} be the dimensions of the corresponding Jordan blocks. Then we say α is an invariant zero of Σ_* with multiplicity structure $S_\alpha^\star(\Sigma_*)$ (see also [109]),

$$S_\alpha^\star(\Sigma_*) = \left\{n_{\alpha,1}, n_{\alpha,2}, \cdots, n_{\alpha,\tau_\alpha}\right\}. \tag{2.4.33}$$

The geometric multiplicity of α is then simply given by τ_α, and the algebraic multiplicity of α is given by $\sum_{i=1}^{\tau_\alpha} n_{\alpha,i}$. Here we should note that the invariant zeros together with their structures of Σ_* are related to the structural invariant indices list $\mathcal{I}_1(\Sigma_*)$ of Morse [94].

The special coordinate basis can also reveal the infinite zero structure of Σ_*. We note that the infinite zero structure of Σ_* can be either defined in association with root-locus theory or as Smith-McMillan zeros of the transfer function at infinity. For the sake of simplicity, we only consider the infinite zeros from the point of view of Smith-McMillan theory here. To define the zero structure of $H_*(\varsigma)$ at infinity, one can use the familiar Smith-McMillan description of the zero structure at finite frequencies of a general not necessarily square but strictly proper transfer function matrix $H_*(\varsigma)$. Namely, a rational matrix $H_*(\varsigma)$ possesses an infinite zero of order k when $H_*(1/z)$ has a finite zero of precisely that order at $z = 0$ (see [42], [104], [107] and [131]). The number of zeros at infinity together with their orders indeed defines an infinite zero structure. Owens [97] related the orders of the infinite zeros of the root-loci of a square system with a nonsingular transfer function matrix to C^* structural invariant indices list $\mathcal{I}_4$ of Morse [94]. This connection reveals that even for general not necessarily strictly proper systems, the *structure at infinity is in fact the topology of inherent integrations between the input and the output variables.* The special coordinate basis of Theorem 2.4.1 explicitly shows this topology of inherent integrations. The following property pinpoints this.

Property 2.4.3. Σ_* has $m_0 =$ rank (D_*) infinite zeros of order 0. The infinite zero structure (of order greater than 0) of Σ_* is given by

$$S_\infty^\star(\Sigma_*) = \left\{ q_1, q_2, \cdots, q_{m_d} \right\}. \tag{2.4.34}$$

That is, each q_i corresponds to an infinite zero of Σ_* of order q_i. Note that for a single-input-single-output system Σ_*, we have $S_\infty^\star(\Sigma_*) = \{q_1\}$, where q_1 is the *relative degree* of Σ_*. ▣

The special coordinate basis can also exhibit the invertibility structure of a given system Σ_*. The formal definitions of right invertibility and left invertibility of a linear system can be found in [95]. Basically, for the usual case when $[\,B_*' \;\; D_*'\,]$ and $[\,C_* \;\; D_*\,]$ are of maximal rank, the system Σ_* or equivalently $H_*(\varsigma)$ is said to be left invertible if there exists a rational matrix function, say $L_*(\varsigma)$, such that

$$L_*(\varsigma) H_*(\varsigma) = I_m. \tag{2.4.35}$$

Σ_* or $H_*(\varsigma)$ is said to be right invertible if there exists a rational matrix function, say $R_*(\varsigma)$, such that

$$H_*(\varsigma) R_*(\varsigma) = I_p. \tag{2.4.36}$$

Σ_* is invertible if it is both left and right invertible, and Σ_* is degenerate if it is neither left nor right invertible.

Property 2.4.4. The given system Σ_* is right invertible if and only if x_b (and hence y_b) are non-existent, left invertible if and only if x_c (and hence u_c) are non-existent, and invertible if and only if both x_b and x_c are non-existent. Moreover, Σ_* is degenerate if and only if both x_b and x_c are present. ▣

The special coordinate basis can also be modified to obtain the structural invariant indices lists $\mathcal{I}_2$ and $\mathcal{I}_3$ of Morse [94] of the given system Σ_*. In order to display $\mathcal{I}_2(\Sigma_*)$, we let X_c and X_i be nonsingular matrices such that the controllable pair (A_{cc}, B_c) is transformed into the controllability structural decomposition (see Theorem 2.3.1), i.e.,

$$X_c^{-1} A_{cc} X_c = \begin{bmatrix} 0 & I_{\ell_1 - 1} & \cdots & 0 & 0 \\ \star & \star & \cdots & \star & \star \\ \vdots & \vdots & \ddots & \vdots & \vdots \\ 0 & 0 & \cdots & 0 & I_{\ell_{m_c} - 1} \\ \star & \star & \cdots & \star & \star \end{bmatrix}, \quad X_c^{-1} B_c X_i = \begin{bmatrix} 0 & \cdots & 0 \\ 1 & \cdots & 0 \\ \vdots & \ddots & \vdots \\ 0 & \cdots & 0 \\ 0 & \cdots & 1 \end{bmatrix}, \tag{2.4.37}$$

where $\star$s denote constant scalars or row vectors. Then we have

$$\mathcal{I}_2(\Sigma_*) = \left\{ \ell_1, \cdots, \ell_{m_c} \right\}, \tag{2.4.38}$$

which is also called the controllability index of (A_{cc}, B_c). Similarly, we have

$$\mathcal{I}_3(\Sigma_*) = \left\{\mu_1, \cdots, \mu_{p_b}\right\}, \tag{2.4.39}$$

where $\left\{\mu_1, \cdots, \mu_{p_b}\right\}$ is the controllability index of the controllable pair (A'_{bb}, C'_b).

By now it is clear that the special coordinate basis decomposes the state-space into several distinct parts. In fact, the state-space $\mathcal{X}$ is decomposed as

$$\mathcal{X} = \mathcal{X}_a^- \oplus \mathcal{X}_a^0 \oplus \mathcal{X}_a^+ \oplus \mathcal{X}_b \oplus \mathcal{X}_c \oplus \mathcal{X}_d. \tag{2.4.40}$$

Here $\mathcal{X}_a^-$ is related to the stable invariant zeros, i.e., the eigenvalues of A_{aa}^- are the stable invariant zeros of Σ_*. Similarly, $\mathcal{X}_a^0$ and $\mathcal{X}_a^+$ are respectively related to the invariant zeros of Σ_* located in the marginally stable and unstable regions. On the other hand, $\mathcal{X}_b$ is related to the right invertibility, i.e., the system is right invertible if and only if $\mathcal{X}_b = \{0\}$, while $\mathcal{X}_c$ is related to left invertibility, i.e., the system is left invertible if and only if $\mathcal{X}_c = \{0\}$. Finally, $\mathcal{X}_d$ is related to zeros of Σ_* at infinity.

There are interconnections between the special coordinate basis and various invariant geometric subspaces. To show these interconnections, we introduce the following geometric subspaces:

Definition 2.4.2. (Geometric Subspaces $\mathcal{V}^\times$ and $\mathcal{S}^\times$). The weakly unobservable subspaces of Σ_*, $\mathcal{V}^\times$, and the strongly controllable subspaces of Σ_*, $\mathcal{S}^\times$, are defined as follows:

1. $\mathcal{V}^\times(\Sigma_*)$ is the maximal subspace of $\mathbb{R}^n$ which is $(A_*+B_*F_*)$-invariant and contained in Ker $(C_* + D_*F_*)$ such that the eigenvalues of $(A_* + B_*F_*)|\mathcal{V}^\times$ are contained in $\mathbb{C}^\times \subseteq \mathbb{C}$ for some constant matrix F_*.

2. $\mathcal{S}^\times(\Sigma_*)$ is the minimal $(A_* + K_*C_*)$-invariant subspace of $\mathbb{R}^n$ containing Im $(B_* + K_*D_*)$ such that the eigenvalues of the map which is induced by $(A_* + K_*C_*)$ on the factor space $\mathbb{R}^n/\mathcal{S}^\times$ are contained in $\mathbb{C}^\times \subseteq \mathbb{C}$ for some constant matrix K_*.

Moreover, we let $\mathcal{V}^- = \mathcal{V}^\times$ and $\mathcal{S}^- = \mathcal{S}^\times$, if $\mathbb{C}^\times = \mathbb{C}^- \cup \mathbb{C}^0$; $\mathcal{V}^+ = \mathcal{V}^\times$ and $\mathcal{S}^+ = \mathcal{S}^\times$, if $\mathbb{C}^\times = \mathbb{C}^+$; $\mathcal{V}^\odot = \mathcal{V}^\times$ and $\mathcal{S}^\odot = \mathcal{S}^\times$, if $\mathbb{C}^\times = \mathbb{C}^\odot \cup \mathbb{C}^\circ$; $\mathcal{V}^\otimes = \mathcal{V}^\times$ and $\mathcal{S}^\otimes = \mathcal{S}^\times$, if $\mathbb{C}^\times = \mathbb{C}^\otimes$; and finally $\mathcal{V}^* = \mathcal{V}^\times$ and $\mathcal{S}^* = \mathcal{S}^\times$, if $\mathbb{C}^\times = \mathbb{C}$. ⊡

Various components of the state vector of the special coordinate basis have the following geometrical interpretations.

Property 2.4.5.

1. $\mathcal{X}_a^- \oplus \mathcal{X}_a^0 \oplus \mathcal{X}_c$ spans $\begin{cases} \mathcal{V}^-(\Sigma_*), & \text{if } \Sigma_* \text{ is of continuous-time,} \\ \mathcal{V}^\odot(\Sigma_*), & \text{if } \Sigma_* \text{ is of discrete-time.} \end{cases}$

2. $\mathcal{X}_a^+ \oplus \mathcal{X}_c$ spans $\begin{cases} \mathcal{V}^+(\Sigma_*), & \text{if } \Sigma_* \text{ is of continuous-time,} \\ \mathcal{V}^\otimes(\Sigma_*), & \text{if } \Sigma_* \text{ is of discrete-time.} \end{cases}$

3. $\mathcal{X}_a^- \oplus \mathcal{X}_a^0 \oplus \mathcal{X}_a^+ \oplus \mathcal{X}_c$ spans $\mathcal{V}^*(\Sigma_*)$.

4. $\mathcal{X}_a^+ \oplus \mathcal{X}_c \oplus \mathcal{X}_d$ spans $\begin{cases} \mathcal{S}^-(\Sigma_*), & \text{if } \Sigma_* \text{ is of continuous-time,} \\ \mathcal{S}^\odot(\Sigma_*), & \text{if } \Sigma_* \text{ is of discrete-time.} \end{cases}$

5. $\mathcal{X}_a^- \oplus \mathcal{X}_a^0 \oplus \mathcal{X}_c \oplus \mathcal{X}_d$ spans $\begin{cases} \mathcal{S}^+(\Sigma_*), & \text{if } \Sigma_* \text{ is of continuous-time,} \\ \mathcal{S}^\otimes(\Sigma_*), & \text{if } \Sigma_* \text{ is of discrete-time.} \end{cases}$

6. $\mathcal{X}_c \oplus \mathcal{X}_d$ spans $\mathcal{S}^*(\Sigma_*)$. ▣P

Finally, for future development on deriving solvability conditions for H_∞ almost disturbance decoupling problems, we introduce two more subspaces of Σ_*. The original definitions of these subspaces were given by Scherer [118,119].

Definition 2.4.3. (Geometric Subspaces $\mathcal{V}_\lambda$ and $\mathcal{S}_\lambda$). For any $\lambda \in \mathbb{C}$, we define

$$\mathcal{V}_\lambda(\Sigma_*) := \left\{ \zeta \in \mathbb{C}^n \;\middle|\; \exists\, \omega \in \mathbb{C}^m \;:\; 0 = \begin{bmatrix} A_* - \lambda I & B_* \\ C_* & D_* \end{bmatrix} \begin{pmatrix} \zeta \\ \omega \end{pmatrix} \right\}, \tag{2.4.41}$$

and

$$\mathcal{S}_\lambda(\Sigma_*) := \left\{ \zeta \in \mathbb{C}^n \;\middle|\; \exists\, \omega \in \mathbb{C}^{n+m} \;:\; \begin{pmatrix} \zeta \\ 0 \end{pmatrix} = \begin{bmatrix} A_* - \lambda I & B_* \\ C_* & D_* \end{bmatrix} \omega \right\}. \tag{2.4.42}$$

$\mathcal{V}_\lambda(\Sigma_*)$ and $\mathcal{S}_\lambda(\Sigma_*)$ are associated with the so-called state zero directions of Σ_* if λ is an invariant zero of Σ_*. ▣D

These subspaces $\mathcal{S}_\lambda(\Sigma_*)$ and $\mathcal{V}_\lambda(\Sigma_*)$ can also be easily obtained using the special coordinate basis. We have the following new property of the special coordinate basis.

Property 2.4.6.

$$\mathcal{S}_\lambda(\Sigma_*) = \operatorname{Im} \left\{ \Gamma_s \begin{bmatrix} \lambda I - A_{aa} & 0 & 0 & 0 \\ 0 & Y_{b\lambda} & 0 & 0 \\ 0 & 0 & I_{n_c} & 0 \\ 0 & 0 & 0 & I_{n_d} \end{bmatrix} \right\}, \tag{2.4.43}$$

where

$$\operatorname{Im}\{Y_{b\lambda}\} = \operatorname{Ker}\left[C_b(A_{bb} + K_b C_b - \lambda I)^{-1}\right], \tag{2.4.44}$$

and where K_b is any appropriately dimensional matrix subject to the constraint that $A_{bb} + K_b C_b$ has no eigenvalue at λ. We note that such a K_b always exists as (A_{bb}, C_b) is completely observable.

$$\mathcal{V}_\lambda(\Sigma_*) = \operatorname{Im}\left\{ \Gamma_s \begin{bmatrix} X_{a\lambda} & 0 \\ 0 & 0 \\ 0 & X_{c\lambda} \\ 0 & 0 \end{bmatrix} \right\}, \tag{2.4.45}$$

where $X_{a\lambda}$ is a matrix whose columns form a basis for the subspace,

$$\left\{ \zeta_a \in \mathbb{C}^{n_a} \;\middle|\; (\lambda I - A_{aa})\zeta_a = 0 \right\}, \tag{2.4.46}$$

and

$$X_{c\lambda} := \left(A_{cc} + B_c F_c - \lambda I \right)^{-1} B_c, \tag{2.4.47}$$

with F_c being any appropriately dimensional matrix subject to the constraint that $A_{cc} + B_c F_c$ has no eigenvalue at λ. Again, we note that the existence of such an F_c is guaranteed by the controllability of (A_{cc}, B_c). 🄿

Clearly, if $\lambda \notin \lambda(A_{aa})$, then we have

$$\mathcal{V}_\lambda(\Sigma_*) \subseteq \mathcal{V}^\times(\Sigma_*), \tag{2.4.48}$$

and

$$\mathcal{S}_\lambda(\Sigma_*) \supseteq \mathcal{S}^\times(\Sigma_*). \tag{2.4.49}$$

Next, we would like to note that the subspaces $\mathcal{V}^\times(\Sigma_*)$ and $\mathcal{S}^\times(\Sigma_*)$ are dual in the sense that $\mathcal{V}^\times(\Sigma_*^\star) = \mathcal{S}^\times(\Sigma_*)^\perp$, where $\Sigma_*^\star$ is characterized by the quadruple (A_*', C_*', B_*', D_*'). Also, $\mathcal{S}_\lambda(\Sigma_*) = \mathcal{V}_{\bar\lambda}(\Sigma_*^\star)^\perp$.

We illustrate the procedure for constructing the special coordinate basis and all its associated properties in the following numerical example.

Example 2.4.1. Consider a linear time-invariant system Σ_* characterized by

$$\begin{cases} \delta(x) = A_*\, x + B_*\, u, \\ \;\; y \;\; = C_*\, x + D_*\, u, \end{cases} \tag{2.4.50}$$

where

$$A_* = \begin{bmatrix} 1 & 2 & 3 & 1 \\ 2 & 3 & 4 & 5 \\ 4 & 5 & 6 & 7 \\ 5 & 6 & 7 & 8 \end{bmatrix}, \quad B_* = \begin{bmatrix} 1 \\ 2 \\ 3 \\ 4 \end{bmatrix}, \tag{2.4.51}$$

and

$$C_* = [\,0 \quad 3 \quad -2 \quad 0\,], \qquad D_* = 0. \tag{2.4.52}$$

The procedure for constructing the special coordinate basis of Σ_* proceeds as follows:

Step 1. Differentiating (shifting) the output of the given system. It involves the following sub-steps.

1. Since $D_* = 0$, we have

$$\delta(y) = C_*\delta(x) = C_*A_*x + C_*B_*u = [\,-2 \quad -1 \quad 0 \quad 1\,]\,x + 0 \cdot u.$$

2. Since $C_*B_* = 0$, we compute

$$\delta^2(y) = C_*A_*^2x + C_*A_*B_*u = [\,1 \quad -1 \quad -3 \quad 1\,]\,x + 0 \cdot u,$$

where $\delta^2(\cdot) = \delta(\delta(\cdot))$.

3. Since $C_*A_*B_* = 0$, we continue on computing

$$\delta^3(y) = C_*A_*^3x + C_*A_*^2B_* = -[\,8 \quad 10 \quad 12 \quad 17\,]\,x - 6 \cdot u,$$

where $\delta^3(\cdot) = \delta(\delta(\delta(\cdot)))$. Step 1 stops here as $C_*A_*^2B_* \neq 0$.

Step 2. Constructing a preliminary state transformation. Let X_0 be an appropriately dimensional matrix such that

$$T = \begin{bmatrix} X_0 \\ C_* \\ C_*A_* \\ C_*A_*^2 \end{bmatrix}, \tag{2.4.53}$$

is nonsingular. Then, define a new set of state variables $\check{x}$,

$$\check{x} = \begin{pmatrix} \check{x}_1 \\ \check{x}_2 \\ \check{x}_3 \\ \check{x}_4 \end{pmatrix} := Tx = \begin{bmatrix} X_0 \\ C_* \\ C_*A_* \\ C_*A_*^2 \end{bmatrix} x = \begin{pmatrix} X_0x \\ y \\ \delta(y) \\ \delta^2(y) \end{pmatrix}. \tag{2.4.54}$$

It is simple to verify that T with $X_0 = [\,1 \quad 0 \quad 0 \quad 0\,]$ is a nonsingular matrix. Furthermore,

$$\delta(\check{x}_1) = 8\check{x}_1 + \check{x}_2 + \frac{8}{3}\check{x}_3 - \frac{5}{3}\check{x}_4 + u, \tag{2.4.55}$$

$$\delta(\check{x}_2) = \check{x}_3, \tag{2.4.56}$$

$$\delta(\check{x}_3) = \check{x}_4, \tag{2.4.57}$$

$$\delta(\check{x}_4) = -72\check{x}_1 - 9\check{x}_2 - 27\check{x}_3 + 10\check{x}_4 - 6u. \tag{2.4.58}$$

Step 3. Eliminating u in $\delta(\check{x}_1)$. (2.4.58) implies that

$$u = -12\check{x}_1 - \frac{3}{2}\check{x}_2 - \frac{9}{2}\check{x}_3 + \frac{5}{3}\check{x}_4 - \frac{1}{6}\delta(\check{x}_4). \tag{2.4.59}$$

Substituting this into (2.4.55), we obtain

$$\delta(\breve{x}_1) = -4\breve{x}_1 - \frac{1}{2}\breve{x}_2 - \frac{11}{6}\breve{x}_3 - \frac{1}{6}\delta(\breve{x}_4). \tag{2.4.60}$$

We have got rid of u in $\delta(\breve{x}_1)$. Unfortunately, we have also introduced an additional $\delta(\breve{x}_4)$ in (2.4.60).

Step 4. Eliminating $\delta(\breve{x}_4)$ in $\delta(\breve{x}_1)$. Define a new variable $\bar{x}_1$ as follows,

$$\bar{x}_1 := \breve{x}_1 + \frac{1}{6}\breve{x}_4. \tag{2.4.61}$$

We have

$$\delta(\bar{x}_1) = -4\bar{x}_1 - \frac{1}{2}\breve{x}_2 - \frac{11}{6}\breve{x}_3 + \frac{2}{3}\breve{x}_4, \tag{2.4.62}$$

and

$$\delta(\breve{x}_4) = -72\bar{x}_1 - 9\breve{x}_2 - 27\breve{x}_3 + 22\breve{x}_4 - 6u. \tag{2.4.63}$$

Step 5. Eliminating $\breve{x}_3$ and $\breve{x}_4$ in $\delta(\bar{x}_1)$. This step involves two sub-steps.

1. Letting

$$\hat{x}_1 := \bar{x}_1 - \frac{2}{3}\breve{x}_3, \tag{2.4.64}$$

we have

$$\delta(\hat{x}_1) = -4\hat{x}_1 - \frac{1}{2}\breve{x}_2 - \frac{9}{2}\breve{x}_3, \tag{2.4.65}$$

and

$$\delta(\breve{x}_4) = -72\hat{x}_1 - 9\breve{x}_2 - 75\breve{x}_3 + 22\breve{x}_4 - 6u. \tag{2.4.66}$$

2. Letting

$$\tilde{x}_1 := \hat{x}_1 + \frac{9}{2}\breve{x}_2, \tag{2.4.67}$$

we have

$$\delta(\tilde{x}_1) = -4\tilde{x}_1 + \frac{35}{2}\breve{x}_2, \tag{2.4.68}$$

and

$$\delta(\breve{x}_4) = -72\tilde{x}_1 + 315\breve{x}_2 - 75\breve{x}_3 + 22\breve{x}_4 - 6u. \tag{2.4.69}$$

Step 6. Forming the nonsingular state, output and input transformations. Let

$$\tilde{x}_2 = \breve{x}_2, \quad \tilde{x}_3 = \breve{x}_3, \quad \tilde{x}_3 = \breve{x}_3, \tag{2.4.70}$$

or equivalently let

$$x = \Gamma_s \tilde{x} = \Gamma_s \begin{pmatrix} \tilde{x}_1 \\ \tilde{x}_2 \\ \tilde{x}_3 \\ \tilde{x}_4 \end{pmatrix}, \tag{2.4.71}$$

with

$$\Gamma_s = \left\{ \begin{bmatrix} 1 & 9/2 & -2/3 & 1/6 \\ 0 & 1 & 0 & 0 \\ 0 & 0 & 1 & 0 \\ 0 & 0 & 0 & 1 \end{bmatrix} \begin{bmatrix} 1 & 0 & 0 & 0 \\ 0 & 3 & -2 & 0 \\ -2 & -1 & 0 & 1 \\ 1 & -1 & -3 & 1 \end{bmatrix} \right\}^{-1}. \tag{2.4.72}$$

Also, let

$$u = \Gamma_i \tilde{u} = -\frac{1}{6}\tilde{u}, \quad y = \Gamma_o \tilde{y} = 1 \cdot \tilde{y}. \tag{2.4.73}$$

Finally, we obtain the dynamic equations of the transformed system,

$$\delta(\tilde{x}_1) = -4\tilde{x}_1 + \frac{35}{2}\tilde{x}_2, \tag{2.4.74}$$

$$\delta(\tilde{x}_2) = \tilde{x}_3, \quad \tilde{y} = \tilde{x}_2, \tag{2.4.75}$$

$$\delta(\tilde{x}_3) = \tilde{x}_4, \tag{2.4.76}$$

$$\delta(\tilde{x}_4) = -72\tilde{x}_1 + 315\tilde{x}_2 - 75\tilde{x}_3 + 22\tilde{x}_4 + \tilde{u}. \tag{2.4.77}$$

The above structure is now in the standard form of the special coordinate basis. $\tilde{x}_1$ is associated with $\mathcal{X}_a$ and $\tilde{x}_2$, $\tilde{x}_3$ and $\tilde{x}_4$ are associated with $\mathcal{X}_d$. Both $\mathcal{X}_b$ and $\mathcal{X}_c$ are non-existent for the given Σ_*.

Let us now examine the properties of Σ_*. Following Properties 2.4.1 to 2.4.6 of the special coordinate basis, it is simple to verify that Σ_* is controllable and observable, and has an invariant zero at -4 as well as an infinite zero (relative degree) of order 3. It is obvious that the given system is invertible as both x_c and x_b are non-existent.

The geometric subspaces $\mathcal{V}_\lambda(\Sigma_*)$ and $\mathcal{S}_\lambda(\Sigma_*)$ can be obtained as follows: for $\lambda = -4$,

$$\mathcal{V}_\lambda(\Sigma_*) = \operatorname{Im}\left\{ \begin{bmatrix} 3 \\ 2 \\ 3 \\ 8 \end{bmatrix} \right\}, \tag{2.4.78}$$

$$\mathcal{S}_\lambda(\Sigma_*) = \operatorname{Im}\left\{ \begin{bmatrix} 1 & 2 & 27 \\ 2 & 2 & 16 \\ 3 & 3 & 27 \\ 4 & 9 & 70 \end{bmatrix} \right\}, \tag{2.4.79}$$

and for $\lambda \neq -4$,

$$\mathcal{V}_\lambda(\Sigma_*) = \{0\}, \quad \mathcal{S}_\lambda(\Sigma_*) = \mathbb{R}^4. \tag{2.4.80}$$

The geometric subspaces $\mathcal{V}^\times(\Sigma_*)$ and $\mathcal{S}^\times(\Sigma_*)$ of Σ_* can also be easily computed:

1. If Σ_* is a continuous-time system, then

$$\mathcal{V}^-(\Sigma_*) = \mathcal{V}^*(\Sigma_*) = \operatorname{Im}\left\{\begin{bmatrix}3\\2\\3\\8\end{bmatrix}\right\}, \quad \mathcal{V}^+(\Sigma_*) = \{0\}, \tag{2.4.81}$$

and

$$\mathcal{S}^-(\Sigma_*) = \mathcal{S}^*(\Sigma_*) = \operatorname{Im}\left\{\begin{bmatrix}1 & 2 & 27\\2 & 2 & 16\\3 & 3 & 27\\4 & 9 & 70\end{bmatrix}\right\}, \quad \mathcal{S}^+(\Sigma_*) = \mathbb{R}^4. \tag{2.4.82}$$

2. If Σ_* is a discrete-time system, then

$$\mathcal{V}^{\otimes}(\Sigma_*) = \mathcal{V}^*(\Sigma_*) = \operatorname{Im}\left\{\begin{bmatrix}3\\2\\3\\8\end{bmatrix}\right\}, \quad \mathcal{V}^{\odot}(\Sigma_*) = \{0\}, \tag{2.4.83}$$

and

$$\mathcal{S}^{\otimes}(\Sigma_*) = \mathcal{S}^*(\Sigma_*) = \operatorname{Im}\left\{\begin{bmatrix}1 & 2 & 27\\2 & 2 & 16\\3 & 3 & 27\\4 & 9 & 70\end{bmatrix}\right\}, \quad \mathcal{S}^{\odot}(\Sigma_*) = \mathbb{R}^4. \tag{2.4.84}$$

Here we would like to note that the computation of the special coordinate basis for a multiple-input-multiple-output system is of course much more complicated than that for a single-input-single-output system, but the idea is basically the same. 🄴

Finally, we conclude this section by summarizing in a graphical form in Figure 2.4.2 some major properties of the tools of linear systems, which combines the mechanisms of the special coordinate basis, the Jordan canonical form and the controllability structural decomposition (CSD). Such tools has been used in the literature to solve many system and control problems such as the squaring down and decoupling of linear systems (see e.g., Sannuti and Saberi [116]), linear system factorizations (see e.g., Chen *et al.* [35], and Lin *et al.* [85]), blocking zeros and strong stabilizability (see e.g., Chen *et al.* [36]), zero placements (see e.g., Chen and Zheng [41]), loop transfer recovery (see e.g., Chen [12], Chen and Chen [20], and Saberi *et al.* [110]), H_2 optimal control (see e.g., Chen *et al.* [37,39], and Saberi *et al.* [114]), disturbance decoupling with static measurement feedback (see e.g., Chen [16] and Chen *et al.* [27]), and control with saturations (see e.g., Lin [80,82]), to name a few. These tools will be used intensively throughout this book to solve problems related to H_∞ control.

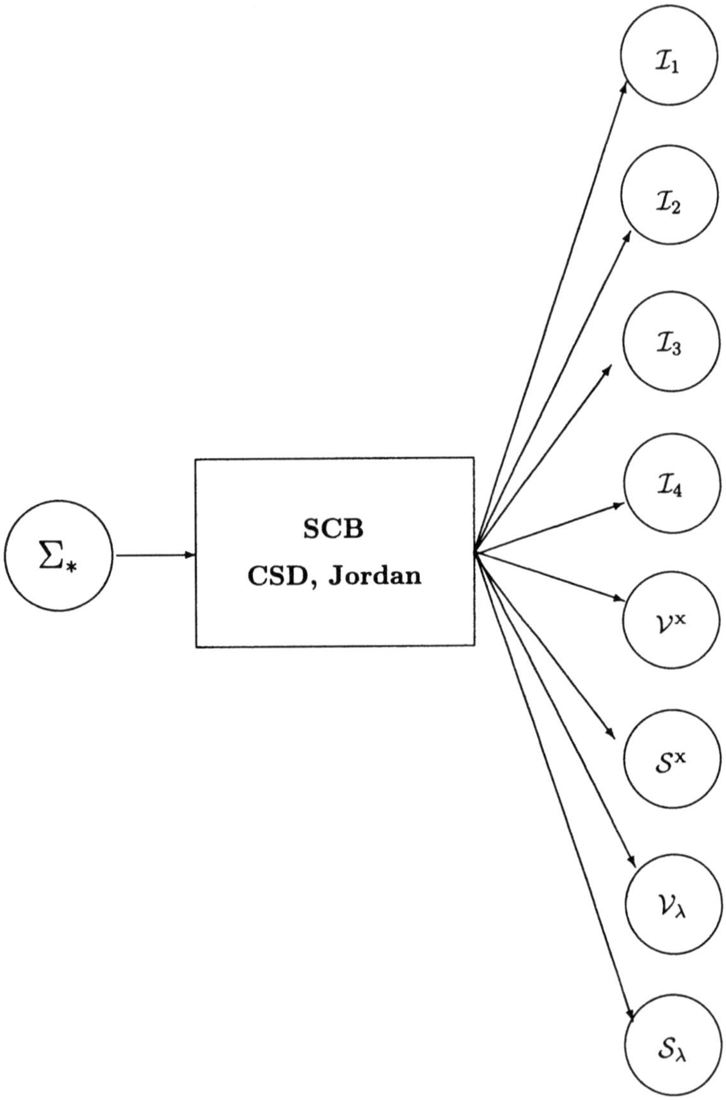

Figure 2.4.2: Tools and structural properties of linear time-invariant systems.

2.5. Proofs of Properties of Special Coordinate Basis

In this section, we provide detailed proofs for all the properties of the special coordinate basis listed in the previous section. Somehow, these proofs were missing in the original work of Sannuti and Saberi [116]. We would like to note that although some of the properties of the special coordinate basis, e.g., the controllability and observability, are quite obvious, some of them, e.g., the interconnections between the geometric subspaces and the subsystems of the special coordinate basis, are not transparent at all to general readers. The results of this section were reported in Chen [17]. It is to give rigorous proofs to all these properties.

We recall the following two lemmas whose results are quite well-known in the literature. The first lemma is about the effects of state feedback laws.

Lemma 2.5.1. Consider a given system Σ_* characterized by a constant matrix quadruple (A_*, B_*, C_*, D_*) or in the state space form of (2.4.1). Also, consider a constant state feedback gain matrix $F_* \in \mathbb{R}^{m \times n}$. Then, $\Sigma_{*\mathrm{F}}$ as characterized by the quadruple $(A_* + B_*F_*, B_*, C_* + D_*F_*, D_*)$ has the following properties:

1. $\Sigma_{*\mathrm{F}}$ is a controllable (stabilizable) system if and only if Σ_* is a controllable (stabilizable) system;
2. The normal rank of $\Sigma_{*\mathrm{F}}$ is equal to that of Σ_*;
3. The invariant zero structure of $\Sigma_{*\mathrm{F}}$ is the same as that of Σ_*;
4. The infinite zero structure of $\Sigma_{*\mathrm{F}}$ is the same as that of Σ_*;
5. $\Sigma_{*\mathrm{F}}$ is (left- or right- or non-) invertible if and only if Σ_* is (left- or right- or non-) invertible;
6. $\mathcal{V}^\times(\Sigma_{*\mathrm{F}}) = \mathcal{V}^\times(\Sigma_*)$ and $\mathcal{S}^\times(\Sigma_{*\mathrm{F}}) = \mathcal{S}^\times(\Sigma_*)$; and
7. $\mathcal{V}_\lambda(\Sigma_{*\mathrm{F}}) = \mathcal{V}_\lambda(\Sigma_*)$ and $\mathcal{S}_\lambda(\Sigma_{*\mathrm{F}}) = \mathcal{S}_\lambda(\Sigma_*)$. ◻L

Proof. Item 1 is obvious. Items 3, 4 and 5 are well-known as all the lists of Morse, i.e., $\mathcal{I}_1$ to $\mathcal{I}_4$, are invariant under any state feedback laws. Furthermore, Items 2 and 5 can be seen from the following simple manipulations:

$$\begin{aligned}
H_{*\mathrm{F}}(\varsigma) &:= (C_* + D_*F_*)(\varsigma I - A_* - B_*F_*)^{-1}B_* + D_* \\
&= (C_* + D_*F_*)(\varsigma I - A_*)^{-1}[I - B_*F_*(\varsigma I - A_*)^{-1}]^{-1}B_* + D_* \\
&= (C_* + D_*F_*)(\varsigma I - A_*)^{-1}B_*[I - F_*(\varsigma I - A_*)^{-1}B_*]^{-1} + D_* \\
&= [C_*(\varsigma I - A_*)^{-1}B_* + D_*][I - F_*(\varsigma I - A_*)^{-1}B_*]^{-1} \\
&= H_*(\varsigma)[I - F_*(\varsigma I - A_*)^{-1}B_*]^{-1}. \qquad (2.5.1)
\end{aligned}$$

Since $[I - F_*(\varsigma I - A_*)^{-1}B_*]^{-1}$ is well-defined almost everywhere on the complex plane, the results of Items 2 and 5 follow.

For Item 6, it is obvious from the definition of $\mathcal{V}^\times$, it is invariant under any state feedback laws. Next, for any subspace $\mathcal{S}$ that satisfies the following conditions:

$$(A_* + K_*C_*)\mathcal{S} \subseteq \mathcal{S}, \tag{2.5.2}$$

$$\mathrm{Im}\,(B_* + K_*D_*) \subseteq \mathcal{S}, \tag{2.5.3}$$

we have

$$(A_* + K_*C_* + B_*F_* + K_*D_*F_*)\mathcal{S} = (A_* + K_*C_*)\mathcal{S} + (B_* + K_*D_*)F_*\mathcal{S} \subseteq \mathcal{S}.$$

Thus, $\mathcal{S}^\times$ is invariant under any state feedback laws as well.

Let us now prove Item 7. Recalling the definition of $\mathcal{V}_\lambda$, we have

$$\mathcal{V}_\lambda(\Sigma_{*\mathrm{F}}) = \left\{ \zeta \in \mathbb{C}^n \;\middle|\; \exists\, \omega \in \mathbb{C}^m \,:\, 0 = \begin{bmatrix} A_* + B_*F_* - \lambda I & B_* \\ C_* + D_*F_* & D_* \end{bmatrix} \begin{pmatrix} \zeta \\ \omega \end{pmatrix} \right\}.$$

Then, for any $\zeta \in \mathcal{V}_\lambda(\Sigma_{*\mathrm{F}})$, there exist an $\omega \in \mathbb{C}^m$ such that

$$0 = \begin{bmatrix} A_* + B_*F_* - \lambda I & B_* \\ C_* + D_*F_* & D_* \end{bmatrix} \begin{pmatrix} \zeta \\ \omega \end{pmatrix} = \begin{bmatrix} A_* - \lambda I & B_* \\ C_* & D_* \end{bmatrix} \begin{bmatrix} I & 0 \\ F_* & I \end{bmatrix} \begin{pmatrix} \zeta \\ \omega \end{pmatrix},$$

or

$$0 = \begin{bmatrix} A_* - \lambda I & B_* \\ C_* & D_* \end{bmatrix} \begin{pmatrix} \zeta \\ \tilde{\omega} \end{pmatrix},$$

where $\tilde{\omega} = F_*\zeta + \omega$. Thus, $\zeta \in \mathcal{V}_\lambda(\Sigma_*)$ and hence $\mathcal{V}_\lambda(\Sigma_{*\mathrm{F}}) \subseteq \mathcal{V}_\lambda(\Sigma_*)$. Similarly, one can show that $\mathcal{V}_\lambda(\Sigma_*) \subseteq \mathcal{V}_\lambda(\Sigma_{*\mathrm{F}})$, and hence $\mathcal{V}_\lambda(\Sigma_*) = \mathcal{V}_\lambda(\Sigma_{*\mathrm{F}})$. The result that $\mathcal{S}_\lambda(\Sigma_{*\mathrm{F}}) = \mathcal{S}_\lambda(\Sigma_*)$ can be shown using the similar arguments. ⊠

The following lemma is about the effects of output injection laws.

Lemma 2.5.2. Consider a given system Σ_* characterized by a constant matrix quadruple (A_*, B_*, C_*, D_*) or in the state space form of (2.4.1). Also, consider a constant output injection gain matrix $K_* \in \mathbb{R}^{n \times p}$. Then, $\Sigma_{*\mathrm{K}}$ as characterized by the quadruple $(A_* + K_*C_*, B_* + K_*D_*, C_*, D_*)$ has the following properties:

1. $\Sigma_{*\mathrm{K}}$ is an observable (detectable) system if and only if Σ_* is an observable (detectable) system;

2. The normal rank of $\Sigma_{*\mathrm{K}}$ is equal to that of Σ_*;

3. The invariant zero structure of $\Sigma_{*\mathrm{K}}$ is the same as that of Σ_*;

4. The infinite zero structure of $\Sigma_{*\mathrm{K}}$ is the same as that of Σ_*;

5. $\Sigma_{*\mathrm{K}}$ is (left- or right- or non-) invertible if and only if Σ_* is (left- or right- or non-) invertible;

6. $\mathcal{V}^\times(\Sigma_{*\mathrm{K}}) = \mathcal{V}^\times(\Sigma_*)$ and $\mathcal{S}^\times(\Sigma_{*\mathrm{K}}) = \mathcal{S}^\times(\Sigma_*)$; and

7. $\mathcal{V}_\lambda(\Sigma_{*\mathrm{K}}) = \mathcal{V}_\lambda(\Sigma_*)$ and $\mathcal{S}_\lambda(\Sigma_{*\mathrm{K}}) = \mathcal{S}_\lambda(\Sigma_*)$. ◻

Proof. It is the dual version of Lemma 2.5.1. ◻

Now, we are ready to prove the properties of the special coordinate basis. Without loss of any generality but for simplicity of presentation, we assume throughout the rest of this section that the given system Σ_* has already been transformed into the special coordinate basis of Theorem 2.4.1 or into the compact form of (2.4.20) to (2.4.23), i.e.,

$$A_* = \begin{bmatrix} A_{aa} & L_{ab}C_b & 0 & L_{ad}C_d \\ 0 & A_{bb} & 0 & L_{bd}C_d \\ B_cE_{ca} & L_{cb}C_b & A_{cc} & L_{cd}C_d \\ B_dE_{da} & B_dE_{db} & B_dE_{dc} & A^*_{dd} + B_dE_{dd} + L_{dd}C_d \end{bmatrix} + B_{*,0}C_{*,0}, \quad (2.5.4)$$

$$B_* = [\, B_{*,0} \quad B_{*,1} \,] = \begin{bmatrix} B_{0a} & 0 & 0 \\ B_{0b} & 0 & 0 \\ B_{0c} & 0 & B_c \\ B_{0d} & B_d & 0 \end{bmatrix}, \quad (2.5.5)$$

and

$$C_* = \begin{bmatrix} C_{*,0} \\ C_{*,1} \end{bmatrix} = \begin{bmatrix} C_{0a} & C_{0b} & C_{0c} & C_{0d} \\ 0 & 0 & 0 & C_d \\ 0 & C_b & 0 & 0 \end{bmatrix}, \quad D_* = \begin{bmatrix} I_{m_0} & 0 & 0 \\ 0 & 0 & 0 \\ 0 & 0 & 0 \end{bmatrix}. \quad (2.5.6)$$

We further note that A^*_{dd}, B_d and C_d have the following forms,

$$A^*_{dd} = \mathrm{blkdiag}\left\{A_{q_1}, \cdots, A_{q_{m_d}}\right\}, \quad (2.5.7)$$

and

$$B_d = \mathrm{blkdiag}\left\{B_{q_1}, \cdots, B_{q_{m_d}}\right\}, \quad C_d = \mathrm{blkdiag}\left\{C_{q_1}, \cdots, C_{q_{m_d}}\right\}, \quad (2.5.8)$$

where A_{q_i}, B_{q_i} and C_{q_i}, $i = 1, 2, \cdots, m_d$, are defined as in (2.4.16).

Proof of Property 2.4.1. Let us define a state feedback gain matrix F_* as follows:

$$F_* = -\begin{bmatrix} C_{0a} & C_{0b} & C_{0c} & C_{0d} \\ E_{da} & E_{db} & E_{dc} & E_{dd} \\ E_{ca} & 0 & 0 & 0 \end{bmatrix}. \quad (2.5.9)$$

Then, we have

$$A_* + B_*F_* = \begin{bmatrix} A_{aa} & L_{ab}C_b & 0 & L_{ad}C_d \\ 0 & A_{bb} & 0 & L_{bd}C_d \\ 0 & L_{cb}C_b & A_{cc} & L_{cd}C_d \\ 0 & 0 & 0 & A_{dd}^* + L_{dd}C_d \end{bmatrix}. \tag{2.5.10}$$

Noting that (A_{cc}, B_c) is completely controllable, we have for any $\lambda \in \mathbb{C}$,

$$\begin{aligned} &\text{rank}\begin{bmatrix} A_* + B_*F_* - \lambda I & B_* \end{bmatrix} \\ &= \text{rank}\begin{bmatrix} A_{aa}-\lambda I & L_{ab}C_b & 0 & L_{ad}C_d & B_{0a} & 0 & 0 \\ 0 & A_{bb}-\lambda I & 0 & L_{bd}C_d & B_{0b} & 0 & 0 \\ 0 & L_{cb}C_b & A_{cc}-\lambda I & L_{cd}C_d & B_{0c} & 0 & B_c \\ 0 & 0 & 0 & A_{dd}^*+L_{dd}C_d-\lambda I & B_{0d} & B_d & 0 \end{bmatrix} \\ &= \text{rank}\begin{bmatrix} A_{aa}-\lambda I & L_{ab}C_b & 0 & L_{ad}C_d & B_{0a} & 0 & 0 \\ 0 & A_{bb}-\lambda I & 0 & L_{bd}C_d & B_{0b} & 0 & 0 \\ 0 & 0 & A_{cc}-\lambda I & 0 & 0 & 0 & B_c \\ 0 & 0 & 0 & A_{dd}^*+L_{dd}C_d-\lambda I & B_{0d} & B_d & 0 \end{bmatrix} \\ &= \text{rank}\begin{bmatrix} A_{\rm con}-\lambda I & 0 & B_{\rm con1}C_d & B_{\rm con0} & 0 & 0 \\ 0 & A_{cc}-\lambda I & 0 & 0 & 0 & B_c \\ 0 & 0 & A_{dd}^*+L_{dd}C_d-\lambda I & B_{0d} & B_d & 0 \end{bmatrix} \end{aligned} \tag{2.5.11}$$

where

$$A_{\rm con} = \begin{bmatrix} A_{aa} & L_{ab}C_b \\ 0 & A_{bb} \end{bmatrix}, \quad B_{\rm con} = [\, B_{\rm con0} \quad B_{\rm con1} \,] = \begin{bmatrix} B_{0a} & L_{ad} \\ B_{0b} & L_{bd} \end{bmatrix}. \tag{2.5.12}$$

Also, noting the special structure of (A_{dd}^*, B_d, C_d), it is simple to verify that $[\, A_*+B_*F_*-\lambda I \quad B_* \,]$ is of maximal rank if and only if $[\, A_{\rm con}-\lambda I \quad B_{\rm con} \,]$ is of maximal rank. By Lemma 2.5.1, we have that (A, B) is controllable (stabilizable) if and only if $(A_{\rm con}, B_{\rm con})$ is controllable (stabilizable).

Similarly, one can show that (A, C) is observable (detectable) if and only if $(A_{\rm obs}, C_{\rm obs})$ is observable (detectable). ⊠

Proof of Property 2.4.2. Let us define a state feedback gain matrix F_* as in (2.5.9) and an output injection gain matrix K_* as follows:

$$K_* = -\begin{bmatrix} B_{0a} & L_{ad} & L_{ab} \\ B_{0b} & L_{bd} & 0 \\ B_{0c} & L_{cd} & L_{cb} \\ B_{0d} & L_{dd} & 0 \end{bmatrix}. \tag{2.5.13}$$

We have

$$\check{A}_* = A_* + B_*F_* + K_*C_* + K_*D_*F_* = \begin{bmatrix} A_{aa} & 0 & 0 & 0 \\ 0 & A_{bb} & 0 & 0 \\ 0 & 0 & A_{cc} & 0 \\ 0 & 0 & 0 & A^*_{dd} \end{bmatrix}, \tag{2.5.14}$$

$$\check{B}_* = B_* + K_*D_* = \begin{bmatrix} 0 & 0 & 0 \\ 0 & 0 & 0 \\ 0 & 0 & B_c \\ 0 & B_d & 0 \end{bmatrix}, \tag{2.5.15}$$

$$\check{C}_* = C_* + D_*F_* = \begin{bmatrix} 0 & 0 & 0 & 0 \\ 0 & 0 & 0 & C_d \\ 0 & C_b & 0 & 0 \end{bmatrix}, \tag{2.5.16}$$

and

$$\check{D}_* = D_* = \begin{bmatrix} I_{m_0} & 0 & 0 \\ 0 & 0 & 0 \\ 0 & 0 & 0 \end{bmatrix}. \tag{2.5.17}$$

Let $\check{\Sigma}_*$ be characterized by the quadruple $(\check{A}_*, \check{B}_*, \check{C}_*, \check{D}_*)$. It is simple to verify that the transfer function of $\check{\Sigma}_*$ is given by

$$\check{H}_*(\varsigma) = \check{C}_*(\varsigma I - \check{A}_*)^{-1}\check{B}_* + \check{D}_* = \begin{bmatrix} I_{m_0} & 0 & 0 \\ 0 & C_d(\varsigma I - A^*_{dd})^{-1}B_d & 0 \\ 0 & 0 & 0 \end{bmatrix}. \tag{2.5.18}$$

Furthermore, we can show that

$$C_d(\varsigma I - A^*_{dd})^{-1}B_d = \begin{bmatrix} \frac{1}{\varsigma^{q_1}} & & \\ & \ddots & \\ & & \frac{1}{\varsigma^{q_{m_d}}} \end{bmatrix}. \tag{2.5.19}$$

By Lemmas 2.5.1 and 2.5.2, we have

$$\text{normrank}\,\{H_*(\varsigma)\} = \text{normrank}\,\{\check{H}_*(\varsigma)\} = m_0 + m_d. \tag{2.5.20}$$

Next, it follows from Lemmas 2.5.1 and 2.5.2 that the invariant zeros of Σ_* and $\check{\Sigma}_*$ are equivalent. By the definition of the invariant zeros of a linear system, i.e., a complex scalar α is an invariant zero of $\check{\Sigma}_*$ if

$$\text{rank}\begin{bmatrix} \check{A}_* - \alpha I & \check{B}_* \\ \check{C}_* & \check{D}_* \end{bmatrix} < n + \text{normrank}\,\{\check{H}_*(\varsigma)\} = n + m_0 + m_d, \tag{2.5.21}$$

and also noting the special structure of (A_{dd}^*, B_d, C_d) and the facts that (A_{bb}, C_b) is observable, and (A_{cc}, B_c) is controllable, we have

$$\begin{aligned}\operatorname{rank}\{P_{\check{\Sigma}_*}(\alpha)\} &= \operatorname{rank}\begin{bmatrix}\check{A}_*-\alpha I & \check{B}_* \\ \check{C}_* & \check{D}_*\end{bmatrix}\\ &= \operatorname{rank}\begin{bmatrix} A_{aa}-\alpha I & 0 & 0 & 0 & 0 & 0 & 0\\ 0 & A_{bb}-\alpha I & 0 & 0 & 0 & 0 & 0\\ 0 & 0 & A_{cc}-\alpha I & 0 & 0 & 0 & B_c\\ 0 & 0 & 0 & A_{dd}^*-\alpha I & 0 & B_d & 0\\ 0 & 0 & 0 & 0 & I_{m_0} & 0 & 0\\ 0 & 0 & 0 & C_d & 0 & 0 & 0\\ 0 & C_b & 0 & 0 & 0 & 0 & 0\end{bmatrix}\\ &= n_b+n_c+n_d+m_0+m_d+\operatorname{rank}\{A_{aa}-\alpha I\}. \end{aligned} \tag{2.5.22}$$

Clearly, the rank of $P_{\check{\Sigma}_*}(\alpha)$ drops below $n+m_0+m_d$ if and only if $\alpha \in \lambda(A_{aa})$. Hence, the invariant zeros of $\check{\Sigma}_*$, or equivalently the invariant zeros of Σ_*, are given by the eigenvalues of A_{aa}, which are the union of $\lambda(A_{aa}^-)$, $\lambda(A_{aa}^0)$, and $\lambda(A_{aa}^+)$. This completes the proof of Property 2.4.2. ⊠

Proof of Property 2.4.3. It follows from Lemmas 2.5.1 and 2.5.2 that the infinite zeros of Σ_* and $\check{\Sigma}_*$ are equivalent. It is clear to see from (2.5.18) and (2.5.19) that the infinite zeros of $\check{\Sigma}_*$, or equivalently the infinite zeros of Σ_*, of order higher than 0, are given by

$$S_\infty^\star(\Sigma_*) = S_\infty^\star(\check{\Sigma}_*) = \left\{q_1, q_2, \cdots, q_{m_d}\right\}. \tag{2.5.23}$$

Furthermore, $\check{\Sigma}_*$ or Σ_* has m_0 infinite zeros of order 0. ⊠

Proof of Property 2.4.4. Again, it follows from Lemmas 2.5.1 and 2.5.2 that Σ_* or $H_*(\varsigma)$ is (left- or right- or non-) invertible if and only if $\check{\Sigma}_*$ or $\check{H}_*(\varsigma)$ is (left- or right- or non-) invertible. The results of Property 2.4.4 can be seen from the transfer function $\check{H}_*(\varsigma)$ in (2.5.18). ⊠

Proof of Property 2.4.5. We will only prove the geometric subspace $\mathcal{V}^*(\Sigma_*)$, i.e.,

$$\mathcal{V}^*(\Sigma_*) = \mathcal{X}_a \oplus \mathcal{X}_c = \operatorname{Im}\left\{\Gamma_s \begin{bmatrix} I_{n_a} & 0\\ 0 & 0\\ 0 & I_{n_c}\\ 0 & 0\end{bmatrix}\right\}. \tag{2.5.24}$$

Here $\Gamma_s = I_n$ as the given system Σ_* is assumed to be already in the form of the special coordinate basis. It follows from Lemma 2.5.2 that $\mathcal{V}^*$ is invariant under any output injection laws. Let us choose an output injection gain matrix K_* as in (2.5.13). Then, we have

$$\hat{A}_* = A_* + K_* C_* = \begin{bmatrix} A_{aa} & 0 & 0 & 0 \\ 0 & A_{bb} & 0 & 0 \\ B_c E_{ca} & 0 & A_{cc} & 0 \\ B_d E_{da} & B_d E_{db} & B_d E_{dc} & A_{dd}^* + B_d E_{dd} \end{bmatrix}, \tag{2.5.25}$$

and

$$\hat{B}_* = B_* + K_* D_* = \check{B}_* = \begin{bmatrix} 0 & 0 & 0 \\ 0 & 0 & 0 \\ 0 & 0 & B_c \\ 0 & B_d & 0 \end{bmatrix}. \tag{2.5.26}$$

Let $\hat{\Sigma}_*$ be a system characterized by $(\hat{A}_*, \hat{B}_*, C_*, D_*)$. Then it is sufficient to show the property of $\mathcal{V}^*(\Sigma_*)$ by showing that

$$\mathcal{V}^*(\hat{\Sigma}_*) = \operatorname{Im}\left\{\begin{bmatrix} I_{n_a} & 0 \\ 0 & 0 \\ 0 & I_{n_c} \\ 0 & 0 \end{bmatrix}\right\}. \tag{2.5.27}$$

First, let us choose a matrix F_* as given in (2.5.9). Then, we have

$$\hat{A}_* + \hat{B}_* F_* = \begin{bmatrix} A_{aa} & 0 & 0 & 0 \\ 0 & A_{bb} & 0 & 0 \\ 0 & 0 & A_{cc} & 0 \\ 0 & 0 & 0 & A_{dd}^* \end{bmatrix}, \tag{2.5.28}$$

and

$$C_* + D_* F_* = \begin{bmatrix} 0 & 0 & 0 & 0 \\ 0 & 0 & 0 & C_d \\ 0 & C_b & 0 & 0 \end{bmatrix}. \tag{2.5.29}$$

It is now simple to see that for any

$$\zeta \in \mathcal{X}_a \oplus \mathcal{X}_c = \operatorname{Im}\left\{\begin{bmatrix} I_{n_a} & 0 \\ 0 & 0 \\ 0 & I_{n_c} \\ 0 & 0 \end{bmatrix}\right\}, \tag{2.5.30}$$

we have

$$\zeta = \begin{pmatrix} \zeta_a \\ 0 \\ \zeta_c \\ 0 \end{pmatrix}, \tag{2.5.31}$$

and

$$(\hat{A}_* + \hat{B}_* F_*)\zeta = \begin{pmatrix} A_{aa}\zeta_a \\ 0 \\ A_{cc}\zeta_c \\ 0 \end{pmatrix} \in \operatorname{Im}\left\{ \begin{bmatrix} I_{n_a} & 0 \\ 0 & 0 \\ 0 & I_{n_c} \\ 0 & 0 \end{bmatrix} \right\} = \mathcal{X}_a \oplus \mathcal{X}_c, \tag{2.5.32}$$

and

$$(C_* + D_* F_*)\zeta = 0. \tag{2.5.33}$$

Clearly, $\mathcal{X}_a \oplus \mathcal{X}_c$ is a $(\hat{A}_* + \hat{B}_* F_*)$-invariant subspace of $\mathbb{R}^n$ and is contained in $\operatorname{Ker}(C_* + D_* F_*)$. By the definition of $\mathcal{V}^*$, we have

$$\mathcal{X}_a \oplus \mathcal{X}_c \subseteq \mathcal{V}^*(\hat{\Sigma}_*). \tag{2.5.34}$$

Conversely, for any $\zeta \in \mathcal{V}^*(\hat{\Sigma}_*)$, by Definition 2.4.2, there exists a gain matrix $\hat{F}_* \in \mathbb{R}^{m\times n}$ such that

$$(\hat{A}_* + \hat{B}_* \hat{F}_*)\zeta \in \mathcal{V}^*(\hat{\Sigma}_*), \tag{2.5.35}$$

and

$$(C_* + D_* \hat{F}_*)\zeta = 0. \tag{2.5.36}$$

(2.5.35) and (2.5.36) imply that for any $\zeta \in \mathcal{V}^*(\hat{\Sigma}_*)$,

$$(C_* + D_* \hat{F}_*)(\hat{A}_* + \hat{B}_* \hat{F}_*)^k \zeta = 0, \quad k = 0, 1, \cdots, n-1. \tag{2.5.37}$$

Thus, (2.5.34) and (2.5.37) imply that

$$(C_* + D_* \hat{F}_*)(\hat{A}_* + \hat{B}_* \hat{F}_*)^k \begin{bmatrix} I_{n_a} & 0 \\ 0 & 0 \\ 0 & I_{n_c} \\ 0 & 0 \end{bmatrix} = 0, \quad k = 0, 1, \cdots, n-1. \tag{2.5.38}$$

Next, let us partition this $\hat{F}_*$ as follows:

$$\hat{F}_* = \begin{bmatrix} F_{a0} - C_{0a} & F_{b0} - C_{0b} & F_{c0} - C_{0c} & F_{d0} - C_{0d} \\ F_{ad} - E_{da} & F_{bd} - E_{db} & F_{cd} - E_{dc} & F_{dd} - E_{dd} \\ F_{ac} - E_{ca} & F_{bc} & F_{cc} & F_{dc} \end{bmatrix}. \tag{2.5.39}$$

We have

$$C_* + D_* \hat{F}_* = \begin{bmatrix} F_{a0} & F_{b0} & F_{c0} & F_{d0} \\ 0 & 0 & 0 & C_d \\ 0 & C_b & 0 & 0 \end{bmatrix}, \tag{2.5.40}$$

and

$$\hat{A}_* + \hat{B}_* \hat{F}_* = \begin{bmatrix} A_{aa} & 0 & 0 & 0 \\ 0 & A_{bb} & 0 & 0 \\ B_c F_{ac} & B_c F_{bc} & A_{cc} + B_c F_{cc} & B_c F_{dc} \\ B_d F_{ad} & B_d F_{bd} & B_d F_{cd} & A_{dd}^{**} \end{bmatrix}. \tag{2.5.41}$$

where $A_{dd}^{**} = A_{dd}^* + B_d F_{dd}$. Then, using (2.5.38) with $k = 0$, we have

$$(C_* + D_* \hat{F}_*) \begin{bmatrix} I_{n_a} & 0 \\ 0 & 0 \\ 0 & I_{n_c} \\ 0 & 0 \end{bmatrix} = 0, \tag{2.5.42}$$

which implies

$$F_{a0} = 0, \quad F_{c0} = 0, \tag{2.5.43}$$

and

$$C_* + D_* \hat{F}_* = \begin{bmatrix} 0 & \star & 0 & \star \\ 0 & 0 & 0 & C_d \\ 0 & C_b & 0 & 0 \end{bmatrix}, \tag{2.5.44}$$

where $\star$s are some matrices of not much interest. Using (2.5.38) with $k = 1$ together with (2.5.44), we have

$$C_d B_d F_{ad} = 0, \quad C_d B_d F_{cd} = 0, \tag{2.5.45}$$

and

$$(C_* + D_* \hat{F}_*)(\hat{A}_* + \hat{B}_* \hat{F}_*) = \begin{bmatrix} 0 & \star & 0 & \star \\ 0 & C_d B_d F_{bd} & 0 & C_d A_{dd}^{**} \\ 0 & C_b A_{bb} & 0 & 0 \end{bmatrix}. \tag{2.5.46}$$

In general, one can show that for any positive integer k,

$$C_d (A_{dd}^{**})^{k-1} B_d F_{ad} = 0, \quad C_d (A_{dd}^{**})^{k-1} B_d F_{cd} = 0, \tag{2.5.47}$$

and

$$(C_* + D_* \hat{F}_*)(\hat{A}_* + \hat{B}_* \hat{F}_*)^k = \begin{bmatrix} 0 & \star & 0 & \star \\ 0 & \star & 0 & C_d (A_{dd}^{**})^k \\ 0 & C_b (A_{bb})^k & 0 & 0 \end{bmatrix}. \tag{2.5.48}$$

As a by-product, we can easily show that $F_{ad} = 0$ and $F_{cd} = 0$, because of the fact that (A_{dd}^{**}, B_d, C_d) is controllable, observable, invertible and is free of invariant zeros. Now, for any

$$\zeta = \begin{pmatrix} \zeta_a \\ \zeta_b \\ \zeta_c \\ \zeta_d \end{pmatrix} \in \mathcal{V}^*(\hat{\Sigma}_*), \tag{2.5.49}$$

it follows from (2.5.37) and (2.5.48) that

$$C_b(A_{bb})^k\zeta_b = 0, \quad k = 0, 1, \cdots, n-1, \tag{2.5.50}$$

which implies $\zeta_b = 0$ because (A_{bb}, C_b) is completely observable, and

$$C_d(A_{dd}^{**})^k\zeta_d + \star\cdot\zeta_b = C_d(A_{dd}^{**})^k\zeta_d = 0, \quad k = 0, 1, \cdots, n-1, \tag{2.5.51}$$

which implies $\zeta_d = 0$ because (A_{dd}^{**}, C_d) is also completely observable. Hence,

$$\zeta = \begin{pmatrix} \zeta_a \\ 0 \\ \zeta_c \\ 0 \end{pmatrix} \in \operatorname{Im}\left\{\begin{bmatrix} I_{n_a} & 0 \\ 0 & 0 \\ 0 & I_{n_c} \\ 0 & 0 \end{bmatrix}\right\} = \mathcal{X}_a \oplus \mathcal{X}_c, \tag{2.5.52}$$

and

$$\mathcal{V}^*(\hat{\Sigma}_*) \subseteq \mathcal{X}_a \oplus \mathcal{X}_c. \tag{2.5.53}$$

Obviously, (2.5.34) and (2.5.53) imply the result.

Similarly, one can follow the same procedure as in the above to show the properties of the other subspaces in Property 2.4.5. ⊠

Proof of Property 2.4.6. Let us prove the property of $\mathcal{V}_\lambda(\Sigma_*)$. It follows from Lemmas 2.5.1 and 2.5.2 that $\mathcal{V}_\lambda$ is invariant under any state feedback and output injection laws. Thus, it is sufficient to prove the property of $\mathcal{V}_\lambda(\Sigma_*)$ by showing that

$$\mathcal{V}_\lambda(\check{\Sigma}_*) = \operatorname{Im}\left\{\begin{bmatrix} X_{a\lambda} & 0 \\ 0 & 0 \\ 0 & X_{c\lambda} \\ 0 & 0 \end{bmatrix}\right\}, \tag{2.5.54}$$

where $\check{\Sigma}_*$ is as defined in the proof of Property 2.4.2, $X_{a\lambda}$ is a matrix whose columns form a basis for the subspace,

$$\left\{\zeta_a \in \mathbb{C}^{n_a} \,\middle|\, (\lambda I - A_{aa})\zeta_a = 0\right\}, \tag{2.5.55}$$

and

$$X_{c\lambda} = (A_{cc} + B_c F_c - \lambda I)^{-1} B_c, \tag{2.5.56}$$

with F_c being an appropriately dimensional matrix such that $A_{cc} + B_c F_c - \lambda I$ is invertible.

For any $\zeta \in \mathcal{V}_\lambda(\check{\Sigma}_*)$, by Definition 2.4.3, there exists a vector $\omega \in \mathbb{C}^m$ such that

$$\begin{bmatrix} \check{A}_* - \lambda I & \check{B}_* \\ \check{C}_* & \check{D}_* \end{bmatrix} \begin{pmatrix} \zeta \\ \omega \end{pmatrix} = 0, \tag{2.5.57}$$

or equivalently,

$$\begin{bmatrix} A_{aa}-\lambda I & 0 & 0 & 0 & 0 & 0 & 0 \\ 0 & A_{bb}-\lambda I & 0 & 0 & 0 & 0 & 0 \\ 0 & 0 & A_{cc}-\lambda I & 0 & 0 & 0 & B_c \\ 0 & 0 & 0 & A_{dd}^{*}-\lambda I & 0 & B_d & 0 \\ 0 & 0 & 0 & 0 & I_{m_0} & 0 & 0 \\ 0 & 0 & 0 & C_d & 0 & 0 & 0 \\ 0 & C_b & 0 & 0 & 0 & 0 & 0 \end{bmatrix} \begin{pmatrix} \zeta_a \\ \zeta_b \\ \zeta_c \\ \zeta_d \\ \omega_0 \\ \omega_d \\ \omega_c \end{pmatrix} = 0. \tag{2.5.58}$$

Hence, we have

$$(A_{aa}-\lambda I)\zeta_a = 0, \tag{2.5.59}$$

which implies that $\zeta_a \in \operatorname{Im}\{X_{a\lambda}\}$,

$$\begin{bmatrix} A_{bb}-\lambda I \\ C_b \end{bmatrix} \zeta_b = 0, \tag{2.5.60}$$

which implies that $\zeta_b = 0$ as (A_{bb}, C_b) is completely observable, and

$$\begin{bmatrix} A_{dd}^{*}-\lambda I & B_d \\ C_d & 0 \end{bmatrix} \begin{pmatrix} \zeta_d \\ \omega_d \end{pmatrix} = 0, \tag{2.5.61}$$

which implies that $\zeta_d = 0$ and $\omega_d = 0$ as (A_{dd}^{*}, B_d, C_d) is square invertible and free of invariant zeros. We also have

$$(A_{cc}-\lambda I)\zeta_c + B_c\omega_c = 0, \tag{2.5.62}$$

which implies that

$$(A_{cc}+B_cF_c-\lambda I)\zeta_c + B_c(\omega_c - F_c\zeta_c) = 0, \tag{2.5.63}$$

or

$$\zeta_c = (A_{cc}+B_cF_c-\lambda I)^{-1}B_c(F_c\zeta_c-\omega_c) = X_{c\lambda}(F_c\zeta_c-\omega_c). \tag{2.5.64}$$

Hence $\zeta_c \in \operatorname{Im}\{X_{c\lambda}\}$. Clearly,

$$\zeta \in \operatorname{Im}\left\{ \begin{bmatrix} X_{a\lambda} & 0 \\ 0 & 0 \\ 0 & X_{c\lambda} \\ 0 & 0 \end{bmatrix} \right\} \implies \mathcal{V}_\lambda(\check{\Sigma}_*) \subseteq \operatorname{Im}\left\{ \begin{bmatrix} X_{a\lambda} & 0 \\ 0 & 0 \\ 0 & X_{c\lambda} \\ 0 & 0 \end{bmatrix} \right\}. \tag{2.5.65}$$

Conversely, for any

$$\zeta = \begin{pmatrix} \zeta_a \\ \zeta_b \\ \zeta_c \\ \zeta_d \end{pmatrix} \in \operatorname{Im}\left\{ \begin{bmatrix} X_{a\lambda} & 0 \\ 0 & 0 \\ 0 & X_{c\lambda} \\ 0 & 0 \end{bmatrix} \right\}, \tag{2.5.66}$$

we have $\zeta_b = 0$, $\zeta_d = 0$, $\zeta_a \in \operatorname{Im}\{X_{a\lambda}\}$, which implies that $(\lambda I - A_{aa})\zeta_a = 0$, and $\zeta_c \in \operatorname{Im}\{X_{c\lambda}\}$, which implies that there exists a vector $\tilde{\omega}_c$ such that

$$\zeta_c = X_{c\lambda}\tilde{\omega}_c = (A_{cc} + B_cF_c - \lambda I)^{-1}B_c\tilde{\omega}_c. \tag{2.5.67}$$

Thus, we have

$$(A_{cc} + B_cF_c - \lambda I)\zeta_c = B_c\tilde{\omega}_c, \tag{2.5.68}$$

or

$$(A_{cc} - \lambda I)\zeta_c + B_c(F_c\zeta_c - \tilde{\omega}_c) = 0. \tag{2.5.69}$$

Let

$$\omega = \begin{pmatrix} \omega_0 \\ \omega_d \\ \omega_c \end{pmatrix} = \begin{pmatrix} 0 \\ 0 \\ F_c\zeta_c - \tilde{\omega}_c \end{pmatrix}. \tag{2.5.70}$$

It is now straightforward to verify using (2.5.58) that

$$\begin{bmatrix} \check{A}_* - \lambda I & \check{B}_* \\ \check{C}_* & \check{D}_* \end{bmatrix} \begin{pmatrix} \zeta \\ \omega \end{pmatrix} = 0. \tag{2.5.71}$$

By Definition 2.4.3, we have

$$\zeta \in \mathcal{V}_\lambda(\check{\Sigma}_*) \quad \Longrightarrow \quad \operatorname{Im}\left\{ \begin{bmatrix} X_{a\lambda} & 0 \\ 0 & 0 \\ 0 & X_{c\lambda} \\ 0 & 0 \end{bmatrix} \right\} \subseteq \mathcal{V}_\lambda(\check{\Sigma}_*). \tag{2.5.72}$$

Finally, (2.5.65) and (2.5.72) imply the result.

The proof of $\mathcal{S}_\lambda(\Sigma_*)$ follows from the same lines of reasoning. ⊠

Chapter 3

Structural Mappings of Bilinear Transformations

3.1. Introduction

WE RECALL IN this chapter the work of Chen and Weller [40] on bilinear and inverse bilinear transformations of linear time-invariant systems. Their result presents a comprehensive picture of the mapping of structural properties associated with general linear multivariable systems under bilinear and inverse bilinear transformations. They have completely investigated the problem of how the finite and infinite zero structures, as well as invertibility structures of a general continuous-time (discrete-time) linear time-invariant multivariable system are mapped to those of its discrete-time (continuous-time) counterpart under the bilinear (inverse bilinear) transformation. It is worth noting that we have added in this chapter some new results on the mapping of geometric subspaces under the bilinear (inverse bilinear) transformation.

The bilinear and inverse bilinear transformations have widespread use in digital control and signal processing. As will be seen shortly, the bilinear transformation actually plays a crucial role in the computation of infima for discrete-time systems as well as in finding the solutions to discrete-time Riccati equations. The results presented in this section were first reported in Chen and Weller [40]. In fact, the need to perform continuous-time to discrete-time model conversions arises in a range of engineering contexts, including sampled-data control system design, and digital signal processing. As a consequence, numerous discretization procedures exist, including zero- and first-order hold input approximations, impulse invariant transformation, and bilinear transformation (see, for example [2] and [55]). Despite the widespread use of the bilinear transform, however, a comprehensive treatment detailing how key structural properties of continuous-time systems, such as the finite and infinite zero structures, and invertibility properties, are inherited by their discrete-time counterparts is lacking in the literature. Given the important role played by the infinite and finite zero structures in control system design, a clear understanding of the zero structures under bilinear transformation would be useful in the design of sampled-data control systems, and would complement existing results on the mapping of finite and infinite zero structures under zero-order hold sampling (see, for example, [1] and [60]).

In this chapter, we present a comprehensive study of how the structures, i.e., the finite and infinite zero structures, invertibility structures, as well as geometric subspaces of a general continuous-time (discrete-time) linear time-invariant system are mapped to those of its discrete-time (continuous-time) counterpart under the well known bilinear (inverse bilinear) transformations

$$s = a\left(\frac{z-1}{z+1}\right) \quad \text{and} \quad z = \frac{a+s}{a-s}, \tag{3.1.1}$$

respectively.

3.2. Mapping of Continuous-time to Discrete-time

In this section, we will consider a continuous-time linear time-invariant system Σ_c characterized by

$$\Sigma_c : \begin{cases} \dot{x} = A\,x + B\,u, \\ y = C\,x + D\,u, \end{cases} \tag{3.2.1}$$

where $x \in \mathbb{R}^n$, $y \in \mathbb{R}^p$, $u \in \mathbb{R}^m$ and A, B, C and D are matrices of appropriate dimensions. Without loss of any generality, we assume that both matrices $[\,C \;\; D\,]$ and $[\,B' \;\; D'\,]$ are of full rank. Σ_c has a transfer function

$$G_c(s) = C(sI - A)^{-1}B + D. \tag{3.2.2}$$

Let us apply a bilinear transformation to the above continuous-time system, by replacing s in (3.2.2) with

$$s = \frac{2}{T}\left(\frac{z-1}{z+1}\right) = a\left(\frac{z-1}{z+1}\right), \tag{3.2.3}$$

where $T = 2/a$ is the sampling period. As presented in (3.2.3), the bilinear transformation is often called Tustin's approximation [2], while the choice

$$a = \frac{\omega_1}{\tan(\omega_1 T/2)} \tag{3.2.4}$$

yields the pre-warped Tustin approximation, in which the frequency responses of the continuous-time system and its discrete-time counterpart are matched at frequency ω_1. In this way, we obtain a discrete-time system

$$G_d(z) = C\left(a\frac{z-1}{z+1}I - A\right)^{-1} B + D. \tag{3.2.5}$$

The following lemma provides a direct state-space realization of $G_d(z)$. While this result is well known (see for example [55]), the proof is included as it is brief and self-contained.

Lemma 3.2.1. A state-space realization of $G_d(z)$, the discrete-time counterpart of the continuous-time system Σ_c of (3.2.1) under the bilinear transformation (3.2.3), is given by

$$\Sigma_d : \begin{cases} x(k+1) = \tilde{A}\, x(k) + \tilde{B}\, u(k), \\ \quad y(k) \;\; = \tilde{C}\, x(k) + \tilde{D}\, u(k), \end{cases} \tag{3.2.6}$$

where

$$\left.\begin{aligned} \tilde{A} &= (aI + A)(aI - A)^{-1}, \\ \tilde{B} &= \sqrt{2a}\,(aI - A)^{-1}B, \\ \tilde{C} &= \sqrt{2a}\,C(aI - A)^{-1}, \\ \tilde{D} &= D + C(aI - A)^{-1}B, \end{aligned}\right\} \tag{3.2.7}$$

or

$$\left.\begin{aligned} \tilde{A} &= (aI + A)(aI - A)^{-1}, \\ \tilde{B} &= B, \\ \tilde{C} &= 2a\,C(aI - A)^{-2}, \\ \tilde{D} &= D + C(aI - A)^{-1}B. \end{aligned}\right\} \tag{3.2.8}$$

Here we clearly assume that matrix A has no eigenvalue at a. ◻

Proof. First, it is straightforward to verify that

$$\begin{aligned} G_d(z) &= C\left(a\frac{z-1}{z+1}I - A\right)^{-1} B + D \\ &= (z+1)C\big[a(z-1)I - (z+1)A\big]^{-1}B + D \\ &= (z+1)C(aI-A)^{-1}\big[zI - (aI+A)(aI-A)^{-1}\big]^{-1}B + D \\ &= zC(aI-A)^{-1}\left(zI-\tilde{A}\right)^{-1}B + \left[C(aI-A)^{-1}\left(zI-\tilde{A}\right)^{-1}B + D\right]. \end{aligned} \tag{3.2.9}$$

If we introduce $\tilde{G}_d(z) = zC(aI - A)^{-1}\left(zI - \tilde{A}\right)^{-1}B$, it follows that

$$\begin{cases} \tilde{x}(k+1) = \tilde{A}'\tilde{x}(k) + (aI - A')^{-1}C'\tilde{u}(k), \\ \quad \tilde{y}(k) \;\; = B'\tilde{x}(k+1) = B'\tilde{A}'\tilde{x}(k) + B'(aI - A')^{-1}C'\tilde{u}(k), \end{cases} \tag{3.2.10}$$

is a state-space realization of $\tilde{G}'_d(z)$, from which

$$\tilde{G}_d(z) = C(aI - A)^{-1} \left(zI - \tilde{A}\right)^{-1} \tilde{A}B + C(aI - A)^{-1}B. \tag{3.2.11}$$

Substituting (3.2.11) into (3.2.9), we obtain

$$\begin{aligned} G_d(z) &= C(aI - A)^{-1} \left(zI - \tilde{A}\right)^{-1} (\tilde{A} + I)B + [C(aI - A)^{-1}B + D] \\ &= \tilde{C} \left(zI - \tilde{A}\right)^{-1} \tilde{B} + \tilde{D}, \end{aligned}$$

and the rest of Lemma 3.2.1 follows. ⊠

The following theorem establishes the interconnection of the structural properties of Σ_c and Σ_d, and forms the major contribution of this chapter.

Theorem 3.2.1. Consider the continuous-time system Σ_c of (3.2.1) characterized by the quadruple (A, B, C, D) with matrix A having no eigenvalue at a, and its discrete-time counterpart under the bilinear transformation (3.2.3), i.e., Σ_d of (3.2.6) characterized by the quadruple $(\tilde{A}, \tilde{B}, \tilde{C}, \tilde{D})$ of (3.2.7). We have the following properties:

1. Controllability (stabilizability) and observability (detectability) of Σ_d:

   (a) The pair $(\tilde{A}, \tilde{B})$ is controllable (stabilizable) if and only if the pair (A, B) is controllable (stabilizable).

   (b) The pair $(\tilde{A}, \tilde{C})$ is observable (detectable) if and only if the pair (A, C) is observable (detectable).

2. Effects of nonsingular state, output and input transformations, together with state feedback and output injection laws:

   (a) For any given nonsingular state, output and input transformations T_s, T_o and T_i, the quadruple

   $$(T_s^{-1}\tilde{A}T_s, T_s^{-1}\tilde{B}T_i, T_o^{-1}\tilde{C}T_s, T_o^{-1}\tilde{D}T_i), \tag{3.2.12}$$

   is the discrete-time counterpart under the bilinear transformation (3.2.3), of the continuous time system

   $$(T_s^{-1}AT_s, T_s^{-1}BT_i, T_o^{-1}CT_s, T_o^{-1}DT_i). \tag{3.2.13}$$

(b) For any $F \in \mathbb{R}^{m\times n}$ with $A + BF$ having no eigenvalue at a, define a nonsingular matrix

$$\begin{aligned}\tilde{T}_i &:= I + F(aI - A - BF)^{-1}B \\ &= [I - F(aI - A)^{-1}B]^{-1} \in \mathbb{R}^{m\times m},\end{aligned} \tag{3.2.14}$$

and a constant matrix

$$\tilde{F} := \sqrt{2a}\, F(aI - A - BF)^{-1} \in \mathbb{R}^{m\times n}. \tag{3.2.15}$$

Then a continuous-time system $\Sigma_{\rm cF}$ characterized by

$$(A + BF, B, C + DF, D), \tag{3.2.16}$$

is mapped to a discrete-time system $\Sigma_{\rm dF}$, characterized by

$$(\tilde{A} + \tilde{B}\tilde{F}, \tilde{B}\tilde{T}_i, \tilde{C} + \tilde{D}\tilde{F}, \tilde{D}\tilde{T}_i), \tag{3.2.17}$$

under the bilinear transformation (3.2.3). Here we note that $\Sigma_{\rm cF}$ is the closed-loop system comprising Σ_c and a state feedback law with gain matrix F, and $\Sigma_{\rm dF}$ is the closed-loop system comprising Σ_d and a state feedback law with gain matrix $\tilde{F}$, together with a nonsingular input transformation $\tilde{T}_i$.

(c) For any $K \in \mathbb{R}^{n\times p}$ with $A + KC$ having no eigenvalue at a, define a nonsingular matrix

$$\tilde{T}_o := [I + C(aI - A - KC)^{-1}K]^{-1} \in \mathbb{R}^{p\times p}, \tag{3.2.18}$$

and a constant matrix

$$\tilde{K} := \sqrt{2a}\,(aI - A - KC)^{-1}K. \tag{3.2.19}$$

Then a continuous-time system $\Sigma_{\rm cK}$ characterized by

$$(A + KC, B + KD, C, D), \tag{3.2.20}$$

is mapped to a discrete-time system $\Sigma_{\rm dK}$, characterized by

$$(\tilde{A} + \tilde{K}\tilde{C}, \tilde{B} + \tilde{K}\tilde{D}, \tilde{T}_o^{-1}\tilde{C}, \tilde{T}_o^{-1}\tilde{D}), \tag{3.2.21}$$

under the bilinear transformation (3.2.3). We note that $\Sigma_{\rm cK}$ is the closed-loop system comprising Σ_c and an output injection law with gain matrix K, and $\Sigma_{\rm dK}$ is the closed-loop system comprising Σ_d and an output injection law with gain matrix $\tilde{K}$, together with a nonsingular output transformation $\tilde{T}_o$.

3. Invertibility and structural invariant indices lists $\mathcal{I}_2$ and $\mathcal{I}_3$ of Σ_d:

   (a) $\mathcal{I}_2(\Sigma_d) = \mathcal{I}_2(\Sigma_c)$, and $\mathcal{I}_3(\Sigma_d) = \mathcal{I}_3(\Sigma_c)$.

   (b) Σ_d is left (right) invertible if and only if Σ_c is left (right) invertible.

   (c) Σ_d is invertible (degenerate) if and only if Σ_c is invertible (degenerate).

4. The invariant zeros of Σ_d and their associated structures consist of the following two parts:

   (a) Let the infinite zero structure (of order greater than 0) of Σ_c be given by $S^\star_\infty(\Sigma_c) = \{q_1, q_2, \cdots, q_{m_d}\}$. Then $z = -1$ is an invariant zero of Σ_d with the multiplicity structure $S^\star_{-1}(\Sigma_d) = \{q_1, q_2, \cdots, q_{m_d}\}$.

   (b) Let $s = \alpha \neq a$ be an invariant zero of Σ_c with the multiplicity structure $S^\star_\alpha(\Sigma_c) = \{n_{\alpha,1}, n_{\alpha,2}, \cdots, n_{\alpha,\tau_\alpha}\}$. Then $z = \beta = (a+\alpha)/(a-\alpha)$ is an invariant zero of its discrete-time counterpart Σ_d with the multiplicity structure $S^\star_\beta(\Sigma_d) = \{n_{\alpha,1}, n_{\alpha,2}, \cdots, n_{\alpha,\tau_\alpha}\}$.

5. The infinite zero structure of Σ_d consists of the following two parts:

   (a) Let $m_0 =$ rank (D), and let m_d be the total number of infinite zeros of Σ_c of order greater than 0. Also, let τ_a be the geometric multiplicity of the invariant zero of Σ_c at $s = a$. Then we have rank $(\tilde{D}) = m_0 + m_d - \tau_a$.

   (b) Let $s = a$ be an invariant zero of the given continuous-time system Σ_c with a multiplicity structure $S^\star_a(\Sigma_c) = \{n_{a,1}, n_{a,2}, \cdots, n_{a,\tau_a}\}$. Then the discrete-time counterpart Σ_d has an infinite zero (of order greater than 0) structure $S^\star_\infty(\Sigma_d) = \{n_{a,1}, n_{a,2}, \cdots, n_{a,\tau_a}\}$.

6. The mappings of geometric subspaces:

   (a) $\mathcal{V}^+(\Sigma_c) = \mathcal{S}^\odot(\Sigma_d)$.

   (b) $\mathcal{S}^+(\Sigma_c) = \mathcal{V}^\odot(\Sigma_d)$. ⊤

Proof. See Section 3.4. ⊠

We have the following two interesting observations. The first is with regard to the minimum phase and nonminimum phase properties of Σ_d, while the second concerns the asymptotic behavior of Σ_d as the sampling period T tends to zero (or, equivalently, as $a \to \infty$).

Observation 3.2.1. Consider a general continuous-time system Σ_c and its discrete-time counterpart Σ_d under the bilinear transformation (3.2.3). Then it follows from 4(a) and 4(b) of Theorem 3.2.1 that

1. Σ_d has all its invariant zeros inside the unit circle if and only if Σ_c has all its invariant zeros in the open left-half plane and has no infinite zero of order greater than 0;
2. Σ_d has invariant zeros on the unit circle if and only if Σ_c has invariant zeros on the imaginary axis, and/or Σ_c has at least one infinite zero of order greater than 0;
3. Σ_d has invariant zeros outside the unit circle if and only if Σ_c has invariant zeros in the open right-half plane. ⊡

Observation 3.2.2. Consider a general continuous-time system Σ_c and its discrete-time counterpart Σ_d under the bilinear transformation (3.2.3). Then a consequence of Theorem 3.2.1, Σ_d has the following asymptotic properties as the sampling period T tends to zero (but not equal to zero):

1. Σ_d has no infinite zero of order greater than 0, i.e., no delays from the input to the output;
2. Σ_d has one invariant zero at $z = -1$ with an appropriate multiplicity structure if Σ_c has any infinite zero of order greater than 0; and
3. The remaining invariant zeros of Σ_d, if any, tend to the point $z = 1$. More interestingly, the invariant zeros of Σ_d corresponding to the stable invariant zeros of Σ_c are always stable, and approach the point $z = 1$ from inside the unit circle. Conversely, the invariant zeros of Σ_d corresponding to the unstable invariant zeros of Σ_c are always unstable, and approach the point $z = 1$ from outside the unit circle. Finally, those associated with the imaginary axis invariant zeros of Σ_c are always mapped onto the unit circle and move towards to the point $z = 1$. ⊡

The following example illustrates the results in Theorem 3.2.1.

Example 3.2.1. Consider a continuous-time system Σ_c characterized by the quadruple (A,B,C,D) with

$$A = \begin{bmatrix} 1 & 1 & 0 & 0 & 1 & 0 \\ 0 & 1 & 1 & 0 & 1 & 0 \\ 0 & 0 & 1 & 0 & 1 & 0 \\ 0 & 0 & 0 & 3 & 1 & 0 \\ 0 & 0 & 0 & 0 & 0 & 1 \\ 1 & 1 & 1 & 1 & 1 & 1 \end{bmatrix}, \quad B = \begin{bmatrix} 0 & 0 \\ 0 & 0 \\ 0 & 0 \\ 1 & 0 \\ 0 & 0 \\ 0 & 1 \end{bmatrix}, \tag{3.2.22}$$

and

$$C=\begin{bmatrix}0&0&0&1&0&0\\0&0&0&0&1&0\end{bmatrix},\quad D=\begin{bmatrix}1&0\\0&0\end{bmatrix}. \tag{3.2.23}$$

We note that the above system Σ_c is already in the form of the special coordinate basis as in Theorem 2.4.1. Furthermore, Σ_c is controllable, observable and invertible with one infinite zero of order 0, and one infinite zero of order 2, i.e., $S_\infty^\star(\Sigma_c)=\{2\}$. The system Σ_c also has two invariant zeros at $s=2$ and $s=1$, respectively, with structures $S_2^\star(\Sigma_c)=\{1\}$ and $S_1^\star(\Sigma_c)=\{3\}$.

1. If $a=1$, we obtain a discrete-time system Σ_d characterized by the quadruple $(\tilde{A},\tilde{B},\tilde{C},\tilde{D})$, with

$$\tilde{A}=\begin{bmatrix}1&2&-3&1&0&-2\\-2&-1&2&0&0&0\\0&-2&1&0&0&0\\0&0&1&-2&0&0\\0&0&-2&0&-1&0\\0&0&-2&0&-2&-1\end{bmatrix},\quad \tilde{B}=\frac{\sqrt{2}}{2}\begin{bmatrix}1&-2\\0&0\\0&0\\-1&0\\0&0\\0&0\end{bmatrix},$$

$$\tilde{C}=\frac{\sqrt{2}}{2}\begin{bmatrix}0&0&1&-1&0&0\\0&0&-2&0&0&0\end{bmatrix},\quad \tilde{D}=\frac{1}{2}\begin{bmatrix}1&0\\0&0\end{bmatrix}.$$

Utilizing either the toolbox of Chen [11] or that of Lin [79], we find that Σ_d is indeed controllable, observable and invertible, with one infinite zero of order 0 and one infinite zero of order 3, i.e., $S_\infty^\star(\Sigma_d)=\{3\}$. Σ_d also has two invariant zeros at $z=-3$ and $z=-1$ respectively, with structures $S_{-3}^\star(\Sigma_d)=\{1\}$ and $S_{-1}^\star(\Sigma_d)=\{2\}$.

2. If $a=2$, we obtain another discrete-time system Σ_d, characterized by

$$\tilde{A}=\begin{bmatrix}0&-2&-5&3&-3&-3\\-2&-1&-2&2&-2&-2\\-1&-2&0&1&-1&-1\\1&2&3&-6&1&1\\-1&-2&-3&1&-2&-1\\-2&-4&-6&2&-6&-3\end{bmatrix},\quad \tilde{B}=\frac{1}{2}\begin{bmatrix}3&-3\\2&-2\\1&-1\\-5&1\\1&-1\\2&-2\end{bmatrix},$$

and

$$\tilde{C}=\frac{1}{2}\begin{bmatrix}1&2&3&-5&1&1\\-1&-2&-3&1&-1&-1\end{bmatrix},\quad \tilde{D}=\frac{1}{4}\begin{bmatrix}-1&1\\1&-1\end{bmatrix},$$

which is controllable, observable and invertible with one infinite zero of order 0 and one infinite zero of order 1, i.e., $S_\infty^\star(\Sigma_d)=\{1\}$. It also has two invariant zeros at $z=3$ and $z=-1$ respectively, with structures $S_3^\star(\Sigma_d)=\{3\}$ and $S_{-1}^\star(\Sigma_d)=\{2\}$, in accordance with Theorem 3.2.1. ◾

3.3. Mapping of Discrete-time to Continuous-time

We present in this section a similar result as in the previous section, but for the inverse bilinear transformation mapping a discrete-time system to a continuous-time system. We begin with a discrete-time linear time-invariant system $\tilde{\Sigma}_d$ characterized by

$$\tilde{\Sigma}_d \; : \; \begin{cases} x(k+1) = \tilde{A}\, x(k) + \tilde{B}\, u(k), \\ \quad y(k) \quad = \tilde{C}\, x(k) + \tilde{D}\, u(k), \end{cases} \tag{3.3.1}$$

where $x \in \mathbb{R}^n$, $y \in \mathbb{R}^p$, $u \in \mathbb{R}^m$ and $\tilde{A}$, $\tilde{B}$, $\tilde{C}$ and $\tilde{D}$ are matrices of appropriate dimensions. Without loss of any generality, we assume that both matrices $[\tilde{C} \;\; \tilde{D}]$ and $[\tilde{B}' \;\; \tilde{D}']$ are of full rank. Σ_d has a transfer function

$$H_d(z) = \tilde{C}(zI - \tilde{A})^{-1}\tilde{B} + \tilde{D}. \tag{3.3.2}$$

The inverse bilinear transformation corresponding to (3.2.3) replaces z in the above equation (3.3.2) with

$$z = \frac{a+s}{a-s}, \tag{3.3.3}$$

to obtain the following continuous-time system:

$$H_c(s) = \tilde{C}\left(\frac{a+s}{a-s}I - \tilde{A}\right)^{-1}\tilde{B} + \tilde{D}. \tag{3.3.4}$$

The following lemma is analogous to Lemma 3.2.1, and provides a state-space realization of $H_c(s)$.

Lemma 3.3.1. A state-space realization of $H_c(s)$, the continuous-time counterpart of the discrete-time system $\tilde{\Sigma}_d$ of (3.3.1) under the inverse bilinear transformation (3.3.3), is given by

$$\tilde{\Sigma}_c \; : \; \begin{cases} \dot{x} = A\, x + B\, u, \\ y = C\, x + D\, u, \end{cases} \tag{3.3.5}$$

where

$$\left. \begin{aligned} A &= a(\tilde{A}+I)^{-1}(\tilde{A}-I), \\ B &= \sqrt{2a}\,(\tilde{A}+I)^{-1}\tilde{B}, \\ C &= \sqrt{2a}\,\tilde{C}(\tilde{A}+I)^{-1}, \\ D &= \tilde{D} - \tilde{C}(\tilde{A}+I)^{-1}\tilde{B}, \end{aligned} \right\} \tag{3.3.6}$$

or

$$\left. \begin{aligned} A &= a(\tilde{A}+I)^{-1}(\tilde{A}-I), \\ B &= \tilde{B}, \\ C &= 2a\,\tilde{C}(\tilde{A}+I)^{-2}, \\ D &= \tilde{D} - \tilde{C}(\tilde{A}+I)^{-1}\tilde{B}. \end{aligned} \right\} \tag{3.3.7}$$

Here we clearly assume that the matrix $\tilde{A}$ has no eigenvalue at -1. ◻

The following theorem is analogous to Theorem 3.2.1.

Theorem 3.3.1. Consider the discrete-time system $\tilde{\Sigma}_d$ of (3.3.1) characterized by the quadruple $(\tilde{A}, \tilde{B}, \tilde{C}, \tilde{D})$ with matrix $\tilde{A}$ having no eigenvalue at -1, and its continuous-time counterpart under the inverse bilinear transformation (3.3.3), i.e., $\tilde{\Sigma}_c$ of (3.3.5) characterized by the quadruple (A, B, C, D) of (3.3.6). We have the following properties:

1. Controllability (stabilizability) and observability (detectability) of $\tilde{\Sigma}_c$:

   (a) The pair (A, B) is controllable (stabilizable) if and only if the pair $(\tilde{A}, \tilde{B})$ is controllable (stabilizable).

   (b) The pair (A, C) is observable (detectable) if and only if the pair $(\tilde{A}, \tilde{C})$ is observable (detectable).

2. Effects of nonsingular state, output and input transformations, together with state feedback and output injection laws:

   (a) For any given nonsingular state, output and input transformations T_s, T_o and T_i, the quadruple

$$(T_s^{-1}AT_s, T_s^{-1}BT_i, T_o^{-1}CT_s, T_o^{-1}DT_i), \tag{3.3.8}$$

   is the continuous-time counterpart of the inverse bilinear transformation, i.e., (3.3.3), of the discrete-time system

$$(T_s^{-1}\tilde{A}T_s, T_s^{-1}\tilde{B}T_i, T_o^{-1}\tilde{C}T_s, T_o^{-1}\tilde{D}T_i). \tag{3.3.9}$$

   (b) For any $\tilde{F} \in \mathbb{R}^{m \times n}$ with $\tilde{A} + \tilde{B}\tilde{F}$ having no eigenvalue at -1, define a nonsingular matrix

$$T_i := I - \tilde{F}(I + \tilde{A} + \tilde{B}\tilde{F})^{-1}\tilde{B} \in \mathbb{R}^{m \times m}, \tag{3.3.10}$$

   and a constant matrix

$$F := \sqrt{2a}\, \tilde{F}(I + \tilde{A} + \tilde{B}\tilde{F})^{-1} \in \mathbb{R}^{m \times n}. \tag{3.3.11}$$

   Then a discrete-time system $\tilde{\Sigma}_{d\text{F}}$ characterized by

$$(\tilde{A} + \tilde{B}\tilde{F}, \tilde{B}, \tilde{C} + \tilde{D}\tilde{F}, \tilde{D}), \tag{3.3.12}$$

   is mapped to a continuous-time counterpart $\tilde{\Sigma}_{c\text{F}}$ characterized by

$$(A + BF, BT_i, C + DF, DT_i), \tag{3.3.13}$$

under the inverse bilinear transformation (3.3.3). Note that $\tilde{\Sigma}_{dF}$ is the closed-loop system comprising $\tilde{\Sigma}_d$ and a state feedback law with gain matrix $\tilde{F}$, and $\tilde{\Sigma}_{dF}$ is the closed-loop system comprising $\tilde{\Sigma}_d$ and a state feedback law with gain matrix F, together with a nonsingular input transformation T_i.

(c) For any $\tilde{K} \in \mathbb{R}^{n\times p}$ with $\tilde{A}+\tilde{K}\tilde{C}$ having no eigenvalue at -1, define a nonsingular matrix

$$T_o := [I - \tilde{C}(I + \tilde{A} + \tilde{K}\tilde{C})^{-1}\tilde{K}]^{-1} \in \mathbb{R}^{p\times p}, \tag{3.3.14}$$

and a constant matrix

$$K := \sqrt{2a}\,(I + \tilde{A} + \tilde{K}\tilde{C})^{-1}\tilde{K}. \tag{3.3.15}$$

Then a discrete-time system $\tilde{\Sigma}_{dK}$ characterized by

$$(\tilde{A} + \tilde{K}\tilde{C}, \tilde{B} + \tilde{K}\tilde{D}, \tilde{C}, \tilde{D}), \tag{3.3.16}$$

is mapped to a continuous-time $\tilde{\Sigma}_{cK}$, characterized by

$$(A + KC, B + KD, T_o^{-1}C, T_o^{-1}D), \tag{3.3.17}$$

under the inverse bilinear transformation (3.3.3). We note that $\tilde{\Sigma}_{dK}$ is the closed-loop system comprising $\tilde{\Sigma}_d$ and an output injection law with gain matrix $\tilde{K}$, and $\tilde{\Sigma}_{cK}$ is the closed-loop system comprising $\tilde{\Sigma}_c$ and an output injection law with gain matrix K, together with a nonsingular output transformation T_o.

3. Invertibility and structural invariant indices lists $\mathcal{I}_2$ and $\mathcal{I}_3$ of $\tilde{\Sigma}_c$:

   (a) $\mathcal{I}_2(\tilde{\Sigma}_c) = \mathcal{I}_2(\tilde{\Sigma}_d)$, and $\mathcal{I}_3(\tilde{\Sigma}_c) = \mathcal{I}_3(\tilde{\Sigma}_d)$.

   (b) $\tilde{\Sigma}_c$ is left (right) invertible if and only if $\tilde{\Sigma}_d$ is left (right) invertible.

   (c) $\tilde{\Sigma}_c$ is invertible (degenerate) if and only if $\tilde{\Sigma}_d$ is invertible (degenerate).

4. Invariant zeros of $\tilde{\Sigma}_c$ and their structures consist of the following two parts:

   (a) Let the infinite zero structure (of order greater than 0) of $\tilde{\Sigma}_d$ be given by $S_\infty^\star(\tilde{\Sigma}_d) = \{q_1, q_2, \cdots, q_{m_d}\}$. Then $s = a$ is an invariant zero of $\tilde{\Sigma}_c$ with the multiplicity structure $S_a^\star(\tilde{\Sigma}_c) = \{q_1, q_2, \cdots, q_{m_d}\}$.

(b) Let $z = \alpha \neq -1$ be an invariant zero of $\tilde{\Sigma}_d$ with the multiplicity structure $S^\star_\alpha(\tilde{\Sigma}_d) = \{n_{\alpha,1}, n_{\alpha,2}, \cdots, n_{\alpha,\tau_\alpha}\}$. Then $s = \beta = a\frac{\alpha-1}{\alpha+1}$ is an invariant zero of its continuous-time counterpart $\tilde{\Sigma}_c$ with the multiplicity structure $S^\star_\beta(\tilde{\Sigma}_c) = \{n_{\alpha,1}, n_{\alpha,2}, \cdots, n_{\alpha,\tau_\alpha}\}$.

5. The infinite zero structure of $\tilde{\Sigma}_c$ consists of the following two parts:

(a) Let $m_0 =$ rank $(\tilde{D})$, and let m_d be the total number of infinite zeros of $\tilde{\Sigma}_d$ of order greater than 0. Also, let τ_{-1} be the geometric multiplicity of the invariant zero of $\tilde{\Sigma}_d$ at $z = -1$. Then we have rank $(D) = m_0 + m_d - \tau_{-1}$.

(b) Let $z = -1$ be an invariant zero of the given discrete-time system $\tilde{\Sigma}_d$ with the multiplicity structure $S^\star_{-1}(\tilde{\Sigma}_d) = \{n_{-1,1}, n_{-1,2}, \cdots, n_{-1,\tau_{-1}}\}$. Then $\tilde{\Sigma}_c$ has an infinite zero (of order greater than 0) structure $S^\star_\infty(\tilde{\Sigma}_c) = \{n_{-1,1}, n_{-1,2}, \cdots, n_{-1,\tau_{-1}}\}$.

6. The mappings of geometric subspaces:

(a) $\mathcal{V}^\odot(\tilde{\Sigma}_d) = \mathcal{S}^+(\tilde{\Sigma}_c)$.

(b) $\mathcal{S}^\odot(\tilde{\Sigma}_d) = \mathcal{V}^+(\tilde{\Sigma}_c)$. ⊤

Proof. The proof of this theorem is similar to that of Theorem 3.2.1. ⊠

We illustrate the result above with the following example.

Example 3.3.1. Consider a discrete-time linear time-invariant system $\tilde{\Sigma}_d$ characterized by the quadruple $(\tilde{A}, \tilde{B}, \tilde{C}, \tilde{D})$ with

$$\tilde{A} = \begin{bmatrix} -1 & 0 & 0 & 1 & 0 & 1 & 0 \\ 0 & -1 & 1 & 1 & 0 & 1 & 1 \\ 0 & 0 & -1 & 1 & 0 & 1 & 1 \\ 0 & 0 & 0 & 1 & 0 & 1 & 1 \\ 1 & 1 & 1 & 1 & 1 & 1 & 1 \\ 1 & 1 & 1 & 1 & 1 & 1 & 1 \\ 0 & 1 & 1 & 1 & 1 & 1 & 1 \end{bmatrix}, \quad \tilde{B} = \begin{bmatrix} 0 & 0 & 0 \\ 0 & 0 & 0 \\ 0 & 0 & 0 \\ 0 & 0 & 0 \\ 0 & 0 & 1 \\ 1 & 0 & 0 \\ 0 & 1 & 0 \end{bmatrix}, \tag{3.3.18}$$

and

$$\tilde{C} = \begin{bmatrix} 0 & 0 & 0 & 1 & 0 & 0 & 0 \\ 0 & 0 & 0 & 0 & 0 & 1 & 0 \\ 0 & 0 & 0 & 0 & 0 & 0 & 1 \end{bmatrix}, \quad \tilde{D} = \begin{bmatrix} 0 & 0 & 0 \\ 0 & 0 & 0 \\ 0 & 0 & 0 \end{bmatrix}. \tag{3.3.19}$$

Again the above system is already in the form of the special coordinate basis. It is simple to verify that $\tilde{\Sigma}_d$ is controllable, observable and is degenerate, i.e., neither left nor right invertible, with two infinite zeros of order 1, i.e., $S_\infty^\star(\tilde{\Sigma}_d) = \{1,1\}$, $\mathcal{I}_2(\tilde{\Sigma}_d) = \{1\}$ and $\mathcal{I}_3(\tilde{\Sigma}_d) = \{1\}$. It also has one invariant zero at $z = -1$ with a structure of $S_{-1}^\star(\tilde{\Sigma}_d) = \{1,2\}$. Applying the result in Lemma 3.3.1 (with $a = 1$), we obtain $\tilde{\Sigma}_c$ which is characterized by (A, B, C, D) with

$$A = \begin{bmatrix} 5 & 0 & 0 & -2 & 0 & -2 & 2 \\ 0 & 3 & 4 & -2 & 2 & -2 & -2 \\ 0 & -2 & 3 & 0 & 0 & 0 & 0 \\ 0 & 0 & 2 & -1 & 0 & 0 & 0 \\ -2 & 0 & -2 & 2 & -1 & 2 & 0 \\ -2 & 0 & -2 & 2 & 0 & 1 & 0 \\ 2 & 0 & -2 & 0 & 0 & 0 & 1 \end{bmatrix}, \quad B = \sqrt{2} \begin{bmatrix} 1 & -1 & 0 \\ 1 & 1 & -1 \\ 0 & 0 & 0 \\ 0 & 0 & 0 \\ -1 & 0 & 1 \\ 0 & 0 & 0 \\ 0 & 0 & 0 \end{bmatrix},$$

$$C = \sqrt{2} \begin{bmatrix} 0 & 0 & -1 & 1 & 0 & 0 & 0 \\ 1 & 0 & 1 & -1 & 0 & 0 & 0 \\ -1 & 0 & 1 & 0 & 0 & 0 & 0 \end{bmatrix}, \qquad D = \begin{bmatrix} 0 & 0 & 0 \\ 0 & 0 & 0 \\ 0 & 0 & 0 \end{bmatrix}.$$

Then, it is straightforward to verify, using the software toolboxes of Chen [11] or Lin [79], for example, that $\tilde{\Sigma}_c$ is controllable, observable and degenerate with an infinite zero structure of $S_\infty^\star(\tilde{\Sigma}_c) = \{1,2\}$, $\mathcal{I}_2(\tilde{\Sigma}_c) = \{1\}$ and $\mathcal{I}_3(\tilde{\Sigma}_c) = \{1\}$. Furthermore, $\tilde{\Sigma}_c$ has one invariant zero at $s = 1$ with associated structure $S_1^\star(\tilde{\Sigma}_c) = \{1,1\}$, in accordance with Theorem 3.3.1. ▣

Finally, we conclude this section by summarizing in a graphical form in Figures 3.3.1 the structural mappings associated with the bilinear and inverse bilinear transformations.

3.4. Proof of Theorem 3.2.1

We present in this section the detailed proof of Theorem 3.2.1. For the sake of simplicity in presentation, and without loss of any generality, we assume that the constant a in (3.2.3) is equal to unity, i.e., $a = 2/T = 1$, throughout this proof. We will prove this theorem item-by-item.

1(a). Let β be an eigenvalue of $\tilde{A}$, i.e., $\beta \in \lambda(\tilde{A})$. It is straightforward to verify that $\beta \neq -1$, provided A has no eigenvalue at $a = 1$ and $\alpha = (\beta - 1)/(\beta + 1)$ is an eigenvalue of A, i.e., $\alpha \in \lambda(A)$. Next, we consider the matrix pencil

$$\begin{aligned} [\,\beta I - \tilde{A} \quad \tilde{B}\,] &= [\,\beta I - (I - A)^{-1}(I + A) \quad \sqrt{2}(I - A)^{-1}B\,] \\ &= (I - A)^{-1}\,[\,\beta(I - A) - (I + A) \quad \sqrt{2}\,B\,] \\ &= (I - A)^{-2}\,[\,(\beta - 1)I - (\beta + 1)A \quad \sqrt{2}\,B\,] \\ &= (I - A)^{-2}\,[\,\alpha I - A \quad B\,] \begin{bmatrix} (\beta + 1)I_n & 0 \\ 0 & \sqrt{2}\,I_m \end{bmatrix}. \end{aligned}$$

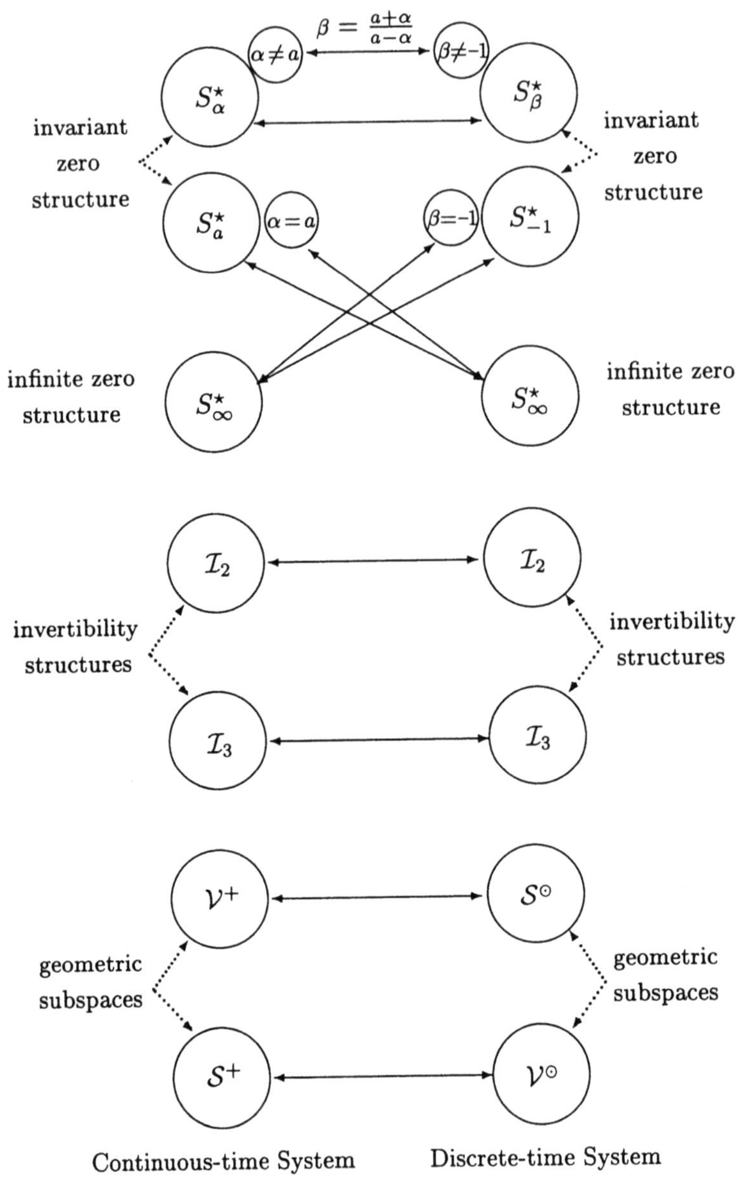

Figure 3.3.1: Structural mappings of bilinear transformations.

Clearly, rank $[\,\beta I - \tilde{A} \;\; \tilde{B}\,] = $ rank $[\,\alpha I - A \;\; B\,]$, and the result 1(a) follows. ☒

1(b). Dual of 1(a). ☒

2(a). It is trivial. ☒

2(b). It follows from Lemma 3.2.1 that the discrete-time counterpart $\Sigma_{d\mathrm{F}}$ of the bilinear transformation of $\Sigma_{c\mathrm{F}}$, characterized by $(A + BF, B, C + DF, D)$, is given by $(\tilde{A}_{\mathrm{F}}, \tilde{B}_{\mathrm{F}}, \tilde{C}_{\mathrm{F}}, \tilde{D}_{\mathrm{F}})$ with

$$\left.\begin{aligned} \tilde{A}_{\mathrm{F}} &= (I + A + BF)(I - A - BF)^{-1}, \\ \tilde{B}_{\mathrm{F}} &= \sqrt{2}\,(I - A - BF)^{-1}B, \\ \tilde{C}_{\mathrm{F}} &= \sqrt{2}\,(C + DF)(I - A - BF)^{-1}, \\ \tilde{D}_{\mathrm{F}} &= D + (C + DF)(I - A - BF)^{-1}B. \end{aligned}\right\} \tag{3.4.1}$$

We first recall from the Appendix of Kailath [64] the following matrix identities that are frequently used in the derivation of our result:

$$(I + XY)^{-1}X = X(I + YX)^{-1}, \tag{3.4.2}$$

and

$$\left[I + X(sI - Z)^{-1}Y\right]^{-1} = I - X(sI - Z + YX)^{-1}Y. \tag{3.4.3}$$

Next, we note that

$$\begin{aligned} \tilde{A}_{\mathrm{F}} &= (I + A + BF)(I - A - BF)^{-1} \\ &= (I + A + BF)(I - A)^{-1}[I - BF(I - A)^{-1}]^{-1} \\ &= [\tilde{A} + BF(I - A)^{-1}][I - BF(I - A)^{-1}]^{-1} \\ &= [\tilde{A} + BF(I - A)^{-1}][I + BF(I - A - BF)^{-1}] \\ &= \tilde{A} + \tilde{A}BF(I - A - BF)^{-1} + BF(I - A)^{-1}[I + BF(I - A - BF)^{-1}] \\ &= \tilde{A} + \tilde{A}BF(I - A - BF)^{-1} + BF(I - A)^{-1}(I - A)(I - A - BF)^{-1} \\ &= \tilde{A} + \tilde{A}BF(I - A - BF)^{-1} + BF(I - A - BF)^{-1} \\ &= \tilde{A} + (\tilde{A} + I)BF(I - A - BF)^{-1} \\ &= \tilde{A} + 2(I - A)^{-1}BF(I - A - BF)^{-1} \\ &= \tilde{A} + \tilde{B}\tilde{F}, \end{aligned}$$

and

$$\begin{aligned} \tilde{B}_{\mathrm{F}} &= \sqrt{2}\,(I - A - BF)^{-1}B \\ &= \sqrt{2}\,[I - (I - A)^{-1}BF]^{-1}(I - A)^{-1}B \\ &= \sqrt{2}\,(I - A)^{-1}B\left[I - F(I - A)^{-1}B\right]^{-1} = \tilde{B}\tilde{T}_i. \end{aligned}$$

Also, we have

$$\begin{aligned}
\tilde{C}_{\mathrm{F}} &= \sqrt{2}\,(C+DF)(I-A-BF)^{-1} \\
&= \sqrt{2}\,(C+DF)(I-A)^{-1}[I-BF(I-A)^{-1}]^{-1} \\
&= \sqrt{2}\,(C+DF)(I-A)^{-1}[I+BF(I-A-BF)^{-1}] \\
&= \sqrt{2}\,C(I-A)^{-1}+\sqrt{2}\,DF(I-A)^{-1} \\
&\quad +\sqrt{2}\,(C+DF)(I-A)^{-1}BF(I-A-BF)^{-1} \\
&= \tilde{C}+\sqrt{2}\,\big[DF(I-A)^{-1}(I-A-BF) \\
&\quad +(C+DF)(I-A)^{-1}BF\big](I-A-BF)^{-1} \\
&= \tilde{C}+\sqrt{2}\,\big[DF-DF(I-A)^{-1}BF+C(I-A)^{-1}BF+DF(I-A)^{-1}BF\big] \\
&\quad \times (I-A-BF)^{-1} \\
&= \tilde{C}+[D+C(I-A)^{-1}B]\sqrt{2}\,F(I-A-BF)^{-1} \\
&= \tilde{C}+\tilde{D}\tilde{F},
\end{aligned}$$

and

$$\begin{aligned}
\tilde{D}_{\mathrm{F}} &= D+(C+DF)(I-A-BF)^{-1}B \\
&= D+(C+DF)\left[I-(I-A)^{-1}BF\right]^{-1}(I-A)^{-1}B \\
&= D+(C+DF)(I-A)^{-1}B\left[I-F(I-A)^{-1}B\right]^{-1} \\
&= \left\{D\left[I-F(I-A)^{-1}B\right]+(C+DF)(I-A)^{-1}B\right\}\tilde{T}_i \\
&= \left\{D-DF(I-A)^{-1}B+C(I-A)^{-1}B+DF(I-A)^{-1}B\right\}\tilde{T}_i \\
&= \tilde{D}\tilde{T}_i,
\end{aligned}$$

which completes the proof of 2(b). ⊠

2(c). Dual of 2(b). ⊠

With the benefit of properties of 2(a)–2(c), the remainder of the proof is considerably simplified. It is well known that the structural invariant indices lists of Morse, which correspond precisely to the structures of finite and infinite zeros as well as invertibility, are invariant under nonsingular state, output and input transformations, state feedback laws and output injections. We can thus apply appropriate nonsingular state, output and input transformations, as well as state feedback and output injection, to Σ_c and so obtain a new system, say Σ_c^*. If this new system has Σ_d^* as its discrete-time counterpart under bilinear transformation, then from Properties 2(a)–2(c), it follows that Σ_d^* and Σ_d have the same structural invariant properties. It is therefore sufficient for the remainder of the proof that we show that 3(a)–6(b) are indeed properties of Σ_d^*.

Let us first apply nonsingular state, output and input transformations Γ_s, Γ_o and Γ_i to Σ_c such that the resulting system is in the form of the special coordinate basis as in Theorem 2.4.1, or, equivalently, the compact form in (2.4.20)–(2.4.23) with A_{aa} and C_{0a} being given by (2.4.25), E_{da} and E_{ca} being given by (2.4.26), and B_{0a}, L_{ab} and L_{ad} being given by (2.4.28). We will further assume that A_{aa} is already in the Jordan form of (2.2.1) and (2.4.32), and that matrices A_{aa}, L_{ad}, B_{a0}, E_{da}, C_{0a}, E_{ca} and L_{ab} are partitioned as follows:

$$A_{aa} = \begin{bmatrix} A_{aa}^a & 0 \\ 0 & A_{aa}^* \end{bmatrix}, \quad L_{ad} = \begin{bmatrix} L_{ad}^a \\ L_{ad}^* \end{bmatrix}, \quad B_{a0} = \begin{bmatrix} B_{a0}^a \\ B_{a0}^* \end{bmatrix}, \quad L_{ab} = \begin{bmatrix} L_{ab}^a \\ L_{ab}^* \end{bmatrix}, \tag{3.4.4}$$

$$E_{da} = [\, E_{da}^a \quad E_{da}^* \,], \quad C_{0a} = [\, C_{0a}^a \quad C_{0a}^* \,], \quad E_{ca} = [\, E_{ca}^a \quad E_{ca}^* \,], \tag{3.4.5}$$

where matrix A_{aa}^a has all its eigenvalues at $a = 1$, i.e.,

$$A_{aa}^a = I + \begin{bmatrix} 0 & I_{n_{a,1}-1} & \cdots & 0 & 0 \\ 0 & 0 & \cdots & 0 & 0 \\ \vdots & \vdots & \ddots & \vdots & \vdots \\ 0 & 0 & \cdots & 0 & I_{n_{a,\tau_a}-1} \\ 0 & 0 & \cdots & 0 & 0 \end{bmatrix}, \tag{3.4.6}$$

and A_{aa}^* contains the remaining invariant zeros of Σ_c. Furthermore, we assume that the pair (A_{cc}, B_c) is in the controllability structural decomposition of (2.4.37), as is the pair (A_{bb}', C_b'). Next, define a state feedback gain matrix

$$F = -\Gamma_i \begin{bmatrix} C_{0a}^a - C_2^a & C_{0a}^* & C_{0b} & C_{0c} & C_{0d} \\ E_{da}^a - C_1^a & E_{da}^* & E_{db} & E_{dc} & E_{dd} \\ E_{ca}^a & E_{ca}^* & 0 & E_{cc} & 0 \end{bmatrix} \Gamma_s^{-1}, \tag{3.4.7}$$

and an output injection gain matrix

$$K = -\Gamma_s \begin{bmatrix} B_{a0}^a - B_2^a & L_{ad}^a - B_1^a & L_{ab}^a \\ B_{a0}^* & L_{ad}^* & L_{ab}^* \\ B_{b0} & L_{bd} & L_{bb} \\ B_{c0} & L_{cd} & L_{cb} \\ B_{d0} & L_{dd} & 0 \end{bmatrix} \Gamma_o^{-1}. \tag{3.4.8}$$

Here, E_{cc} is chosen such that all $\star$s in (2.4.37) are cleaned out, i.e.,

$$A_{cc}^* := A_{cc} - B_c E_{cc}, \tag{3.4.9}$$

is in Jordan form with all diagonal elements equal to 0. Similarly, L_{bb} is chosen such that

$$(A_{bb}^*)' := (A_{bb} - L_{bb} C_b)', \tag{3.4.10}$$

is in Jordan form with with all diagonal elements equal to 0. Likewise, E_{dd} and L_{dd} are chosen such that

$$A_{dd}^* := A_{dd} - L_{dd}C_d - B_dE_{dd}, \tag{3.4.11}$$

is in Jordan form with all diagonal elements equal to 0, which in turn implies

$$C_d(I - A_{dd}^*)^{-1}B_d = I_{m_d}. \tag{3.4.12}$$

The matrices B_1^a, B_2^a, C_1^a and C_2^a are chosen in conformity with A_{aa}^a of (3.4.6) as follows:

$$B^a := [\, B_2^a \quad B_1^a \,] := \begin{bmatrix} 0 & 0 & \cdots & 0 \\ 0 & 1 & \cdots & 0 \\ \vdots & \ddots & \vdots & \vdots \\ 0 & 0 & \cdots & 0 \\ 0 & 0 & \cdots & 1 \end{bmatrix}, \tag{3.4.13}$$

and

$$C^a := \begin{bmatrix} C_2^a \\ C_1^a \end{bmatrix} := \begin{bmatrix} 0 & 0 & \cdots & 0 & 0 \\ 1 & 0 & \cdots & 0 & 0 \\ \vdots & \vdots & \ddots & \vdots & \vdots \\ 0 & 0 & \cdots & 1 & 0 \end{bmatrix}. \tag{3.4.14}$$

This can always be done, as a consequence of the assumption that the matrix A has no eigenvalue at $a = 1$, which implies that the invariant zero at $a = 1$ of Σ_c is completely controllable and observable.

Finally, we obtain a continuous-time system Σ_c^* characterized by the quadruple (A^*,B^*,C^*,D^*), where

$$\begin{aligned} A^* &= P^{-1}\Gamma_s^{-1}(A + BF + KC + KDF)\Gamma_s P \\ &= \begin{bmatrix} A_{aa}^* & 0 & 0 & 0 & 0 \\ 0 & A_{bb}^* & 0 & 0 & 0 \\ 0 & 0 & A_{cc}^* & 0 & 0 \\ 0 & 0 & 0 & A_{dd}^* & B_dC_1^a \\ 0 & 0 & 0 & B_1^aC_d & A_{aa}^a + B_2^aC_2^a \end{bmatrix}, \end{aligned} \tag{3.4.15}$$

$$B^* = P^{-1}\Gamma_s^{-1}(B + KD)\Gamma_i = \begin{bmatrix} 0 & 0 & 0 \\ 0 & 0 & 0 \\ 0 & 0 & B_c \\ 0 & B_d & 0 \\ B_2^a & 0 & 0 \end{bmatrix}, \tag{3.4.16}$$

$$C^* = \Gamma_o^{-1}(C + DF)\Gamma_s P = \begin{bmatrix} 0 & 0 & 0 & 0 & C_2^a \\ 0 & 0 & 0 & C_d & 0 \\ 0 & C_b & 0 & 0 & 0 \end{bmatrix}, \tag{3.4.17}$$

and

$$D^* = \Gamma_o^{-1} D \Gamma_i = \begin{bmatrix} I_{m_0} & 0 & 0 \\ 0 & 0 & 0 \\ 0 & 0 & 0 \end{bmatrix}, \tag{3.4.18}$$

where P is a permutation matrix that transforms A_{aa}^a from its original position, i.e., Block $(1,1)$, to Block $(5,5)$ in (3.4.15).

Next, define a subsystem (A_s, B_s, C_s, D_s) with

$$A_s := \begin{bmatrix} A_{dd}^* & B_d C_1^a \\ B_1^a C_d & A_{aa}^a + B_2^a C_2^a \end{bmatrix}, \quad B_s := \begin{bmatrix} 0 & B_d \\ B_2^a & 0 \end{bmatrix}, \tag{3.4.19}$$

and

$$C_s := \begin{bmatrix} 0 & C_2^a \\ C_d & 0 \end{bmatrix}, \quad D_s := \begin{bmatrix} I_{m_0} & 0 \\ 0 & 0 \end{bmatrix}. \tag{3.4.20}$$

It is straightforward to verify that with the choice of B^a and C^a as in (3.4.13) and (3.4.14), A_s has no eigenvalue at $a = 1$. Hence A^* has no eigenvalue at $a = 1$ either, since both A_{bb}^* and A_{cc}^* have all eigenvalues at 0, and A_{aa}^* contains only the invariant zeros of Σ_c which are not equal to $a = 1$. Applying the bilinear transformation (3.2.3) to Σ_c^*, it follows from Lemma 3.2.1 that we obtain a discrete-time system Σ_d^*, characterized by $(\tilde{A}^*, \tilde{B}^*, \tilde{C}^*, \tilde{D}^*)$, with

$$\tilde{A}^* = \begin{bmatrix} (I+A_{aa}^*)(I-A_{aa}^*)^{-1} & 0 & 0 & 0 \\ 0 & (I+A_{bb}^*)(I-A_{bb}^*)^{-1} & 0 & 0 \\ 0 & 0 & (I+A_{cc}^*)(I-A_{cc}^*)^{-1} & 0 \\ 0 & 0 & 0 & (I+A_s)(I-A_s)^{-1} \end{bmatrix}, \tag{3.4.21}$$

$$\tilde{B}^* = \sqrt{2} \begin{bmatrix} 0 & 0 \\ 0 & 0 \\ 0 & (I-A_{cc}^*)^{-1} B_c \\ (I-A_s)^{-1} B_s & 0 \end{bmatrix}, \tag{3.4.22}$$

$$\tilde{C}^* = \sqrt{2} \begin{bmatrix} 0 & 0 & 0 & C_s(I-A_s)^{-1} \\ 0 & C_b(I-A_s)^{-1} & 0 & 0 \end{bmatrix}, \tag{3.4.23}$$

and

$$\tilde{D}^* = \begin{bmatrix} D_s + C_s(I-A_s)^{-1} B_s & 0 \\ 0 & 0 \end{bmatrix}. \tag{3.4.24}$$

Our next task is to find appropriate transformations, state feedback, and output injection laws, so as to transform the above system into the form of the special coordinate basis displaying the Properties 3(a)–6(b).

To simplify the presentation, we first focus on the subsystem $(\tilde{A}_s, \tilde{B}_s, \tilde{C}_s, \tilde{D}_s)$ with

$$\tilde{A}_s := (I + A_s)(I - A_s)^{-1}, \quad \tilde{B}_s := \sqrt{2}\,(I - A_s)^{-1} B_s, \tag{3.4.25}$$

and

$$\tilde{C}_s := \sqrt{2}C_s(I - A_s)^{-1}, \quad \tilde{D}_s := D_s + C_s(I - A_s)^{-1}B_s. \tag{3.4.26}$$

Using (3.4.12) in conjunction with Appendix A.22 of Kailath [64], it is straightforward to compute $(I - A_s)^{-1} =$

$$\begin{bmatrix} X_1 & (I-A_{dd}^*)^{-1}B_dC_1^a(I-A_{aa}-B^aC^a)^{-1} \\ (I-A_{aa}-B^aC^a)^{-1}B_1^aC_d(I-A_{dd}^*)^{-1} & (I-A_{aa}-B^aC^a)^{-1} \end{bmatrix}, \tag{3.4.27}$$

where

$$X_1 = (I - A_{dd}^*)^{-1} + (I - A_{dd}^*)^{-1}B_dC_1^a(I - A_{aa} - B^aC^a)^{-1}B_1^aC_d(I - A_{dd}^*)^{-1},$$

and hence

$$\tilde{A}_s = \begin{bmatrix} X_2 & 2(I-A_{dd}^*)^{-1}B_dC_1^a(I-A_{aa}^a-B^aC^a)^{-1} \\ 2(I-A_{aa}^a-B^aC^a)^{-1}B_1^aC_d(I-A_{dd}^*)^{-1} & (I+A_{aa}^a+B^aC^a)(I-A_{aa}^a-B^aC^a)^{-1} \end{bmatrix}, \tag{3.4.28}$$

where

$$X_2 = (I+A_{dd}^*)(I-A_{dd}^*)^{-1}+2(I-A_{dd}^*)^{-1}B_dC_1^a(I-A_{aa}^a-B^aC^a)^{-1}B_1^aC_d(I-A_{dd}^*)^{-1},$$

$$\tilde{B}_s = \sqrt{2}\begin{bmatrix} (I-A_{dd}^*)^{-1}B_dC_1^a(I-A_{aa}^a-B^aC^a)^{-1}B_2^a & (I-A_{dd}^*)^{-1}B_d[I+C_1^a(I-A_{aa}^a-B^aC^a)^{-1}B_1^a] \\ (I-A_{aa}^a-B^aC^a)^{-1}B_2^a & (I-A_{aa}^a-B^aC^a)^{-1}B_1^a \end{bmatrix}, \tag{3.4.29}$$

$$\tilde{C}_s = \sqrt{2}\begin{bmatrix} C_2^a(I-A_{aa}^a-B^aC^a)^{-1}B_1^aC_d(I-A_{dd}^*)^{-1} & C_2^a(I-A_{aa}^a - B^aC^a)^{-1} \\ [I+C_1^a(I-A_{aa}^a-B^aC^a)^{-1}B_1^a]C_d(I-A_{dd}^*)^{-1} & C_1^a(I-A_{aa}^a-B^aC^a)^{-1} \end{bmatrix}, \tag{3.4.30}$$

and

$$\tilde{D}_s = \begin{bmatrix} I+C_2^a(I-A_{aa}^a-B^aC^a)^{-1}B_2^a & C_2^a(I-A_{aa}^a-B^aC^a)^{-1}B_1^a \\ C_1^a(I-A_{aa}^a-B^aC^a)^{-1}B_2^a & I+C_1^a(I-A_{aa}^a-B^aC^a)^{-1}B_1^a \end{bmatrix}. \tag{3.4.31}$$

Noting the structure of A_{aa}^a in (3.4.6), and the structures of B^a and C^a in (3.4.13) and (3.4.14), we have

$$(I - A_{aa} - B^aC^a)^{-1} = \begin{bmatrix} 0 & -1 & \cdots & 0 & 0 \\ -I_{n_{a,1}-1} & 0 & \cdots & 0 & 0 \\ \vdots & \vdots & \ddots & \vdots & \vdots \\ 0 & 0 & \cdots & 0 & -1 \\ 0 & 0 & \cdots & -I_{n_{a,r_a}-1} & 0 \end{bmatrix}, \tag{3.4.32}$$

$$C_1^a(I - A_{aa} - B^a C^a)^{-1} B_2^a = 0, \quad C_2^a(I - A_{aa} - B^a C^a)^{-1} B_1^a = 0, \tag{3.4.33}$$

and

$$C^a(I - A_{aa} - B^a C^a)^{-1} B^a = \begin{bmatrix} 0 & 0 \\ 0 & -I_{\tau_a} \end{bmatrix}. \tag{3.4.34}$$

Thus, $\tilde{B}_s$, $\tilde{C}_s$ and $\tilde{D}_s$ reduce to the following forms:

$$\tilde{B}_s = \sqrt{2} \begin{bmatrix} 0 & (I - A_{dd}^*)^{-1} B_d [I + C_1^a (I - A_{aa}^a - B^a C^a)^{-1} B_1^a] \\ (I - A_{aa}^a - B^a C^a)^{-1} B_2^a & (I - A_{aa}^a - B^a C^a)^{-1} B_1^a \end{bmatrix}, \tag{3.4.35}$$

$$\tilde{C}_s = \sqrt{2} \begin{bmatrix} 0 & C_2^a (I - A_{aa}^a - B^a C^a)^{-1} \\ [I + C_1^a (I - A_{aa}^a - B^a C^a)^{-1} B_1^a] C_d (I - A_{dd}^*)^{-1} & C_1^a (I - A_{aa}^a - B^a C^a)^{-1} \end{bmatrix}, \tag{3.4.36}$$

and

$$\tilde{D}_s = \begin{bmatrix} I + C_2^a (I - A_{aa}^a - B^a C^a)^{-1} B_2^a & 0 \\ 0 & I + C_1^a (I - A_{aa}^a - B^a C^a)^{-1} B_1^a \end{bmatrix}. \tag{3.4.37}$$

Next, define

$$\tilde{F}_s := \sqrt{2} \begin{bmatrix} 0 & 0 \\ -C_d (I - A_{dd}^*)^{-1} & 0 \end{bmatrix}, \tag{3.4.38}$$

and

$$\tilde{K}_s := \sqrt{2} \begin{bmatrix} 0 & -(I - A_{dd}^*)^{-1} B_d \\ 0 & 0 \end{bmatrix}, \tag{3.4.39}$$

from which it follows that

$$\begin{aligned} \tilde{A}_{sc} &= \tilde{A}_s + \tilde{B}_s \tilde{F}_s + \tilde{K}_s \tilde{C}_s + \tilde{K}_s \tilde{D}_s \tilde{F}_s \\ &= \begin{bmatrix} \tilde{A}_{aa}^{**} & 0 \\ 0 & (I + A_{aa}^a + B^a C^a)(I - A_{aa}^a - B^a C^a)^{-1} \end{bmatrix}, \end{aligned}$$

where

$$\tilde{A}_{aa}^{**} := (I + A_{dd}^*)(I - A_{dd}^*)^{-1} - 2(I - A_{dd}^*)^{-1} B_d C_d (I - A_{dd}^*)^{-1}, \tag{3.4.40}$$

$$\tilde{B}_{sc} = \tilde{B}_s + \tilde{K}_s \tilde{D}_s = \sqrt{2} \begin{bmatrix} 0 & 0 \\ (I - A_{aa}^a - B^a C^a)^{-1} B_2^a & (I - A_{aa}^a - B^a C^a)^{-1} B_1^a \end{bmatrix},$$

and

$$\tilde{C}_{sc} = \tilde{C}_s + \tilde{D}_s \tilde{F}_s = \sqrt{2} \begin{bmatrix} 0 & C_2^a (I - A_{aa}^a - B^a C^a)^{-1} \\ 0 & C_1^a (I - A_{aa}^a - B^a C^a)^{-1} \end{bmatrix}.$$

Next, repartition B^a and C^a of (3.4.13) and (3.4.14) as follows:

$$B^a = [0 \quad \tilde{B}_a] \quad \text{and} \quad C^a = \begin{bmatrix} 0 \\ \tilde{C}_a \end{bmatrix}, \tag{3.4.41}$$

where both $\tilde{B}_a$ and $\tilde{C}_a$ are of maximal rank. We thus obtain

$$\tilde{A}_{sc} = \begin{bmatrix} \tilde{A}^{**}_{aa} & 0 \\ 0 & (I+A^a_{aa}+\tilde{B}_a\tilde{C}_a)(I-A^a_{aa}-\tilde{B}_a\tilde{C}_a)^{-1} \end{bmatrix},$$

$$\tilde{B}_{sc} = \sqrt{2}\begin{bmatrix} 0 & 0 \\ 0 & (I - A_{aa} - \tilde{B}_a\tilde{C}_a)^{-1}\tilde{B}_a \end{bmatrix},$$

and

$$\tilde{C}_{sc} = \sqrt{2}\begin{bmatrix} 0 & 0 \\ 0 & \tilde{C}_a(I - A_{aa} - \tilde{B}_a\tilde{C}_a)^{-1} \end{bmatrix}, \quad \tilde{D}_{sc} = \tilde{D}_s = \begin{bmatrix} I_{m_0+m_d-\tau_a} & 0 \\ 0 & 0 \end{bmatrix}.$$

Using (3.4.6) and (3.4.32), straightforward manipulations yield

$$(I + A^a_{aa} + \tilde{B}_a\tilde{C}_a)(I - A^a_{aa} - \tilde{B}_a\tilde{C}_a)^{-1}$$
$$= \begin{bmatrix} \begin{bmatrix} 0 & -2 \\ -2I_{n_{a,1}-1} & 0 \end{bmatrix} - I_{n_{a,1}} & \cdots & 0 \\ \vdots & \ddots & \vdots \\ 0 & \cdots & \begin{bmatrix} 0 & -2 \\ -2I_{n_{a,\tau_a}-1} & 0 \end{bmatrix} - I_{n_{a,\tau_a}} \end{bmatrix},$$

$$(I-A^a_{aa}-\tilde{B}_a\tilde{C}_a)^{-1}\tilde{B}_a = -\begin{bmatrix} 1 & \cdots & 0 \\ 0 & \cdots & 0 \\ \vdots & \ddots & \vdots \\ 0 & \cdots & 1 \\ 0 & \cdots & 0 \end{bmatrix},$$

and

$$\tilde{C}_a(I-A^a_{aa}-\tilde{B}_a\tilde{C}_a)^{-1} = -\begin{bmatrix} 0 & 1 & \cdots & 0 & 0 \\ \vdots & \vdots & \ddots & \vdots & \vdots \\ 0 & 0 & \cdots & 0 & 1 \end{bmatrix}.$$

Moreover, it can be readily verified that each subsystem $(\tilde{A}_{ai}, \tilde{B}_{ai}, \tilde{C}_{ai})$, $i = 1, \cdots, \tau_a$, with

$$\tilde{A}_{ai} = -I_{n_{a,i}} + \begin{bmatrix} 0 & -2 \\ -2I_{n_{a,i}-1} & 0 \end{bmatrix}, \quad \tilde{B}_{ai} = \begin{bmatrix} -1 \\ 0 \end{bmatrix}, \quad \tilde{C}_{ai} = [0 \quad -1],$$

has the following properties:

$$\tilde{C}_{ai}\tilde{B}_{ai} = \tilde{C}_{ai}\tilde{A}_{ai}\tilde{B}_{ai} = \cdots = \tilde{C}_{ai}(\tilde{A}_{ai})^{n_{a,i}-2}\tilde{B}_{ai} = 0,$$

and

$$\tilde{C}_{ai}(\tilde{A}_{ai})^{n_{a,i}-1}\tilde{B}_{ai} \neq 0.$$

It follows from Theorem 2.4.1 that there exist nonsingular transformations Γ_{sa}, Γ_{oa} and Γ_{ia} such that

$$\tilde{A}_d = \Gamma_{sa}^{-1}\left[(I + A_{aa}^a + \tilde{B}_a\tilde{C}_a)(I - A_{aa}^a - \tilde{B}_a\tilde{C}_a)^{-1}\right]\Gamma_{sa} = \begin{bmatrix} \star & I_{n_{a,1}-1} & \cdots & 0 & 0 \\ \star & \star & \cdots & 0 & 0 \\ \vdots & \vdots & \ddots & \vdots & \vdots \\ 0 & 0 & \cdots & \star & I_{n_{a,r_a}-1} \\ 0 & 0 & \cdots & \star & \star \end{bmatrix}, \tag{3.4.42}$$

$$\tilde{B}_d = \Gamma_{sa}^{-1}\left[(I - A_{aa}^a - \tilde{B}_a\tilde{C}_a)^{-1}\tilde{B}_a\right]\Gamma_{ia} = \begin{bmatrix} 0 & \cdots & 0 \\ 1 & \cdots & 0 \\ \vdots & \ddots & \vdots \\ 0 & \cdots & 0 \\ 0 & \cdots & 1 \end{bmatrix}, \tag{3.4.43}$$

and

$$\tilde{C}_d = \Gamma_{oa}^{-1}\left[\tilde{C}_a(I - A_{aa}^a - \tilde{B}_a\tilde{C}_a)^{-1}\right]\Gamma_{sa} = \begin{bmatrix} 1 & 0 & \cdots & 0 & 0 \\ \vdots & \vdots & \ddots & \vdots & \vdots \\ 0 & 0 & \cdots & 1 & 0 \end{bmatrix}. \tag{3.4.44}$$

Now, let us return to Σ_d^* characterized by $(\tilde{A}^*, \tilde{B}^*, \tilde{C}^*, \tilde{D}^*)$ as in (3.4.21) to (3.4.24). Using the properties of the subsystem $(\tilde{A}_s, \tilde{B}_s, \tilde{C}_s, \tilde{D}_s)$ just derived, we are in a position to define appropriate state feedback and output injection gain matrices, say $\tilde{F}^*$ and $\tilde{K}^*$, together with nonsingular state, output and input transformations $\tilde{\Gamma}_s^*$, $\tilde{\Gamma}_o^*$ and $\tilde{\Gamma}_i^*$, such that

$$\tilde{A}_{\text{SCB}}^* := (\tilde{\Gamma}_s^*)^{-1}\left(\tilde{A}^* + \tilde{B}^*\tilde{F}^* + \tilde{K}^*\tilde{C}^* + \tilde{K}^*\tilde{D}^*\tilde{F}^*\right)\tilde{\Gamma}_s^*$$
$$= \begin{bmatrix} (I+A_{aa}^*)(I-A_{aa}^*)^{-1} & 0 & 0 & 0 & 0 \\ 0 & (I+A_{bb}^*)(I-A_{bb}^*)^{-1} & 0 & 0 & 0 \\ 0 & 0 & (I+A_{cc}^*)(I-A_{cc}^*)^{-1} & 0 & 0 \\ 0 & 0 & 0 & \tilde{A}_{aa}^{**} & 0 \\ 0 & 0 & 0 & 0 & \tilde{A}_d \end{bmatrix}, \tag{3.4.45}$$

with $\tilde{A}_{aa}^{**}$ given by (3.4.40), and

$$\tilde{B}_{\text{SCB}}^* := (\tilde{\Gamma}_s^*)^{-1}\left(\tilde{B}^* + \tilde{K}^*\tilde{D}^*\right)\tilde{\Gamma}_i^* = \begin{bmatrix} 0 & 0 & 0 \\ 0 & 0 & 0 \\ 0 & 0 & (I - A_{cc}^*)^{-1}B_c \\ 0 & 0 & 0 \\ 0 & \tilde{B}_d & 0 \end{bmatrix}, \tag{3.4.46}$$

$$\tilde{C}_{\text{SCB}}^* := (\tilde{\Gamma}_o^*)^{-1}\left(\tilde{C}^* + \tilde{D}^*\tilde{F}^*\right)\tilde{\Gamma}_s^* = \begin{bmatrix} 0 & 0 & 0 & 0 & 0 \\ 0 & C_b(I - A_{bb}^*)^{-1} & 0 & 0 & 0 \\ 0 & 0 & 0 & 0 & \tilde{C}_d \end{bmatrix}, \tag{3.4.47}$$

and

$$\tilde{D}^*_{\rm SCB} := (\tilde{\Gamma}^*_o)^{-1}\tilde{D}^*\tilde{\Gamma}^*_i = \begin{bmatrix} I_{m_0+m_d-\tau_a} & 0 & 0 \\ 0 & 0 & 0 \\ 0 & 0 & 0 \end{bmatrix}. \tag{3.4.48}$$

Clearly, $\Sigma^*_{\rm SCB}$ characterized by $(\tilde{A}^*_{\rm SCB}, \tilde{B}^*_{\rm SCB}, \tilde{C}^*_{\rm SCB}, \tilde{D}^*_{\rm SCB})$ has the same structural invariant indices lists as Σ^*_d does, which in turn has the same structural invariant indices lists as Σ_d. Most importantly, $\Sigma^*_{\rm SCB}$ is in the form of the special coordinate basis, and we are now ready to prove Properties 3(a)–6(b) of the theorem.

3(a). First, we note that $\mathcal{I}_2(\Sigma_d) = \mathcal{I}_2(\Sigma^*_{\rm SCB})$. From (3.4.45) to (3.4.48) and the properties of the special coordinate basis, we know that $\mathcal{I}_2(\Sigma^*_{\rm SCB})$ is given by the controllability index of the pair

$$\Big((I+A^*_{cc})(I-A^*_{cc})^{-1},\ (I-A^*_{cc})^{-1}B_c\Big) \quad \text{or} \quad \Big((I+A^*_{cc})(I-A^*_{cc})^{-1},\ B_c\Big).$$

Recalling the definitions of A^*_{cc} and B_c:

$$A^*_{cc} = \begin{bmatrix} 0 & I_{\ell_1-1} & \cdots & 0 & 0 \\ 0 & 0 & \cdots & 0 & 0 \\ \vdots & \vdots & \ddots & \vdots & \vdots \\ 0 & 0 & \cdots & 0 & I_{\ell_{m_c}-1} \\ 0 & 0 & \cdots & 0 & 0 \end{bmatrix}, \quad B_c = \begin{bmatrix} 0 & \cdots & 0 \\ 1 & \cdots & 0 \\ \vdots & \ddots & \vdots \\ 0 & \cdots & 0 \\ 0 & \cdots & 1 \end{bmatrix},$$

it is straightforward to verify that the controllability index of

$$\Big((I+A^*_{cc})(I-A^*_{cc})^{-1},\ B_c\Big)$$

is also given by $\{\ell_1, \cdots, \ell_{m_c}\}$, and thus $\mathcal{I}_2(\Sigma_d) = \mathcal{I}_2(\Sigma_c)$.

Likewise, the proof that $\mathcal{I}_3(\Sigma_d) = \mathcal{I}_3(\Sigma_c)$ follows along similar lines. ☒

3(b)–3(c). These follow directly from 3(a). ☒

4(a). It follows from the properties of the special coordinate basis that the invariant zero structure of $\tilde{\Sigma}^*_{\rm SCB}$, or equivalently Σ_d, is given by the eigenvalues of $\tilde{A}^{**}_{aa}$ and $(I+A^*_{aa})(I-A^*_{aa})^{-1}$, together with their associated Jordan blocks. Property 4(a) corresponds with the eigenvalues of $\tilde{A}^{**}_{aa}$ of (3.4.40), together with their associated Jordan blocks. First, we note that for any $z \in \mathbb{C}$,

$$zI - \tilde{A}^{**}_{aa} = \left[(z-1)I - (z+1)A^*_{dd} + 2(I-A^*_{dd})^{-1}B_dC_d\right](I-A^*_{dd})^{-1}. \tag{3.4.49}$$

Recall the definitions of A^*_{dd}, B_d and C_d:

$$A^*_{dd} = \begin{bmatrix} 0 & I_{n_{q_1}-1} & \cdots & 0 & 0 \\ 0 & 0 & \cdots & 0 & 0 \\ \vdots & \vdots & \ddots & \vdots & \vdots \\ 0 & 0 & \cdots & 0 & I_{q_{m_d}-1} \\ 0 & 0 & \cdots & 0 & 0 \end{bmatrix}, \quad B_d = \begin{bmatrix} 0 & \cdots & 0 \\ 1 & \cdots & 0 \\ \vdots & \ddots & \vdots \\ 0 & \cdots & 0 \\ 0 & \cdots & 1 \end{bmatrix},$$

and

$$C_d = \begin{bmatrix} 1 & 0 & \cdots & 0 & 0 \\ \vdots & \vdots & \ddots & \vdots & \vdots \\ 0 & 0 & \cdots & 1 & 0 \end{bmatrix}.$$

It can be shown that

$$(z-1)I - (z+1)A_{dd}^{*} + 2(I - A_{dd}^{*})^{-1}B_d C_d = \text{blkdiag}\{Q_1(z), \cdots, Q_i(z)\},$$

where $Q_i(z) \in \mathbb{C}^{n_{q_i} \times n_{q_i}}$ is given by

$$Q_i(z) = \begin{bmatrix} z+1 & -(z+1) & 0 & \cdots & 0 & 0 \\ 2 & z-1 & -(z+1) & \cdots & 0 & 0 \\ 2 & 0 & z-1 & \ddots & 0 & 0 \\ \vdots & \vdots & \vdots & \ddots & \ddots & \vdots \\ 2 & 0 & 0 & \cdots & z-1 & -(z+1) \\ 2 & 0 & 0 & \cdots & 0 & z-1 \end{bmatrix}, \quad (3.4.50)$$

for $i = 1, \cdots, m_d$. It follows from (3.4.49) that the eigenvalue of $\tilde{A}_{aa}^{**}$ is the scalar z that causes the rank of

$$\text{blkdiag}\{Q_1(z), \cdots, Q_{m_d}(z)\},$$

to drop below $n_d = \sum_{i=1}^{m_d} q_i$. Using the particular form of $Q_i(z)$, it is clear that the only such scalar $z \in \mathbb{C}$ which causes $Q_i(z)$ to drop rank is $z = -1$. Moreover, rank $\{Q_i(-1)\} = n_{q_i} - 1$, i.e., $Q_i(-1)$ has only one linearly independent eigenvector. Hence, $z = -1$ is the eigenvalue of $\tilde{A}_{aa}^{**}$, or equivalently the invariant zero of Σ_d, with the multiplicity structure

$$S_{-1}^{\star}(\Sigma_d) = \{q_1, \cdots, q_{m_d}\} = S_{\infty}^{\star}(\Sigma_c),$$

thereby proving 4(a). ☒

4(b). This part of the infinite zero structure corresponds to the invariant zeros of the matrix $(I + A_{aa}^{*})(I - A_{aa}^{*})^{-1}$. With A_{aa}^{*} in Jordan form, Property 4(b) follows by straightforward manipulations. ☒

5(a). It follows directly from (3.4.48). ☒

5(b). This follows from the structure of $(\tilde{A}_d, \tilde{B}_d, \tilde{C}_d)$ in (3.4.42) to (3.4.44), in conjunction with Property 2.4.3 of the special coordinate basis. ☒

6(a)-6(b). We let the state space of the system (3.2.1) be $\mathcal{X}$ and be partitioned in its SCB subsystems as follows:

$$\mathcal{X} = \mathcal{X}_a^- \oplus \mathcal{X}_a^0 \oplus \mathcal{X}_a^+ \oplus \mathcal{X}_b \oplus \mathcal{X}_c \oplus \mathcal{X}_d. \quad (3.4.51)$$

We further partition $\mathcal{X}_a^+$ as

$$\mathcal{X}_a^+ = \mathcal{X}_{a1}^+ \oplus \mathcal{X}_{a*}^+, \tag{3.4.52}$$

where $\mathcal{X}_{a1}^+$ is associated with the zero dynamics of the unstable zero of (3.2.1) at $s = a = 1$ and $\mathcal{X}_{a*}^+$ is associated with the rest of unstable zero dynamics of (3.2.1). Similarly, we let the state space of the transformed system (3.2.6) be $\tilde{\mathcal{X}}$ and be partitioned in its SCB subsystems as follows:

$$\tilde{\mathcal{X}} = \tilde{\mathcal{X}}_a^- \oplus \tilde{\mathcal{X}}_a^0 \oplus \tilde{\mathcal{X}}_a^+ \oplus \tilde{\mathcal{X}}_b \oplus \tilde{\mathcal{X}}_c \oplus \tilde{\mathcal{X}}_d \tag{3.4.53}$$

with $\tilde{\mathcal{X}}_a^0$ being further partitioned as

$$\tilde{\mathcal{X}}_a^0 = \tilde{\mathcal{X}}_{a1}^0 \oplus \tilde{\mathcal{X}}_{a*}^0, \tag{3.4.54}$$

where $\tilde{\mathcal{X}}_{a1}^0$ is associated with the zero dynamics of the invariant zero of (3.2.6) at $z = -1$ and $\tilde{\mathcal{X}}_{a*}^0$ is associated the rest of the zero dynamics of the zeros of (3.2.6) on the unit circle. Then, from the above derivations of 1(a) to 5(b), we have the following mappings between the subsystems of Σ_c of (3.2.1) and those of Σ_d of (3.2.6):

$$\left.\begin{array}{ccc}
\mathcal{X}_a^- & \Longleftrightarrow & \tilde{\mathcal{X}}_a^-, \\
\mathcal{X}_d & \Longleftrightarrow & \tilde{\mathcal{X}}_{a1}^0, \\
\mathcal{X}_a^0 & \Longleftrightarrow & \tilde{\mathcal{X}}_{a*}^0, \\
\mathcal{X}_{a*}^+ & \Longleftrightarrow & \tilde{\mathcal{X}}_a^+, \\
\mathcal{X}_b & \Longleftrightarrow & \tilde{\mathcal{X}}_b, \\
\mathcal{X}_c & \Longleftrightarrow & \tilde{\mathcal{X}}_c, \\
\mathcal{X}_{a1}^+ & \Longleftrightarrow & \tilde{\mathcal{X}}_d.
\end{array}\right\} \tag{3.4.55}$$

Noting that both geometric subspaces $\mathcal{V}^\times$ and $\mathcal{S}^\times$ are invariant under any nonsingular output and input transformations, as well as any state feedback and output injection laws, we have

$$\mathcal{V}^+(\Sigma_c) = \mathcal{X}_{a*}^+ \oplus \mathcal{X}_{a1}^+ \oplus \mathcal{X}_c = \tilde{\mathcal{X}}_a^+ \oplus \tilde{\mathcal{X}}_d \oplus \tilde{\mathcal{X}}_c = \mathcal{S}^\odot(\Sigma_d), \tag{3.4.56}$$

and

$$\mathcal{S}^+(\Sigma_c) = \mathcal{X}_a^- \oplus \mathcal{X}_a^0 \oplus \mathcal{X}_c \oplus \mathcal{X}_d = \tilde{\mathcal{X}}_a^- \oplus \tilde{\mathcal{X}}_{a*}^0 \oplus \tilde{\mathcal{X}}_c \oplus \tilde{\mathcal{X}}_{a1}^0 = \mathcal{V}^\odot(\Sigma_d). \tag{3.4.57}$$

Unfortunately, other geometric subspaces do not have such clear relationships as above. ⊠

This concludes the proof of Theorem 3.2.1 and this chapter. ⊛

Chapter 4

Existence Conditions of H_∞ Suboptimal Controllers

4.1. Introduction

THE FIRST FUNDAMENTAL issue one faces in an H_∞ optimization problem, is when, or under what conditions a γ suboptimal controller exists. Fortunately, the problem regarding the existence conditions of γ-suboptimal controllers for either the regular or singular type of continuous-time or discrete-time systems has almost been completely solved in the literature. As it was mentioned in the introduction, there were four main different approaches developed in early years, which include: 1) Interpolation approach (see e.g., Limbeer and Anderson [77]); 2) Frequency domain approach (see e.g., Doyle [47], Francis [54] and Glover [57]); 3) Polynomial approach (see e.g., Kwakernaak [69]); and 4) J-spectral factorization approach (see e.g., Kimura [67]). All these techniques mainly deal with the regular problem.

We recall in this chapter the existence conditions of γ suboptimal controllers for the H_∞ optimization problem derived from the pure time-domain methods based on algebraic Riccati equations or linear matrix inequalities. For the regular continuous-time systems, the problem was solved by Doyle, Glover, Khargonekar and Francis [49], i.e., DGKF, and Tadmor [129]. For general singular continuous-time systems with no invariant zero on the imaginary axis, the problem was solved by Stoorvogel and Trentelman [127] and Stoorvogel [124]. In the situation when systems have invariant zeros on the imaginary axis, the result was derived by Scherer [117–119]. The existence conditions of γ-suboptimal controllers for discrete-time systems were reported in Stoorvogel

[124] and Stoorvogel, Saberi and Chen [125]. These results will form a base for the results reported in the coming chapters.

4.2. Continuous-time Systems

We consider in this section a general continuous-time linear time-invariant (LTI) system Σ with a state-space description,

$$\Sigma : \begin{cases} \dot{x} = A\ x + B\ u + E\ w, \\ y = C_1\ x + D_1\ w, \\ h = C_2\ x + D_2\ u + D_{22}\ w, \end{cases} \tag{4.2.1}$$

where $x \in \mathbb{R}^n$ is the state, $u \in \mathbb{R}^m$ is the control input, $w \in \mathbb{R}^q$ is the external disturbance input, $y \in \mathbb{R}^p$ is the measurement output, and $h \in \mathbb{R}^\ell$ is the controlled output of Σ. We also consider the following proper measurement feedback control law,

$$\Sigma_{\rm cmp} : \begin{cases} \dot{v} = A_{\rm cmp}\ v + B_{\rm cmp}\ y, \\ u = C_{\rm cmp}\ v + D_{\rm cmp}\ y. \end{cases} \tag{4.2.2}$$

For simplicity of presentation, we will first set the direct feedthrough term from the disturbance w to controlled output h in (4.2.1) to be equal to zero, i.e., $D_{22} = 0$. For easy reference, we define $\Sigma_{\rm P}$ to be the subsystem characterized by the matrix quadruple (A, B, C_2, D_2), and $\Sigma_{\rm Q}$ to be the subsystem characterized by the matrix quadruple (A, E, C_1, D_1), which respectively have transfer functions:

$$G_{\rm P}(s) = C_2(sI - A)^{-1}B + D_2, \tag{4.2.3}$$

and

$$G_{\rm Q}(s) = C_1(sI - A)^{-1}E + D_1. \tag{4.2.4}$$

We recall in this section some important results in the literature regarding the existence conditions of γ-suboptimal control laws for the continuous-time H_∞ optimization problem.

The first result given below is due to [124]. Before we introduce the theorem, let us define the following quadratic matrices,

$$F_\gamma(P) := \begin{bmatrix} A'P + PA + C_2'C_2 + \gamma^{-2}PEE'P & PB + C_2'D_2 \\ B'P + D_2'C_2 & D_2'D_2 \end{bmatrix}, \tag{4.2.5}$$

and

$$G_\gamma(Q) := \begin{bmatrix} AQ + QA' + EE' + \gamma^{-2}QC_2'C_2Q & QC_1' + ED_1' \\ C_1Q + D_1E' & D_1D_1' \end{bmatrix}. \tag{4.2.6}$$

It should be noted that the above matrices are dual of each other. In addition to these two matrices, we define two polynomial matrices whose roles are again completely dual:

$$L_\gamma(P,s) := [\, sI - A - \gamma^{-2}EE'P \quad -B\,], \tag{4.2.7}$$

and

$$M_\gamma(Q,s) := \begin{bmatrix} sI - A - \gamma^{-2}QC_2'C_2 \\ -C_1 \end{bmatrix}. \tag{4.2.8}$$

Now we are ready to introduce the following theorem which gives a set of necessary and sufficient conditions for the existence of a γ-suboptimal controller for the continuous-time system (4.2.1) with $D_{22}=0$ and with both subsystems $\Sigma_{\rm P}$ and $\Sigma_{\rm Q}$ having no invariant zero on the imaginary axis.

Theorem 4.2.1. Consider the continuous-time linear time-invariant system of (4.2.1) with $D_{22}=0$. Assume that $\Sigma_{\rm P}$ and $\Sigma_{\rm Q}$ have no invariant zero on the imaginary axis. Then the following statements are equivalent:

1. There exists a linear time-invariant and proper dynamic compensator $\Sigma_{\rm cmp}$ of (4.2.2) such that when it is applied to (4.2.1), the resulting closed-loop system is internally stable. Moreover, the H_∞-norm of the closed-loop transfer function from the disturbance input w to the controlled output h is less than γ.

2. There exist positive semi-definite matrices P and Q such that the following conditions are satisfied:

   (a) $F_\gamma(P) \geq 0$.

   (b) $\text{rank}\{F_\gamma(P)\} = \text{normrank}\,\{G_{\rm P}(s)\}$.

   (c) $\text{rank}\begin{bmatrix} L_\gamma(P,s) \\ F_\gamma(P) \end{bmatrix} = n + \text{normrank}\,\{G_{\rm P}(s)\},\ \forall s \in \mathbb{C}^0 \cup \mathbb{C}^+$.

   (d) $G_\gamma(Q) \geq 0$.

   (e) $\text{rank}\{G_\gamma(Q)\} = \text{normrank}\,\{G_{\rm Q}(s)\}$.

   (f) $\text{rank}[M_\gamma(Q,s),\ G_\gamma(Q)] = n + \text{normrank}\,\{G_{\rm Q}(s)\},\ \forall s \in \mathbb{C}^0 \cup \mathbb{C}^+$.

   (g) $\rho(PQ) < \gamma^2$.

   Here $G_{\rm P}(s)$ and $G_{\rm Q}(s)$ are respectively the transfer function of $\Sigma_{\rm P}$ and $\Sigma_{\rm Q}$, and "normrank" denotes the rank of a matrix with entries in the field of rational functions. 🅃

The following remark concerns the full information feedback and full state feedback cases. It turns out that for the system with $D_{22} = 0$, the existence conditions of γ-suboptimal controllers for the full information feedback case and for the full state feedback case are identical.

Remark 4.2.1. For the special cases of full information and full state feedback, the solution to the linear matrix inequality (LMI), i.e., Condition 2.(d) of Theorem 4.2.1, which satisfies Conditions 2.(e) and 2.(f), is identically zero. This implies that Condition 2.(g) is automatically satisfied. Hence, the existence conditions of γ-suboptimal controllers for both the full information and the full state feedback cases reduce to Conditions 2.(a)-2.(c). Moreover, it can be shown that a γ-suboptimal static control law exists. ®

The following corollary deals with the regular systems or regular case. It was first reported in Doyle *et al.* [49] and Tadmor [129].

Corollary 4.2.1. Consider the continuous-time linear time-invariant system of (4.2.1) with $D_{22} = 0$. Assume that Σ_P and Σ_Q have no invariant zero on the imaginary axis, D_2 is of full column rank and D_1 is of full row rank. Then the following statements are equivalent:

1. There exists a linear time-invariant and proper dynamic compensator Σ_{cmp} of (4.2.2) such that when it is applied to (4.2.1), the resulting closed-loop system is internally stable. Moreover, the H_∞-norm of the closed-loop transfer function from the disturbance input w to the controlled output h is less than γ.

2. There exist positive semi-definite matrices P and Q such that the following conditions are satisfied:

   (a) P is the solution of the Riccati equation:

$$A'P + PA + C_2'C_2 + \gamma^2 PEE'P \\ -(PB + C_2'D_2)(D_2'D_2)^{-1}(B'P + D_2'C_2) = 0. \quad (4.2.9)$$

   (b) A_{clP} is asymptotically stable, where

$$A_{clP} := A + \gamma^{-2}EE'P - B(D_2'D_2)^{-1}(B'P + D_2'C_2). \quad (4.2.10)$$

   (c) Q is the solution of the Riccati equation:

$$AQ + QA' + EE' + \gamma^2 QC_2'C_2Q \\ -(QC_1' + ED_1')(D_1D_1')^{-1}(C_1Q + D_1E') = 0. \quad (4.2.11)$$

(d) A_{clQ} is asymptotically stable, where

$$A_{\mathrm{clQ}} := A + \gamma^{-2} Q C_2' C_2 - (Q C_1' + E D_1')(D_1 D_1')^{-1} C_1. \quad (4.2.12)$$

(e) $\rho(PQ) < \gamma^2$. ▣

If the given system (4.2.1) with nonzero D_{22} term, then the general conditions for the existence of γ-suboptimal controllers are rather complicated. We will derive these conditions later in Chapter 6. In what follows, we recall a corollary that deals with a special full information feedback case when D_2 is of full column rank and Σ_{P} has no invariant zero on the imaginary axis.

Corollary 4.2.2. Consider the continuous-time linear time-invariant system of (4.2.1) with $y = (\, x' \;\; w' \,)'$ and D_2 being of full column rank. Assume that Σ_{P} has no invariant zero on the imaginary axis. Then the following statements are equivalent:

1. There exist constant gain matrices F_1 and F_2 such that when the control law $u = F_1 x + F_2 w$ is applied to (4.2.1), the resulting closed-loop system is internally stable. Moreover, the H_∞-norm of the closed-loop transfer function from the disturbance input w to the controlled output h is less than γ.

2. The following conditions are satisfied:

   (a) $D_{22}' \left(I - D_2 (D_2' D_2)^{-1} D_2'\right) D_{22} < \gamma^2 I$.

   (b) There exists a positive semi-definite solution P to the Riccati equation:

   $$0 = PA + A'P + C_2' C_2 - \begin{bmatrix} B'P + D_2' C_2 \\ E'P + D_{22}' C_2 \end{bmatrix}' G^{-1} \begin{bmatrix} B'P + D_2' C \\ E'P + D_{22}' C \end{bmatrix},$$

   where

   $$G := \begin{bmatrix} D_2' D_2 & D_2' D_{22} \\ D_{22}' D_2 & D_{22}' D_{22} - \gamma^2 I \end{bmatrix},$$

   such that the matrix,

   $$A_{\mathrm{clP}} := A - [\, B \quad E \,] G^{-1} \begin{bmatrix} B'P + D_1' C \\ E'P + D_2' C \end{bmatrix},$$

   is asymptotically stable. ▣

Note that the existence conditions of a γ-suboptimal controller for the full state feedback case with D_2 being of full column rank and $\Sigma_{\rm P}$ having no invariant zero on the imaginary axis, are similar to those in Item 2 of Corollary 4.2.2 except that one has to replace 2.(a) by $D_{22}'D_{22} < \gamma^2 I$.

Next, we will remove the restrictions on the invariant zeros of the subsystems $\Sigma_{\rm P}$ and $\Sigma_{\rm Q}$, i.e., we will allow both $\Sigma_{\rm P}$ and $\Sigma_{\rm Q}$ to have invariant zeros on the imaginary axis. The following theorem is due to Scherer [119].

Theorem 4.2.2. Consider the continuous-time linear time-invariant system of (4.2.1) with $D_{22} = 0$. Then the following statements are equivalent:

1. There exists a linear time-invariant and proper dynamic compensator $\Sigma_{\rm cmp}$ of (4.2.2) such that when it is applied to to (4.2.1), the resulting closed-loop system is internally stable. Moreover, the H_∞-norm of the closed-loop transfer function from the disturbance input w to the controlled output h is less than γ.

2. There exist appropriate dimensional constant matrices F and K, and positive definite matrices $P > 0$ and $Q > 0$ such that the following conditions are satisfied:

   (a) $(A+BF)'P+P(A+BF)+\gamma^{-2}PEE'P+(C_2+D_2F)'(C_2+D_2F) < 0$.

   (b) $(A+KC_1)Q+Q(A+KC_1)'+\gamma^{-2}QC_2'C_2Q+(E+KD_1)(E+KD_1)' < 0$.

   (c) $\rho(PQ) < \gamma^2$. ◻T

The above Conditions 2.(a) and 2.(b) in Theorem 4.2.2 can be converted into conditions of the existences of positive definite solutions for some reduced order algebraic Riccati inequalities, which are independent of F and K. This can be done by transforming the subsystems $\Sigma_{\rm P}$ and $\Sigma_{\rm Q}$ of the given system into the special coordinate basis as in Chapter 2.

4.3. Discrete-time Systems

We now consider in this section a general discrete-time linear time-invariant (LTI) system Σ with a state-space description

$$\Sigma : \begin{cases} x(k+1) = A\ x(k) + B\ u(k) + E\ w(k), \\ y(k) \ = C_1\ x(k) + D_1\ w(k), \\ h(k) \ = C_2\ x(k) + D_2\ u(k) + D_{22}\ w(k), \end{cases} \tag{4.3.1}$$

where $x \in \mathbb{R}^n$ is the state, $u \in \mathbb{R}^m$ is the control input, $w \in \mathbb{R}^q$ is the disturbance input, $y \in \mathbb{R}^p$ is the measurement output, and $h \in \mathbb{R}^\ell$ is the controlled output of Σ. The following $\Sigma_{\rm cmp}$ is the controller considered:

$$\Sigma_{\rm cmp} : \begin{cases} v(k+1) = A_{\rm cmp}\, v(k) + B_{\rm cmp}\, y(k), \\ \quad u(k) \;\; = C_{\rm cmp}\, v(k) + D_{\rm cmp}\, y(k). \end{cases} \tag{4.3.2}$$

We would like to note that in principle, one could transfer all the results in the continuous-time case to the discrete-time one using the bilinear and the inverse bilinear transformations of Chapter 3. For the regular full information feedback case, the interconnection between the continuous-time and discrete-time H_∞ optimization problems as well as the relationship between the continuous-time and discrete-time H_∞ algebraic Riccati equations will be explicitly established later in Chapter 5.

Again, as in the continuous-time case, we define $\Sigma_{\rm P}$ to be the subsystem characterized by the matrix quadruple (A, B, C_2, D_2), and $\Sigma_{\rm Q}$ to be the subsystem characterized by the matrix quadruple (A, E, C_1, D_1), which respectively have transfer functions:

$$G_{\rm P}(z) = C_2(zI - A)^{-1}B + D_2, \tag{4.3.3}$$

and

$$G_{\rm Q}(z) = C_1(zI - A)^{-1}E + D_1. \tag{4.3.4}$$

The following result is due to Stoorvogel, Saberi and Chen [125].

Theorem 4.3.1. Consider the system (4.3.1). Assume that the subsystems $\Sigma_{\rm P}$ and $\Sigma_{\rm Q}$ have no invariant zero on the unit circle. Then the following two statements are equivalent:

1. There exists a linear time-invariant and causal dynamic compensator $\Sigma_{\rm cmp}$ of (4.3.2) such that when it is applied to (4.3.1), the resulting closed loop system is internally stable and the closed loop transfer matrix from the disturbance input w to the controlled output h is less than γ.

2. There exist symmetric matrices $P \geq 0$ and $Q \geq 0$ such that

   (a) The following matrix R is positive definite,

$$\begin{aligned} R := \gamma^2 I - D_{22}' D_{22} - E'PE \\ + \big(E'PB + D_{22}'D_2\big)V^\dagger\big(B'PE + D_2'D_{22}\big) > 0, \end{aligned} \tag{4.3.5}$$

   where

$$V := B'PB + D_2'D_2. \tag{4.3.6}$$

(b) P satisfies the discrete algebraic Riccati equation:

$$P = A'PA + C_2'C_2 - \begin{bmatrix} B'PA + D_2'C_2 \\ E'PA + D_{22}'C_2 \end{bmatrix}' G^\dagger \begin{bmatrix} B'PA + D_2'C_2 \\ E'PA + D_{22}'C_2 \end{bmatrix}, \tag{4.3.7}$$

where

$$G := \begin{bmatrix} D_2'D_2 + B'PB & D_2'D_{22} + B'PE \\ D_{22}'D_2 + E'PB & E'PE + D_{22}'D_{22} - \gamma^2 I \end{bmatrix}. \tag{4.3.8}$$

(c) For all $z \in \mathbb{C}$ with $|z| \geq 1$, we have

$$\operatorname{rank} \begin{bmatrix} zI - A & -B & -E \\ B'PA + D_2'C_2 & B'PB + D_2'D_2 & B'PE + D_2'D_{22} \\ E'PA + D_{22}'C_2 & E'PB + D_{22}'D_2 & E'PE + D_{22}'D_{22} - \gamma^2 I \end{bmatrix}$$
$$= n + q + \operatorname{normrank}\{G_{\mathrm{P}}(z)\}.$$

(d) The following matrix S is positive definite,

$$S := \gamma^2 I - D_{22}D_{22}' - C_2QC_2'$$
$$+ (C_2QC_1' + D_{22}D_1')W^\dagger(C_1QC_2' + D_1D_{22}') > 0, \tag{4.3.9}$$

where

$$W := D_1D_1' + C_1QC_1'. \tag{4.3.10}$$

(e) Q satisfies the following discrete algebraic Riccati equation:

$$Q = AQA' + EE' - \begin{bmatrix} C_1QA' + D_1E' \\ C_2QA' + D_{22}E' \end{bmatrix}' H^\dagger \begin{bmatrix} C_1QA' + D_1E' \\ C_2QA' + D_{22}E' \end{bmatrix}, \tag{4.3.11}$$

where

$$H := \begin{bmatrix} D_1D_1' + C_1QC_1' & D_1D_{22}' + C_1QC_2' \\ D_{22}D_1' + C_2QC_1' & C_2QC_2' + D_{22}D_{22}' - \gamma^2 I \end{bmatrix}. \tag{4.3.12}$$

(f) For all $z \in \mathbb{C}$ with $|z| \geq 1$, we have

$$\operatorname{rank} \begin{bmatrix} zI - A & AQC_1' + ED_1' & AQC_2' + ED_{22}' \\ -C_1 & C_1QC_1' + D_1D_1' & C_1QC_2' + D_1D_{22}' \\ -C_2 & C_2QC_1' + D_{22}D_1' & C_2QC_2' + D_{22}D_{22}' - \gamma^2 I \end{bmatrix}$$
$$= n + \ell + \operatorname{normrank}\{G_{\mathrm{Q}}(z)\}.$$

(g) $\rho(PQ) < \gamma^2$. 🅃

Here we should note that Condition 2.(b) is the standard Riccati equation used in discrete-time H_∞ optimization except that the inverse is replaced by a generalized inverse. Condition 2.(c) is nothing other than the requirement that P must be a stabilizing solution of the Riccati equation. Conditions 2.(b) and 2.(c) uniquely determine, if it exists, the matrix P. A similar comment can be made about Conditions 2.(d)-2.(f). Condition 2.(g) is as usual the coupling condition. The solutions to the above mentioned P and Q can be obtained by transforming the subsystems $\Sigma_{\rm P}$ and $\Sigma_{\rm Q}$ into the special coordinate basis as in Chapter 2 and then solving two standard discrete-time Riccati equations without generalized inverses. These will be given later in Chapter 10.

The following remark concerns the full information feedback and full state feedback cases.

Remark 4.3.1. For the special cases of full information and full state feedback we can dispense with the second Riccati equation. More specifically:

1. *Full information feedback case:* In this case we know both the state and the disturbance of the system at time k. It is easy to check that $Q = 0$ satisfies Conditions 2.(d)-2.(f). Moreover this guarantees that the coupling Condition 2.(g) is automatically satisfied. Therefore there exists a stabilizing controller which yields a closed loop system with the H_∞ norm strictly less than γ if and only if there exists a positive semi-definite matrix P satisfying Conditions 2.(a)-2.(c).

2. *Full state feedback case:* In this case, it is easy to see that a necessary condition for the existence of a positive semi-definite matrix Q satisfying Conditions 2.(d)-2.(f) is that $|D_{22}| < \gamma$. It is also easy to check that for the full state feedback case,

$$Q = E(I - \gamma^{-2} D_{22} D_{22}')^{-1} E', \tag{4.3.13}$$

satisfies Conditions 2.(d)-2.(f). Condition 2.(g) then reduces to

$$\gamma^2 I - D_{22} D_{22}' - E'PE > 0. \tag{4.3.14}$$

Moreover, Condition (4.3.14) implies that Condition 2.(a) is automatically satisfied. Therefore there exists a stabilizing controller which yields a closed loop system with the H_∞ norm strictly less than γ if and only if there exists a positive semi-definite matrix P satisfying Conditions 2.(b), 2.(c) and additionally Condition (4.3.14).

Furthermore, it can be shown that either in the full information case or in the full state feedback case, there always exists a γ-suboptimal static control law whenever the above-mentioned conditions are satisfied. 🅁

The following corollary deals with the regular case in discrete-time H_∞ optimization and is due to [124].

Corollary 4.3.1. Consider the system (4.3.1). Assume that the subsystem $\Sigma_{\rm P}$ is left invertible and has no invariant zero on the unit circle, and the subsystem $\Sigma_{\rm Q}$ is right invertible and has no invariant zero on the unit circle. Then the following two statements are equivalent:

1. There exists a linear time-invariant and causal dynamic compensator $\Sigma_{\rm cmp}$ of (4.3.2) such that when it is applied to (4.3.1), the resulting closed loop system is internally stable and the closed loop transfer matrix from the disturbance input w to the controlled output h is less than γ.

2. There exist symmetric matrices $P \geq 0$ and $Q \geq 0$ such that

   (a) The following matrices V and R are positive definite,

$$V := B'PB + D_2'D_2 > 0, \tag{4.3.15}$$

   and

$$R := \gamma^2 I - D_{22}'D_{22} - E'PE + (E'PB + D_{22}'D_2)V^{-1}(B'PE + D_2'D_{22}) > 0. \tag{4.3.16}$$

   (b) P satisfies the discrete algebraic Riccati equation:

$$P = A'PA + C_2'C_2 - \begin{bmatrix} B'PA + D_2'C_2 \\ E'PA + D_{22}'C_2 \end{bmatrix}' G(P)^{-1} \begin{bmatrix} B'PA + D_2'C_2 \\ E'PA + D_{22}'C_2 \end{bmatrix}, \tag{4.3.17}$$

   where

$$G(P) := \begin{bmatrix} D_2'D_2 + B'PB & D_2'D_{22} + B'PE \\ D_{22}'D_2 + E'PB & E'PE + D_{22}'D_{22} - \gamma^2 I \end{bmatrix}. \tag{4.3.18}$$

   (c) The following matrix $A_{\rm clP}$ is asymptotically stable,

$$A_{\rm clP} := A - [\,B \quad E\,]\,G(P)^{-1} \begin{bmatrix} B'PA + D_2'C_2 \\ E'PA + D_{22}'C_2 \end{bmatrix}. \tag{4.3.19}$$

(d) The following matrices W and S are positive definite,

$$W := D_1 D_1' + C_1 Q C_1' > 0, \tag{4.3.20}$$

and

$$S := \gamma^2 I - D_{22} D_{22}' - C_2 Q C_2' + (C_2 Q C_1' + D_{22} D_1') W^{-1} (C_1 Q C_2' + D_1 D_{22}') > 0. \tag{4.3.21}$$

(e) Q satisfies the following discrete algebraic Riccati equation:

$$Q = AQA' + EE' - \begin{bmatrix} C_1 QA' + D_1 E' \\ C_2 QA' + D_{22} E' \end{bmatrix}' H(Q)^{-1} \begin{bmatrix} C_1 QA' + D_1 E' \\ C_2 QA' + D_{22} E' \end{bmatrix}, \tag{4.3.22}$$

where

$$H(Q) := \begin{bmatrix} D_1 D_1' + C_1 Q C_1' & D_1 D_{22}' + C_1 Q C_2' \\ D_{22} D_1' + C_2 Q C_1' & C_2 Q C_2' + D_{22} D_{22}' - \gamma^2 I \end{bmatrix}. \tag{4.3.23}$$

(f) The following matrix A_{clQ} is asymptotically stable,

$$A_{\mathrm{clQ}} := A - \begin{bmatrix} C_1 QA' + D_1 E' \\ C_2 QA' + D_{22} E' \end{bmatrix}' H(Q)^{-1} \begin{bmatrix} C_1 \\ C_2 \end{bmatrix}. \tag{4.3.24}$$

(g) $\rho(PQ) < \gamma^2$. ▣

It is interesting to note that all the conditions in Corollary 4.3.1 are related to those in Corollary 4.2.1 by a properly defined bilinear transformation. This will be shown later in Chapter 5 in more details. In fact, following the result of Glover [57], we can show that the continuous-time H_∞ optimization problem and the discrete-time H_∞ optimization problem are equivalent under the bilinear transformation (see the detailed properties of the bilinear transformation in Chapter 3). Thus, all the results for the discrete-time case can be derived from those of its continuous-time counterpart.

Chapter 5

Solutions to Discrete-time Riccati Equations

5.1. Introduction

THE DISCRETE-TIME algebraic Riccati equation (DARE) has been investigated extensively in the literature (see, for example [9,68,72,101,105,123]). Here, most of the work was based on the discrete-time algebraic Riccati equation appearing in a linear quadratic control problem (hereafter we will refer to such a DARE as the H_2-DARE). Recently, the problem of H_∞ control and that of differential games for discrete-time systems, have been studied by a number of researchers including [4,63,78]. This work gives rise to a different kind of algebraic Riccati equation (hereafter we call it an H_∞-DARE). Analyzing and solving such an H_∞-DARE are very difficult primarily because of an indefinite nonlinear term and because we cannot *a priori* guarantee the existence of solutions. In this chapter, we recall the results of Chen *et al.* [38] on non-recursive methods for solving general DAREs, as well as H_2-DAREs and H_∞-DAREs. In particular, we will cast the problem of solving a given H_∞-DARE to the problem of solving an auxiliary continuous-time algebraic Riccati equation associated with the continuous-time H_∞ control problem (H_∞-CARE) for which the well known non-recursive solving methods are available. The advantages of this approach are: it reduces the computation involved in the recursive algorithms while giving much more accurate solutions, and it readily provides the properties of the general H_∞-DARE. More importantly, the results given in this chapter build an interconnection between the discrete-time and continuous-time H_∞ optimization problems.

5.2. Solution to a General DARE

We first introduce in this section a non-recursive method for solving the following discrete-time algebraic Riccati equation, which is even more general than the H_∞-DARE and which plays a critical role in solving the H_∞-DARE,

$$P = A'PA - (A'PM + N)(R + M'PM)^{-1}(M'PA + N') + Q, \qquad (5.2.1)$$

where A, M, N, R and Q are real matrices of dimensions $n \times n$, $n \times m$, $n \times m$, $m \times m$ and $n \times n$, respectively, and with Q and R being symmetric matrices. We will show that the DARE of (5.2.1) can be converted to a continuous-time Riccati equation. Assume that matrix A has no eigenvalue at -1. We define

$$\left.\begin{aligned} F &:= (A+I)^{-1}(A-I), \\ G &:= 2(A+I)^{-2}M, \\ W &:= R + M'(A'+I)^{-1}Q(A+I)^{-1}M \\ &\qquad -N'(A+I)^{-1}M - M'(A'+I)^{-1}N, \\ H &:= -Q(A+I)^{-1}M + N. \end{aligned}\right\} \qquad (5.2.2)$$

Note that matrices F, G, W and H are in fact defined using the inverse bilinear transformation.

We have the following theorem.

Theorem 5.2.1. Assume that matrix A has no eigenvalue at -1. Then the following two statements are equivalent.

1. P is a symmetric solution to the DARE (5.2.1) and W is nonsingular.
2. $\tilde{P}$ is a symmetric solution to the continuous algebraic Riccati equation,

$$\tilde{P}F + F'\tilde{P} - (\tilde{P}G + H)W^{-1}(\tilde{P}G + H)' + Q = 0, \qquad (5.2.3)$$

and $R + 2G'(I - F')^{-1}\tilde{P}(I - F)^{-1}G$ is nonsingular.

Moreover, P and $\tilde{P}$ are related by $P = 2(A' + I)^{-1}\tilde{P}(A + I)^{-1}$. ◻

Proof. See Subsection 5.4.A. ◻

We would like to note that Theorem 5.2.1 can be regarded as a bridge connecting discrete-time algebraic Riccati equations and continuous-time algebraic Riccati equations. The result of Theorem 5.2.1 shows that any discrete-time Riccati equation of the form (5.2.1) can be converted into an equivalent continuous-time Riccati equation of (5.2.3) for which many numerically stable non-recursive solving methods are available. Thus, in our opinion, there is no need to develop separate techniques for solving discrete-time algebraic Riccati equations.

5.3. Solution to an H_∞-DARE

In this section we present a non-recursive procedure that generates symmetric positive semi-definite matrices P such that

$$V := B'PB + D_2'D_2 > 0, \tag{5.3.1}$$

$$R := \gamma^2 I - D_{22}'D_{22} - E'PE + (E'PB + D_{22}'D_2)V^{-1}(B'PE + D_2'D_{22}) > 0, \tag{5.3.2}$$

and such that the following discrete-time algebraic Riccati equation (DARE) is satisfied:

$$P = A'PA + C_2'C_2 - \begin{bmatrix} B'PA + D_2'C_2 \\ E'PA + D_{22}'C_2 \end{bmatrix}' G^{-1} \begin{bmatrix} B'PA + D_2'C_2 \\ E'PA + D_{22}'C_2 \end{bmatrix}, \tag{5.3.3}$$

where

$$G := \begin{bmatrix} D_2'D_2 + B'PB & D_2'D_{22} + B'PE \\ D_{22}'D_2 + E'PB & E'PE + D_{22}'D_{22} - \gamma^2 I \end{bmatrix}. \tag{5.3.4}$$

The conditions of (5.3.1) and (5.3.2) guarantee that the matrix G is invertible. We are particularly interested in solutions P of (5.3.1), (5.3.2) and (5.3.3) such that all the eigenvalues of the matrix $A_{\rm cl}$ are inside the unit circle, where

$$A_{\rm cl} := A - [B \quad E] G^{-1} \begin{bmatrix} B'PA + D_2'C_2 \\ E'PA + D_{22}'C_2 \end{bmatrix}. \tag{5.3.5}$$

The interest in this particular Riccati equation stems from the discrete-time H_∞ control theory (see Corollary 4.3.1). Also, it is simple to see that by letting $E = 0$ and $D_{22} = 0$, (5.3.1), (5.3.2) and (5.3.3) reduce to the well-known Riccati equation from linear quadratic control theory. For clarity, we first recall the relation between the above Riccati equation and the discrete-time full information feedback H_∞ control problem. Let us define a system $\Sigma_{\rm FI}$ by

$$\Sigma_{\rm FI} : \begin{cases} x(k+1) = A\, x(k) + B\, u(k) + E\, w(k), \\ y(k) = \begin{pmatrix} I \\ 0 \end{pmatrix} x(k) + \begin{pmatrix} 0 \\ I \end{pmatrix} w(k), \\ h(k) = C_2\, x(k) + D_2\, u(k) + D_{22}\, w(k), \end{cases} \tag{5.3.6}$$

where $x \in \mathbb{R}^n$ is the state, $u \in \mathbb{R}^m$ is the control input, $w \in \mathbb{R}^q$ the disturbance input, $h \in \mathbb{R}^\ell$ the controlled output and $y \in \mathbb{R}^{n+q}$ the measurement. Then the following lemma follows from Corollary 4.3.1.

Lemma 5.3.1. Consider a given system (5.3.6). Assume that (A, B, C_2, D_2) is left invertible and has no invariant zero on the unit circle. Then the following two statements are equivalent:

1. There exists a static feedback $u = K_1 x + K_2 w$, which stabilizes $\Sigma_{\rm FI}$ and makes the H_∞ norm of the closed-loop transfer function from w to h less than γ.

2. There exists a symmetric positive semi-definite solution P to (5.3.1), (5.3.2) and (5.3.3) such that matrix $A_{\rm cl}$ of (5.3.5) has all its eigenvalues inside the unit circle. ▢

In what follows, we provide a non-recursive method for computing the stabilizing solution to the H_∞-DARE for the full information problem, i.e., (5.3.1), (5.3.2) and (5.3.3). We first define an auxiliary H_∞-CARE from the given system data and we connect the stabilizing solution for the given H_∞-DARE to the stabilizing solution for the auxiliary H_∞-CARE, for which non-recursive methods of obtaining solutions are available.

We first choose any constant matrix F such that $A + BF$ has no eigenvalue at -1. We note that this can always be done as (A, B) is stabilizable with respect to $\mathbb{C}^\circ \cup \mathbb{C}^\otimes$. Next, define an auxiliary H_∞-CARE,

$$0 = \tilde{P}\tilde{A} + \tilde{A}'\tilde{P} + \tilde{C}_2'\tilde{C}_2 - \begin{bmatrix} \tilde{B}'\tilde{P} + \tilde{D}_2'\tilde{C}_2 \\ \tilde{E}'\tilde{P} + \tilde{D}_{22}'\tilde{C}_2 \end{bmatrix}' \tilde{G}^{-1} \begin{bmatrix} \tilde{B}'\tilde{P} + \tilde{D}_2'\tilde{C}_2 \\ \tilde{E}'\tilde{P} + \tilde{D}_{22}'\tilde{C}_2 \end{bmatrix}, \tag{5.3.7}$$

with the associated condition

$$\tilde{D}_{22}'\left(I - \tilde{D}_2(\tilde{D}_2'\tilde{D}_2)^{-1}\tilde{D}_2'\right)\tilde{D}_{22} < \gamma^2 I, \tag{5.3.8}$$

where

$$\left.\begin{aligned} \tilde{A} &:= (A + BF + I)^{-1}(A + BF - I), \\ \tilde{B} &:= 2(A + BF + I)^{-2}B, \\ \tilde{E} &:= 2(A + BF + I)^{-2}E, \\ \tilde{C}_2 &:= C_2 + D_2F, \\ \tilde{D}_2 &:= D_2 - (C_2 + D_2F)(A + BF + I)^{-1}B, \\ \tilde{D}_{22} &:= D_2 - (C_2 + D_2F)(A + BF + I)^{-1}E, \end{aligned}\right\} \tag{5.3.9}$$

and

$$\tilde{G} := \begin{bmatrix} \tilde{D}_2'\tilde{D}_2 & \tilde{D}_2'\tilde{D}_{22} \\ \tilde{D}_{22}'\tilde{D}_2 & \tilde{D}_{22}'\tilde{D}_{22} - \gamma^2 I \end{bmatrix}. \tag{5.3.10}$$

If matrix $\tilde{D}_2$ is injective, then Condition (5.3.8) implies $\tilde{G}$ in (5.3.10) is invertible. Again, we are particularly interested in solution $\tilde{P}$ of (5.3.7) such that the eigenvalues of $\tilde{A}_{\rm cl}$ are in the open-left plane, where

$$\tilde{A}_{\rm cl} := \tilde{A} - [\tilde{B} \quad \tilde{E}]\tilde{G}^{-1} \begin{bmatrix} \tilde{B}'\tilde{P} + \tilde{D}_2'\tilde{C}_2 \\ \tilde{E}'\tilde{P} + \tilde{D}_{22}'\tilde{C}_2 \end{bmatrix}. \tag{5.3.11}$$

We note that under the conditions when $\tilde{D}_2$ is injective, $(\tilde{A}, \tilde{B}, \tilde{C}_2, \tilde{D}_2)$ has no invariant zero on the $j\omega$ axis, and (5.3.8), the above H_∞-CARE (5.3.7) is related to the continuous-time H_∞ γ-suboptimal full information feedback control problem for the following system,

$$\tilde{\Sigma}_{\rm FI}: \begin{cases} \dot{x} = \tilde{A}\, x + \tilde{B}\, u + \tilde{E}\, w, \\ y = \begin{pmatrix} I \\ 0 \end{pmatrix} x + \begin{pmatrix} 0 \\ I \end{pmatrix} w, \\ h = \tilde{C}_2\, x + \tilde{D}_2\, u + \tilde{D}_{22}\, w. \end{cases} \tag{5.3.12}$$

The following lemma follows from Corollary 4.2.2.

Lemma 5.3.2. Consider a given system (5.3.12). Assume that $\tilde{D}_2$ is injective and $(\tilde{A}, \tilde{B}, \tilde{C}_2, \tilde{D}_2)$ has no invariant zero on the $j\omega$ axis. Then the following two statements are equivalent:

1. There exists a static feedback law $u = \tilde{K}_1 x + \tilde{K}_2 w$, which stabilizes $\tilde{\Sigma}_{\rm FI}$ and makes the H_∞ norm of the closed-loop transfer function from w to h less than γ.

2. Condition (5.3.8) holds and there exists a symmetric $\tilde{P} \geq 0$ such that (5.3.7) is satisfied and such that the matrix $\tilde{A}_{\rm cl}$ of (5.3.11) has all its eigenvalues in the open left-half plane. ⓛ

Now, we are ready to present our main results.

Theorem 5.3.1. The following two statements are equivalent:

1. (A, B) is stabilizable and (A, B, C_2, D_2) is left invertible with no invariant zero on the unit circle. Moreover, there exists a symmetric positive semi-definite matrix P such that (5.3.1), (5.3.2) and (5.3.3) are satisfied along with the matrix $A_{\rm cl}$ of (5.3.5) having all its eigenvalues inside the unit circle.

2. $(\tilde{A}, \tilde{B})$ is stabilizable, $\tilde{D}_2$ is injective and $(\tilde{A}, \tilde{B}, \tilde{C}_2, \tilde{D}_2)$ has no invariant zero on the $j\omega$ axis, and (5.3.8) holds. Moreover, there exists a symmetric positive semi-definite solution $\tilde{P}$ of the H_∞-CARE (5.3.7) such that the eigenvalues of $\tilde{A}_{\rm cl}$, where $\tilde{A}_{\rm cl}$ is as in (5.3.11), are in the open left-half complex plane.

Moreover, P and $\tilde{P}$ are related by $P = 2(A' + F'B' + I)^{-1} \tilde{P} (A + BF + I)^{-1}$. Ⓣ

Proof. See Subsection 5.4.B. ⊠

Remark 5.3.1. We should point out that the left invertibility of (A, B, C_2, D_2) is a necessary condition for the existence of the stabilizing solution to the H_∞-DARE for the full information problem (see [124]). Moreover, following the proof of Theorem 5.3.1 and the properties of the continuous-time algebraic Riccati equation, it is easy to show that the condition that (A, B, C_2, D_2) has no invariant zero on the unit circle is also necessary for the existence of the stabilizing solution to the H_∞-DARE for the full information problem. Ⓡ

Remark 5.3.2. From Theorem 5.3.1, a non-iterative method of obtaining the stabilizing solution P to the H_∞-DARE for the full information problem can be established as follows:

1. Obtain the auxiliary H_∞-CARE;

2. Obtain the stabilizing solution $\tilde{P}$ to the H_∞-CARE using some well-known non-iterative methods. For clarity, we recall in the following a so-called Schur method (see e.g., [73,114]): Define a Hamiltonian matrix

$$H_{\rm m} = \begin{bmatrix} H_{11} & H_{12} \\ H_{21} & H_{22} \end{bmatrix}, \tag{5.3.13}$$

where

$$\left.\begin{aligned} H_{11} &= \tilde{A} - [\,\tilde{B} \quad \tilde{E}\,]\tilde{G}^{-1}[\,\tilde{D}_2 \quad \tilde{D}_{22}\,]'\tilde{C}_2, \\ H_{12} &= -[\,\tilde{B} \quad \tilde{E}\,]\tilde{G}^{-1}[\,\tilde{B} \quad \tilde{E}\,]', \\ H_{21} &= -\tilde{C}_2'\{I - [\,\tilde{D}_2 \quad \tilde{D}_{22}\,]\tilde{G}^{-1}[\,\tilde{D}_2 \quad \tilde{D}_{22}\,]'\}\tilde{C}_2, \\ H_{22} &= -\{\tilde{A} - [\,\tilde{B} \quad \tilde{E}\,]\tilde{G}^{-1}[\,\tilde{D}_2 \quad \tilde{D}_{22}\,]'\tilde{C}_2\}'. \end{aligned}\right\} \tag{5.3.14}$$

Find an orthogonal matrix $T_{\rm m} \in \mathbb{R}^{2n\times 2n}$ that puts $H_{\rm m}$ in the real Schur form

$$T_{\rm m}' H_{\rm m} T_{\rm m} = \begin{bmatrix} S_{11} & S_{12} \\ 0 & S_{22} \end{bmatrix}, \tag{5.3.15}$$

where $S_{11} \in \mathbb{R}^{n\times n}$ is a stable matrix and $S_{22} \in \mathbb{R}^{n\times n}$ is an anti-stable matrix. Partition $T_{\rm m}$ into four $n \times n$ blocks:

$$T_{\rm m} = \begin{bmatrix} T_{11} & T_{12} \\ T_{21} & T_{22} \end{bmatrix}. \tag{5.3.16}$$

Then $\tilde{P}$ is given by $\tilde{P} = T_{21}T_{11}^{-1}$.

3. The stabilizing solution to the H_∞-DARE for the full information problem is given by $P = 2(A' + F'B' + I)^{-1}\tilde{P}(A + BF + I)^{-1}$. Ⓡ

It is well-known that the H_∞-DARE is the generalization of the H_2-DARE. Namely, by letting $\gamma = \infty$, or equivalently $E = 0$ and $D_{22} = 0$, we obtain the general H_2-DARE. For completeness, we give the following corollary that provides a non-iterative method of solving the general H_2-DARE.

Corollary 5.3.1. The following two statements are equivalent:

1. (A, B) is stabilizable and (A, B, C_2, D_2) is left invertible with no invariant zero on the unit circle. Moreover, there exists a positive semi-definite matrix P such that

$$B'PB + D_2'D_2 > 0, \tag{5.3.17}$$

$$P = A'PA + C_2'C_2 - (A'PB + C_2'D_2)(D_2'D_2 + B'PB)^{-1}(A'PB + C_2'D_2)' \tag{5.3.18}$$

and such that the eigenvalues of the matrix $A_{\rm cl}$ are inside the unit circle, where

$$A_{\rm cl} = A - B(D_2'D_2 + B'PB)^{-1}(A'PB + C_2'D_2)'. \tag{5.3.19}$$

2. $(\tilde{A}, \tilde{B})$ is stabilizable, $\tilde{D}_2$ is injective and $(\tilde{A}, \tilde{B}, \tilde{C}_2, \tilde{D}_2)$ has no invariant zero on the $j\omega$ axis. Moreover, there exists a positive semi-definite solution $\tilde{P}$ of the following CARE

$$0 = \tilde{P}\tilde{A} + \tilde{A}'\tilde{P} + \tilde{C}_2'\tilde{C}_2 - (\tilde{P}\tilde{B} + \tilde{C}_2'\tilde{D}_2)(\tilde{D}_2'\tilde{D}_2)^{-1}(\tilde{P}\tilde{B} + \tilde{C}_2'\tilde{D}_2)', \tag{5.3.20}$$

such that the eigenvalues of $\tilde{A}_{\rm cl}$ are in the open left-half complex plane, where

$$\tilde{A}_{\rm cl} = \tilde{A} - \tilde{B}(\tilde{D}_2'\tilde{D}_2)^{-1}(\tilde{P}\tilde{B} + \tilde{C}_2'\tilde{D}_2)'. \tag{5.3.21}$$

Moreover, P and $\tilde{P}$ are related by $P = 2(A' + F'B' + I)^{-1}\tilde{P}(A + BF + I)^{-1}$. ©

Lemmas 5.3.1 and 5.3.2, and Theorem 5.3.1 show the interconnection between the H_∞ γ-suboptimal control problem for the discrete-time system $\Sigma_{\rm FI}$ and the continuous-time system $\tilde{\Sigma}_{\rm FI}$. This connection is formalized in the following lemma.

Lemma 5.3.3. Assume that (A, B) is stabilizable and (A, B, C_2, D_2) is left invertible with no invariant zero on the unit circle. Then the following statements are equivalent:

1. The full information feedback discrete-time system $\Sigma_{\rm FI}$ of (5.3.6) has at least one γ-suboptimal control law. Namely, for a given γ, there exists a static full information feedback $u = K_1x + K_2w$ such that the closed-loop transfer function from w to h has an H_∞-norm less than γ.

2. The full information feedback continuous-time system $\tilde{\Sigma}_{\text{FI}}$ of (5.3.12) has at least one γ-suboptimal control law. Namely, for a given γ, there exists a static full information feedback $u = \tilde{K}_1 x + \tilde{K}_2 w$ such that the closed-loop transfer function from w to h has an H_∞-norm less than γ. ▣

Remark 5.3.3. The results of Lemma 5.3.3 can easily be obtained by a different route. It is well known that the Hankel norm and the H_∞ norm of a transfer function are invariant under a bilinear transformation (see e.g., Glover [57]). Hence one can recast the H_∞ γ-suboptimal control problem for the discrete-time system Σ_{FI} into an equivalent H_∞ γ-suboptimal control problem for an auxiliary continuous-time system obtained by performing bilinear transformation on Σ_{FI}. It can be shown that one of the state space realizations of this auxiliary continuous-time system, Σ_{BL}, is given by

$$\Sigma_{\text{BL}} : \begin{cases} \dot{x} = \tilde{A} \; x + \tilde{B} \; u + \tilde{E} \; w, \\ y = \begin{pmatrix} I \\ 0 \end{pmatrix} x + \begin{pmatrix} \tilde{D}_3 \\ 0 \end{pmatrix} u + \begin{pmatrix} \tilde{D}_4 \\ I \end{pmatrix} w, \\ z = \tilde{C}_2 \; x + \tilde{D}_2 \; u + \tilde{D}_{22} \; w, \end{cases} \tag{5.3.22}$$

where $\tilde{D}_3 = -(A + BF + I)^{-1}B$, $\tilde{D}_4 = -(A + BF + I)^{-1}E$, and $\tilde{A}$, $\tilde{B}$, $\tilde{E}$, $\tilde{C}_2$, $\tilde{D}_2$ and $\tilde{D}_{22}$ are as defined in (5.3.9). Consequently the H_∞ γ-suboptimal control problem for the discrete-time Σ_{FI} has a solution if and only if the H_∞ γ-suboptimal control problem for the continuous-time system Σ_{BL} has a solution. However, we note that Σ_{BL} is not completely in the full information form. This difficulty can easily be removed by redefining the measurement output in Σ_{BL} as

$$\tilde{y} := \begin{bmatrix} I & -\tilde{D}_4 \\ 0 & I \end{bmatrix} \left(y - \begin{bmatrix} \tilde{D}_3 \\ 0 \end{bmatrix} u \right) = \begin{pmatrix} I \\ 0 \end{pmatrix} x + \begin{pmatrix} 0 \\ I \end{pmatrix} w, \tag{5.3.23}$$

It is now obvious that Σ_{BL} with the new measurement output $\tilde{y}$ is in fact the same as $\tilde{\Sigma}_{\text{FI}}$. Also, it is easy to show that the H_∞ γ-suboptimal problem for Σ_{BL} has a solution if and only if the H_∞ γ-suboptimal problem for Σ_{FI} has a solution and hence the result of Lemma 5.3.3 follows. It is important to note that the bilinear transformation approach does not establish a relationship between the stabilizing solution of the H_∞-CARE associated with the continuous-time system $\tilde{\Sigma}_{\text{FI}}$, obtained by performing a bilinear transformation on discrete-time system Σ_{FI} and defining the new measurement as in (5.3.23), and the H_∞-DARE associated with the given discrete-time system Σ_{FI}. In fact, the main contribution of Theorem 5.3.1 is to establish such a relationship. ▣

We present in the following a numerical example to illustrate our results.

Example 5.3.1. Let us consider a discrete-time H_∞-DARE for the full information problem with

$$A = \begin{bmatrix} 1 & 1 & 0 & 1 & 0 \\ 1 & 1 & 1 & 0 & 0 \\ 0 & 1 & 0 & 1 & 0 \\ 0 & 0 & 1 & 1 & 1 \\ 0 & -1 & 0 & 1 & 1 \end{bmatrix}, \; B = \begin{bmatrix} 1 & 0 \\ 0 & 1 \\ 1 & 0 \\ 0 & 1 \\ 1 & 0 \end{bmatrix}, \; E = \begin{bmatrix} 1 \\ 0 \\ -1 \\ 0 \\ 1 \end{bmatrix}, \tag{5.3.24}$$

$$C_2 = \begin{bmatrix} 0 & 0 & 0 & 0 & 0 \\ 1 & 0 & 1 & 0 & 1 \\ 0 & 1 & 0 & 1 & 0 \end{bmatrix}, \; D_2 = \begin{bmatrix} 1 & 0 \\ 0 & 0 \\ 0 & 0 \end{bmatrix}, \; D_{22} = \begin{bmatrix} 0 \\ 0 \\ 0.5 \end{bmatrix}, \tag{5.3.25}$$

and $\gamma = 1$. It is simple to verify that (A, B, C_2, D_2) is left invertible with an invariant zero at 0. Following (5.3.9), we obtain the auxiliary H_∞-CARE with

$$\tilde{A} = \begin{bmatrix} 1 & -2 & 6 & -4 & 2 \\ -1 & 3 & -8 & 6 & -3 \\ 2 & -4 & 11 & -8 & 4 \\ -1 & 2 & -4 & 3 & -1 \\ 0 & 0 & -2 & 2 & -1 \end{bmatrix}, \; \tilde{B} = \begin{bmatrix} 68 & -50 \\ -92 & 68 \\ 128 & -94 \\ -52 & 38 \\ -18 & 14 \end{bmatrix}, \; \tilde{E} = \begin{bmatrix} -20 \\ 28 \\ -40 \\ 16 \\ 6 \end{bmatrix},$$

$$\tilde{C}_2 = \begin{bmatrix} 0 & 0 & 0 & 0 & 0 \\ 1 & 0 & 1 & 0 & 1 \\ 0 & 1 & 0 & 1 & 0 \end{bmatrix}, \quad \tilde{D}_2 = \begin{bmatrix} 1 & 0 \\ 10 & -8 \\ -9 & 6 \end{bmatrix}, \quad \tilde{D}_{22} = \begin{bmatrix} 0.0 \\ -4.0 \\ 3.5 \end{bmatrix}.$$

Solving (5.2.3) in MATLAB, we obtain the stabilizing solution to the auxiliary H_∞-CARE as

$$\tilde{P} = 10^3 \times \begin{bmatrix} 0.767767 & 1.110081 & 0.180720 & -0.307296 & -0.617828 \\ 1.110081 & 1.607297 & 0.260775 & -0.448623 & -0.897322 \\ 0.180720 & 0.260775 & 0.046343 & -0.064704 & -0.139318 \\ -0.307296 & -0.448623 & -0.064704 & 0.143150 & 0.264285 \\ -0.617828 & -0.897322 & -0.139318 & 0.264285 & 0.511644 \end{bmatrix},$$

and the stabilizing solution to the H_∞-DARE for the full information problem is given by,

$$P = \begin{bmatrix} 127.143494 & 187.057481 & 1 & -84.671880 & -134.864680 \\ 187.057481 & 278.730887 & 0 & -124.061419 & -201.396153 \\ 1 & 0 & 1 & 0 & 1 \\ -84.671880 & -124.061419 & 0 & 61.078015 & 92.569717 \\ -134.864680 & -201.396153 & 1 & 92.569717 & 147.982935 \end{bmatrix}.$$

It is straightforward to verify that the above P satisfies (5.3.1), (5.3.2) and (5.3.3). Moreover, the eigenvalues of $A_{\rm cl}$ are given by $\{0.4125 \pm j0.0733, 0, 0, 0\}$, which are inside the unit circle. ⓔ

5.4. Proofs of Main Results

The proofs of the main results of this chapter are given in the next.

5.4.A. Proof of Theorem 5.2.1

First, let us consider the following reductions:

$$\begin{aligned}
A'PA - P + Q &= 2A'(A'+I)^{-1}\tilde{P}(A+I)^{-1}A - 2(A'+I)^{-1}\tilde{P}(A+I)^{-1} + Q \\
&= 2(A'+I)^{-1}A'\tilde{P}A(A+I)^{-1} - 2(A'+I)^{-1}\tilde{P}(A+I)^{-1} + Q \\
&= (A'+I)^{-1}(2A'\tilde{P}A - 2\tilde{P})(A+I)^{-1} + Q \\
&= (A'+I)^{-1}[(A'+I)\tilde{P}(A-I) + (A'-I)\tilde{P}(A+I)](A+I)^{-1} + Q \\
&= \tilde{P}(A-I)(A+I)^{-1} + (A'+I)^{-1}(A'-I)\tilde{P} + Q \\
&= \tilde{P}F + F'\tilde{P} + Q. \qquad (5.4.1)
\end{aligned}$$

(1. ⇒ 2.) Let us start with the following trivial equality,

$$A'PA - P + (A'+I)P(A+I) - (A'+I)PA - A'P(A+I) = 0,$$

which implies that

$$\begin{aligned}
&P - PA(A+I)^{-1} - (A'+I)^{-1}A'P \\
&\qquad + (A'+I)^{-1}A'PA(A+I)^{-1} - (A'+I)^{-1}P(A+I)^{-1} = 0.
\end{aligned}$$

Then we have

$$\begin{aligned}
W &= R + M'(A'+I)^{-1}Q(A+I)^{-1}M - N'(A+I)^{-1}M - M'(A'+I)^{-1}N \\
&= R + M'(A'+I)^{-1}Q(A+I)^{-1}M - N'(A+I)^{-1}M - M'(A'+I)^{-1}N \\
&\quad + M'PM - M'PA(A+I)^{-1}M - M'(A'+I)^{-1}A'PM \\
&\quad + M'(A'+I)^{-1}A'PA(A+I)^{-1}M - M'(A'+I)^{-1}P(A+I)^{-1}M \\
&= R + M'PM - (M'PA+N')(A+I)^{-1}M - M'(A'+I)^{-1}(A'PM+N) \\
&\quad + M'(A'+I)^{-1}(A'PA+Q-P)(A+I)^{-1}M \qquad (5.4.2) \\
&= R + M'PM - (M'PA+N')(A+I)^{-1}M - M'(A'+I)^{-1}(A'PM+N) \\
&\quad + M'(A'+I)^{-1}(A'PM+N)(R+M'PM)^{-1}(M'PA+N') \\
&\quad \times (A+I)^{-1}M \qquad (5.4.3) \\
&= [I - M'(A'+I)^{-1}(A'PM+N)(R+M'PM)^{-1}] \\
&\quad \times (R+M'PM)[I - (R+M'PM)^{-1}(M'PA+N')(A+I)^{-1}M]. \qquad (5.4.4)
\end{aligned}$$

Here we note that we have used (5.2.1) to get (5.4.3) from (5.4.2). By the assumption that W is nonsingular, we have

$$R + M'PM = [I - M'(A' + I)^{-1}(A'PM + N)(R + M'PM)^{-1}]^{-1}W \\ \times [I - (R + M'PM)^{-1}(M'PA + N')(A + I)^{-1}M]^{-1}.$$

Hence,

$$\begin{aligned}(A'PM+N)&(R+M'PM)^{-1}(M'PA+N') \\ &= (A'PM+N)[I-(R+M'PM)^{-1}(M'PA+N')(A+I)^{-1}M]W^{-1} \\ &\quad \times [I-(R+M'PM)^{-1}(M'PA+N')(A+I)^{-1}M]'(M'PA+N') \\ &= [A'PM-(A'PM+N)(R+M'PM)^{-1}(M'PA+N')(A+I)^{-1}M+N]W^{-1} \\ &\quad \times [A'PM-(A'PM+N)(R+M'PM)^{-1}(M'PA+N') \\ &\quad \times (A+I)^{-1}M+N]' \qquad (5.4.5) \\ &= [A'PM+(P-A'PA-Q)(A+I)^{-1}M+N]W^{-1} \\ &\quad \times [A'PM+(P-A'PA-Q)(A+I)^{-1}M+N]' \qquad (5.4.6) \\ &= [(A'P+P-Q)(A+I)^{-1}M+N]W^{-1}[(A'P+P-Q)(A+I)^{-1}M+N]' \\ &= [(A'+I)P(A+I)(A+I)^{-2}M-Q(A+I)^{-1}M+N]W^{-1} \\ &\quad \times [(A'+I)P(A+I)(A+I)^{-2}M-Q(A+I)^{-1}M+N]' \\ &= (\tilde{P}G+H)W^{-1}(\tilde{P}G+H)'. \qquad (5.4.7)\end{aligned}$$

Again, we have used (5.2.1) to get (5.4.6) from (5.4.5). Finally, (5.2.1), (5.4.1) and (5.4.7) imply that

$$\tilde{P}F + F'\tilde{P} - (\tilde{P}G + H)W^{-1}(\tilde{P}G + H)' + Q = 0.$$

(2. $\Rightarrow$ 1.) It follows from (5.2.2) that

$$\left.\begin{aligned} A &= (I + F)(I - F)^{-1}, \\ M &= 2(I - F)^{-2}G, \\ H &= -Q(I - F)^{-1}G + N, \\ P &= (I - F')\tilde{P}(I - F)/2, \\ W &= R + G'(I - F')^{-1}Q(I - F)^{-1}G \\ &\quad - N'(I - F)^{-1}G - G'(I - F')^{-1}N, \\ R + M'PM &= R + 2G'(I - F')^{-1}\tilde{P}(I - F)^{-1}G. \end{aligned}\right\} \qquad (5.4.8)$$

Then we have

$$\begin{aligned}
R+M'PM &= R+G'(I-F')^{-1}[Q+(\tilde{P}-\tilde{P}F-Q) \\
&\quad + (\tilde{P}-F'\tilde{P}-Q)+(\tilde{P}F+F'\tilde{P}+Q)](I-F)^{-1}G \\
&= R+G'(I-F')^{-1}Q(I-F)^{-1}G-N'(I-F)^{-1}G-G'(I-F')^{-1}N \\
&\quad + G'(I-F')^{-1}[\tilde{P}G-Q(I-F)^{-1}G+N]+[\tilde{P}G-Q(I-F)^{-1}G+N]' \\
&\quad \times (I-F)^{-1}G+G'(I-F')^{-1}(\tilde{P}F+F'\tilde{P}+Q)(I-F)^{-1}G \qquad (5.4.9) \\
&= W+G'(I-F')^{-1}(\tilde{P}G+H)+(\tilde{P}G+H)'(I-F)^{-1}G \\
&\quad + G'(I-F')^{-1}(\tilde{P}G+H)W^{-1}(\tilde{P}G+H)'(I-F)^{-1}G \qquad (5.4.10) \\
&= [I+W^{-1}(\tilde{P}G+H)'(I-F)^{-1}G]'W[I+W^{-1}(\tilde{P}G+H)'(I-F)^{-1}G]. \qquad (5.4.11)
\end{aligned}$$

Here we note that we have used (5.2.3) to get (5.4.10) from (5.4.9).

By assumption, we have $R + M'PM$ nonsingular. Thus, we can rewrite (5.4.11) as,

$$\begin{aligned}
W = [I + G'(I - F')^{-1}(\tilde{P}G + H)W^{-1}]^{-1}(R + M'PM) \\
\times [I + W^{-1}(\tilde{P}G + H)'(I - F)^{-1}G]^{-1}.
\end{aligned}$$

We have the following reductions,

$$\begin{aligned}
&(\tilde{P}G+H)W^{-1}(\tilde{P}G+H)' \\
&\quad = (\tilde{P}G+H)[I+W^{-1}(\tilde{P}G+H)'(I-F)^{-1}G] \\
&\qquad \times (R+M'PM)^{-1}[I+W^{-1}(\tilde{P}G+H)'(I-F)^{-1}G]'(\tilde{P}G+H)' \\
&\quad = [\tilde{P}G+H+(\tilde{P}G+H)W^{-1}(\tilde{P}G+H)'(I-F)^{-1}G](R+M'PM)^{-1} \\
&\qquad \times [\tilde{P}G+H+(\tilde{P}G+H)W^{-1}(\tilde{P}G+H)'(I-F)^{-1}G]' \qquad (5.4.12) \\
&\quad = [\tilde{P}G-Q(I-F)^{-1}G+(\tilde{P}F+F'\tilde{P}+Q)(I-F)^{-1}G+N](R+M'PM)^{-1} \\
&\qquad \times [\tilde{P}G-Q(I-F)^{-1}G+(\tilde{P}F+F'\tilde{P}+Q)(I-F)^{-1}G+N]' \qquad (5.4.13) \\
&\quad = [(I+F')\tilde{P}(I-F)^{-1}G+N](R+M'PM)^{-1}[G'(I-F')^{-1}\tilde{P}(I+F)+N'] \\
&\quad = (A'PM+N)(R+M'PM)^{-1}(M'PA+N'). \qquad (5.4.14)
\end{aligned}$$

Again, we have used (5.2.3) to get (5.4.13) from (5.4.12). Finally, it follows from (5.2.3), (5.4.1) and (5.4.14) that

$$A'PA - (A'PM + N)(R + M'PM)^{-1}(M'PA + N') + Q - P = 0.$$

This completes the proof of Theorem 5.2.1. ⊠

5.4.B. Proof of Theorem 5.3.1

We note that the constant matrix F, a pre-state feedback, is introduced merely to overcome the situation when A has eigenvalues at -1. It is well-known in the literature that a pre-state feedback law does not affect the solution of the Riccati equation (5.3.3). Hence, for simplicity of presentation, we prove Theorem 5.3.1 for the case that $F = 0$ and $\gamma = 1$.

($1. \Rightarrow 2.$) It follows from Lemma 3.3.1 that the quadruple $(\tilde{A}, \tilde{B}, \tilde{C}_2, \tilde{D}_2)$ is an inverse bilinear transformation of the quadruple (A, B, C_2, D_2) with $a = 1$. Hence, it follows from Theorem 3.3.1 that $(\tilde{A}, \tilde{B})$ is stabilizable (see Item 1.a of Theorem 3.3.1) and $(\tilde{A}, \tilde{B}, \tilde{C}_2, \tilde{D}_2)$ is left invertible (see Item 3.b of Theorem 3.3.1) with no invariant zero on the $j\omega$ axis (see Item 4 of Theorem 3.3.1) and with no infinite zero of order higher than 0 (see Item 5 of Theorem 3.3.1). Hence, $\tilde{D}_2$ is injective as (A, B, C_2, D_2) has no invariant zero at -1.

Next, we will show that (5.3.8) holds. Let

$$\left.\begin{aligned}
M &:= [\,B \quad E\,], \\
N &:= C_2'\,[\,D_2 \quad D_{22}\,], \\
R &:= \begin{bmatrix} D_2'D_2 & D_2'D_{22} \\ D_{22}'D_2 & D_{22}'D_{22} - I \end{bmatrix}, \\
Q &:= C_2'C_2, \\
F &:= \tilde{A}, \\
G &:= 2(A+I)^{-2}M, \\
H &:= -Q(A+I)^{-1}M + N, \\
W &:= R + M'(A'+I)^{-1}Q(A+I)^{-1}M - N'(A+I)^{-1}M \\
&\qquad - M'(A'+I)^{-1}N, \\
X &:= I - (R + M'PM)^{-1}(M'PA + N')(A+I)^{-1}M.
\end{aligned}\right\} \tag{5.4.15}$$

It is simple to verify that

$$W = \begin{bmatrix} \tilde{D}_2'\tilde{D}_2 & \tilde{D}_2'\tilde{D}_{22} \\ \tilde{D}_{22}'\tilde{D}_2 & \tilde{D}_{22}'\tilde{D}_{22} - I \end{bmatrix}.$$

Then, (5.3.3) and (5.3.7) reduce to (5.2.1) and (5.2.3), respectively, and (5.3.5) and (5.3.11) can be written, respectively, as

$$A_{\rm cl} = A - M(R + M'PM)^{-1}(M'PA + N'), \tag{5.4.16}$$

and

$$\tilde{A}_{\rm cl} = F - GW^{-1}(\tilde{P}G + H)'. \tag{5.4.17}$$

Noting that

$$\begin{aligned}\det[X] &= \det[I - (R + M'PM)^{-1}(M'PA + N')(A + I)^{-1}M] \\ &= \det[I - M(R + M'PM)^{-1}(M'PA + N')(A + I)^{-1}] \\ &= \det[I + A_{\text{cl}}] \cdot \det[(A + I)^{-1}],\end{aligned}$$

it follows that X is nonsingular provided that the eigenvalues of A_{cl} are inside the unit circle. Recalling (5.4.4) in the proof of Theorem 5.2.1, we have W nonsingular and

$$W^{-1} = X^{-1}(R + M'PM)^{-1}(X^{-1})', \tag{5.4.18}$$

which implies that the inertia of W^{-1} is equal to the inertia of $(R + M'PM)^{-1}$ (see e.g., Theorem 4.9 of [3]). Again, noting that

$$W^{-1} = \begin{bmatrix} I & Y \\ 0 & I \end{bmatrix} \begin{bmatrix} (\tilde{D}_2'\tilde{D}_2)^{-1} & 0 \\ 0 & \left[\tilde{D}_{22}'\left(I - \tilde{D}_2(\tilde{D}_2'\tilde{D}_2)^{-1}\tilde{D}_2'\right)\tilde{D}_{22} - I\right]^{-1} \end{bmatrix} \begin{bmatrix} I & Y \\ 0 & I \end{bmatrix}',$$

and

$$(R + M'PM)^{-1} = \begin{bmatrix} I & Z \\ 0 & I \end{bmatrix} \begin{bmatrix} V^{-1} & 0 \\ 0 & -R^{-1} \end{bmatrix} \begin{bmatrix} I & Z \\ 0 & I \end{bmatrix}',$$

where $Y = -(\tilde{D}_2'\tilde{D}_2)^{-1}\tilde{D}_2'\tilde{D}_{22}$ and $Z = -V^{-1}B'PE$, together with (5.4.18) and the facts that $V > 0$ and $R > 0$, it follows that

$$\tilde{D}_{22}'\left(I - \tilde{D}_2(\tilde{D}_2'\tilde{D}_2)^{-1}\tilde{D}_2'\right)\tilde{D}_{22} < I.$$

Using the fact that W is nonsingular, it follows from Theorem 5.2.1 that $\tilde{P}$ is a positive semi-definite solution of (5.3.7).

Finally, we are ready to prove that $\tilde{A}_{\text{cl}}$ has all its eigenvalues in the open left-half complex plane. It follows from (5.4.7) in the proof of Theorem 5.2.1 that

$$\begin{aligned}\tilde{A}_{\text{cl}} &= F - GW^{-1}(\tilde{P}G + H)' = F - GX^{-1}(R + M'PM)^{-1}(M'PA + N') \\ &= (A + I)^{-1}(A - I) - 2(A + I)^{-2}M[I - (R + M'PM)^{-1}(M'PA + N') \\ &\qquad \times (A + I)^{-1}M]^{-1}(R + M'PM)^{-1}(M'PA + N') \\ &= (A + I)^{-1}\{A - I - 2[I - (A + I)^{-1}M(R + M'PM)^{-1}(M'PA + N')]^{-1} \\ &\qquad \times (A + I)^{-1}M(R + M'PM)^{-1}(M'PA + N')\} \\ &= (A + I)^{-1}\{A - I - 2[I + A - M(R + M'PM)^{-1}(M'PA + N')]^{-1} \\ &\qquad \times M(R + M'PM)^{-1}(M'PA + N')\}\end{aligned}$$

$$
\begin{aligned}
&= (A+I)^{-1}\,(A_{\rm cl}+I)^{-1}\,\{[I+A-M(R+M'PM)^{-1}(M'PA+N')] \\
&\qquad \times (A-I) - 2M(R+M'PM)^{-1}(M'PA+N')\} \\
&= (A+I)^{-1}\,(A_{\rm cl}+I)^{-1}\,(A_{\rm cl}-I)\,(A+I), \qquad (5.4.19)
\end{aligned}
$$

which implies that the eigenvalues of $\tilde{A}_{\rm cl}$ are in the open left-half plane provided that the eigenvalues of $A_{\rm cl}$ are inside the unit circle.

(2. $\Rightarrow$ 1.) First, following the results of Theorem 3.2.1, it is straightforward to show that (A, B) is stabilizable and (A, B, C_2, D_2) is left invertible with no invariant zero on the unit circle, provided that $(\tilde{A}, \tilde{B})$ is stabilizable, $\tilde{D}_2$ is injective and $(\tilde{A}, \tilde{B}, \tilde{C}_2, \tilde{D}_2)$ has no invariant zero on the $j\omega$ axis. Next, noting that

$$
\begin{aligned}
\det&[I+W^{-1}(\tilde{P}G+H)'(I-F)^{-1}G] \\
&= \det[I+GW^{-1}(\tilde{P}G+H)'(I-F)^{-1}] \\
&= \det[I-F+GW^{-1}(\tilde{P}G+H)']\cdot\det[(I-F)^{-1}] \\
&= \det\left[I-\tilde{A}_{\rm cl}\right]\cdot\det[(I-F)^{-1}],
\end{aligned}
$$

and $\tilde{A}_{\rm cl}$ has all its eigenvalue in the open left-half plane, it follows from (5.4.11) that $R+M'PM$ is nonsingular. Thus, the condition in Part 2 of Theorem 5.2.1 holds. The rest of the proof in the reverse direction of Theorem 5.3.1 follows from an almost identical procedure as (1. $\Rightarrow$ 2.). This completes our proof. ⊠

Chapter 6

Infima in Continuous-time H_∞ Optimization

6.1. Introduction

IN THIS CHAPTER, we address the problem of computing infima in H_∞ optimization for continuous-time systems. The H_∞-CARE based approach to this problem simply provides an iterative scheme of approximating the infimum, γ^*, of the H_∞-norm of the closed-loop transfer function. For example, in the regular measurement feedback case and utilizing the results of Doyle *et al.* [49] (see also Corollary 4.2.1), an iterative procedure for approximating γ^* would proceed as follows: one starts with a value of γ and determines whether $\gamma > \gamma^*$ by solving two "indefinite" algebraic Riccati equations and checking the positive semi-definiteness and stabilizing properties of these solutions. In the case when such positive semi-definite solutions exist and satisfy a *coupling condition*, then we have $\gamma > \gamma^*$ and one simply repeats the above steps using a smaller value of γ. In principle, one can approximate the infimum γ^* to within any degree of accuracy in this manner. However this search procedure is exhaustive and can be very costly. More significantly, due to the possible high-gain occurrence as γ gets close to γ^*, numerical solutions for these H_∞-CAREs can become highly sensitive and ill-conditioned. This difficulty also arises in the *coupling condition*. Namely, as γ decreases, evaluation of the *coupling condition* would generally involve finding eigenvalues of stiff matrices. These numerical difficulties are likely to be more severe for problems associated with the singular case. Thus, in general, the iterative procedure for the computation of γ^* based on AREs is not reliable.

Our goal here is to develop a non-iterative procedure to compute exactly the value of γ^* for a fairly large class of systems, which are associated with the singular case and satisfy certain geometric conditions. The computation of γ^* in our procedure involves solving two well-defined Riccati and two Lyapunov equations, which are independent of γ. The algorithm has been implemented efficiently in a MATLAB-software environment for numerical solutions. The results of this chapter are based on those reported in Chen [15] and Chen *et al.* [28,30–32].

The outline of this chapter is as follows: In Section 6.2, we will present a non-iterative algorithm that computes the infimum, γ^*, for the continuous-time H_∞ optimization problem under full information feedback, which is equivalent to that under full state feedback if the direct feedthrough term from the disturbance to the controlled output is equal to zero. Section 6.3 deals with the computation of γ^* for the measurement feedback case. Both Sections 6.2 and 6.3 require the given systems to have no invariant zero on the imaginary axis and satisfying certain geometric conditions. Finally, in Section 6.4, we will remove the constraints on the imaginary axis invariant zeros, i.e., we will present a non-iterative computational algorithm for finding γ^* for systems with invariant zeros on the imaginary axis.

6.2. Full Information Feedback Case

We consider in this section the H_∞ optimization problem for the class of continuous-time systems characterized by

$$\Sigma : \begin{cases} \dot{x} = A\ x + B\ u + E\ w, \\ y = \begin{pmatrix} I \\ 0 \end{pmatrix} x + \begin{pmatrix} 0 \\ I \end{pmatrix} w, \\ h = C_2\ x + D_2\ u + D_{22}\ w, \end{cases} \tag{6.2.1}$$

where $x \in \mathbb{R}^n$ is the state, $u \in \mathbb{R}^m$ is the control input, $w \in \mathbb{R}^q$ is the external disturbance input, $y \in \mathbb{R}^{n+q}$ is the measurement output, and $h \in \mathbb{R}^\ell$ is the controlled output of Σ. It is labelled a full information problem in the literature because all information about the system, i.e., both x and w, are available for feedback. For the purpose of easy reference in future developments, we define Σ_P to be the subsystem characterized by the matrix quadruple (A, B, C_2, D_2).

We first make the following assumptions:

Assumption 6.F.1: (A, B) is stabilizable;

Assumption 6.F.2: Σ_P has no invariant zero on the imaginary axis;

Assumption 6.F.3: $\operatorname{Im}(E) \subset \mathcal{V}^-(\Sigma_P) + \mathcal{S}^-(\Sigma_P)$; and

Assumption 6.F.4: $D_{22} = 0$. ▣

Remark 6.2.1. Here we note that the first assumption, i.e., (A, B) is stabilizable, is necessary for the existence of any stabilizing controller. The second assumption will be removed in Section 6.4. Also, Assumption 6.F.3 will be automatically satisfied if Σ_P is right invertible. In fact, in this case, Assumption 6.F.4 will no longer be necessary. This will be treated as a special case at the end of this section (see Remark 6.2.4). ▣

We have the following non-iterative algorithm for computing the infimum, γ^*, of the full information system (6.2.1).

Step 6.F.1. Without loss of generality, we assume that (A, B, C_2, D_2), i.e., Σ_P, has been partitioned in the form of (2.4.4). Then, transform Σ_P into the special coordinate basis as described in Chapter 2 (see also (2.4.20) to (2.4.23) for the compact form of the special coordinate basis). In this algorithm, for easy reference in future developments, we introduce an additional permutation matrix to the state transformation Γ_s such that the new state variables are ordered as follows:

$$\tilde{x} = \begin{pmatrix} x_a^+ \\ x_b \\ x_a^- \\ x_c \\ x_d \end{pmatrix}. \tag{6.2.2}$$

We also choose the output transformation Γ_o to have the following form:

$$\Gamma_o = \begin{bmatrix} I_{m_0} & 0 \\ 0 & \Gamma_{or} \end{bmatrix}, \tag{6.2.3}$$

where $m_0 = \operatorname{rank}(D_2)$. Next, we compute

$$\Gamma_s^{-1} E = \begin{bmatrix} E_a^+ \\ E_b \\ E_a^- \\ E_c \\ E_d \end{bmatrix}. \tag{6.2.4}$$

It is simple to verify from the properties of the special coordinate basis that Assumption 6.F.3 is equivalent to $E_b = 0$. Also, for economy of notation, we denote n_x the dimension of $\mathbb{R}^n / \mathcal{S}^+(\Sigma_P)$, which is equivalent to $n_a^+ + n_b$. We note that $n_x = 0$ if and only if the system Σ_P is right invertible and is of minimum phase.

Step 6.F.2. Next, we define

$$A_{11} := \begin{bmatrix} A_{aa}^+ & L_{ab}^+ C_b \\ 0 & A_{bb} \end{bmatrix}, \; B_{11} := \begin{bmatrix} B_{0a}^+ \\ B_{0b} \end{bmatrix}, \; A_{13} := \begin{bmatrix} L_{ad}^+ \\ L_{bd} \end{bmatrix}, \tag{6.2.5}$$

$$C_{21} := \Gamma_{or} \begin{bmatrix} 0 & 0 \\ 0 & C_b \end{bmatrix}, \quad C_{23} := \Gamma_{or} \begin{bmatrix} C_d C_d' \\ 0 \end{bmatrix}, \tag{6.2.6}$$

and

$$A_x := A_{11} - A_{13}(C_{23}'C_{23})^{-1}C_{23}'C_{21}, \tag{6.2.7}$$

$$B_x B_x' := B_{11}B_{11}' + A_{13}(C_{23}'C_{23})^{-1}A_{13}', \tag{6.2.8}$$

$$C_x'C_x := C_{21}'C_{21} - C_{21}'C_{23}(C_{23}'C_{23})^{-1}C_{23}'C_{21}. \tag{6.2.9}$$

Then we solve for the positive definite solution S_x of the algebraic Riccati equation,

$$A_x S_x + S_x A_x' - B_x B_x' + S_x C_x' C_x S_x = 0, \tag{6.2.10}$$

together with the matrix T_x defined by

$$T_x := \begin{bmatrix} T_{ax} & 0 \\ 0 & 0 \end{bmatrix}, \tag{6.2.11}$$

where T_{ax} is the unique solution of the algebraic Lyapunov equation,

$$A_{aa}^+ T_{ax} + T_{ax}(A_{aa}^+)' = E_a^+(E_a^+)'. \tag{6.2.12}$$

Here we should note that $(-A_x, C_x)$ is detectable since $-A_{aa}^+$ is stable and (A_{bb}, C_b) is observable. Furthermore, Assumption 6.F.1 implies that (A_x, B_x) is stabilizable. Hence the existence and uniqueness of the solutions S_x and T_{ax} follow from the results of Richardson and Kwong [106].

Step 6.F.3. The infimum, γ^*, is given by

$$\gamma^* = \sqrt{\lambda_{\max}(T_x S_x^{-1})}. \tag{6.2.13}$$

It can be shown using the result of Wielandt [135] that all the eigenvalues of $T_x S_x^{-1}$ are real and nonnegative. ⓐ

We have the following theorem.

Theorem 6.2.1. Consider the full information system given by (6.2.1). Then under Assumptions 6.F.1 to 6.F.4,

1. γ^* given by (6.2.13) is indeed its infimum, and

2. for $\gamma > \gamma^*$, the positive semi-definite matrix $P(\gamma)$ given by

$$P(\gamma) = (\Gamma_s^{-1})' \begin{bmatrix} (S_x - T_x/\gamma^2)^{-1} & 0 \\ 0 & 0 \end{bmatrix} \Gamma_s^{-1}, \tag{6.2.14}$$

is the unique solution that satisfies Conditions 2.(a)-2.(c) of Theorem 4.2.1. Moreover, such a solution $P(\gamma)$ does not exist when $\gamma < \gamma^*$. 🆃

Proof. As stated in Step 6.F.1 of the algorithm, we assume that Σ_P has been partitioned as in (2.4.4). Hence, the full information system of (6.2.1) can be rewritten as

$$\begin{cases} \dot{x} = A\; x + [\,B_0 \;\; B_1\,] \begin{pmatrix} u_0 \\ u_1 \end{pmatrix} + E\; w, \\ \begin{pmatrix} h_0 \\ h_1 \end{pmatrix} = \begin{bmatrix} C_{2,0} \\ C_{2,1} \end{bmatrix} x + \begin{bmatrix} I_{m_0} & 0 \\ 0 & 0 \end{bmatrix} \begin{pmatrix} u_0 \\ u_1 \end{pmatrix} + \begin{bmatrix} D_{22,0} \\ D_{22,1} \end{bmatrix} w, \end{cases} \tag{6.2.15}$$

where in this proof, we consider both $D_{22,0} = 0$ and $D_{22,1} = 0$. Let us apply a pre-feedback law,

$$u_0 = -C_{2,0}\, x + v_0, \tag{6.2.16}$$

to the above system. Then it is trivial to write the new system as,

$$\begin{cases} \dot{x} = (A - B_0 C_{2,0})\, x + [\,B_0 \;\; B_1\,] \begin{pmatrix} v_0 \\ u_1 \end{pmatrix} + E\, w, \\ \begin{pmatrix} h_0 \\ h_1 \end{pmatrix} = \begin{bmatrix} 0 \\ C_{2,1} \end{bmatrix} x + \begin{bmatrix} I_{m_0} & 0 \\ 0 & 0 \end{bmatrix} \begin{pmatrix} v_0 \\ u_1 \end{pmatrix}. \end{cases} \tag{6.2.17}$$

It follows from the theorem of the special coordinate basis, i.e., Theorem 2.4.1, that there exist nonsingular transformations, Γ_s, Γ_o and Γ_i such that

$$\begin{pmatrix} v_0 \\ u_1 \end{pmatrix} = \Gamma_i \begin{pmatrix} v_0 \\ u_d \\ u_c \end{pmatrix}, \quad x = \Gamma_s \begin{pmatrix} x_a^+ \\ x_b \\ x_a^- \\ x_c \\ x_d \end{pmatrix}, \quad \begin{pmatrix} h_0 \\ h_1 \end{pmatrix} = \Gamma_o \begin{pmatrix} h_0 \\ h_d \\ h_b \end{pmatrix}.$$

By Assumption 6.F.2, i.e., Σ_P has no invariant zero on the imaginary axis, the state component x_a^0 is nonexistent and the transformed system is given by

$$\begin{pmatrix} \dot{x}_a^+ \\ \dot{x}_b \\ \dot{x}_a^- \\ \dot{x}_c \\ \dot{x}_d \end{pmatrix} = \begin{bmatrix} A_{aa}^+ & L_{ab}^+ C_b & 0 & 0 & L_{ad}^+ C_d \\ 0 & A_{bb} & 0 & 0 & L_{bd} C_d \\ 0 & L_{ab}^- C_b & A_{aa}^- & 0 & L_{ad}^- C_d \\ B_c E_{ca}^+ & L_{cb} C_b & B_c E_{ca}^- & A_{cc} & L_{cd} C_d \\ B_d E_{da}^+ & B_d E_{db} & B_d E_{da}^- & B_d E_{dc} & A_{dd} \end{bmatrix} \begin{pmatrix} x_a^+ \\ x_b \\ x_a^- \\ x_c \\ x_d \end{pmatrix}$$

$$
+\begin{bmatrix} B_{0a}^{+} & 0 & 0 \\ B_{0b} & 0 & 0 \\ B_{0a}^{-} & 0 & 0 \\ B_{0c} & 0 & B_c \\ B_{0d} & B_d & 0 \end{bmatrix} \begin{pmatrix} v_0 \\ u_d \\ u_c \end{pmatrix} + \begin{bmatrix} E_a^{+} \\ E_b \\ E_a^{-} \\ E_c \\ E_d \end{bmatrix} w, \tag{6.2.18}
$$

where $E_b = 0$, and

$$
\begin{pmatrix} h_0 \\ h_d \\ h_b \end{pmatrix} = \begin{bmatrix} I_{m_0} & 0 \\ 0 & \Gamma_{or} \end{bmatrix} \begin{bmatrix} 0 & 0 & 0 & 0 & 0 \\ 0 & 0 & 0 & 0 & C_d \\ 0 & C_b & 0 & 0 & 0 \end{bmatrix} \begin{pmatrix} x_a^{+} \\ x_b \\ x_a^{-} \\ x_c \\ x_d \end{pmatrix} + \begin{bmatrix} I_{m_0} & 0 & 0 \\ 0 & 0 & 0 \\ 0 & 0 & 0 \end{bmatrix} \begin{pmatrix} v_0 \\ u_d \\ u_c \end{pmatrix}. \tag{6.2.19}
$$

The above transformation of the system with a pre-state feedback law,

$$
u_0 = -C_{2,0}\, x + v_0,
$$

along with the nonsingular state and control input transformations does not change our solution since it does not affect the value of γ^*. We need to introduce the following lemmas in order to prove the theorem.

Lemma 6.2.1. Given the system of (6.2.1), which satisfies Assumptions 6.F.1, 6.F.2 and 6.F.4, and $\gamma > 0$, then there exists a full information feedback control law $u = F_1 x + F_2 w$ such that when it is applied to (6.2.1), the resulting $\|T_{hw}\|_\infty < \gamma$ and $\lambda(A + BF) \subset \mathbb{C}^-$, if and only if there exists a real symmetric solution $P_x > 0$ to the algebraic Riccati equation

$$
P_x A_x + A_x' P_x + P_x E_x E_x' P_x / \gamma^2 - P_x B_x B_x' P_x + C_x' C_x = 0, \tag{6.2.20}
$$

where A_x, B_x and C_x are as defined in (6.2.7) to (6.2.9), and

$$
E_x = \begin{bmatrix} E_a^{+} \\ E_b \end{bmatrix}, \tag{6.2.21}
$$

with no restriction on E_b. Note that $E_b = 0$ if Assumption 6.F.3 holds. ⓛ

Proof. Without loss of generality, we assume that the given system has been transformed into the form of (6.2.18) and (6.2.19). Now let us define the new state variables,

$$
x_1 := \begin{pmatrix} x_a^{+} \\ x_b \end{pmatrix}, \quad \begin{pmatrix} x_2 \\ x_3 \end{pmatrix} := \begin{pmatrix} x_a^{-} \\ x_c \\ x_d \end{pmatrix}, \tag{6.2.22}
$$

where x_3 contains only the m_d states of x_d which are directly associated with the controlled output h_d while x_2 contains x_a^{-}, x_c and the remaining states of

x_d. Hence, the dynamics of the transformed system in (6.2.18) and (6.2.19) can be partitioned as follows,

$$\dot{x}_1 = A_{11}x_1 + [\,B_{11} \quad A_{13}\,]\begin{pmatrix} v_0 \\ x_3 \end{pmatrix} + E_x w, \tag{6.2.23}$$

$$\begin{pmatrix} \dot{x}_2 \\ \dot{x}_3 \end{pmatrix} = \begin{bmatrix} A_{22} & A_{23} \\ A_{32} & A_{33} \end{bmatrix}\begin{pmatrix} x_2 \\ x_3 \end{pmatrix} + \begin{bmatrix} B_{22} \\ B_{32} \end{bmatrix} u_1 + \begin{bmatrix} B_{21} & A_{21} \\ B_{31} & A_{31} \end{bmatrix}\begin{pmatrix} v_0 \\ x_1 \end{pmatrix} + \begin{bmatrix} E_2 \\ E_3 \end{bmatrix} w, \tag{6.2.24}$$

$$\begin{pmatrix} h_0 \\ h_1 \end{pmatrix} = \begin{bmatrix} 0 \\ C_{21} \end{bmatrix} x_1 + \begin{bmatrix} I_{m_0} & 0 \\ 0 & C_{23} \end{bmatrix}\begin{pmatrix} v_0 \\ x_3 \end{pmatrix}, \tag{6.2.25}$$

where A_{11}, B_{11}, A_{13}, C_{21} and C_{23} are as defined in (6.2.5) to (6.2.6), while A_{22}, A_{23}, $\cdots$, E_3 are the matrices with appropriate dimensions. It is now straightforward to verify using the properties of the special coordinate basis that the quadruple characterized by

$$\left(\begin{bmatrix} A_{22} & A_{23} \\ A_{32} & A_{33} \end{bmatrix}, \begin{bmatrix} B_{22} \\ B_{32} \end{bmatrix}, [\,0 \quad I\,], 0 \right), \tag{6.2.26}$$

is right invertible and of minimum phase. Moreover, the state space $\mathcal{X}_2 \oplus \mathcal{X}_3$ spans the strongly controllable subspace $\mathcal{S}^+(\Sigma_P)$. On the other hand, the subsystem characterized by the quadruple

$$\left(A_{11}, [\,B_{11} \quad A_{13}\,], \begin{bmatrix} 0 \\ C_{21} \end{bmatrix}, \begin{bmatrix} I_{m_0} & 0 \\ 0 & C_{23} \end{bmatrix} \right), \tag{6.2.27}$$

is left invertible with no infinite zero and with no stable invariant zero. The result of Lemma 6.2.1 follows from Corollary 5.2 and Theorem 6.2 of [127]. ⊠

Lemma 6.2.2. Given the system of (6.2.1) which satisfies Assumptions 6.F.1 to 6.F.4, then the algebraic Riccati equation of (6.2.20) has a symmetric solution $P_x > 0$ if and only if $S_x > T_x/\gamma^2$, where S_x and T_x are respectively given by (6.2.10) and (6.2.11). 🄻

Proof. First, we note that T_x of (6.2.11) is in fact the solution to the following Lyapunov equation

$$A_x T_x + T_x A_x' = E_x E_x, \tag{6.2.28}$$

where

$$E_x = \begin{bmatrix} E_a^+ \\ 0 \end{bmatrix},$$

since Assumption 6.F.3 holds. Also note that

$$T_x C_{21} = 0 \quad \text{and} \quad T_x C_x' C_x T_x = 0. \tag{6.2.29}$$

Now, suppose that $S_x > T_x/\gamma^2$ and define a positive definite matrix,

$$X := S_x - T_x/\gamma^2.$$

It follows from (6.2.10), (6.2.28) and (6.2.29) that

$$A_x X + X A_x' + E_x E_x'/\gamma^2 - B_x B_x' + X C_x' C_x X = 0. \tag{6.2.30}$$

Now, let us pre- and post-multiply (6.2.30) by $P_x := X^{-1}$, we obtain

$$P_x A_x + A_x' P_x + P_x E_x E_x' P_x/\gamma^2 - P_x B_x B_x' P_x + C_x' C_x = 0. \tag{6.2.31}$$

Hence, $P_x > 0$ is a solution to (6.2.20).

Conversely, suppose that (6.2.20) has a solution $P_x > 0$. Let $X := P_x^{-1} > 0$. We have

$$A_x X + X A_x' + E_x E_x'/\gamma^2 - B_x B_x' + X C_x' C_x X = 0. \tag{6.2.32}$$

Also, let T_x be the solution to the Lyapunov equation

$$A_x T_x + T_x A_x' = E_x E_x', \tag{6.2.33}$$

which has the special form as in (6.2.11). Thus, (6.2.29) holds. Next, we define $\bar{S}_x = T_x/\gamma^2 + X$. Clearly, we have $\bar{S}_x > T_x/\gamma^2$ and $\bar{S}_x \geq X > 0$. Then, we have

$$\begin{aligned} A_x \bar{S}_x + \bar{S}_x A_x' - B_x B_x' + \bar{S}_x C_x' C_x \bar{S}_x &= A_x (T_x/\gamma^2 + X) \\ &\quad + (T_x/\gamma^2 + X) A_x' - B_x B_x' + (T_x/\gamma^2 + X) C_x' C_x (T_x/\gamma^2 + X) \\ &= (A_x T_x + T_x A_x' - E_x E_x')/\gamma^2 \\ &\quad + A_x X + X A_x' + E_x E_x'/\gamma^2 - B_x B_x' + X C_x' C_x X \\ &= 0, \end{aligned}$$

which implies that $\bar{S}_x > 0$ is a solution of the Riccati equation (6.2.10). Since (6.2.10) can only have one positive definite solution, thus we have $\bar{S}_x = S_x$ and $S_x > T_x/\gamma^2$. This completes our proof of Lemma 6.2.2. ⊠

Now, let us get back to the proof of Theorem 6.2.1. Suppose that $\gamma > \gamma^*$. It is easy to verify that

$$P(\gamma) = (\Gamma_s^{-1})' \begin{bmatrix} (S_x - T_x/\gamma^2)^{-1} & 0 \\ 0 & 0 \end{bmatrix} \Gamma_s^{-1},$$

satisfies Conditions 2.(a)-2.(c) of Theorem 4.2.1. Hence, there exists a state feedback law $u = Fx$ with $F \in \mathbb{R}^{m \times n}$ (and obviously there exists a full information feedback law $u = F_1 x + F_2 w$) such that the H_∞-norm of the resulting closed-loop system from the disturbance w to the controlled output h, $T_{hw}(s)$, is less than γ and $\lambda(A+BF) \subset \mathbb{C}^-$.

The converse part of the theorem follows immediately from Lemmas 6.2.1 and 6.2.2 since the condition $\gamma > \{\lambda_{\max}(T_x S_x^{-1})\}^{\frac{1}{2}}$ is equivalent to $S_x > T_x/\gamma^2$. This completes our proof of Theorem 6.2.1. ⊠

The following remarks are in order.

Remark 6.2.2. For the continuous-time systems, the infimum for the full information system of (6.2.1) with $D_{22}=0$ is equivalent to the infimum for the full state feedback system, i.e.,

$$\begin{cases} \dot{x} = A\ x + B\ u + E\ w, \\ y = \quad x \\ h = C_2\ x + D_2\ u + 0\ w. \end{cases} \tag{6.2.34}$$

Thus, the infimum for the above full state feedback system is also given by γ^* in (6.2.13). Ⓡ

Remark 6.2.3. If Assumption 6.F.3, i.e., the geometric condition, is not satisfied, then an iterative scheme might be used to determine the infimum. This can be done by finding the smallest scalar, say $\tilde{\gamma}^*$, such that the Riccati equation

$$\tilde{P}_x A_x + A_x' \tilde{P}_x + \tilde{P}_x E_x E_x' \tilde{P}_x/(\tilde{\gamma}^*)^2 - \tilde{P}_x B_x B_x' \tilde{P}_x + C_x' C_x = 0, \tag{6.2.35}$$

has a positive definite solution $\tilde{P}_x > 0$. One could also apply the result of Scherer [117] directly to the Riccati equation (6.2.20) to develop an iterative algorithm of the Newton type to compute an approximation of γ^*. The algorithm of Scherer has a quadratic convergent rate. Ⓡ

Remark 6.2.4. If Σ_P is right invertible, then Assumption 6.F.3 is automatically satisfied. Moreover, Assumption 6.F.4 is no longer necessary and the infimum γ^* for the full information feedback system (6.2.1) can be obtained as follows:

$$\gamma^* = \left(\lambda_{\max}\left\{\begin{bmatrix} D_{22,1}' D_{22,1} & 0 \\ 0 & \tilde{T}_x \tilde{S}_x^{-1} \end{bmatrix}\right\}\right)^{\frac{1}{2}}, \tag{6.2.36}$$

where $\tilde{T}_x$ and $\tilde{S}_x$ are the positive semi-definite and positive definite solutions of the following Lyapunov equations,

$$\begin{aligned} A_{aa}^+ \tilde{T}_x + \tilde{T}_x (A_{aa}^+)' &= \left(E_a^+ - B_{0a}^+ D_{22,0} - L_{ad}^+ \Gamma_{or}^{-1} D_{22,1}\right) \\ &\quad \times \left(E_a^+ - B_{0a}^+ D_{22,0} - L_{ad}^+ \Gamma_{or}^{-1} D_{22,1}\right)', \end{aligned} \tag{6.2.37}$$

$$A_{aa}^+ \tilde{S}_x + \tilde{S}_x (A_{aa}^+)' = B_{0a}^+ (B_{0a}^+)' + L_{ad}^+ \Gamma_{or}^{-1} (L_{ad}^+ \Gamma_{or}^{-1})', \tag{6.2.38}$$

respectively, and $D_{22,0}$ and $D_{22,1}$ are as defined in (6.2.15) but for nonzero D_{22}. On the other hand, the infimum for the full state feedback system (6.2.34) is different from (6.2.36) and is given by

$$\gamma^* = \left(\lambda_{\max}\left\{\begin{bmatrix} D_{22}'D_{22} & 0 \\ 0 & \tilde{T}_x\tilde{S}_x^{-1} \end{bmatrix}\right\}\right)^{\frac{1}{2}}, \tag{6.2.39}$$

where $\tilde{T}_x$ and $\tilde{S}_x$ are again the positive semi-definite and positive definite solutions of the Lyapunov equations (6.2.37) and (6.2.38), respectively. These claims can be verified using similar arguments as in the proof of Theorem 6.2.1. The detailed proofs can be found in Chen [15]. ®

We conclude this section with the following illustrative examples.

Example 6.2.1. Consider a full information system (6.2.1) and a full state feedback system (6.2.34) characterized by

$$A = \begin{bmatrix} 1 & 1 & 1 & 0 & 1 \\ 0 & 1 & 0 & 0 & 1 \\ 0 & 1 & 1 & 0 & 1 \\ 1 & 1 & 1 & 1 & 1 \\ 1 & 1 & 1 & 1 & 0 \end{bmatrix}, \quad B = \begin{bmatrix} 0 & 0 & 0 \\ 0 & 0 & 0 \\ 1 & 0 & 0 \\ 0 & 0 & 1 \\ 0 & 1 & 0 \end{bmatrix}, \quad E = \begin{bmatrix} 5 & 1 \\ 0 & 0 \\ 0 & 0 \\ 2 & 3 \\ 1 & 4 \end{bmatrix}, \tag{6.2.40}$$

and

$$C_2 = \begin{bmatrix} 0 & 0 & 1 & 0 & 0 \\ 0 & 0 & 0 & 0 & 1 \\ 0 & 1 & 0 & 0 & 0 \\ 0 & 0 & 1 & 0 & 0 \end{bmatrix}, \quad D_2 = \begin{bmatrix} 1 & 0 & 0 \\ 0 & 0 & 0 \\ 0 & 0 & 0 \\ 0 & 0 & 0 \end{bmatrix}, \quad D_{22} = 0. \tag{6.2.41}$$

It is simple to verify that the subsystem (A, B, C_2, D_2) is neither left- nor right-invertible with one unstable invariant zero at $s = 1$. Moreover, it is already in the form of special coordinate basis with

$$\Gamma_s = I_5, \quad \Gamma_{or} = I_3, \quad n_x = 3,$$

$$A_x = \begin{bmatrix} 1 & 1 & 1 \\ 0 & 1 & 0 \\ 0 & 1 & 0 \end{bmatrix}, \quad B_xB_x' = \begin{bmatrix} 1 & 1 & 1 \\ 1 & 1 & 1 \\ 1 & 1 & 2 \end{bmatrix}, \quad C_x'C_x = \begin{bmatrix} 0 & 0 & 0 \\ 0 & 1 & 0 \\ 0 & 0 & 1 \end{bmatrix},$$

and

$$A_{aa}^+ = 1, \quad E_a^+ = [5 \quad 1].$$

Then solving equations (6.2.10) and (6.2.12), we obtain

$$S_x = \begin{bmatrix} 0.556281 & 0.185427 & -0.305593 \\ 0.185427 & 0.395142 & 0.231469 \\ -0.305593 & 0.231469 & 1.217984 \end{bmatrix}, \quad T_x = \begin{bmatrix} 13 & 0 & 0 \\ 0 & 0 & 0 \\ 0 & 0 & 0 \end{bmatrix},$$

and for both systems (6.2.1) and (6.2.34), the infima are given by

$$\gamma^* = \sqrt{\lambda_{\max}(T_x S_x^{-1})} = 6.4679044. \qquad \boxed{\text{E}}$$

Example 6.2.2. Consider a full information system (6.2.1) and a full state feedback system (6.2.34) characterized by

$$A = \begin{bmatrix} 3 & 0 & 0 & 1 \\ 1 & 1 & 0 & 1 \\ 1 & 1 & 0 & 0 \\ 0 & 0 & 1 & 0 \end{bmatrix}, \quad B = \begin{bmatrix} 1 & 0 & 0 \\ 0 & 0 & 1 \\ 0 & 1 & 0 \\ 0 & 0 & 0 \end{bmatrix}, \quad E = \begin{bmatrix} 4 \\ 3 \\ 2 \\ 1 \end{bmatrix}, \tag{6.2.42}$$

and

$$C_2 = \begin{bmatrix} 1 & 0 & 0 & 0 \\ 0 & 0 & 0 & 1 \end{bmatrix}, \quad D_2 = \begin{bmatrix} 1 & 0 & 0 \\ 0 & 0 & 0 \end{bmatrix}, \quad D_{22} = \begin{bmatrix} 2 \\ 1 \end{bmatrix}. \tag{6.2.43}$$

It is simple to verify that the subsystem (A, B, C_2, D_2) or Σ_P is controllable and right invertible with one unstable invariant zero at 2 and one infinite zero of order 2. Following Remark 6.2.4, we obtain

$$\Gamma_s = I_4, \quad \Gamma_{or} = 1, \quad n_x = 1, \quad A_{aa}^+ = 2, \quad B_{0a}^+ = 1,$$

$$L_{ad}^+ = 1, \quad E_a^+ = 4, \quad D_{22,0} = 2, \quad D_{22,1} = 1,$$

and

$$\tilde{S}_x = 0.5, \quad \tilde{T}_x = 0.25.$$

Then, the infimum for the full information feedback system is given

$$\gamma^* = \left(\lambda_{\max}\left\{\begin{bmatrix} D_{22,1}' D_{22,1} & 0 \\ 0 & \tilde{T}_x \tilde{S}_x^{-1} \end{bmatrix}\right\}\right)^{\frac{1}{2}} = \left(\lambda_{\max}\left\{\begin{bmatrix} 1 & 0 \\ 0 & 0.5 \end{bmatrix}\right\}\right)^{\frac{1}{2}} = 1,$$

and the infimum for the full state feedback system is

$$\gamma^* = \left(\lambda_{\max}\left\{\begin{bmatrix} D_{22}' D_{22} & 0 \\ 0 & \tilde{T}_x \tilde{S}_x^{-1} \end{bmatrix}\right\}\right)^{\frac{1}{2}} = \left(\lambda_{\max}\left\{\begin{bmatrix} 5 & 0 \\ 0 & 0.5 \end{bmatrix}\right\}\right)^{\frac{1}{2}} = \sqrt{5}.$$

Clearly, they are different. $\boxed{\text{E}}$

Finally, we conclude this section by posting an open problem related to the exact computation of the infimum, γ^*, for the full information feedback system of (6.2.1). The algorithm that yields the exact value of γ^* for this type of problem was built based on the following crucial assumption,

$$\operatorname{Im}(E) \subset \mathcal{V}^-(\Sigma_P) + \mathcal{S}^-(\Sigma_P), \tag{6.2.44}$$

and some minor ones. As will be seen shortly in an example, the assumption of (6.2.44) is not a necessary condition for obtaining the exact value of γ^*. Here is the open problem.

Open Problem. How to compute the exact value of the infimum, i.e., γ^*, associated with the full information feedback system of (6.2.1) without imposing the condition as given in (6.2.44)?

We believe that the above problem is solvable or at least partially solvable. The following is an example for which we are able to obtain the exact value of γ^* without imposing the condition of (6.2.44).

Example 6.2.3. Consider a full information feedback system of (6.2.1) with

$$A = \begin{bmatrix} 1 & 0 \\ 0 & 1 \end{bmatrix}, \quad B = \begin{bmatrix} 1 & 0 \\ 0 & 1 \end{bmatrix}, \quad E = \begin{bmatrix} 1 & 0 \\ 0 & 1 \end{bmatrix}, \tag{6.2.45}$$

and

$$C_2 = \begin{bmatrix} 0 & 0 \\ 0 & 0 \\ 1 & 0 \\ 0 & 2 \end{bmatrix}, \quad D_2 = \begin{bmatrix} 1 & 0 \\ 0 & 1 \\ 0 & 0 \\ 0 & 0 \end{bmatrix}, \quad D_{22} = 0. \tag{6.2.46}$$

It is simple to check using the linear system tools of Chapter 2 that

$$\mathcal{V}^-(\Sigma_{\rm P}) + \mathcal{S}^-(\Sigma_{\rm P}) = \{0\}, \tag{6.2.47}$$

and hence the condition of (6.2.44) is not valid. It is also straightforward to verify that the existence of a γ-suboptimal control law with $\gamma > \gamma^* \geq 0$ for (6.2.1) is equivalent to the existence of a positive definite solution P for the following algebraic Riccati equation,

$$PA + A'P + PEE'P/\gamma^2 - PBB'P + C_2'C_2 = 0. \tag{6.2.48}$$

Let

$$P := \begin{bmatrix} P_1 & P_0 \\ P_0 & P_2 \end{bmatrix} \quad \text{and} \quad \frac{1}{\alpha} := \frac{1}{\gamma^2} - 1. \tag{6.2.49}$$

Then (6.2.48) is equivalent to

$$\begin{bmatrix} P_0^2 + P_1^2 + 2\alpha P_1 + \alpha & P_0(P_1 + P_2 + 2\alpha) \\ P_0(P_1 + P_2 + 2\alpha) & P_0^2 + P_2^2 + 2\alpha P_2 + 4\alpha \end{bmatrix} = 0, \tag{6.2.50}$$

or

$$P_0(P_1 + P_2 + 2\alpha) = 0, \tag{6.2.51}$$

$$P_0^2 + P_1^2 + 2\alpha P_1 + \alpha = 0, \tag{6.2.52}$$

$$P_0^2 + P_2^2 + 2\alpha P_2 + 4\alpha = 0. \tag{6.2.53}$$

(6.2.51) implies that either

$$P_0 = 0 \quad \text{or} \quad P_1 + P_2 + 2\alpha = 0. \tag{6.2.54}$$

If we choose $P_1 + P_2 + 2\alpha = 0$, then we have

$$P_1 = -P_2 - 2\alpha, \tag{6.2.55}$$

which together with (6.2.52) imply that

$$P_0^2 + P_2^2 + 2\alpha P_2 + \alpha = 0. \tag{6.2.56}$$

Clearly, (6.2.53) and (6.2.56) imply that $\alpha = 0$ or equivalently $\gamma = 0$. Note that $\gamma > \gamma^* \geq 0$. Hence, it is a contradiction. Thus, we will have to choose $P_0 = 0$. Then (6.2.52) and (6.2.53) reduce to

$$P_1^2 + 2\alpha P_1 + \alpha = 0, \tag{6.2.57}$$

$$P_2^2 + 2\alpha P_2 + 4\alpha = 0. \tag{6.2.58}$$

It can be readily verified that the above equations have positive solutions P_1 and P_2 if and only if $\alpha < 0$, or equivalently $\gamma > 1$. Therefore, the exact value of the infimum is given by $\gamma^* = 1$. Moreover, the positive definite solution P of (6.2.48) is given by

$$P = \begin{bmatrix} \frac{\gamma}{\gamma^2 - 1}\left(\sqrt{2\gamma^2 - 1} + \gamma\right) & 0 \\ 0 & \frac{\gamma}{\gamma^2 - 1}\left(\sqrt{5\gamma^2 - 4} + \gamma\right) \end{bmatrix}, \tag{6.2.59}$$

for any given $\gamma > \gamma^* = 1$. 🄴

In general, we feel that there is a large class of systems that do not necessarily satisfy the geometric condition (6.2.44) but their infima are exactly computable. It is an interesting and of course very challenging problem.

6.3. Output Feedback Case

We present in this section an elegant well-conditioned non-iterative algorithm for the exact computation of γ^* of the following measurement feedback system,

$$\Sigma : \begin{cases} \dot{x} = A\ x + B\ u + E\ w, \\ y = C_1\ x + D_1\ w, \\ h = C_2\ x + D_2\ u + D_{22}\ w, \end{cases} \tag{6.3.1}$$

where $x \in \mathbb{R}^n$ is the state, $u \in \mathbb{R}^m$ is the control input, $w \in \mathbb{R}^q$ is the external disturbance input, $y \in \mathbb{R}^p$ is the measurement output, and $h \in \mathbb{R}^\ell$ is the controlled output of Σ. Again, for the purpose of easy reference, we define $\Sigma_{\rm P}$ to be the subsystem characterized by the matrix quadruple (A, B, C_2, D_2) and $\Sigma_{\rm Q}$ to be the subsystem characterized by the matrix quadruple (A, E, C_1, D_1). We first make the following assumptions:

Assumption 6.M.1: (A, B) is stabilizable;

Assumption 6.M.2: $\Sigma_{\rm P}$ has no invariant zero on the imaginary axis;

Assumption 6.M.3: $\mathrm{Im}\,(E) \subset \mathcal{V}^-(\Sigma_{\rm P}) + \mathcal{S}^-(\Sigma_{\rm P})$;

Assumption 6.M.4: (A, C_1) is detectable;

Assumption 6.M.5: $\Sigma_{\rm Q}$ has no invariant zero on the imaginary axis;

Assumption 6.M.6: $\mathrm{Ker}\,(C_2) \supset \mathcal{V}^-(\Sigma_{\rm Q}) \cap \mathcal{S}^-(\Sigma_{\rm Q})$; and

Assumption 6.M.7: $D_{22} = 0$. ▣A

Remark 6.3.1. Here we note that Assumptions 6.M.1 and 6.M.4, i.e., (A, B) is stabilizable and (A, C_1) is detectable, are necessary for the existence of any stabilizing controller. Assumptions 6.M.2 and 6.M.5 will be removed later in Section 6.4. Also, Assumptions 6.M.3 and 6.M.6 will be automatically satisfied if $\Sigma_{\rm P}$ is right invertible and if $\Sigma_{\rm Q}$ is left invertible. Moreover, in this case, $D_{22} = 0$, i.e., Assumption 6.M.7, can be removed without any difficulties (see Remark 6.3.3 later in this section). ▣R

We have the following non-iterative algorithm for computing the infimum, γ^*, of the general measurement feedback system (6.3.1).

Step 6.M.1. Define an auxiliary full information system

$$\begin{cases} \dot{x} = A\ x + B\ u + E\ w, \\ y = \begin{pmatrix} I \\ 0 \end{pmatrix} x + \begin{pmatrix} 0 \\ I \end{pmatrix} w, \\ h = C_2\ x + D_2\ u + D_{22}\ w, \end{cases} \tag{6.3.2}$$

and perform Steps 6.F.1 and 6.F.2 of the algorithm as given in Section 6.2. For easy reference in future development, we append a subscript '${\rm P}$' to all sub-matrices and transformations in the special coordinate basis associated with the system (6.3.2). In particular, we rename the state transformation of the special coordinate basis for $\Sigma_{\rm P}$ as $\Gamma_{s\rm P}$, and the dimension of $\mathbb{R}^n/\mathcal{S}^+(\Sigma_{\rm P})$ as $n_{x\rm P}$. Furthermore, S_x of (6.2.10) and T_x of (6.2.11) are respectively renamed to $S_{x\rm P}$ and $T_{x\rm P}$.

Step 6.M.2. Define another auxiliary full information system

$$\begin{cases} \dot{x} = A' \; x + C_1' \; u + C_2' \; w, \\ y = \begin{pmatrix} I \\ 0 \end{pmatrix} x + \begin{pmatrix} 0 \\ I \end{pmatrix} w, \\ z = E' \; x + D_1' \; u + D_{22}' \; w, \end{cases} \tag{6.3.3}$$

and again perform Steps 6.F.1 and 6.F.2 of the algorithm as given in Section 6.2 one more time but for this auxiliary system. To all sub-matrices and transformations in the special coordinate basis of $\Sigma_{\mathrm{Q}}^{\star}$, where $\Sigma_{\mathrm{Q}}^{\star}$ is the dual system of Σ_{Q} and is characterized by quadruple (A', C_1', E', D_1'), we append a subscript 'Q' to signify their relation to the system $\Sigma_{\mathrm{Q}}^{\star}$. In particular, we rename the state transformation of the special coordinate basis for this case as $\Gamma_{s\mathrm{Q}}$, and the dimension of $\mathbb{R}^n/\mathcal{S}^+(\Sigma_{\mathrm{Q}}^{\star})$ as $n_{x\mathrm{Q}}$. As in Step 6.M.1, we also rename S_x of (6.2.10) and T_x of (6.2.11) as $S_{x\mathrm{Q}}$ and $T_{x\mathrm{Q}}$, respectively.

Step 6.M.3. Partition

$$\Gamma_{s\mathrm{P}}^{-1}(\Gamma_{s\mathrm{Q}}^{-1})' = \begin{bmatrix} \Gamma & \star \\ \star & \star \end{bmatrix}, \tag{6.3.4}$$

where Γ is a $n_{x\mathrm{P}} \times n_{x\mathrm{Q}}$ matrix, and define a constant matrix

$$M = \begin{bmatrix} T_{x\mathrm{P}} S_{x\mathrm{P}}^{-1} + \Gamma S_{x\mathrm{Q}}^{-1} \Gamma' S_{x\mathrm{P}}^{-1} & -\Gamma S_{x\mathrm{Q}}^{-1} \\ - T_{x\mathrm{Q}} S_{x\mathrm{Q}}^{-1} \Gamma' S_{x\mathrm{P}}^{-1} & T_{x\mathrm{Q}} S_{x\mathrm{Q}}^{-1} \end{bmatrix}. \tag{6.3.5}$$

Step 6.M.4. The infimum γ^* for the measurement feedback system (6.3.1) is then given by

$$\gamma^* = \sqrt{\lambda_{\max}(M)}. \tag{6.3.6}$$

It will be shown later in Proposition 6.3.4 that the matrix M of (6.3.5) has only real and nonnegative eigenvalues. 🄰

The proof of the above algorithm is rather involved. We would have to introduce several lemmas before proceeding to its final proof. Let us first define

$$\gamma_{\mathrm{P}}^* := \left\{\lambda_{\max}(T_{x\mathrm{P}} S_{x\mathrm{P}}^{-1})\right\}^{\frac{1}{2}} \quad \text{and} \quad \gamma_{\mathrm{Q}}^* := \left\{\lambda_{\max}(T_{x\mathrm{Q}} S_{x\mathrm{Q}}^{-1})\right\}^{\frac{1}{2}}, \tag{6.3.7}$$

$$P(\gamma) := (\Gamma_{s\mathrm{P}}^{-1})' \begin{bmatrix} (S_{x\mathrm{P}} - T_{x\mathrm{P}}/\gamma^2)^{-1} & 0 \\ 0 & 0 \end{bmatrix} \Gamma_{s\mathrm{P}}^{-1}, \tag{6.3.8}$$

and

$$Q(\gamma) := (\Gamma_{s\mathrm{Q}}^{-1})' \begin{bmatrix} (S_{x\mathrm{Q}} - T_{x\mathrm{Q}}/\gamma^2)^{-1} & 0 \\ 0 & 0 \end{bmatrix} \Gamma_{s\mathrm{Q}}^{-1}. \tag{6.3.9}$$

We have the following lemma.

Lemma 6.3.1. Consider the system (6.3.1), which satisfies Assumptions 6.M.1 to 6.M.7. Then we have

1. For $\gamma > \gamma_{\text{P}}^*$, the positive semi-definite matrix $P(\gamma)$ given by (6.3.8) is the unique solution to the matrix inequality $F_\gamma(P) \geq 0$, i.e., Condition 2.(a) of Theorem 4.2.1, and satisfies both rank Conditions 2.(b) and 2.(c) of Theorem 4.2.1. Moreover, such a solution $P(\gamma)$ does not exist when $\gamma < \gamma_{\text{P}}^*$.

2. For $\gamma > \gamma_{\text{Q}}^*$, the positive semi-definite matrix $Q(\gamma)$ given by (6.3.9) is the unique solution to the matrix inequality $G_\gamma(Q) \geq 0$, i.e., Condition 2.(d) of Theorem 4.2.1, and satisfies both rank Conditions 2.(e) and 2.(f) of Theorem 4.2.1. Moreover, such a solution $Q(\gamma)$ does not exist when $\gamma < \gamma_{\text{Q}}^*$. ⊡

Proof. It follows from Theorem 6.2.1. ⊠

The next lemma gives an equivalence of the infimum, γ^*, for the measurement feedback system (6.3.1).

Lemma 6.3.2. Let $\gamma_{\text{PQ}}^* := \max\{\gamma_{\text{P}}^*, \gamma_{\text{Q}}^*\}$. Then the infimum for the given measurement feedback system (6.3.1) is equivalent to

$$\gamma^* = \inf \left\{ \gamma \in (\gamma_{\text{PQ}}^*, \infty) \mid f(\gamma) < \gamma^2 \right\}, \tag{6.3.10}$$

where the scalar function

$$f(\gamma) := \rho\{P(\gamma)Q(\gamma)\}, \tag{6.3.11}$$

and $P(\gamma)$ and $Q(\gamma)$ are given by (6.3.8) and (6.3.9) respectively. ⊡

Proof. It follows Lemma 6.3.2 that $\gamma^* \geq \gamma_{\text{PQ}}^*$. Next, for any $\hat{\gamma} \in (\gamma_{\text{PQ}}^*, \infty)$ such that $f(\hat{\gamma}) < \hat{\gamma}^2$, i.e., $\rho\{P(\hat{\gamma})Q(\hat{\gamma})\} < \hat{\gamma}^2$, then the corresponding $P(\hat{\gamma})$ and $Q(\hat{\gamma})$ as given in (6.3.8) and (6.3.9) satisfy the conditions of Theorem 4.2.1. Hence, $\hat{\gamma} > \gamma^*$ and γ^* is equivalent to that of (6.3.10). ⊠

It is then straightforward to show that the scalar function $f(\gamma)$ of (6.3.11) is given by

$$f(\gamma) = \lambda_{\max}\left\{ (S_{x\text{P}} - \gamma^{-2} T_{x\text{P}})^{-1} \Gamma (S_{x\text{Q}} - \gamma^{-2} T_{x\text{Q}})^{-1} \Gamma' \right\}. \tag{6.3.12}$$

The function $f(\gamma)$ of (6.3.12) is a well-defined mapping from $(\gamma_{\text{PQ}}^*, \infty)$ to $[0, \infty)$. Its evaluation involves the computation of the maximum eigenvalue of a matrix of dimension $n_{x\mathrm{P}} \times n_{x\mathrm{P}}$, which is normally of a much smaller dimension than the original product $P(\gamma)Q(\gamma)$. We establish some important properties of the function $f(\gamma)$ in the following propositions.

Proposition 6.3.1. $f(\gamma)$ is a continuous, nonnegative and non-increasing function of γ on $(\gamma^*_{\mathrm{PQ}}, \infty)$. 🄿

Proof. We first show that $P_x(\gamma) := (S_{x\mathrm{P}} - \gamma^{-2}T_{x\mathrm{P}})^{-1}$ is non-increasing, i.e., if $\gamma_2 > \gamma_1$ then $P_x(\gamma_2) \le P_x(\gamma_1)$. Recall that $S_{x\mathrm{P}} > 0$ and $T_{x\mathrm{P}} \ge 0$, we have for all $\gamma_2 > \gamma_1 > \gamma^*_{\mathrm{PQ}}$

$$(\gamma_1^{-2} - \gamma_2^{-2})T_{x\mathrm{P}} \ge 0,$$

which implies that

$$S_{x\mathrm{P}} - \gamma_1^{-2}T_{x\mathrm{P}} \le S_{x\mathrm{P}} - \gamma_2^{-2}T_{x\mathrm{P}}.$$

Hence,

$$P_x(\gamma_2) \le P_x(\gamma_1), \quad \text{for } \gamma_2 > \gamma_1.$$

Similarly, one can show that $Q_x(\gamma) := (S_{x\mathrm{Q}} - \gamma^{-2}T_{x\mathrm{Q}})^{-1}$ is non-increasing. This implies that $\Gamma Q_x(\gamma)\Gamma'$ is also non-increasing. Then clearly $f(\gamma)$ is a continuous, nonnegative and non-increasing function of γ on $(\gamma^*_{\mathrm{PQ}}, \infty)$. ⊠

The function $f(\gamma)$ defined above can be extended as a mapping from $[\gamma^*_{\mathrm{PQ}}, \infty)$ to $[0, \infty)$ by setting

$$f(\gamma^*_{\mathrm{PQ}}) = \lim_{\gamma \to \gamma^*_{\mathrm{PQ}}} f(\gamma). \tag{6.3.13}$$

It follows from Proposition 6.3.1 that the limit $f(\gamma^*_{\mathrm{PQ}})$ exists and could be finite or infinite.

Proposition 6.3.2. $f(\gamma) = \gamma^2$ has either no solution or a unique solution in the interval $(\gamma^*_{\mathrm{PQ}}, \infty)$. 🄿

Proof. The result follows from Proposition 6.3.1 and the fact that γ^2 is strictly increasing for positive γ. ⊠

Proposition 6.3.3. If $f(\gamma) = \gamma^2$ has no solution in the interval $(\gamma^*_{\mathrm{PQ}}, \infty)$ then γ^* is equal to γ^*_{PQ}. Otherwise, γ^* is equal to the unique solution of $f(\gamma) = \gamma^2$ in the interval $(\gamma^*_{\mathrm{PQ}}, \infty)$. 🄿

Proof. If $f(\gamma) = \gamma^2$ has no solution in the interval $(\gamma^*_{\mathrm{PQ}}, \infty)$, then $f(\gamma) < \gamma^2$ for all $\gamma \in (\gamma^*_{\mathrm{PQ}}, \infty)$ and hence according to Lemma 6.3.2, $\gamma^* = \gamma^*_{\mathrm{PQ}}$. On the other hand, it is obvious that γ^* is equal to the unique solution of $f(\gamma) = \gamma^2$ when such a solution exists. ⊠

At first glance, it seems that the solution of $f(\gamma) = \gamma^2$ would involve the rooting of a highly nonlinear algebraic equation in γ. Actually its solution can be achieved in one step. Namely the problem of solving $f(\gamma) = \gamma^2$, if such a solution exists in the interval (γ^*_{PQ}, ∞), can be converted to the problem of calculating the maximum eigenvalue of a constant matrix, i.e., M of (6.3.5). In fact, we would also show that, when $f(\gamma) = \gamma^2$ has no solution in the interval (γ^*_{PQ}, ∞), the maximum eigenvalue of this matrix M is equal to γ^*_{PQ}, which is γ^* as well. To prove this, we would have to introduce a matrix function of γ,

$$N(\gamma) := (S_{xP} - \gamma^{-2}T_{xP})^{-1}\Gamma(S_{xQ} - \gamma^{-2}T_{xQ})^{-1}\Gamma' - \gamma^2 I. \tag{6.3.14}$$

We have the following propositions on the properties of the matrices M and $N(\gamma)$.

Proposition 6.3.4. The eigenvalues of the matrix M of (6.3.5) are real and nonnegative. Ⓟ

Proof. First, we have

$$\begin{aligned}
\lambda\{M\} &= \lambda\left\{\begin{bmatrix} I & 0 \\ 0 & T_{xQ}\end{bmatrix}\begin{bmatrix} T_{xP} + \Gamma S_{xQ}^{-1}\Gamma' & -\Gamma S_{xQ}^{-1} \\ -S_{xQ}^{-1}\Gamma' & S_{xQ}^{-1}\end{bmatrix}\begin{bmatrix} S_{xP}^{-1} & 0 \\ 0 & I\end{bmatrix}\right\} \\
&= \lambda\left\{\begin{bmatrix} S_{xP}^{-1} & 0 \\ 0 & I\end{bmatrix}\begin{bmatrix} I & 0 \\ 0 & T_{xQ}\end{bmatrix}\begin{bmatrix} T_{xP} + \Gamma S_{xQ}^{-1}\Gamma' & -\Gamma S_{xQ}^{-1} \\ -S_{xQ}^{-1}\Gamma' & S_{xQ}^{-1}\end{bmatrix}\right\} \\
&= \lambda\left\{\begin{bmatrix} S_{xP}^{-1} & 0 \\ 0 & T_{xQ}\end{bmatrix}\begin{bmatrix} T_{xP} + \Gamma S_{xQ}^{-1}\Gamma' & -\Gamma S_{xQ}^{-1} \\ -S_{xQ}^{-1}\Gamma' & S_{xQ}^{-1}\end{bmatrix}\right\}.
\end{aligned} \tag{6.3.15}$$

Now, it is trivial to verify that both sub-matrices in (6.3.15) are symmetric and positive semi-definite. Then, using the result of Wielandt [135] (i.e., Theorem 3), it is simple to show that the eigenvalues of M are real and nonnegative. ⊠

Proposition 6.3.5.

1. $N(\gamma)$ has real eigenvalues for all $\gamma \in (\gamma^*_{PQ}, \infty)$.

2. $\lambda_{\max}\{N(\gamma)\} = f(\gamma) - \gamma^2$ is a continuous and strictly decreasing function of γ in (γ^*_{PQ}, ∞). Ⓟ

Proof. Note that both $(S_{xP} - \gamma^{-2}T_{xP})^{-1}$ and $(S_{xQ} - \gamma^{-2}T_{xQ})^{-1}$ are symmetric and positive definite for all $\gamma \in (\gamma^*_{PQ}, \infty)$. Hence, all the eigenvalues of $N(\gamma)$ are real for $\gamma \in (\gamma^*_{PQ}, \infty)$. The second item follows from Proposition 6.3.1. ⊠

Proposition 6.3.6. The roots of $\det[N(\gamma)] = 0$ are real. Moreover, the largest root of $\det[N(\gamma)] = 0$ in the interval (γ^*_{PQ}, ∞) is equal to $\{\lambda_{\max}(M)\}^{\frac{1}{2}}$. Ⓟ

Proof. Using the definition of $N(\gamma)$ in (6.3.14), we have

$$\begin{aligned}\det[N(\gamma)] &= (-1)^{n_{xP}}\cdot\det\left[\gamma^2 I-(S_{xP}-\gamma^{-2}T_{xP})^{-1}\Gamma(S_{xQ}-\gamma^{-2}T_{xQ})^{-1}\Gamma'\right]\\ &= \frac{(-1)^{n_{xP}}}{\det[S_{xP}-\gamma^{-2}T_{xP}]}\cdot\det\left[\gamma^2 S_{xP}-T_{xP}-\gamma^2\Gamma(\gamma^2 S_{xQ}-T_{xQ})^{-1}\Gamma'\right]\\ &= \frac{(-1)^{n_{xP}}}{\det[S_{xP}-\gamma^{-2}T_{xP}]\cdot\det[\gamma^2 S_{xQ}-T_{xQ}]}\cdot\det\begin{bmatrix}\gamma^2 S_{xP}-T_{xP} & \Gamma\\ \gamma^2\Gamma' & \gamma^2 S_{xQ}-T_{xQ}\end{bmatrix}\\ &= \frac{(-1)^{n_{xP}}\cdot\det[S_{xP}]\cdot\det[S_{xQ}]}{\det[S_{xP}-\gamma^{-2}T_{xP}]\cdot\det[\gamma^2 S_{xQ}-T_{xQ}]}\cdot\det\left[\gamma^2 I-M\right].\end{aligned}\tag{6.3.16}$$

Now it is simple to see that the roots of $\det[N(\gamma)]=0$ are real since all the roots of $\det[\gamma^2 S_{xP}-T_{xP}]=0$, $\det[\gamma^2 S_{xQ}-T_{xQ}]=0$ and $\det[\gamma^2 I-M]=0$ are real. Clearly, $\det[S_{xP}-\gamma^{-2}T_{xP}]\neq 0$ and $\det[\gamma^2 S_{xQ}-T_{xQ}]\neq 0$ for all $\gamma\in(\gamma^*_{PQ},\infty)$. Hence the largest root of $\det[N(\gamma)]=0$ in (γ^*_{PQ},∞) is equal to the largest root of $\det[\gamma^2 I-M]=0$, which is equal to $\{\lambda_{\max}(M)\}^{\frac{1}{2}}$. ⊠

Finally, we are ready to prove our algorithm for computing the infimum γ^* for measurement feedback systems. We have the following theorem.

Theorem 6.3.1. Consider the measurement feedback system (6.3.1), which satisfies Assumptions 6.M.1 to 6.M.7. Then

$$\gamma^* = \sqrt{\lambda_{\max}(M)},\tag{6.3.17}$$

where M as defined in (6.3.5), is indeed its infimum. 🅃

Proof. First, we will show that γ^* is equal to the largest root of $\det[N(\gamma)]=0$ when $f(\gamma)=\gamma^2$ has a unique solution in (γ^*_{PQ},∞). It is simple to observe that $\det[N(\gamma^*)]=0$ since $\lambda_{\max}[N(\gamma^*)]=f(\gamma^*)-(\gamma^*)^2=0$. Now suppose that there exists a γ_1 such that $\det[N(\gamma_1)]=0$ and $\gamma_1>\gamma^*$. This implies that there exists an eigenvalue of $N(\gamma_1)$, say $\lambda_i[N(\gamma_1)]$, such that $\lambda_i[N(\gamma_1)]\neq\lambda_{\max}[N(\gamma_1)]$ and $\lambda_i[N(\gamma_1)]=0$. Thus, we have

$$\lambda_{\max}[N(\gamma_1)] > \lambda_i[N(\gamma_1)] = 0 = \lambda_{\max}[N(\gamma^*)],\tag{6.3.18}$$

contradicting the findings in Proposition 6.3.5 that $\lambda_{\max}[N(\gamma)]$ must be a non-increasing function. Hence, γ^* is the largest root of $\det[N(\gamma)]=0$ and it is equal to $\{\lambda_{\max}(M)\}^{\frac{1}{2}}$ as shown in Proposition 6.3.6.

Now we consider the situation when $f(\gamma)=\gamma^2$ has no solution in the interval (γ^*_{PQ},∞). In this case, clearly we have $\gamma^*=\gamma^*_{PQ}$ and $0\le f(\gamma^*_{PQ})\le(\gamma^*_{PQ})^2$. The last inequality and the definition of $N(\gamma)$ in (6.3.14) imply that

$$-(\gamma^*_{PQ})^2 \le \lambda_i[N(\gamma^*_{PQ})] \le 0.\tag{6.3.19}$$

Thus, the determinant of $N(\gamma^*_{PQ})$ is bounded. Evaluating equation (6.3.16) at $\gamma = \gamma^*_{PQ}$, we have

$$\begin{aligned}\det\left[N(\gamma^*_{PQ})\right]\cdot\det\left[S_{xP}-(\gamma^*_{PQ})^{-2}T_{xP}\right]\cdot\det\left[(\gamma^*_{PQ})^2S_{xQ}-T_{xQ}\right]\\ =(-1)^{n_{xP}}\cdot\det\left[S_{xP}\right]\cdot\det\left[S_{xQ}\right]\cdot\det\left[(\gamma^*_{PQ})^2I-M\right].\end{aligned} \quad (6.3.20)$$

Note that from (6.3.7) and the definition of γ^*_{PQ}, we have

$$\det\left[S_{xP}-(\gamma^*_{PQ})^{-2}T_{xP}\right]\cdot\det\left[(\gamma^*_{PQ})^2S_{xQ}-T_{xQ}\right]=0, \quad (6.3.21)$$

and since $\det\left[N(\gamma^*_{PQ})\right]$ is bounded, it follows from (6.3.20) that

$$\det\left[(\gamma^*_{PQ})^2I-M\right]=0, \quad (6.3.22)$$

or $(\gamma^*_{PQ})^2$ is an eigenvalue of M. Furthermore since $\det\left[N(\gamma)\right]=0$ and similarly $\det\left[\gamma^2I-M\right]=0$ do not have a root in (γ^*_{PQ},∞), hence $\gamma^*_{PQ}=\{\lambda_{\max}(M)\}^{\frac{1}{2}}$. This completes the proof of Theorem 6.3.1. ⊠

The following remarks are in order.

Remark 6.3.2. If Assumptions 6.M.3 and 6.M.6, i.e., the geometric conditions, are not satisfied, then an iterative scheme might be used to determine the infimum. This can be done by finding the smallest scalar, say $\tilde{\gamma}^*$, such that the Riccati equation

$$\tilde{P}_xA_{xP}+A'_{xP}\tilde{P}_x+\tilde{P}_xE_{xP}E'_{xP}\tilde{P}_x/(\tilde{\gamma}^*)^2-\tilde{P}_xB_{xP}B'_{xP}\tilde{P}_x+C'_{xP}C_{xP}=0, \quad (6.3.23)$$

has a positive definite solution $\tilde{P}_x>0$, the Riccati equation

$$\tilde{Q}_xA_{xQ}+A'_{xQ}\tilde{Q}_x+\tilde{Q}_xE_{xQ}E'_{xQ}\tilde{Q}_x/(\tilde{\gamma}^*)^2-\tilde{Q}_xB_{xQ}B'_{xQ}\tilde{Q}_x+C'_{xQ}C_{xQ}=0, \quad (6.3.24)$$

has a positive definite solution $\tilde{Q}_x>0$, and

$$\lambda_{\max}\left\{\tilde{P}_x\Gamma\tilde{Q}_x\Gamma'\right\}<(\tilde{\gamma}^*)^2. \quad (6.3.25)$$

Here Γ is as defined in (6.3.4). Also, all sub-matrices with subscript 'P' are related to the special coordinate basis decomposition of Σ_P and the system (6.3.2), and all sub-matrices with subscript 'Q' are related to the special coordinate basis decomposition of $\Sigma^\star_Q$ and the system (6.3.3). Ⓡ

Remark 6.3.3. If Σ_P is right invertible and Σ_Q is left invertible, then Assumptions 6.M.3 and 6.M.6, i.e., the geometric conditions, are automatically satisfied. Moreover, Assumption 6.M.7, $D_{22}=0$, is no longer necessary and the infimum γ^* for the measurement feedback system (6.3.1) can be obtained as follows:

$$\gamma^* = \left(\lambda_{\max} \left\{ \begin{bmatrix} D_{22,1P}' D_{22,1P} & 0 & 0 & 0 \\ 0 & \tilde{T}_{xP}\tilde{S}_{xP}^{-1} + \Gamma \tilde{S}_{xQ}^{-1} \Gamma' \tilde{S}_{xP}^{-1} & -\Gamma \tilde{S}_{xQ}^{-1} & 0 \\ 0 & -\tilde{T}_{xQ}\tilde{S}_{xQ}^{-1} \Gamma' \tilde{S}_{xP}^{-1} & \tilde{T}_{xQ}\tilde{S}_{xQ}^{-1} & 0 \\ 0 & 0 & 0 & D_{22,1Q}' D_{22,1Q} \end{bmatrix} \right\} \right)^{\frac{1}{2}},$$

where Γ is as defined in (6.3.4), $\tilde{T}_{xP}$ and $\tilde{S}_{xP}$ are the positive semi-definite and positive definite solutions of the following Lyapunov equations,

$$A_{aaP}^{+}\tilde{T}_{xP} + \tilde{T}_{xP}(A_{aaP}^{+})' = (E_{aP}^{+} - B_{0aP}^{+}D_{22,0P} - L_{adP}^{+}\Gamma_{orP}^{-1}D_{22,1P}) \times (E_{aP}^{+} - B_{0aP}^{+}D_{22,0P} - L_{adP}^{+}\Gamma_{orP}^{-1}D_{22,1P})', \quad (6.3.26)$$

$$A_{aaP}^{+}\tilde{S}_{xP} + \tilde{S}_{xP}(A_{aaP}^{+})' = B_{0aP}^{+}(B_{0aP}^{+})' + L_{adP}^{+}\Gamma_{orP}^{-1}(L_{adP}^{+}\Gamma_{orP}^{-1})', \quad (6.3.27)$$

and $\tilde{T}_{xQ}$ and $\tilde{S}_{xQ}$ are the positive semi-definite and positive definite solutions of the following Lyapunov equations,

$$A_{aaQ}^{+}\tilde{T}_{xQ} + \tilde{T}_{xQ}(A_{aaQ}^{+})' = (E_{aQ}^{+} - B_{0aQ}^{+}D_{22,0Q} - L_{adQ}^{+}\Gamma_{orQ}^{-1}D_{22,1Q}) \times (E_{aQ}^{+} - B_{0aQ}^{+}D_{22,0Q} - L_{adQ}^{+}\Gamma_{orQ}^{-1}D_{22,1Q})', \quad (6.3.28)$$

$$A_{aaQ}^{+}\tilde{S}_{xQ} + \tilde{S}_{xQ}(A_{aaQ}^{+})' = B_{0aQ}^{+}(B_{0aQ}^{+})' + L_{adQ}^{+}\Gamma_{orQ}^{-1}(L_{adQ}^{+}\Gamma_{orQ}^{-1})'. \quad (6.3.29)$$

Here again all sub-matrices with subscript 'P' are related to the special coordinate basis decomposition of Σ_P and the system (6.3.2), while all sub-matrices with subscript 'Q' are related to the special coordinate basis decomposition of $\Sigma_Q^\star$ and the system (6.3.3). The detailed proof of the above claim is similar to that of Theorem 6.3.1. It can be found in Chen [15]. ®

We illustrate our results in the following examples.

Example 6.3.1. We consider a measurement feedback system (6.3.1) with A, B, E, C_2, D_2, D_{22} being given as in Example 6.2.1 of Section 6.2 and

$$C_1 = \begin{bmatrix} 0 & -2 & -3 & -2 & -1 \\ 1 & 2 & 3 & 2 & 1 \end{bmatrix}, \quad D_1 = \begin{bmatrix} 1 & 0 \\ 0 & 0 \end{bmatrix}. \quad (6.3.30)$$

Step 6.M.1. It was computed in Example 6.2.1 that $\Gamma_{sP} = I_5$, $n_{xP} = 3$ and

$$S_{xP} = \begin{bmatrix} 0.556281 & 0.185427 & -0.305593 \\ 0.185427 & 0.395142 & 0.231469 \\ -0.305593 & 0.231469 & 1.217984 \end{bmatrix}, \quad T_{xP} = \begin{bmatrix} 13 & 0 & 0 \\ 0 & 0 & 0 \\ 0 & 0 & 0 \end{bmatrix}.$$

Step 6.M.2. The subsystem (A, E, C_1, D_1) is invertible and of nonminimum phase with invariant zeros at $\{-1.630662, -3.593415, 0.521129 \pm j0.363042\}$. Following our algorithm, we obtain

$$\Gamma_{sQ} = \begin{bmatrix} -0.011218 & -0.106028 & -0.906482 & -0.212184 & 0.090909 \\ 0.185213 & -0.745725 & 0.194520 & -0.119195 & 0.181818 \\ -0.919232 & 0.096732 & 0.326906 & -0.603079 & 0.272727 \\ 0.279141 & 0.532936 & 0.087364 & -0.581308 & 0.181818 \\ -0.206551 & -0.373195 & 0.161098 & 0.489027 & 0.090909 \end{bmatrix},$$

$$\Gamma_{orQ} = 1, \quad A_Q = A^+_{aaQ} = \begin{bmatrix} 0.433179 & -0.253237 \\ 0.551005 & 0.609080 \end{bmatrix}, \quad n_{xQ} = 2,$$

$$B_Q B'_Q = \begin{bmatrix} 0.033508 & -0.018630 \\ -0.018630 & 0.030289 \end{bmatrix}, \quad C'_Q C_Q = \begin{bmatrix} 0 & 0 \\ 0 & 0 \end{bmatrix},$$

$$E^+_{aQ} = \begin{bmatrix} -0.769496 & 0.010023 & 0.448951 & -0.769496 \\ -0.090061 & 0.655677 & -1.044466 & -0.090061 \end{bmatrix},$$

and

$$S_{xQ} = \begin{bmatrix} 0.026333 & -0.021114 \\ -0.021114 & 0.043965 \end{bmatrix}, \quad T_{xQ} = \begin{bmatrix} 1.274771 & -0.555799 \\ -0.555799 & 1.764580 \end{bmatrix}.$$

Step 6.M.3. The $n_{xP} \times n_{xQ}$ matrix Γ is then given by

$$\Gamma = \begin{bmatrix} -0.011218 & -0.106028 \\ 0.185213 & -0.745725 \\ -0.919232 & 0.096732 \end{bmatrix},$$

and

$$M = 10^2 \times \begin{bmatrix} 0.500695 & -0.334250 & 0.245016 & 0.082332 & 0.052125 \\ -0.442374 & 0.992368 & -0.260321 & 0.032515 & 0.253182 \\ 0.616882 & -0.513348 & 0.588766 & 0.501907 & 0.261525 \\ 1.074941 & -1.295698 & 0.921909 & 0.622391 & 0.172484 \\ -0.583103 & 1.526365 & -0.286520 & 0.180099 & 0.487850 \end{bmatrix}.$$

Step 6.M.4. Finally, the infimum for the measurement feedback system is given by

$$\gamma^* = 13.638725.$$

▣

Example 6.3.2. We consider a measurement feedback system (6.3.1) with A, B, E, C_2, D_2, D_{22} being given as in Example 6.2.2 of Section 6.2 and

$$C_1 = [\,1 \quad -2 \quad 3 \quad -4\,], \quad D_1 = 0. \tag{6.3.31}$$

It is again simple to verify that the subsystem (A, E, C_1, D_1), i.e., Σ_Q, is observable and invertible with two unstable invariant zeros at $0.5 \pm j0.5916$ and one infinite zero of order two. Hence, all assumptions are satisfied. Following Remark 6.3.3, we obtain

$$n_{x\mathrm{P}} = 1, \quad \tilde{S}_{x\mathrm{P}} = 0.5, \quad \tilde{T}_{x\mathrm{P}} = 0.25,$$

$$\Gamma_{or\mathrm{Q}} = 1, \quad n_{x\mathrm{Q}} = 2,$$

$$E^+_{a\mathrm{Q}} = \begin{bmatrix} -1.2230247 & -0.5241535 \\ 1.1679942 & 0.9408842 \end{bmatrix}, \quad L^+_{ad\mathrm{Q}} = \begin{bmatrix} -0.6289841 \\ 1.3756377 \end{bmatrix},$$

$$A^+_{aa\mathrm{Q}} = \begin{bmatrix} 0.8842105 & -0.5101735 \\ 0.9753892 & 0.1157895 \end{bmatrix},$$

$$B^+_{0a\mathrm{Q}} = \emptyset, \quad D_{22,0\mathrm{Q}} = \emptyset, \quad D_{22,1\mathrm{Q}} = [\,2 \quad 1\,],$$

$$\tilde{S}_{x\mathrm{Q}} = \begin{bmatrix} 0.5274947 & 0.5264991 \\ 0.5264991 & 3.7365053 \end{bmatrix}, \quad \tilde{T}_{x\mathrm{Q}} = \begin{bmatrix} 0.5810175 & 0.9950273 \\ 0.9950273 & 3.2589825 \end{bmatrix},$$

$$\Gamma = [\,-1.2230247 \quad 1.1679942\,],$$

$$M = \begin{bmatrix} 1 & 0 & 0 & 0 & 0 \\ 0 & 9.7252904 & 3.0610640 & -0.7439148 & 0 \\ 0 & 2.0766328 & 0.9724337 & 0.1292764 & 0 \\ 0 & 1.2428740 & 1.1820112 & 0.7056473 & 0 \\ 0 & 0 & 0 & 0 & 5 \end{bmatrix},$$

and finally the infimum for the given system,

$$\gamma^* = 3.2088448.$$

Ⓔ

6.4. Plants with Imaginary Axis Zeros

We present in this section a non-iterative algorithm for computing γ^* of the measurement feedback system (6.3.1) whose subsystems Σ_P and/or Σ_Q have invariant zeros on the imaginary axis. The procedure is similar to the algorithm of the previous section, although it is slightly more complicated. It involves finding eigenspaces for the imaginary axis invariant zeros of Σ_P and Σ_Q and finding solutions to two extra Sylvester equations. We consider the system (6.3.1) which satisfies the following assumptions:

Assumption 6.Z.1: (A, B) is stabilizable;

Assumption 6.Z.2: $\mathrm{Im}\,(E) \subset \mathcal{V}^-(\Sigma_\mathrm{P}) + \mathcal{S}^-(\Sigma_\mathrm{P})$;

Assumption 6.Z.3: (A, C_1) is detectable;

Assumption 6.Z.4: $\mathrm{Ker}\,(C_2) \supset \mathcal{V}^-(\Sigma_\mathrm{Q}) \cap \mathcal{S}^-(\Sigma_\mathrm{Q})$; and

Assumption 6.Z.5: $D_{22} = 0$.

Ⓐ

We have the following step-by-step algorithm for computing γ^*. We note that it has some overlaps with that in the previous section. However, this is merely for completeness and to properly define matrices required in the computation of the infimum γ^*.

Step 6.Z.1. Transform the subsystem system $\Sigma_{\rm P}$, i.e., (A, B, C_2, D_2) into the special coordinate basis described in Theorem 2.4.1. To all sub-matrices and transformations in the special coordinate basis of $\Sigma_{\rm P}$, we append the subscript '$_{\rm P}$' to signify their relation to the system $\Sigma_{\rm P}$. We also introduce an additional permutation matrix to the original state transformation such that the transformed state variables are arranged as

$$\tilde{x}_{\rm P} = \begin{pmatrix} x_{a{\rm P}}^+ \\ x_{b{\rm P}} \\ x_{a{\rm P}}^0 \\ x_{a{\rm P}}^- \\ x_{c{\rm P}} \\ x_{d{\rm P}} \end{pmatrix}. \tag{6.4.1}$$

Next, we compute

$$\Gamma_{s{\rm P}}^{-1} E = \begin{bmatrix} E_{a{\rm P}}^+ \\ E_{b{\rm P}} \\ E_{a{\rm P}}^0 \\ E_{a{\rm P}}^- \\ E_{c{\rm P}} \\ E_{d{\rm P}} \end{bmatrix}. \tag{6.4.2}$$

Note that Assumption 6.Z.2 implies $E_{b{\rm P}} = 0$. Then define the following matrices:

$$A_{\rm P} := \begin{bmatrix} A_{aa{\rm P}}^+ & L_{ab{\rm P}}^+ C_{b{\rm P}} & 0 \\ 0 & A_{bb{\rm P}} & 0 \\ 0 & L_{ab{\rm P}}^0 C_{b{\rm P}} & A_{aa{\rm P}}^0 \end{bmatrix}, \quad B_{\rm P} := \begin{bmatrix} B_{0a{\rm P}}^+ & L_{ad{\rm P}}^+ \\ B_{0b{\rm P}} & L_{bd{\rm P}} \\ B_{0a{\rm P}}^0 & L_{ad{\rm P}}^0 \end{bmatrix}, \tag{6.4.3}$$

$$E_{\rm P} := \begin{bmatrix} E_{a{\rm P}}^+ \\ E_{b{\rm P}} \\ E_{a{\rm P}}^0 \end{bmatrix}, \tag{6.4.4}$$

and

$$C_{\rm P} := \Gamma_{o{\rm P}} \begin{bmatrix} 0 & 0 & 0 \\ 0 & 0 & 0 \\ 0 & C_{b{\rm P}} & 0 \end{bmatrix}, \quad D_{\rm P} := \Gamma_{o{\rm P}} \begin{bmatrix} I_{m_{0{\rm P}}} & 0 \\ 0 & C_{d{\rm P}} C_{d{\rm P}}' \\ 0 & 0 \end{bmatrix}. \tag{6.4.5}$$

By some simple algebra, it is straightforward to show that

$$C'_{\mathrm{P}}[I - D_{\mathrm{P}}(D'_{\mathrm{P}}D_{\mathrm{P}})^{-1}D'_{\mathrm{P}}]C_{\mathrm{P}} = \begin{bmatrix} 0 & 0 & 0 \\ 0 & \tilde{C}'_{b\mathrm{P}}\tilde{C}_{b\mathrm{P}} & 0 \\ 0 & 0 & 0 \end{bmatrix}, \tag{6.4.6}$$

for some full row rank $\tilde{C}_{b\mathrm{P}}$,

$$A_{\mathrm{P}} - B_{\mathrm{P}}(D'_{\mathrm{P}}D_{\mathrm{P}})^{-1}D'_{\mathrm{P}}C_{\mathrm{P}} = \begin{bmatrix} A^{+}_{aa\mathrm{P}} & \tilde{L}^{+}_{ab\mathrm{P}}\tilde{C}_{b\mathrm{P}} & 0 \\ 0 & \tilde{A}_{bb\mathrm{P}} & 0 \\ 0 & \tilde{L}^{0}_{ab\mathrm{P}}\tilde{C}_{b\mathrm{P}} & A^{0}_{aa\mathrm{P}} \end{bmatrix}, \tag{6.4.7}$$

and

$$B_{\mathrm{P}}(D'_{\mathrm{P}}D_{\mathrm{P}})^{-1}B'_{\mathrm{P}} = \begin{bmatrix} B^{+}_{0a\mathrm{P}} & \tilde{L}^{+}_{ad\mathrm{P}} \\ B_{0b\mathrm{P}} & \tilde{L}_{bd\mathrm{P}} \\ B^{0}_{0a\mathrm{P}} & \tilde{L}^{0}_{ad\mathrm{P}} \end{bmatrix} \cdot \begin{bmatrix} B^{+}_{0a\mathrm{P}} & \tilde{L}^{+}_{ad\mathrm{P}} \\ B_{0b\mathrm{P}} & \tilde{L}_{bd\mathrm{P}} \\ B^{0}_{0a\mathrm{P}} & \tilde{L}^{0}_{ad\mathrm{P}} \end{bmatrix}' \tag{6.4.8}$$

for some appropriate $\tilde{L}_{ab\mathrm{P}}$, $\tilde{L}^{0}_{ab\mathrm{P}}$, $\tilde{L}^{+}_{ad\mathrm{P}}$, $\tilde{L}_{bd\mathrm{P}}$ and $\tilde{L}^{0}_{ad\mathrm{P}}$. Here we note that it can easily be verified that the pair $(\tilde{A}_{bb\mathrm{P}}, \tilde{C}_{b\mathrm{P}})$ is observable provided that $(A_{bb\mathrm{P}}, C_{b\mathrm{P}})$ is observable.

Step 6.Z.2. Define

$$A_{x\mathrm{P}} := \begin{bmatrix} A^{+}_{aa\mathrm{P}} & \tilde{L}^{+}_{ab\mathrm{P}}\tilde{C}_{b\mathrm{P}} \\ 0 & \tilde{A}_{bb\mathrm{P}} \end{bmatrix}, \quad B_{x\mathrm{P}} := \begin{bmatrix} B^{+}_{0a\mathrm{P}} & \tilde{L}^{+}_{ad\mathrm{P}} \\ B_{0b\mathrm{P}} & \tilde{L}_{bd\mathrm{P}} \end{bmatrix}, \tag{6.4.9}$$

and

$$C_{x\mathrm{P}} := [\,0 \quad \tilde{C}_{b\mathrm{P}}\,], \quad E_{x\mathrm{P}} := \begin{bmatrix} E^{+}_{a\mathrm{P}} \\ E_{b\mathrm{P}} \end{bmatrix}. \tag{6.4.10}$$

Then we solve for the unique positive definite solution $S_{x\mathrm{P}}$ of the Riccati equation,

$$A_{x\mathrm{P}}S_{x\mathrm{P}} + S_{x\mathrm{P}}A'_{x\mathrm{P}} - B_{x\mathrm{P}}B'_{x\mathrm{P}} + S_{x\mathrm{P}}C'_{x\mathrm{P}}C_{x\mathrm{P}}S_{x\mathrm{P}} = 0, \tag{6.4.11}$$

together with the matrix $T_{x\mathrm{P}}$ defined by

$$T_{x\mathrm{P}} := \begin{bmatrix} T_{ax\mathrm{P}} & 0 \\ 0 & 0 \end{bmatrix},$$

where $T_{ax\mathrm{P}}$ is the unique solution to the Lyapunov equation,

$$A^{+}_{aa\mathrm{P}}T_{ax\mathrm{P}} + T_{ax\mathrm{P}}(A^{+}_{aa\mathrm{P}})' = E^{+}_{a\mathrm{P}}(E^{+}_{a\mathrm{P}})'. \tag{6.4.12}$$

Next, solve the unique solution $Y_{x\mathrm{P}}$ of the following Sylvester equation,

$$(A_{x\mathrm{P}} + S_{x\mathrm{P}}C'_{x\mathrm{P}}C_{x\mathrm{P}})Y_{x\mathrm{P}} + Y_{x\mathrm{P}}(A^{0}_{aa\mathrm{P}})' + S_{x\mathrm{P}}C'_{x\mathrm{P}}(\tilde{L}^{0}_{ab\mathrm{P}})' - B_{x\mathrm{P}}\,[\,B^{0}_{0a\mathrm{P}} \quad \tilde{L}^{0}_{ad\mathrm{P}}\,]' = 0. \tag{6.4.13}$$

Let us denote the set of eigenvalues of $A^0_{aa\mathrm{P}}$ with a nonnegative imaginary part as $\{j\omega_{\mathrm{P}1}, \cdots, j\omega_{\mathrm{P}k_\mathrm{P}}\}$ and for $i = 1, \cdots, k_\mathrm{P}$, choose complex matrices $V_{i\mathrm{P}}$, whose columns form a basis for the eigenspace,

$$\left\{x \in \mathbb{C}^{n^0_{a\mathrm{P}}} \mid x^{\mathrm{H}}(j\omega_{\mathrm{P}i} I - A^0_{aa\mathrm{P}}) = 0\right\}, \tag{6.4.14}$$

where $n^0_{a\mathrm{P}}$ is the dimension of $A^0_{aa\mathrm{P}}$. Then define

$$F_{i\mathrm{P}} := V^{\mathrm{H}}_{i\mathrm{P}}\Big(\big[B^0_{0a\mathrm{P}} \quad \tilde{L}^0_{ad\mathrm{P}}\big]\big[B^0_{0a\mathrm{P}} \quad \tilde{L}^0_{ad\mathrm{P}}\big]' + \tilde{L}^0_{ab\mathrm{P}}(\tilde{L}^0_{ab\mathrm{P}})' - \Big[(\tilde{L}^0_{ab\mathrm{P}})' + C_\mathrm{P} Y_\mathrm{P}\Big]'\Big[(\tilde{L}^0_{ab\mathrm{P}})' + C_\mathrm{P} Y_\mathrm{P}\Big]\Big)V_{i\mathrm{P}}, \tag{6.4.15}$$

for $i = 1, \cdots, k_\mathrm{P}$, and

$$F_\mathrm{P} := \mathrm{blkdiag}\Big\{ F_{1\mathrm{P}}, \cdots, F_{k_\mathrm{P}\mathrm{P}} \Big\}. \tag{6.4.16}$$

It is shown in [118] that $F_\mathrm{P} > 0$. Also, define

$$G_\mathrm{P} := \mathrm{blkdiag}\Big\{ V^{\mathrm{H}}_{1\mathrm{P}} E^0_{a\mathrm{P}}(E^0_{a\mathrm{P}})' V_{1\mathrm{P}}, \cdots, V^{\mathrm{H}}_{k_\mathrm{P}\mathrm{P}} E^0_{a\mathrm{P}}(E^0_{a\mathrm{P}})' V_{k_\mathrm{P}\mathrm{P}} \Big\}. \tag{6.4.17}$$

Step 6.Z.3. Transform the subsystem $\Sigma^\star_\mathrm{Q}$, i.e., (A', C'_1, E', D'_1), into the special coordinate basis described in Theorem 2.4.1. Again we add here the subscript 'Q' to all sub-matrices and transformations in the special coordinate basis of the system $\Sigma^\star_\mathrm{Q}$ and rearrange the transformed state variables as

$$\tilde{x}_\mathrm{Q} = \begin{pmatrix} x^+_{a\mathrm{Q}} \\ x_{b\mathrm{Q}} \\ x^0_{a\mathrm{Q}} \\ x^-_{a\mathrm{Q}} \\ x_{c\mathrm{Q}} \\ x_{d\mathrm{Q}} \end{pmatrix}. \tag{6.4.18}$$

Next, we compute

$$\Gamma^{-1}_{s\mathrm{Q}} C'_2 = \begin{bmatrix} E^+_{a\mathrm{Q}} \\ E_{b\mathrm{Q}} \\ E^0_{a\mathrm{Q}} \\ E^-_{a\mathrm{Q}} \\ E_{c\mathrm{Q}} \\ E_{d\mathrm{Q}} \end{bmatrix}. \tag{6.4.19}$$

Note that Assumption 6.Z.4 implies $E_{b\mathrm{Q}} = 0$. Then define the following matrices:

$$A_\mathrm{Q} := \begin{bmatrix} A^+_{aa\mathrm{Q}} & L^+_{ab\mathrm{Q}} C_{b\mathrm{Q}} & 0 \\ 0 & A_{bb\mathrm{Q}} & 0 \\ 0 & L^0_{ab\mathrm{Q}} C_{b\mathrm{Q}} & A^0_{aa\mathrm{Q}} \end{bmatrix}, \quad B_\mathrm{Q} := \begin{bmatrix} B^+_{0a\mathrm{Q}} & L^+_{ad\mathrm{Q}} \\ B_{0b\mathrm{Q}} & L_{bd\mathrm{Q}} \\ B^0_{0a\mathrm{Q}} & L^0_{ad\mathrm{Q}} \end{bmatrix}, \tag{6.4.20}$$

$$E_Q := \begin{bmatrix} E_{aQ}^+ \\ E_{bQ} \\ E_{aQ}^0 \end{bmatrix}, \tag{6.4.21}$$

and

$$C_Q := \Gamma_{oQ} \begin{bmatrix} 0 & 0 & 0 \\ 0 & 0 & 0 \\ 0 & C_{bQ} & 0 \end{bmatrix}, \quad D_Q := \Gamma_{oQ} \begin{bmatrix} I_{moQ} & 0 \\ 0 & C_{dQ}C_{dQ}' \\ 0 & 0 \end{bmatrix}. \tag{6.4.22}$$

By some simple algebra, it is straightforward to show that

$$C_Q'[I - D_Q(D_Q'D_Q)^{-1}D_Q']C_Q = \begin{bmatrix} 0 & 0 & 0 \\ 0 & \tilde{C}_{bQ}'\tilde{C}_{bQ} & 0 \\ 0 & 0 & 0 \end{bmatrix}, \tag{6.4.23}$$

for some full row rank $\tilde{C}_{bQ}$, and

$$A_Q - B_Q(D_Q'D_Q)^{-1}D_Q'C_Q = \begin{bmatrix} A_{aaQ}^+ & \tilde{L}_{abQ}^+\tilde{C}_{bQ} & 0 \\ 0 & \tilde{A}_{bbQ} & 0 \\ 0 & \tilde{L}_{abQ}^0\tilde{C}_{bQ} & A_{aaQ}^0 \end{bmatrix}, \tag{6.4.24}$$

and

$$B_Q(D_Q'D_Q)^{-1}B_Q' = \begin{bmatrix} B_{0aQ}^+ & \tilde{L}_{adQ}^+ \\ B_{0bQ} & \tilde{L}_{bdQ} \\ B_{0aQ}^0 & \tilde{L}_{adQ}^0 \end{bmatrix} \cdot \begin{bmatrix} B_{0aQ}^+ & \tilde{L}_{adQ}^+ \\ B_{0bQ} & \tilde{L}_{bdQ} \\ B_{0aQ}^0 & \tilde{L}_{adQ}^0 \end{bmatrix}'. \tag{6.4.25}$$

for some appropriate $\tilde{L}_{abQ}$, $\tilde{L}_{abQ}^0$, $\tilde{L}_{adQ}^+$, $\tilde{L}_{bdQ}$ and $\tilde{L}_{adQ}^0$. Here we note that it can easily be verified that the pair $(\tilde{A}_{bbQ}, \tilde{C}_{bQ})$ is observable provided that (A_{bbQ}, C_{bQ}) is observable.

Step 6.Z.4. Define

$$A_{xQ} := \begin{bmatrix} A_{aaQ}^+ & \tilde{L}_{abQ}^+\tilde{C}_{bQ} \\ 0 & \tilde{A}_{bbQ} \end{bmatrix}, \quad B_{xQ} := \begin{bmatrix} B_{0aQ}^+ & \tilde{L}_{adQ}^+ \\ B_{0bQ} & \tilde{L}_{bdQ} \end{bmatrix}, \tag{6.4.26}$$

and

$$C_{xQ} := [0 \quad \tilde{C}_{bQ}], \quad E_{xQ} = \begin{bmatrix} E_{aQ}^+ \\ E_{bQ} \end{bmatrix}. \tag{6.4.27}$$

Then we solve for the unique positive definite solution S_{xQ} of the Riccati equation,

$$A_{xQ}S_{xQ} + S_{xQ}A_{xQ}' - B_{xQ}B_{xQ}' + S_{xQ}C_{xQ}'C_{xQ}S_{xQ} = 0, \tag{6.4.28}$$

together with the matrix $T_{x\mathrm{Q}}$ defined by

$$T_{x\mathrm{Q}} := \begin{bmatrix} T_{ax\mathrm{Q}} & 0 \\ 0 & 0 \end{bmatrix},$$

where $T_{ax\mathrm{Q}}$ is the unique solution to the Lyapunov equation,

$$A^+_{aa\mathrm{Q}} T_{ax\mathrm{Q}} + T_{ax\mathrm{Q}} (A^+_{aa\mathrm{Q}})' = E^+_{a\mathrm{Q}} (E^+_{a\mathrm{Q}})'. \tag{6.4.29}$$

Next, solve the unique solution $Y_{x\mathrm{Q}}$ of the following Sylvester equation,

$$(A_{x\mathrm{Q}} + S_{x\mathrm{Q}} C'_{x\mathrm{Q}} C_{x\mathrm{Q}}) Y_{x\mathrm{Q}} + Y_{x\mathrm{Q}} (A^0_{aa\mathrm{Q}})' + S_{x\mathrm{Q}} C'_{x\mathrm{Q}} (\tilde{L}^0_{ab\mathrm{Q}})' - B_{x\mathrm{Q}} \begin{bmatrix} B^0_{0a\mathrm{Q}} & \tilde{L}^0_{ad\mathrm{Q}} \end{bmatrix}' = 0. \tag{6.4.30}$$

Let us denote the set of eigenvalues of $A^0_{aa\mathrm{Q}}$ with a nonnegative imaginary part as $\{j\omega_{\mathrm{Q}1}, \cdots, j\omega_{\mathrm{Q}k_\mathrm{Q}}\}$ and for $i = 1, \cdots, k_\mathrm{Q}$, choose complex matrices $V_{i\mathrm{Q}}$, whose columns form a basis for the eigenspace,

$$\left\{ x \in \mathbb{C}^{n^0_{a\mathrm{Q}}} \mid x^{\mathrm{H}} (j\omega_{\mathrm{Q}i} I - A^0_{aa\mathrm{Q}}) = 0 \right\}, \tag{6.4.31}$$

where $n^0_{a\mathrm{Q}}$ is the dimension of $A^0_{aa\mathrm{Q}}$. Then define

$$F_{i\mathrm{Q}} := V^{\mathrm{H}}_{i\mathrm{Q}} \Big(\begin{bmatrix} B^0_{0a\mathrm{Q}} & \tilde{L}^0_{ad\mathrm{Q}} \end{bmatrix} \begin{bmatrix} B^0_{0a\mathrm{Q}} & \tilde{L}^0_{ad\mathrm{Q}} \end{bmatrix}' + \tilde{L}^0_{ab\mathrm{Q}} (\tilde{L}^0_{ab\mathrm{Q}})' - \Big[(\tilde{L}^0_{ab\mathrm{Q}})' + C_\mathrm{Q} Y_\mathrm{Q} \Big]' \Big[(\tilde{L}^0_{ab\mathrm{Q}})' + C_\mathrm{Q} Y_\mathrm{Q} \Big] \Big) V_{i\mathrm{Q}}, \tag{6.4.32}$$

for $i = 1, \cdots, k_\mathrm{Q}$, and

$$F_\mathrm{Q} := \mathrm{blkdiag} \Big\{ F_{1\mathrm{Q}}, \cdots, F_{k_\mathrm{Q}\mathrm{Q}} \Big\}. \tag{6.4.33}$$

Again, it can be shown that $F_\mathrm{Q} > 0$. Also, define

$$G_\mathrm{Q} := \mathrm{blkdiag} \Big\{ V^{\mathrm{H}}_{1\mathrm{Q}} E^0_{a\mathrm{Q}} (E^0_{a\mathrm{Q}})' V_{1\mathrm{Q}}, \cdots, V^{\mathrm{H}}_{k_\mathrm{Q}\mathrm{Q}} E^0_{a\mathrm{Q}} (E^0_{a\mathrm{Q}})' V_{k_\mathrm{Q}\mathrm{Q}} \Big\}. \tag{6.4.34}$$

Step 6.Z.5. Define

$$n_{x\mathrm{P}} := \dim \{ \mathbb{R}^n / \mathcal{S}^+(\Sigma_\mathrm{P}) \} - n^0_{a\mathrm{P}}, \tag{6.4.35}$$

and

$$n_{x\mathrm{Q}} := \dim \{ \mathbb{R}^n / \mathcal{S}^+(\Sigma^\star_\mathrm{Q}) \} - n^0_{a\mathrm{Q}}, \tag{6.4.36}$$

and partition

$$\Gamma^{-1}_{s\mathrm{P}} (\Gamma^{-1}_{s\mathrm{Q}})' = \begin{bmatrix} \Gamma & \star \\ \star & \star \end{bmatrix}, \tag{6.4.37}$$

where Γ is of dimension $n_{x\mathrm{P}} \times n_{x\mathrm{Q}}$. Finally, define a constant matrix

$$M := \begin{bmatrix} G_\mathrm{P} F_\mathrm{P}^{-1} & 0 & 0 & 0 \\ 0 & T_{x\mathrm{P}} S^{-1}_{x\mathrm{P}} + \Gamma S^{-1}_{x\mathrm{Q}} \Gamma' S^{-1}_{x\mathrm{P}} & -\Gamma S^{-1}_{x\mathrm{Q}} & 0 \\ 0 & -T_{x\mathrm{Q}} S^{-1}_{x\mathrm{Q}} \Gamma' S^{-1}_{x\mathrm{P}} & T_{x\mathrm{Q}} S^{-1}_{x\mathrm{Q}} & 0 \\ 0 & 0 & 0 & G_\mathrm{Q} F_\mathrm{Q}^{-1} \end{bmatrix}. \tag{6.4.38}$$

Step 6.Z.6. The infimum γ^* is then given by

$$\gamma^* = \sqrt{\lambda_{\max}(M)}. \tag{6.4.39}$$

This will be justified in Theorem 6.4.1 below. Ⓐ

We have the following main theorem.

Theorem 6.4.1. Consider the given measurement feedback system (6.3.1). Then under Assumptions 6.Z.1 to 6.Z.5, its infimum is given by (6.4.39). Ⓣ

Proof. Following the results of Scherer [119], it can be show that

$$\gamma > \gamma_{\mathrm{P}}^* := \max\left\{ \sqrt{\lambda_{\max}(T_{x\mathrm{P}} S_{x\mathrm{P}}^{-1})},\ \sqrt{\lambda_{\max}(G_{\mathrm{P}} F_{\mathrm{P}}^{-1})} \right\}, \tag{6.4.40}$$

if and only if the following algebraic Riccati inequality,

$$\begin{aligned}&[A_{\mathrm{P}} - B_{\mathrm{P}}(D_{\mathrm{P}}'D_{\mathrm{P}})^{-1}D_{\mathrm{P}}C_{\mathrm{P}}]X + X[A_{\mathrm{P}} - B_{\mathrm{P}}(D_{\mathrm{P}}'D_{\mathrm{P}})^{-1}D_{\mathrm{P}}C_{\mathrm{P}}]' \\ &\quad + \gamma^{-2}E_{\mathrm{P}}E_{\mathrm{P}}' + XC_{\mathrm{P}}\left[I - D_{\mathrm{P}}(D_{\mathrm{P}}'D_{\mathrm{P}})^{-1}D_{\mathrm{P}}'\right]C_{\mathrm{P}}X - B_{\mathrm{P}}(D_{\mathrm{P}}'D_{\mathrm{P}})^{-1}B_{\mathrm{P}}' < 0,\end{aligned}$$

has a positive definite solution. Then it follows from the results of [118] and [119] (see also Theorem 4.2.2) and some simple algebraic manipulations that for $\gamma > \gamma_{\mathrm{P}}^*$, the positive semi-definite matrix $P(\gamma)$ given by

$$P(\gamma) = (\Gamma_{s\mathrm{P}}^{-1})' \begin{bmatrix} (S_{x\mathrm{P}} - \gamma^{-2}T_{x\mathrm{P}})^{-1} & 0 \\ 0 & 0 \end{bmatrix} \Gamma_{s\mathrm{P}}^{-1}, \tag{6.4.41}$$

is the lower limit point of the set

$$\Big\{P > 0 \,|\, \exists F : (A + BF)'P + P(A + BF) + \gamma^{-2}PEE'P + (C_2 + D_2F)'(C_2 + D_2F) < 0\Big\}.$$

Moreover, such a $P(\gamma)$ does not exist when $\gamma < \gamma_{\mathrm{P}}^*$. By dual reasoning, one can show that

$$\gamma > \gamma_{\mathrm{Q}}^* := \max\left\{ \sqrt{\lambda_{\max}(T_{x\mathrm{Q}} S_{x\mathrm{Q}}^{-1})},\ \sqrt{\lambda_{\max}(G_{\mathrm{Q}} F_{\mathrm{Q}}^{-1})} \right\}, \tag{6.4.42}$$

if and only if the following algebraic Riccati inequality,

$$\begin{aligned}&[A_{\mathrm{Q}} - B_{\mathrm{Q}}(D_{\mathrm{Q}}'D_{\mathrm{Q}})^{-1}D_{\mathrm{Q}}C_{\mathrm{Q}}]Z + Z[A_{\mathrm{Q}} - B_{\mathrm{Q}}(D_{\mathrm{Q}}'D_{\mathrm{Q}})^{-1}D_{\mathrm{Q}}C_{\mathrm{Q}}]' \\ &\quad + \gamma^{-2}E_{\mathrm{Q}}E_{\mathrm{Q}}' + ZC_{\mathrm{Q}}\left[I - D_{\mathrm{Q}}(D_{\mathrm{Q}}'D_{\mathrm{Q}})^{-1}D_{\mathrm{Q}}'\right]C_{\mathrm{Q}}Z - B_{\mathrm{Q}}(D_{\mathrm{Q}}'D_{\mathrm{Q}})^{-1}B_{\mathrm{Q}}' < 0,\end{aligned}$$

has a positive definite solution. For $\gamma > \gamma_{\mathrm{Q}}^*$, the positive semi-definite matrix $Q(\gamma)$ given by

$$Q(\gamma) = (\Gamma_{s\mathrm{Q}}^{-1})' \begin{bmatrix} (S_{x\mathrm{Q}} - \gamma^{-2} T_{x\mathrm{Q}})^{-1} & 0 \\ 0 & 0 \end{bmatrix} \Gamma_{s\mathrm{Q}}^{-1}, \tag{6.4.43}$$

is the lower limit point of the set

$$\Big\{ Q > 0 \,|\, \exists K : (A + KC_1)Q + Q(A + KC_1)' + \gamma^{-2} Q C_2' C_2 Q \\ + (E + KD_1)(E + KD_1)' < 0 \Big\}.$$

Again, such a $Q(\gamma)$ does not exist when $\gamma < \gamma_{\mathrm{Q}}^*$. Now, let us define

$$\gamma_{\mathrm{PQ}}^* := \max \left\{ \sqrt{\lambda_{\max}(T_{x\mathrm{P}} S_{x\mathrm{P}}^{-1})}, \sqrt{\lambda_{\max}(T_{x\mathrm{Q}} S_{x\mathrm{Q}}^{-1})} \right\}, \tag{6.4.44}$$

and

$$\gamma_{coup}^* := \sup \Big\{ \gamma \in (\gamma_{\mathrm{PQ}}^*, \infty) \mid \rho[P(\gamma) Q(\gamma)] < \gamma^2 \Big\}, \tag{6.4.45}$$

where $P(\gamma)$ and $Q(\gamma)$ are as given in (6.4.41) and (6.4.43), respectively. Then following the results of Scherer [119], it can easily be shown that

$$\gamma^* = \max \left\{ \gamma_{coup}^*, \sqrt{\lambda_{\max}(G_{\mathrm{P}} F_{\mathrm{P}}^{-1})}, \sqrt{\lambda_{\max}(G_{\mathrm{Q}} F_{\mathrm{Q}}^{-1})} \right\}. \tag{6.4.46}$$

Also, it follows from Theorem 6.3.1 that

$$\gamma_{coup}^* = \left\{ \lambda_{\max} \begin{bmatrix} T_{x\mathrm{P}} S_{x\mathrm{P}}^{-1} + \Gamma S_{x\mathrm{Q}}^{-1} \Gamma' S_{x\mathrm{P}}^{-1} & -\Gamma S_{x\mathrm{Q}}^{-1} \\ -T_{x\mathrm{Q}} S_{x\mathrm{Q}}^{-1} \Gamma' S_{x\mathrm{P}}^{-1} & T_{x\mathrm{Q}} S_{x\mathrm{Q}}^{-1} \end{bmatrix} \right\}^{\frac{1}{2}}. \tag{6.4.47}$$

Hence, the result of Theorem 6.4.1 follows. ⊠

We illustrate our main result of this section in the following example.

Example 6.4.1. Consider a given system characterized by

$$A = \begin{bmatrix} 0 & 1 & 1 & -1 & 1 \\ 0 & 0 & 0 & 0 & 1 \\ 0 & 1 & 0 & 0 & 1 \\ 1 & 1 & 1 & 0 & 1 \\ 1 & 1 & 1 & 1 & 0 \end{bmatrix}, \quad B = \begin{bmatrix} 0 & 0 \\ 0 & 0 \\ 1 & 0 \\ 0 & 0 \\ 0 & 1 \end{bmatrix}, \quad E = \begin{bmatrix} 1 & 1 \\ 0 & 0 \\ 0 & 0 \\ 2 & 1 \\ 1 & 2 \end{bmatrix}, \tag{6.4.48}$$

$$C_1 = \begin{bmatrix} -1 & 11 & -21.876238 & -4.2239 & -2.425699 \\ 1 & 2 & 3 & 2 & 1 \end{bmatrix}, \quad D_1 = \begin{bmatrix} 1 & 0 \\ 0 & 0 \end{bmatrix}, \tag{6.4.49}$$

and

$$C_2 = \begin{bmatrix} 0 & 0 & 1 & 0 & 0 \\ 0 & 0 & 0 & 0 & 1 \\ 0 & 1 & 0 & 0 & 0 \\ 0 & 0 & 1 & 0 & 0 \end{bmatrix}, \quad D_2 = \begin{bmatrix} 1 & 0 \\ 0 & 0 \\ 0 & 0 \\ 0 & 0 \end{bmatrix}, \quad D_{22} = 0. \tag{6.4.50}$$

First, it is simple to verify that the subsystem Σ_P is left invertible with two invariant zeros at $\pm j$ and Assumption 6.Z.2 is satisfied. Applying the special coordinate basis transformation to Σ_P, we have

$$\Gamma_{sP} = \begin{bmatrix} 0 & 0 & 0 & -1 & 0 \\ 1.3660254 & 0.3660254 & 0 & 0 & 0 \\ 0.1988066 & 1.9900945 & 0 & 0 & 0 \\ 0 & 0 & 1 & 0 & 0 \\ 0 & 0 & 0 & 0 & 1 \end{bmatrix},$$

$$A_{xP} = \begin{bmatrix} -0.1614784 & 0.2246812 \\ 0.6026457 & -0.8385216 \end{bmatrix}, \quad B_{xP} = \begin{bmatrix} 0.6040578 & -0.1762197 \\ 0.4723969 & 0.4878984 \end{bmatrix},$$

$$C_{xP} = \begin{bmatrix} 1.3544397 & 0.2665382 \\ 0.2665382 & 2.0058434 \end{bmatrix}, \quad E_{bP} = \begin{bmatrix} 0 & 0 \\ 0 & 0 \end{bmatrix},$$

$$A^0_{aaP} = \begin{bmatrix} 0 & -1 \\ 1 & 0 \end{bmatrix}, \quad \tilde{L}^0_{abP} = \begin{bmatrix} 0.9489977 & 1.0485243 \\ -0.9489977 & -1.0485243 \end{bmatrix},$$

and

$$\begin{bmatrix} B^0_{0aP} & \tilde{L}^0_{adP} \end{bmatrix} = \begin{bmatrix} 0 & 1 \\ 0 & -1 \end{bmatrix}, \quad E^0_{aP} = \begin{bmatrix} 2 & 1 \\ -1 & -1 \end{bmatrix}..$$

Following Step 6.Z.2, we obtain

$$S_{xP} = \begin{bmatrix} 0.6180716 & -0.2516670 \\ -0.2516670 & 0.7339429 \end{bmatrix}, \quad T_{xP} = \begin{bmatrix} 0 & 0 \\ 0 & 0 \end{bmatrix},$$

$$Y_{xP} = \begin{bmatrix} -0.6928337 & -0.0822109 \\ -0.3161228 & 0.3068152 \end{bmatrix},$$

and

$$F_P = 2.3885733, \qquad G_P = 3.5.$$

Next, the subsystem Σ_Q is invertible and of nonminimum phase with invariant zeros at $\{0.078944, \pm j2.302011, -4.095803\}$. Hence, Assumption 6.Z.4 is automatically satisfied. Applying the special coordinate basis transformation to $\Sigma_Q^\star$, we obtain

$$\Gamma_{sQ} = \begin{bmatrix} 0.2148444 & 0.0018481 & 0.2169145 & 0.0698280 & 0.2 \\ 0.5503097 & 0.6645646 & -0.6352193 & 0.8023543 & 0.4 \\ -0.7990597 & -0.7456317 & -0.5938518 & -0.5805731 & 0.6 \\ -0.0941402 & -0.0440333 & 0.3437855 & 0.0892284 & 0.4 \\ -0.0603521 & 0.0210926 & -0.2803500 & -0.0795282 & 0.2 \end{bmatrix},$$

$$A_{x\mathrm{Q}} = A^+_{aa\mathrm{Q}} = 0.0789442, \quad B_{x\mathrm{Q}} = [\,2.3596219 \quad -0.1725085\,], \quad C_{x\mathrm{Q}} = 0,$$

$$E^+_{a\mathrm{Q}} = [\,0.1593412 \quad 0.0009204 \quad 0.0116587 \quad 0.1593412\,]$$

and

$$A^0_{aa\mathrm{Q}} = \begin{bmatrix} 0.8733954 & -14.3566212 \\ 0.4222493 & -0.8733953 \end{bmatrix},$$

$$[\,B^0_{0a\mathrm{Q}} \quad \tilde{L}^0_{ad\mathrm{Q}}\,] = \begin{bmatrix} 13.8502316 & -10.8089077 \\ 0.3251762 & -1.3752299 \end{bmatrix},$$

$$E^0_{a\mathrm{Q}} = \begin{bmatrix} -1.9958628 & 6.3511003 & -0.7973732 & -1.9958628 \\ -0.5082606 & 0.0920508 & -0.4908900 & -0.5082606 \end{bmatrix}.$$

Following Step 6.Z.4, we have

$$S_{x\mathrm{Q}} = 35.4527292, \quad T_{x\mathrm{Q}} = 0.3224810, \quad Y_{x\mathrm{Q}} = [\,-5.2529064 \quad 93.6614674\,],$$

and

$$F_{\mathrm{Q}} = 8.4694885, \qquad G_{\mathrm{Q}} = 35.4527292.$$

Finally, evaluate

$$M = \begin{bmatrix} 1.4653098 & 0 & 0 & 0 & 0 \\ 0 & -0.0000103 & -0.0000451 & 0.0003744 & 0 \\ 0 & 0.0000632 & 0.0002763 & -0.0022958 & 0 \\ 0 & -0.0002503 & -0.0010946 & 0.0090961 & 0 \\ 0 & 0 & 0 & 0 & 0.2110284 \end{bmatrix}.$$

We obtain

$$\gamma^* = \sqrt{\lambda_{\max}(M)} = 1.2104998.$$

🅔

Finally, we conclude this chapter by noting that it can readily be verified now that the auxiliary systems associated with the the problem of maximization of complex stability radius in Subsection 1.4.2, the robust stabilization problem for plants with additive perturbations in Subsection 1.4.3 and the robust stabilization problem for plants with multiplicative perturbations in Subsection 1.4.4, all in Chapter 1, satisfy Assumptions 6.Z.1 to 6.Z.5. Hence, their infima are exactly computable.

Chapter 7

Solutions to Continuous-time H_∞ Problem

7.1. Introduction

THE MAIN CONTRIBUTION of this chapter is to provide closed-form solutions to the H_∞ suboptimal control problem for continuous-time systems. Here by closed-form solutions we mean solutions which are explicitly parameterized in terms of γ and are obtained without explicitly requiring a value for γ. Hence one can easily tune the parameter γ to obtain the desired level of disturbance attenuation. Such a design can be called a 'one-shot' design. We provide these closed-form solutions for a class of singular H_∞ suboptimal control problems for which the subsystem from the control input to the controlled output and the subsystem from the disturbance to the measurement output satisfy certain geometric conditions and some other minor assumptions, namely, Assumptions 6.M.1 to 6.M.7 of Chapter 6. Moreover, for this class of systems we also provide conditions under which the H_∞ optimal control problem via state feedback has a solution. Explicit expressions for the solutions will also be given. Finally the issue of pole-zero cancellations in the closed-loop system resulting from the H_∞ optimal or suboptimal state or output feedback control laws is examined.

Some significant attributes of our method of generating the closed-form solutions in the H_∞ suboptimal control problem are as follows:

1. No H_∞-CAREs are solved in generating the closed-form solutions. As a result, all the numerical difficulties associated with the H_∞-CAREs are alleviated.

2. The value for γ can be adjusted on line when the closed-form solution to the H_∞ suboptimal control problem is implemented using either software or hardware. Since the effect of such a 'knob' on the performance and the robustness of the closed-loop system is straightforward, it should be very appealing from a practical point of view.

3. Having closed-form solutions to the H_∞ suboptimal control problem enables us to understand the behavior of the controller (i.e., high-gain, bandwidth, etc.) as the parameter γ approaches the infimum value of the H_∞ norm of T_{hw} over all stabilizing controllers.

The above mentioned results were reported in Saberi, Chen and Lin [108]. In the case when Assumptions 6.M.1 and 6.M.7 are not satisfied, a similar method will also be adapted to compute γ-suboptimal solutions. It is, however, no longer a closed-form one. The outline of this chapter is as follows: Section 7.2 gives a closed-form solution to the H_∞ suboptimal state feedback control problem, while Section 7.3 provides a closed-form solution (full order controller) to the H_∞ suboptimal measurement feedback control problem. A reduced order γ-suboptimal controller design method is introduced in Section 7.4. Finally, all main results are to be proved in Section 7.5.

7.2. Full State Feedback

We consider in this section the H_∞ optimization problem for the following full state feedback systems characterized by

$$\Sigma : \begin{cases} \dot{x} = A\ x + B\ u + E\ w, \\ y = x \\ h = C_2\ x + D_2\ u + D_{22}\ w, \end{cases} \tag{7.2.1}$$

where $x \in \mathbb{R}^n$ is the state, $u \in \mathbb{R}^m$ is the control input, $w \in \mathbb{R}^q$ is the external disturbance input, and $h \in \mathbb{R}^\ell$ is the controlled output of Σ. Again, we let Σ_P be the subsystem characterized by the matrix quadruple (A, B, C_2, D_2). As in Section 6.2 of Chapter 6, we first make the following assumptions:

Assumption 7.F.1: (A, B) is stabilizable;

Assumption 7.F.2: Σ_P has no invariant zero on the imaginary axis;

Assumption 7.F.3: $\mathrm{Im}\,(E) \subset \mathcal{V}^-(\Sigma_P) + \mathcal{S}^-(\Sigma_P)$; and

Assumption 7.F.4: $D_{22} = 0$. Ⓐ

We introduce a procedure for obtaining the closed-form solutions for the H_∞ suboptimal state feedback control problem utilizing an asymptotic time-scale and eigenstructure assignment (ATEA). The concept of the ATEA design procedure was proposed originally in Saberi and Sannuti [112] and its complete time-scale properties and Lyapunov stability analysis were done in Chen [12]. It uses the special coordinate basis of the given system (See Theorem 2.4.1). We also give conditions under which the H_∞ optimal control problem has a solution. Furthermore, explicit expressions for these optimal solutions will be given. The following is a step-by-step algorithm to construct the closed-form of the γ-suboptimal state feedback laws, which are explicitly parameterized by $\gamma > \gamma^*$ and a tuning parameter ε.

Step 7.F.1: Transform the system Σ_{P} into the special coordinate basis as given by Theorem 2.4.1 in Chapter 2. To all sub-matrices and transformations in the special coordinate basis of Σ_{P}, we append a subscript P to signify their relation to the system Σ_{P}. We also choose the output transformation $\Gamma_{o\mathrm{P}}$ to have the following form:

$$\Gamma_{o\mathrm{P}} = \begin{bmatrix} I_{m_{0\mathrm{P}}} & 0 \\ 0 & \Gamma_{or\mathrm{P}} \end{bmatrix}, \tag{7.2.2}$$

where $m_{0\mathrm{P}} =$ rank (D_2). Next, we compute

$$\bar{E} = \Gamma_{s\mathrm{P}}^{-1} E = \begin{bmatrix} E_{a\mathrm{P}}^{+} \\ E_{b\mathrm{P}} \\ E_{a\mathrm{P}}^{-} \\ E_{c\mathrm{P}} \\ E_{d\mathrm{P}} \end{bmatrix}. \tag{7.2.3}$$

Note that Assumption 7.F.3 implies $E_{b\mathrm{P}} \equiv 0$. Also, for economy of notation, we denote $n_{x\mathrm{P}}$ the dimension of $\mathbb{R}^n/\mathcal{S}^+(\Sigma_{\mathrm{P}})$. Note that $n_{x\mathrm{P}} = 0$ if and only if the system Σ_{P} is right invertible and is of minimum phase. Next, define

$$A_{11\mathrm{P}} = \begin{bmatrix} A_{aa\mathrm{P}}^{+} & L_{ab\mathrm{P}}^{+} C_{b\mathrm{P}} \\ 0 & A_{bb\mathrm{P}} \end{bmatrix}, \quad B_{11\mathrm{P}} = \begin{bmatrix} B_{0a\mathrm{P}}^{+} \\ B_{0b\mathrm{P}} \end{bmatrix}, \quad A_{13\mathrm{P}} = \begin{bmatrix} L_{ad\mathrm{P}}^{+} \\ L_{bd\mathrm{P}} \end{bmatrix},$$

$$C_{21\mathrm{P}} = \Gamma_{or\mathrm{P}} \begin{bmatrix} 0 & 0 \\ 0 & C_{b\mathrm{P}} \end{bmatrix}, \quad C_{23\mathrm{P}} = \Gamma_{or\mathrm{P}} \begin{bmatrix} C_{d\mathrm{P}} C_{d\mathrm{P}}' \\ 0 \end{bmatrix}, \quad E_{x\mathrm{P}} = \begin{bmatrix} E_{a\mathrm{P}}^{+} \\ E_{b\mathrm{P}} \end{bmatrix},$$

and

$$A_{x\mathrm{P}} = A_{11\mathrm{P}} - A_{13\mathrm{P}} (C_{23\mathrm{P}}' C_{23\mathrm{P}})^{-1} C_{23\mathrm{P}}' C_{21\mathrm{P}},$$

$$B_{x\mathrm{P}} B_{x\mathrm{P}}' = B_{11\mathrm{P}} B_{11\mathrm{P}}' + A_{13\mathrm{P}} (C_{23\mathrm{P}}' C_{23\mathrm{P}})^{-1} A_{13\mathrm{P}}',$$

$$C_{x\mathrm{P}}' C_{x\mathrm{P}} = C_{21\mathrm{P}}' C_{21\mathrm{P}} - C_{21\mathrm{P}}' C_{21\mathrm{P}} (C_{23\mathrm{P}}' C_{23\mathrm{P}})^{-1} C_{23\mathrm{P}}' C_{21\mathrm{P}}.$$

Step 7.F.2: Solve for the unique positive definite solution $S_{x\text{P}}$ of the algebraic matrix Riccati equation,

$$A_{x\text{P}} S_{x\text{P}} + S_{x\text{P}} A'_{x\text{P}} - B_{x\text{P}} B'_{x\text{P}} + S_{x\text{P}} C'_{x\text{P}} C_{x\text{P}} S_{x\text{P}} = 0, \tag{7.2.4}$$

together with the matrix $T_{x\text{P}}$ defined by

$$T_{x\text{P}} = \begin{bmatrix} T_{aa\text{P}} & 0 \\ 0 & 0 \end{bmatrix}, \tag{7.2.5}$$

where $T_{aa\text{P}}$ is the unique semi-positive solution of the algebraic matrix Lyapunov equation,

$$A^+_{aa\text{P}} T_{aa\text{P}} + T_{aa\text{P}} (A^+_{aa\text{P}})' = E^+_{a\text{P}} (E^+_{a\text{P}})'. \tag{7.2.6}$$

Then it was shown in Section 6.2 of Chapter 6 that the infimum for the given system (7.2.1) is given by

$$\gamma^* = \sqrt{\lambda_{\max}(T_{x\text{P}} S^{-1}_{x\text{P}})}. \tag{7.2.7}$$

Then, for any $\gamma > \gamma^*$, we define

$$F_{11}(\gamma) := \begin{bmatrix} F^+_{a0}(\gamma) & F_{b0}(\gamma) \\ F^+_{a1}(\gamma) & F_{b1}(\gamma) \end{bmatrix} = \begin{bmatrix} B'_{11\text{P}} P_x \\ (C'_{23\text{P}} C_{23\text{P}})^{-1} [A'_{13\text{P}} P_x + C'_{23\text{P}} C_{21\text{P}}] \end{bmatrix}, \tag{7.2.8}$$

where

$$P_x := (S_{x\text{P}} - \gamma^{-2} T_{x\text{P}})^{-1}, \tag{7.2.9}$$

and define

$$A^c_{11\text{P}} := A_{11\text{P}} - [\, B_{11\text{P}} \quad A_{13\text{P}} \,] F_{11}(\gamma).$$

We will show later on that the eigenvalues of $A^c_{11\text{P}}$ are in $\mathbb{C}^-$. Let us partition $[\, F^+_{a1}(\gamma) \quad F_{b1}(\gamma) \,]$ as,

$$[\, F^+_{a1}(\gamma) \quad F_{b1}(\gamma) \,] = \begin{bmatrix} F^+_{a11}(\gamma) & F_{b11}(\gamma) \\ F^+_{a12}(\gamma) & F_{b12}(\gamma) \\ \vdots & \vdots \\ F^+_{a1m_{d\text{P}}}(\gamma) & F_{b1m_{d\text{P}}}(\gamma) \end{bmatrix}, \tag{7.2.10}$$

where $F^+_{a1i}(\gamma)$ and $F_{b1i}(\gamma)$ are of dimensions $1 \times n^+_{a\text{P}}$ and $1 \times n_{b\text{P}}$, respectively.

Step 7.F.3: Let $\Delta_{c\text{P}}$ be any arbitrary $m_{c\text{P}} \times n_{c\text{P}}$ matrix subject to the constraint that

$$A^c_{cc\text{P}} = A_{cc\text{P}} - B_{c\text{P}} \Delta_{c\text{P}}, \tag{7.2.11}$$

is a stable matrix. Note that the existence of such a $\Delta_{c\text{P}}$ is guaranteed by the property that $(A_{cc\text{P}}, B_{c\text{P}})$ is controllable.

Step 7.F.4: This step makes use of subsystems, $i = 1$ to $m_{d\mathrm{P}}$, represented by (2.4.14) of Chapter 2. Let $\Lambda_i = \{\lambda_{i1}, \lambda_{i2}, \cdots, \lambda_{iq_i}\}$, $i = 1$ to $m_{d\mathrm{P}}$, be the sets of q_i elements all in $\mathbb{C}^-$, which are closed under complex conjugation, where q_i and $m_{d\mathrm{P}}$ are as defined in Theorem 2.4.1 but associated with the special coordinate basis of Σ_P. Let $\Lambda_{d\mathrm{P}} := \Lambda_1 \cup \Lambda_2 \cup \cdots \cup \Lambda_{m_{d\mathrm{P}}}$. For $i = 1$ to $m_{d\mathrm{P}}$, we define

$$p_i(s) := \prod_{j=1}^{q_i} (s - \lambda_{ij}) = s^{q_i} + F_{i1} s^{q_i - 1} + \cdots + F_{iq_i - 1} s + F_{iq_i}, \tag{7.2.12}$$

and

$$\tilde{F}_i(\varepsilon, \Lambda_i) := \frac{1}{\varepsilon^{q_i}} \left[F_{iq_i}, \ \varepsilon F_{iq_i - 1}, \ \cdots, \ \varepsilon^{q_i - 1} F_{i1} \right]. \tag{7.2.13}$$

Step 7.F.5: In this step, various gains calculated in Steps 7.F.2 to 7.F.4 are put together to form a composite state feedback gain for the given system Σ_P. Let

$$\tilde{F}^+_{a1}(\gamma, \varepsilon, \Lambda_{d\mathrm{P}}) := \begin{bmatrix} F^+_{a11}(\gamma) F_{1q_1} / \varepsilon^{q_1} \\ F^+_{a12}(\gamma) F_{2q_2} / \varepsilon^{q_2} \\ \vdots \\ F^+_{a1m_{d\mathrm{P}}}(\gamma) F_{m_{d\mathrm{P}} q_{m_{d\mathrm{P}}}} / \varepsilon^{q_{m_{d\mathrm{P}}}} \end{bmatrix},$$

and

$$\tilde{F}_{b1}(\gamma, \varepsilon, \Lambda_{d\mathrm{P}}) := \begin{bmatrix} F_{b11}(\gamma) F_{1q_1} / \varepsilon^{q_1} \\ F_{b12}(\gamma) F_{2q_2} / \varepsilon^{q_2} \\ \vdots \\ F_{b1m_{d\mathrm{P}}}(\gamma) F_{m_{d\mathrm{P}} q_{m_{d\mathrm{P}}}} / \varepsilon^{q_{m_{d\mathrm{P}}}} \end{bmatrix}.$$

Then define the state feedback gain $F(\gamma, \varepsilon, \Lambda_{d\mathrm{P}}, \Delta_{c\mathrm{P}})$ as

$$F(\gamma, \varepsilon, \Lambda_{d\mathrm{P}}, \Delta_{c\mathrm{P}}) = -\Gamma_{i\mathrm{P}} \Big(\tilde{F}(\gamma, \varepsilon, \Lambda_{d\mathrm{P}}, \Delta_{c\mathrm{P}}) + \tilde{F}_0 \Big) \Gamma_{s\mathrm{P}}^{-1}, \tag{7.2.14}$$

where $\tilde{F}(\gamma, \varepsilon, \Lambda_{d\mathrm{P}}, \Delta_{c\mathrm{P}})$ is given by

$$\begin{bmatrix} F^+_{a0}(\gamma) & F_{b0}(\gamma) & 0 & 0 & 0 \\ \tilde{F}^+_{a1}(\gamma, \varepsilon, \Lambda_{d\mathrm{P}}) & \tilde{F}_{b1}(\gamma, \varepsilon, \Lambda_{d\mathrm{P}}) & 0 & 0 & \tilde{F}_d(\varepsilon, \Lambda_{d\mathrm{P}}) \\ 0 & 0 & 0 & \Delta_{c\mathrm{P}} & 0 \end{bmatrix}, \tag{7.2.15}$$

$$\tilde{F}_0 = \begin{bmatrix} C^+_{0a\mathrm{P}} & C_{0b\mathrm{P}} & C^-_{0a\mathrm{P}} & C_{0c\mathrm{P}} & C_{0d\mathrm{P}} \\ E^+_{da\mathrm{P}} & E_{db\mathrm{P}} & E^-_{da\mathrm{P}} & E_{dc\mathrm{P}} & E_{d\mathrm{P}} \\ E^+_{ca\mathrm{P}} & E_{cb\mathrm{P}} & E^-_{ca\mathrm{P}} & 0 & 0 \end{bmatrix}, \tag{7.2.16}$$

and where

$$E_{dP} = \begin{bmatrix} E_{11} & \cdots & E_{1m_{dP}} \\ \vdots & \ddots & \vdots \\ E_{m_{dP}1} & \cdots & E_{m_{dP}m_{dP}} \end{bmatrix}, \tag{7.2.17}$$

and

$$\tilde{F}_d(\varepsilon, \Lambda_{dP}) = \text{diag}\Big[\tilde{F}_1(\varepsilon, \Lambda_1),\ \tilde{F}_2(\varepsilon, \Lambda_2),\ \cdots,\ \tilde{F}_{m_{dP}}(\varepsilon, \Lambda_{m_{dP}})\Big]. \tag{7.2.18}$$

This completes the algorithm. ▣A

We have the following theorem.

Theorem 7.2.1. Consider the full state feedback system (7.2.1) which satisfies Assumptions 7.F.1 to 7.F.4. Then with state feedback gain given by (7.2.14), we have the following properties:

1. For any $\gamma > \gamma^*$, for any $\Lambda_{dP} \subset \mathbb{C}^-$ which is closed under complex conjugation and for any Δ_{cP} subject to the constraints that A^c_{ccP} is stable, there exists an $\varepsilon^* > 0$ such that for all $0 < \varepsilon \leq \varepsilon^*$, the state feedback control law,
$$u = F(\gamma, \varepsilon, \Lambda_{dP}, \Delta_{cP})x, \tag{7.2.19}$$
with $F(\gamma, \varepsilon, \Lambda_{dP}, \Delta_{cP})$ being given as in (7.2.14) is a γ-suboptimal control law for the given system (7.2.1). Namely, the closed-loop system comprising Σ_P and the state feedback law (7.2.19) is internally stable and the H_∞-norm of the closed-loop transfer function from the disturbance w to the controlled output h is less than γ, i.e., $\|T_{hw}\|_\infty < \gamma$.

2. Moreover as $\varepsilon \to 0$, the poles of the closed-loop system, i.e., the eigenvalues of $A + BF(\gamma, \varepsilon, \Lambda_{dP}, \Delta_{cP})$, are given by
$$\lambda(A^-_{aaP}),\quad \lambda(A^c_{ccP}),\quad \lambda(A^c_{11P}) + 0(\varepsilon) \quad \text{and} \quad \frac{\Lambda_{dP}}{\varepsilon} + 0(1),$$
Clearly, there are at least n_{dP} poles of the closed-loop system have infinite negative real parts as $\varepsilon \to 0$. ▣T

Proof. See Subsection 7.5.A. ▣

The following remarks respectively deal with 1) the interpretations of the parameters ε, Δ_{dP} and Δ_{cP}; 2) solutions to the regular problem; and 3) solutions to the general problem when the geometric condition (Assumption 7.F.3) is not satisfied.

Remark 7.2.1. (Interpretations of ε, Λ_{dP} and Δ_{cP}). Theorem 7.2.1 shows that the closed-loop system under H_∞ suboptimal state feedback laws, i.e., T_{hw}, has fast eigenvalues Λ_{dP}/ε. So the set of parameters Λ_{dP} in the H_∞ suboptimal gain $F(\gamma,\varepsilon,\Lambda_{dP},\Delta_{cP})$ of (7.2.14) represents the asymptotes of these fast eigenvalues while ε represents their time-scale. The closed-loop system also has $\lambda(A^c_{ccP})$ as slow eigenvalues. These eigenvalues can be assigned to any desired locations in $\mathbb{C}^-$ by choosing an appropriate Δ_{cP}. Hence, the set of parameters Δ_{cP} in the H_∞ suboptimal state feedback gain prescribes the locations of these slow eigenvalues. Ⓡ

Remark 7.2.2. (Regular Case). If D_2 is injective, it is obvious from our algorithm that $F(\gamma,\varepsilon,\Lambda_{dP},\Delta_{cP}) = F(\gamma)$ does not depend on ε, Λ_{dP} and Δ_{cP}, and is given by

$$F(\gamma) = -\Gamma_{iP}\,[\,C^+_{0aP} + F^+_{a0}(\gamma) \quad C_{0bP} + F_{b0}(\gamma) \quad C^-_{0aP}\,]\,\Gamma^{-1}_{sP}.$$

This corresponds to the regular case, and is the central controller given in Doyle *et al.* [49]. Ⓡ

Remark 7.2.3. Finally, we would like to note that if Assumption 7.F.3, i.e., the geometric condition, is not satisfied, one can use the iterative procedure in Chapter 6 to find an approximation of the infimum, say $\tilde{\gamma}^*$. Moreover, the algorithm for finding the γ-suboptimal state feedback laws can be slightly modified to handle this situation. To be more specific, one only needs to modify Step 7.F.2 slightly as follows:

Step 7.F.2m: For any $\gamma > \tilde{\gamma}^*$, we define

$$F_{11}(\gamma) := \begin{bmatrix} F^+_{a0}(\gamma) & F_{b0}(\gamma) \\ F^+_{a1}(\gamma) & F_{b1}(\gamma) \end{bmatrix} = \begin{bmatrix} B'_{11P}P_x \\ (C'_{23P}C_{23P})^{-1}[A'_{13P}P_x + C'_{23P}C_{21P}] \end{bmatrix},$$

where P_x is the positive definite solution of the Riccati equation,

$$P_xA_{xP} + A'_{xP}P_x + P_xE_{xP}E'_{xP}P_x/\gamma^2 - P_xB_{xP}B'_{xP}P_x + C'_{xP}C_{xP} = 0,$$

and define

$$A^c_{11P} := A_{11P} - [\,B_{11P} \quad A_{13P}\,]\,F_{11}(\gamma).$$

Let us partition $[\,F^+_{a1}(\gamma) \quad F_{b1}(\gamma)\,]$ as,

$$[\,F^+_{a1}(\gamma) \quad F_{b1}(\gamma)\,] = \begin{bmatrix} F^+_{a11}(\gamma) & F_{b11}(\gamma) \\ F^+_{a12}(\gamma) & F_{b12}(\gamma) \\ \vdots & \vdots \\ F^+_{a1m_{dP}}(\gamma) & F_{b1m_{dP}}(\gamma) \end{bmatrix},$$

where $F^+_{a1i}(\gamma)$ and $F_{b1i}(\gamma)$ are of dimensions $1 \times n^+_{aP}$ and $1 \times n_{bP}$.

The rest steps of the algorithm, i.e., Steps 7.F.1, 7.F.3 to 7.F.5, remain unchanged. All results in Theorem 7.2.1 are valid for this situation as well. The only difference is that the control law is no longer of closed-form. Ⓡ

The following theorem deals with pole-zero cancellations in the closed-loop system T_{hw} under the state feedback law.

Theorem 7.2.2. (Pole-zero Cancellations). Consider the full state feedback system (7.2.1) which satisfies Assumptions 7.F.1 to 7.F.4. Then with state feedback gain given by (7.2.14), the resulting closed-loop system has the following property: $\lambda(A_{aa\mathrm{P}}^-)$, the stable invariant zeros of the system Σ_P, and $\lambda(A_{cc\mathrm{P}}^c)$ are the output decoupling zeros of the closed-loop transfer matrix T_{hw}. Hence, they cancel with the poles of T_{hw}. Ⓣ

Proof. See Subsection 7.5.B. ⊠

We illustrate our algorithm in the following example.

Example 7.2.1. Reconsider the system in Example 6.2.1, i.e., a full state feedback system characterized by

$$A = \begin{bmatrix} 1 & 1 & 1 & 0 & 1 \\ 0 & 1 & 0 & 0 & 1 \\ 0 & 1 & 1 & 0 & 1 \\ 1 & 1 & 1 & 1 & 1 \\ 1 & 1 & 1 & 1 & 0 \end{bmatrix}, \quad B = \begin{bmatrix} 0 & 0 & 0 \\ 0 & 0 & 0 \\ 1 & 0 & 0 \\ 0 & 0 & 1 \\ 0 & 1 & 0 \end{bmatrix}, \quad E = \begin{bmatrix} 5 & 1 \\ 0 & 0 \\ 0 & 0 \\ 2 & 3 \\ 1 & 4 \end{bmatrix}, \tag{7.2.20}$$

and

$$C_2 = \begin{bmatrix} 0 & 0 & 1 & 0 & 0 \\ 0 & 0 & 0 & 0 & 1 \\ 0 & 1 & 0 & 0 & 0 \\ 0 & 0 & 1 & 0 & 0 \end{bmatrix}, \quad D_2 = \begin{bmatrix} 1 & 0 & 0 \\ 0 & 0 & 0 \\ 0 & 0 & 0 \\ 0 & 0 & 0 \end{bmatrix}, \quad D_{22} = \begin{bmatrix} 0 & 0 \\ 0 & 0 \\ 0 & 0 \\ 0 & 0 \end{bmatrix}. \tag{7.2.21}$$

It is easy to verify that (A, B) is stabilizable, and the system Σ_P is neither right nor left invertible and is of nonminimum phase with an invariant zero at $s = 1$. Moreover, it is already in the form of the special coordinate basis with $n_{a\mathrm{P}}^+ = 1$, $n_{a\mathrm{P}}^- = n_{a\mathrm{P}}^0 = 0$, $n_{b\mathrm{P}} = 2$ and $n_{c\mathrm{P}} = n_{d\mathrm{P}} = 1$. Also, it is simple to see that $\mathrm{Im}\,(E) \subseteq \mathcal{V}^-(\Sigma_\mathrm{P}) \cup \mathcal{S}^-(\Sigma_\mathrm{P})$ since $E_{b\mathrm{P}} = 0$. Hence, all Assumptions 7.F.1 to 7.F.4 are satisfied. Moreover, it was obtained in Example 6.2.1 that the infimum is given by

$$\gamma^* = 6.4679044.$$

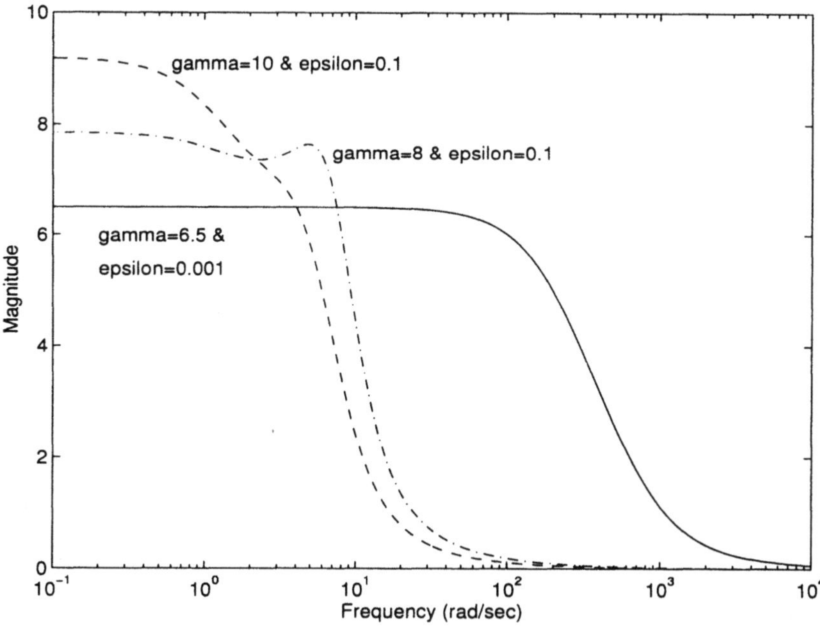

Figure 7.2.1: Maximum singular values of T_{hw} (state feedback case).

Following the algorithm in this section, we obtain the closed-form solution of the γ-suboptimal state feedback gains, $F(\gamma, \varepsilon, \lambda_{d\text{P}}, \Delta_{c\text{P}})$, which is given by

$$\begin{bmatrix} \dfrac{-0.163673\gamma^2}{0.132909\gamma^2 - 5.560084} & -1 + \dfrac{0.294790\gamma^2\lambda_{d\text{P}}}{(0.132909\gamma^2 - 5.560084)\varepsilon} & -1 \\ \dfrac{0.185427\gamma^2 - 3.009097}{0.132909\gamma^2 - 5.560084} & -1 + \dfrac{(0.102145\gamma^2 - 12.824695)\lambda_{d\text{P}}}{(0.132909\gamma^2 - 5.560084)\varepsilon} & -1 \\ \dfrac{-0.318336\gamma^2 + 10.696930}{0.132909\gamma^2 - 5.560084} & -1 + \dfrac{(0.163673\gamma^2 - 2.127749)\lambda_{d\text{P}}}{(0.132909\gamma^2 - 5.560084)\varepsilon} & -1 \\ 0 & -1 & -\Delta_{c\text{P}} \\ 0 & \dfrac{\lambda_{d\text{P}}}{\varepsilon} & 0 \end{bmatrix}', \tag{7.2.22}$$

where the scalars $\lambda_{d\text{P}} < 0$ and $\Delta_{c\text{P}} > 1$ (note that $\Delta_{c\text{P}}$ must be greater than one in order to have stable $A^c_{cc\text{P}}$). We demonstrate our results in Figure 7.2.1 by the plots of maximum singular values of the closed-loop transfer function matrix for several values of γ and ε. Note that in Figure 7.2.1, we choose parameters $\lambda_{d\text{P}} = -1$ and $\Delta_{c\text{P}} = 3$. ▣

7.3. Full Order Output Feedback

This section deals with H_∞ suboptimal and optimal design using full order measurement output feedback laws, i.e., the dynamical order of these control laws will be exactly the same as that of the given system. To be more specific, we consider the following measurement feedback system

$$\Sigma : \begin{cases} \dot{x} = A\ x + B\ u + E\ w, \\ y = C_1\ x \qquad\qquad + D_1\ w, \\ h = C_2\ x + D_2\ u + D_{22}\ w, \end{cases} \tag{7.3.1}$$

where $x \in \mathbb{R}^n$ is the state, $u \in \mathbb{R}^m$ is the control input, $w \in \mathbb{R}^q$ is the external disturbance input, $y \in \mathbb{R}^p$ is the measurement output, and $h \in \mathbb{R}^\ell$ is the controlled output of Σ. Again, we let Σ_P be the subsystem characterized by the matrix quadruple (A, B, C_2, D_2) and Σ_Q be the subsystem characterized by the matrix quadruple (A, E, C_1, D_1). The following assumptions are made first:

Assumption 7.M.1: (A, B) is stabilizable;

Assumption 7.M.2: Σ_P has no invariant zero on the imaginary axis;

Assumption 7.M.3: $\mathrm{Im}\,(E) \subset \mathcal{V}^-(\Sigma_P) + \mathcal{S}^-(\Sigma_P)$;

Assumption 7.M.4: (A, C_1) is detectable;

Assumption 7.M.5: Σ_Q has no invariant zero on the imaginary axis;

Assumption 7.M.6: $\mathrm{Ker}\,(C_2) \supset \mathcal{V}^-(\Sigma_Q) \cap \mathcal{S}^-(\Sigma_Q)$; and

Assumption 7.M.7: $D_{22} = 0$. [A]

The class of output feedback controllers that we consider in this section are basically observer based control laws and can be regarded as an extension of the central output feedback controller that was proposed in Doyle *et al.* [49] for the regular case. We have modified the central output feedback controller of the regular case to deal with the singular case. This modification will be discussed later on. We assume that the infimum γ^* has been obtained using methods given in Section 6.3 of Chapter 6. The procedure for obtaining the closed-form of the H_∞ suboptimal output feedback laws for any $\gamma > \gamma^*$ proceeds as follows.

Step 7.M.1: Define an auxiliary full state feedback system

$$\begin{cases} \dot{x} = A\ x + B\ u + E\ w, \\ y = \quad x \\ h = C_2\ x + D_2\ u + D_{22}\ w, \end{cases}$$

and proceed to perform Steps 7.F.1 to 7.F.5 of Section 7.2 to obtain the gain matrix $F(\gamma,\varepsilon,\Lambda_{dP},\Delta_{cP})$. Also, define

$$P(\gamma) := (\Gamma_{sP}^{-1})' \begin{bmatrix} (S_{xP} - \gamma^{-2}T_{xP})^{-1} & 0 \\ 0 & 0 \end{bmatrix} \Gamma_{sP}^{-1}. \tag{7.3.2}$$

Step 7.M.2: Define another auxiliary full state feedback system as follows,

$$\begin{cases} \dot{x} = A'\,x + C_1'\,u + C_2'\,w, \\ y = \quad x \\ h = E'\,x + D_1'\,u + D_{22}'\,w, \end{cases} \tag{7.3.3}$$

and proceed to perform Steps 7.F.1 to 7.F.5 of Section 7.2 but for this auxiliary system to obtain a gain matrix $F(\gamma,\varepsilon,\Lambda_{dQ},\Delta_{cQ})$. Let us define $K(\gamma,\varepsilon,\Lambda_{dQ},\Delta_{cQ}) := F(\gamma,\varepsilon,\Lambda_{dQ},\Delta_{cQ})'$. Also, define

$$Q(\gamma) := (\Gamma_{sQ}^{-1})' \begin{bmatrix} (S_{xQ} - \gamma^{-2}T_{xQ})^{-1} & 0 \\ 0 & 0 \end{bmatrix} \Gamma_{sQ}^{-1}. \tag{7.3.4}$$

Step 7.M.3: Construct the following full order observer based controller,

$$\Sigma_{\text{cmp}} : \begin{cases} \dot{v} = A_{\text{cmp}}\,v + B_{\text{cmp}}\,y, \\ u = C_{\text{cmp}}\,v + \;0\;\,y, \end{cases} \tag{7.3.5}$$

where

$$\begin{aligned} A_{\text{cmp}} &= A + \gamma^{-2}EE'P(\gamma) + BF(\gamma,\varepsilon,\Lambda_{dP},\Delta_{cP}) \\ &\quad + \big[I - \gamma^{-2}Q(\gamma)P(\gamma)\big]^{-1}\Big\{K(\gamma,\varepsilon,\Lambda_{dQ},\Delta_{cQ})\big[C_1 + \gamma^{-2}D_1E'P(\gamma)\big] \\ &\quad + \gamma^{-2}Q(\gamma)\big[A'P(\gamma) + P(\gamma)A + C_2'C_2 + \gamma^{-2}P(\gamma)EE'P(\gamma)\big] \\ &\quad + \gamma^{-2}Q(\gamma)\big[P(\gamma)B + C_2'D_2\big]F(\gamma,\varepsilon,\Lambda_{dP},\Delta_{cP})\Big\}, \end{aligned} \tag{7.3.6}$$

$$B_{\text{cmp}} = -\big[I - \gamma^{-2}Q(\gamma)P(\gamma)\big]^{-1}K(\gamma,\varepsilon,\Lambda_{dQ},\Delta_{cQ}), \tag{7.3.7}$$

$$C_{\text{cmp}} = F(\gamma,\varepsilon,\Lambda_{dP},\Delta_{cP}). \tag{7.3.8}$$

It is to be shown that Σ_{cmp} is indeed a γ-suboptimal controller. Clearly, it has a dynamical order of n, i.e., it is a full order output feedback controller. ▣

We have the following theorem.

Theorem 7.3.1. Consider the given measurement feedback system (7.3.1) satisfying Assumptions 7.M.1 to 7.M.7. Then for any $\gamma > \gamma^*$, for any $\Lambda_{dP} \subset \mathbb{C}^-$ and $\Lambda_{dQ} \subset \mathbb{C}^-$, which are closed under complex conjugation, and for any Δ_{cP}

and Δ_{cQ} subject to the constraints that A^c_{ccP} and A^c_{ccQ} are stable matrices, there exists an $\varepsilon^* > 0$ such that for all $0 < \varepsilon \leq \varepsilon^*$, the control law $\Sigma_{\rm cmp}$ as given in (7.3.5) is γ-suboptimal controller, namely, the closed-loop system comprising Σ and the output feedback controller $\Sigma_{\rm cmp}$, is internally stable and the H_∞-norm of the closed-loop transfer matrix from the disturbance w to the controlled output h is less than γ, i.e., $\|T_{hw}\|_\infty < \gamma$. 🅃

Proof. See Subsection 7.5.C. ⊠

The following theorem deals with the issue of pole-zero cancellations and the closed-loop eigenvalues in the γ-suboptimal output feedback control.

Theorem 7.3.2. Consider the given measurement feedback system (7.3.1) satisfying Assumptions 7.M.1 to 7.M.7 with the γ-suboptimal control $\Sigma_{\rm cmp}$ as given in (7.3.5). Then the following properties hold:

1. $\lambda(A^-_{aaP})$, the stable invariant zeros of the system Σ_P, and $\lambda(A^c_{ccP})$ are the output decoupling zeros of the closed-loop system T_{hw}. Hence they cancel with the poles of T_{hw}.
2. $\lambda(A^-_{aaQ})$, the stable invariant zeros of the system Σ_Q, and $\lambda(A^c_{ccQ})$ are the input decoupling zeros of the closed-loop system T_{hw}. Hence they cancel with the poles of T_{hw}.
3. As $\varepsilon \to 0$, the fast eigenvalues of the closed-loop system are asymptotically given by $\Lambda_{dP}/\varepsilon + 0(1)$ and $\Lambda_{dQ}/\varepsilon + 0(1)$. 🅃

Proof. See Subsection 7.5.D. ⊠

The following remarks are in order.

Remark 7.3.1. (Interpretations of ε, Λ_{dP}, Λ_{dQ}, Δ_{cP} and Δ_{cQ}). Again, as in Remark 7.2.1, the set of parameters Λ_{dP} and Λ_{dQ} represent the asymptotes of the fast eigenvalues of the closed-loop system while ε represents their time-scale. The set of parameters Δ_{cP} and Δ_{cQ} prescribe the locations of the slow eigenvalues of the closed-loop system corresponding to $\lambda(A^c_{ccP})$ and $\lambda(A^c_{ccQ})$. The eigenvalues can be assigned to any desired locations in $\mathbb{C}^-$ by choosing appropriate Δ_{cP} and Δ_{cQ}. 🅁

Remark 7.3.2. (Regular Case). For the regular problem when D_1 is surjective and D_2 is injective, which implies that Σ_P does not have x_c and x_d and, Σ_Q does not have x_b and x_d in their SCB decompositions, it is straightforward to verify that both $F(\gamma,\varepsilon,\Lambda_{d\text{P}},\Delta_{c\text{P}}) = F(\gamma)$ and $K(\gamma,\varepsilon,\Lambda_{d\text{Q}},\Delta_{c\text{Q}}) = K(\gamma)$ depend only on γ. Moreover, we have

$$\big[P(\gamma)B + C_2'D_2\big]F(\gamma) + \big[A'P(\gamma) + P(\gamma)A + C_2'C_2 + \gamma^{-2}P(\gamma)EE'P(\gamma)\big] = 0.$$

Hence, Σ_{cmp} reduces to

$$\Sigma_{\text{cmp}} \;:\; \begin{cases} \dot{v} = A_{\text{cmp}}\, v + B_{\text{cmp}}\, y, \\ u = C_{\text{cmp}}\, v + \;\; 0 \;\; y, \end{cases}$$

where

$$\begin{aligned} A_{\text{cmp}} &= A + \gamma^{-2}EE'P(\gamma) + BF(\gamma) \\ &\quad + \big[I - \gamma^{-2}Q(\gamma)P(\gamma)\big]^{-1}K(\gamma)\big[C_1 + \gamma^{-2}D_1E'P(\gamma)\big], \\ B_{\text{cmp}} &= -\big[I - \gamma^{-2}Q(\gamma)P(\gamma)\big]^{-1}K(\gamma), \\ C_{\text{cmp}} &= F(\gamma). \end{aligned}$$

This corresponds to the regular case, and is the central controller given in Doyle *et al.* [49]. Ⓡ

Remark 7.3.3. Finally, we would like to note that if Assumptions 7.M.3 and 7.M.6, i.e., the geometric conditions, are not satisfied, one can use the iterative procedure in Chapter 6 to find an approximation of the infimum, say $\tilde{\gamma}^*$. Moreover, the algorithm for finding the γ-suboptimal output feedback laws can also be modified to handle this situation. To be more specific, one only needs to modify Steps 7.M.1 and 8.M.2 slightly as follows:

Step 7.M.1m: Define an auxiliary full state feedback system

$$\begin{cases} \dot{x} = A\ x + B\ u + E\ w, \\ y = \quad x \\ h = C_2\ x + D_2\ u + D_{22}\ w, \end{cases}$$

and proceed to perform Steps 7.F.1, 7.F.2m, and 7.F.3 to 7.F.5 of Section 7.2 to obtain the gain matrix $F(\gamma,\varepsilon,\Lambda_{d\text{P}},\Delta_{c\text{P}})$ and P_x. Let $P_{x\text{P}} := P_x$. Also, define

$$P(\gamma) := (\Gamma_{s\text{P}}^{-1})' \begin{bmatrix} P_{x\text{P}} & 0 \\ 0 & 0 \end{bmatrix} \Gamma_{s\text{P}}^{-1}. \tag{7.3.9}$$

Step 7.M.2m: Define another auxiliary full state feedback system as follows,

$$\Sigma_{\text{Q}} : \begin{cases} \dot{x} = A'\ x + C_1'\ u + C_2'\ w, \\ y = \quad x \\ h = E'\ x + D_1'\ u + D_{22}'\ w, \end{cases}$$

and proceed to perform Steps 7.F.1, 7.F.2m, and 7.F.3 to 7.F.5 of Section 7.2 but for this auxiliary system to obtain $F(\gamma,\varepsilon,\Lambda_{d\mathrm{P}},\Delta_{c\mathrm{P}})$ and P_x. Let $K(\gamma,\varepsilon,\Lambda_{d\mathrm{Q}},\Delta_{c\mathrm{Q}}) := F(\gamma,\varepsilon,\Lambda_{d\mathrm{Q}},\Delta_{c\mathrm{Q}})'$ and $Q_{x\mathrm{Q}} := P_x$. Also, define

$$Q(\gamma) := (\Gamma_{s\mathrm{Q}}^{-1})' \begin{bmatrix} Q_{x\mathrm{Q}} & 0 \\ 0 & 0 \end{bmatrix} \Gamma_{s\mathrm{Q}}^{-1}. \tag{7.3.10}$$

The last step of the algorithm, i.e., Step 7.M.3, remains unchanged. All results in Theorems 7.3.1 and 7.3.2 are valid for this situation as well. However, the output feedback control law is not of closed-form any more. ®

Again, we illustrate our results in the following example.

Example 7.3.1. Consider a given measurement feedback system characterized by matrices A, B, E, C_2, D_2 and D_{22} as given in Example 7.2.1 of the previous section and

$$C_1 = \begin{bmatrix} 0 & -2 & -3 & -2 & -1 \\ 1 & 2 & 3 & 2 & 1 \end{bmatrix}, \quad D_1 = \begin{bmatrix} 1 & 0 \\ 0 & 0 \end{bmatrix}. \tag{7.3.11}$$

We first note that the pair (A, C_1) is detectable, and the system (A, E, C_1, D_1) is invertible (hence Assumption 7.M.6 is satisfied) and of nonminimum phase with invariant zeros at $\{-1.630662,\ -3.593415,\ 0.521129 \pm j0.363042\}$. It was obtained in Example 6.3.1 that

$$\gamma^* = 13.638725.$$

The closed-form to the output feedback suboptimal controllers as in (7.3.5) to (7.3.8) with $F(\gamma,\varepsilon,\lambda_{d\mathrm{P}},\Delta_{c\mathrm{P}})$ given by (7.2.22),

$$K(\gamma,\varepsilon,\lambda_{d\mathrm{Q}},\Delta_{c\mathrm{Q}}) = [\,K_0 \quad K_1\,], \tag{7.3.12}$$

where

$$K_0 = \begin{bmatrix} \dfrac{-43.91\gamma^4 + 4257.86\gamma^2 - 97026.13}{7.12\gamma^4 - 790.42\gamma^2 + 19405.23} \\[2mm] \dfrac{-12.45\gamma^4 + 372.65\gamma^2 - 0.02}{7.12\gamma^4 - 790.42\gamma^2 + 19405.23} \\[2mm] \dfrac{-48.44\gamma^4 + 1803.08\gamma^2 + 0.02}{7.12\gamma^4 - 790.42\gamma^2 + 19405.23} \\[2mm] \dfrac{62.57\gamma^4 - 1212.58\gamma^2 - 38810.46}{7.12\gamma^4 - 790.42\gamma^2 + 19405.23} \\[2mm] \dfrac{17.80\gamma^4 - 83.04\gamma^2 - 19405.21}{7.12\gamma^4 - 790.42\gamma^2 + 19405.23} \end{bmatrix},$$

and

$$K_1 = \begin{bmatrix} -5 + 0.090909\frac{\lambda_{dQ}}{\varepsilon} - \frac{(0.24\gamma^4 - 10.14\gamma^2)\lambda_{dQ}}{(7.12\gamma^4 - 790.42\gamma^2 + 19405.23)\varepsilon} \\ -0.363636 - \frac{(-2.39\gamma^4 + 190.91\gamma^2)\lambda_{dQ}}{(7.12\gamma^4 - 790.42\gamma^2 + 19405.23)\varepsilon} \\ -0.382726 - \frac{(2.04\gamma^4 - 108.95\gamma^2)\lambda_{dQ}}{(7.12\gamma^4 - 790.42\gamma^2 + 19405.23)\varepsilon} \\ -2.545451 + 0.272727\frac{\lambda_{dQ}}{\varepsilon} - \frac{(-1.13\gamma^4 + 14.86\gamma^2)\lambda_{dQ}}{(7.12\gamma^4 - 790.42\gamma^2 + 19405.23)\varepsilon} \\ -1.272726 + 0.363636\frac{\lambda_{dQ}}{\varepsilon} - \frac{(0.69\gamma^4 - 74.56\gamma^2)\lambda_{dQ}}{(7.12\gamma^4 - 790.42\gamma^2 + 19405.23)\varepsilon} \end{bmatrix},$$

with $\lambda_{dQ} < 0$, and

$$P(\gamma) = \frac{1}{0.132909\gamma^2 - 5.560084} \times$$

$$\begin{bmatrix} 0.42770\gamma^2 & -0.29658\gamma^2 & 0.16367\gamma^2 & 0 & 0 \\ -0.29658\gamma^2 & -15.8338 + 0.58415\gamma^2 & 3.0091 - 0.18543\gamma^2 & 0 & 0 \\ 0.16367\gamma^2 & 3.0091 - 0.18543\gamma^2 & -5.1368 + 0.18543\gamma^2 & 0 & 0 \\ 0 & 0 & 0 & 0 & 0 \\ 0 & 0 & 0 & 0 & 0 \end{bmatrix},$$

$$Q(\gamma) = \frac{\gamma^2}{0.071193\gamma^4 - 7.904171\gamma^2 + 194.052288}\left(\gamma^2 Q_1 + Q_0\right),$$

where

$$Q_1 = \begin{bmatrix} 0.083104 & 0.124442 & 0.484459 & -0.768087 & -0.249208 \\ 0.124423 & 1.778706 & 0.340500 & -1.759522 & -1.184163 \\ 0.484459 & 0.340500 & 2.917279 & -4.330299 & -1.256601 \\ -0.768087 & -1.759522 & -4.330299 & 7.332315 & 2.613520 \\ -0.249208 & -1.184163 & -1.256601 & 2.613520 & 1.160281 \end{bmatrix},$$

and

$$Q_0 = \begin{bmatrix} -3.0576430 & -3.7265760 & -18.030782 & 27.934279 & 8.7345960 \\ -3.7265760 & -122.50790 & 6.5460280 & 79.188507 & 70.727376 \\ -18.030781 & 6.5460280 & -113.22255 & 153.81266 & 36.981101 \\ 27.934279 & 79.188509 & 153.81266 & -272.47959 & -102.79025 \\ 8.7345960 & 70.727376 & 36.981101 & -102.79025 & -55.552230 \end{bmatrix}.$$

As in the previous example, we demonstrate our results in Figure 7.3.1 by the plots of maximum singular values of the closed-loop transfer function matrix for several values of γ and ε. Note that in Figure 7.3.1, we choose $\lambda_{dP} = -1$, $\Delta_{cP} = 3$ and $\lambda_{dQ} = -1$. Note that since Σ_Q for this example is left invertible, the gain $K(\gamma, \varepsilon, \lambda_{dQ}, \Delta_{cQ})$ depends only on γ, ε and λ_{dQ}. ⊡

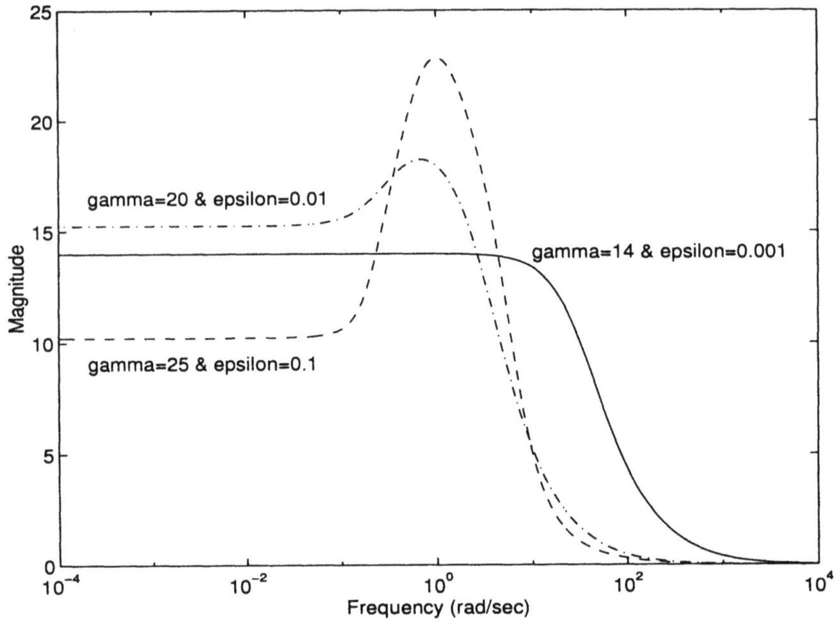

Figure 7.3.1: Maximum singular values of T_{hw} (output feedback case).

7.4. Reduced Order Output Feedback

In this section, the H_∞ control problem with reduced order measurement output feedback is investigated. For the case that some entries of the measurement vector are not noise-corrupted, we show that one can find dynamic compensators of a lower dynamical order. More specifically, we will show that there exists a time-invariant, finite-dimensional dynamic compensator $\Sigma_{\rm cmp}$ of the form

$$\Sigma_{\rm cmp} \ : \ \begin{cases} \dot{v} = A_{\rm cmp} \ v + B_{\rm cmp} \ y, \\ u = C_{\rm cmp} \ v + D_{\rm cmp} \ y, \end{cases} \tag{7.4.1}$$

and with a McMillan degree $n - \text{rank}[C_1, \ D_1] + \text{rank}(D_1) \leq n$ for Σ of (7.3.1) such that the resulting closed loop system is internally stable and the closed loop transfer function from w to h has an H_∞ norm less than $\gamma > \gamma^*$. Moreover, we give an explicit construction of such a reduced order compensator. The result of this section was previously reported in [126] while the original idea for how to construct a reduced order observer for a general system was given by Chen *et al.* [29].

Let γ^* be the infimum for the given system Σ of (7.3.1) and let $\gamma > \gamma^*$ be given. Using the result of the previous section, one can easily find two positive semi-definite matrices P and Q which satisfy

$$F_\gamma(P) := \begin{bmatrix} A'P + PA + C_2'C_2 + PEE'P/\gamma^2 & PB + C_2'D_2 \\ B'P + D_2'C_2 & D_2'D_2 \end{bmatrix} \geq 0,$$

and

$$G_\gamma(Q) := \begin{bmatrix} AQ + QA' + EE' + QC_2'C_2Q/\gamma^2 & QC_1' + ED_1' \\ C_1Q + D_1E' & D_1D_1' \end{bmatrix} \geq 0,$$

respectively, i.e., P and Q are the solutions of the quadratic matrix inequalities $F_\gamma(P) \geq 0$ and $G_\gamma(Q) \geq 0$. Next, we define an auxiliary system,

$$\Sigma_{\mathrm{PQ}} : \begin{cases} \dot{x}_{\mathrm{PQ}} = A_{\mathrm{PQ}}\, x_{\mathrm{PQ}} + B_{\mathrm{PQ}}\, u + E_{\mathrm{PQ}}\, w_{\mathrm{PQ}}, \\ y = C_{1\mathrm{P}}\, x_{\mathrm{PQ}} + D_{1\mathrm{PQ}}\, w_{\mathrm{PQ}}, \\ h_{\mathrm{PQ}} = C_{2\mathrm{P}}\, x_{\mathrm{PQ}} + D_{2\mathrm{P}}\, u, \end{cases} \tag{7.4.2}$$

where

$$\begin{bmatrix} C_{2\mathrm{P}}' \\ D_{2\mathrm{P}}' \end{bmatrix} [C_{2\mathrm{P}} \quad D_{2\mathrm{P}}] := F_\gamma(P), \quad \begin{bmatrix} E_{\mathrm{Q}} \\ D_{1\mathrm{PQ}} \end{bmatrix} [E_{\mathrm{Q}}' \quad D_{1\mathrm{PQ}}'] := G_\gamma(Q)$$

and

$$\left.\begin{aligned} A_{\mathrm{PQ}} &:= A + EE'P/\gamma^2 + (\gamma^2 I - QP)^{-1}QC_{2\mathrm{P}}'C_{2\mathrm{P}}, \\ B_{\mathrm{PQ}} &:= B + (\gamma^2 I - QP)^{-1}QC_{2\mathrm{P}}'D_{2\mathrm{P}}, \\ E_{\mathrm{PQ}} &:= (I - QP/\gamma^2)^{-1}E_{\mathrm{Q}}, \\ C_{1\mathrm{P}} &:= C_1 + D_1E'P/\gamma^2. \end{aligned}\right\} \tag{7.4.3}$$

It can be shown (see e.g., [124]) that i) $(A_{\mathrm{PQ}}, B_{\mathrm{PQ}}, C_{2\mathrm{P}}, D_{2\mathrm{P}})$ is right invertible and of minimum phase; and ii) $(A_{\mathrm{PQ}}, E_{\mathrm{PQ}}, C_{1\mathrm{P}}, D_{1\mathrm{PQ}})$ is left invertible and of minimum phase.

We will build the reduced order compensator upon the above auxiliary system and show later that it works for the original system Σ of (7.3.1) as well. Let us first eliminate states which can be directly observed and concentrate on those states which still need to be observed. In order to do this, we need to choose a suitable basis. Without loss of generality, but for simplicity of presentation, we assume that the matrices $C_{1\mathrm{P}}$ and $D_{2\mathrm{PQ}}$ are transformed in the following form:

$$C_{1\mathrm{P}} = \begin{bmatrix} 0 & C_{1,02} \\ I_k & 0 \end{bmatrix} \quad \text{and} \quad D_{2\mathrm{PQ}} = \begin{bmatrix} D_{1,0} \\ 0 \end{bmatrix}. \tag{7.4.4}$$

Thus, the system Σ_{PQ} as in (7.4.2) can be partitioned as follows,

$$\begin{cases} \begin{pmatrix} \dot{x}_1 \\ \dot{x}_2 \end{pmatrix} = \begin{bmatrix} A_{11} & A_{12} \\ A_{21} & A_{22} \end{bmatrix} \begin{pmatrix} x_1 \\ x_2 \end{pmatrix} + \begin{bmatrix} B_1 \\ B_2 \end{bmatrix} u + \begin{bmatrix} E_1 \\ E_2 \end{bmatrix} w_{PQ}, \\ \begin{pmatrix} y_0 \\ y_1 \end{pmatrix} = \begin{bmatrix} 0 & C_{1,02} \\ I_k & 0 \end{bmatrix} \begin{pmatrix} x_1 \\ x_2 \end{pmatrix} + \begin{bmatrix} D_{1,0} \\ 0 \end{bmatrix} w_{PQ}, \\ h_{PQ} = C_{2P} \quad x_{PQ} + D_{2P} \quad u, \end{cases} \tag{7.4.5}$$

where $(x_1', x_2')' = x_{PQ}$ and $(y_0', y_1')' = y$. We observe that $y_1 = x_1$ is already available and need not be estimated. Thus we need to estimate only the state variable x_2. We first rewrite the state equation for x_1 in terms of the output y_1 and state x_2 as follows,

$$\dot{y}_1 = A_{11} y_1 + A_{12} x_2 + B_1 u + E_1 w_{PQ}, \tag{7.4.6}$$

where y_1 and u are known signals. Then, (7.4.6) can be rewritten as

$$\tilde{y} = A_{12} x_2 + E_1 w_{PQ} = \dot{y}_1 - A_{11} x_1 - B_1 u. \tag{7.4.7}$$

Thus, observation of x_2 is made via (7.4.7) as well as by

$$y_0 = C_{1,02} x_2 + D_{1,0} w_{PQ}.$$

Now, a reduced order system suitable for estimating the state x_2 is given by

$$\begin{cases} \dot{x}_2 = A_{22} \quad x_2 + [\,A_{21} \quad B_2\,] \begin{pmatrix} y_1 \\ u \end{pmatrix} + E_2 \quad w_{PQ}, \\ \begin{pmatrix} y_0 \\ \tilde{y} \end{pmatrix} = \begin{bmatrix} C_{1,02} \\ A_{12} \end{bmatrix} x_2 + \begin{bmatrix} D_{1,0} \\ E_1 \end{bmatrix} w_{PQ}. \end{cases} \tag{7.4.8}$$

Before we proceed to construct the reduced order observer, we present in the following a key lemma which plays an important role in our design.

Lemma 7.4.1. Let Σ_R denote the subsystem characterized by

$$(A_R, B_R, C_R, D_R) := \left(A_{22},\ E_2,\ \begin{bmatrix} C_{1,02} \\ A_{12} \end{bmatrix},\ \begin{bmatrix} D_{1,0} \\ E_1 \end{bmatrix} \right).$$

Then we have

1. Σ_R is (non-)minimum phase if and only if $(A_{PQ}, E_{PQ}, C_{1P}, D_{1PQ})$ is (non-) minimum phase.

2. Σ_R is detectable if and only if $(A_{PQ}, E_{PQ}, C_{1P}, D_{1PQ})$ is detectable.

3. Σ_R is left invertible if and only if $(A_{PQ}, E_{PQ}, C_{1P}, D_{1PQ})$ is left invertible.

4. Invariant zeros of $\Sigma_{\rm R}$ are the same as those of $(A_{\rm PQ}, E_{\rm PQ}, C_{\rm 1P}, D_{\rm 1PQ})$.

5. Orders of infinite zeros of the reduced order system, $\Sigma_{\rm R}$, are reduced by one from those of $(A_{\rm PQ}, E_{\rm PQ}, C_{\rm 1P}, D_{\rm 1PQ})$. ▣

Proof. It follows from Proposition 2.2.1 of Chen [12]. ▣

Now, based on (7.4.8), we can construct a reduced order observer of x_2 as,

$$\dot{\hat{x}}_2 = A_{22}\hat{x}_2 + A_{21}y_1 + B_2u + K_{\rm R}\left(\begin{bmatrix} y_0 \\ \tilde{y} \end{bmatrix} - \begin{bmatrix} C_{1,02} \\ A_{12} \end{bmatrix}\hat{x}_2\right),$$

and

$$\hat{x}_{\rm PQ} = \begin{bmatrix} 0 \\ I_{n-k} \end{bmatrix}\hat{x}_2 + \begin{bmatrix} I_k \\ 0 \end{bmatrix} y_1,$$

where $K_{\rm R}$ is the observer gain matrix for the reduced order system and is chosen such that

$$A_{22} - K_{\rm R}\begin{bmatrix} C_{1,02} \\ A_{12} \end{bmatrix},$$

is asymptotically stable. In order to remove the dependency on $\dot{y}_1$, let us partition $K_{\rm R} = [K_{\rm R0},\ K_{\rm R1}]$ to be compatible with the dimensions of the output $(y_0',\ y_1')'$. Then (see e.g., [71]), one can define a new variable $v := \hat{x}_2 - K_{\rm R1}y_1$ and obtain a new dynamic equation,

$$\begin{aligned}\dot{v} &= (A_{22} - K_{\rm R0}C_{1,02} - K_{\rm R1}A_{12})v + (B_2 - K_{\rm R1}B_1)u \\ &\quad + [K_{\rm R0},\ A_{21} - K_{\rm R1}A_{11} + (A_{22} - K_{\rm R0}C_{1,02} - K_{\rm R1}A_{12})K_{\rm R1}]\begin{pmatrix} y_0 \\ y_1 \end{pmatrix}. \end{aligned} \tag{7.4.9}$$

Thus by implementing (7.4.9), $\hat{x}_2$ can be obtained without generating $\dot{y}_1$.

Theorem 7.4.1. Let $\Sigma_{\rm PQ}$ be given by (7.4.2). Then there exist for every $\varepsilon > 0$, a state feedback gain F and a reduced order observer gain matrix $K_{\rm R}$ such that the following reduced order observer based controller,

$$\Sigma_{\rm cmp}: \begin{cases} \dot{v} = (A_{22} - K_{\rm R0}C_{1,02} - K_{\rm R1}A_{12})v + (B_2 - K_{\rm R1}B_1)u \\ \qquad + [K_{\rm R0},\ A_{21} - K_{\rm R1}A_{11} + (A_{22} - K_{\rm R0}C_{1,02} - K_{\rm R1}A_{12})K_{\rm R1}]\,y, \\ u = -F\hat{x}_{\rm PQ} = -F\begin{bmatrix} 0 \\ I_{n-k} \end{bmatrix} v - F\begin{bmatrix} 0 & I_k \\ 0 & K_{\rm R1} \end{bmatrix} y, \end{cases} \tag{7.4.10}$$

when applied to $\Sigma_{\rm PQ}$ is internally stabilizing and yields an H_∞ norm of the closed-loop transfer matrix from $w_{\rm PQ}$ to $h_{\rm PQ}$ strictly less than ε. Moreover, if $\Sigma_{\rm cmp}$ is applied to the original system Σ of (7.3.1), then the resulting closed-loop system comprising Σ and $\Sigma_{\rm cmp}$ is internally stable and the H_∞ norm of the closed-loop transfer matrix from w to h is less than γ. ▣

Proof. See Subsection 7.5.E. ⊠

Remark 7.4.1. The gain matrix F and $K_{\rm R}$ can be found using a systematic procedure given in Chapter 8. Ⓡ

Remark 7.4.2. In the case that the given system Σ of (7.3.1) is regular, then the controller (7.4.10) reduces to the well-known full order observer based control design for the regular H_∞-optimization as given in [49]. Ⓡ

We illustrate the above result with a numerical example.

Example 7.4.1. We again consider a given measurement feedback system characterized by matrices A, B, E, C_2, D_2 as in Example 7.2.1 and C_1, D_1 as in Example 7.3.1. The infimum for this problem is $\gamma^* = 13.638725$. In what follows, we will construct a reduced order measurement output feedback control law that makes the H_∞ norm of the resulting closed-loop transfer matrix from w to h strictly less that $\gamma = 14$. Following the procedure, we obtain an auxiliary system $\Sigma_{\rm PQ}$ of the form (7.4.2) with

$$A_{\rm PQ} = \begin{bmatrix} 4.2254 & -0.7415 & 4.1946 & 0 & 1.4335 \\ -11.8293 & 7.6804 & -13.7917 & 0 & -0.7102 \\ 19.4695 & -9.0672 & 22.8277 & 0 & 4.0975 \\ -17.4591 & 10.0905 & -19.5135 & 1 & -2.1038 \\ 1.2144 & 0.5197 & 1.4176 & 1 & -0.0983 \end{bmatrix},$$

$$B_{\rm PQ} = \begin{bmatrix} 0.9327 & 0 & 0 \\ -4.4755 & 0 & 0 \\ 7.8569 & 0 & 0 \\ -6.3735 & 0 & 1 \\ 0.1940 & 1 & 0 \end{bmatrix}, \quad E_{\rm PQ} = \begin{bmatrix} 18.5391 & 0.8299 \\ -62.8474 & -29.3560 \\ 102.9481 & 28.5462 \\ -97.9601 & -22.3008 \\ -0.0958 & 3.1029 \end{bmatrix},$$

$$C_{\rm 1P} = \begin{bmatrix} 0.1044 & -2.0724 & -2.9601 & -2 & -1 \\ 1 & 2 & 3 & 2 & 1 \end{bmatrix},$$

$$C_{\rm 2P} = \begin{bmatrix} 3.0616 & -0.9592 & 2.8464 & 0 & 0.6772 \\ -1.0146 & -1.3601 & 0.6330 & 0 & -0.7358 \end{bmatrix},$$

and

$$D_{\rm 1P} = \begin{bmatrix} 0.9409 & -0.3383 \\ 0 & 0 \end{bmatrix}, \quad D_{\rm 2PQ} = \begin{bmatrix} 0.9409 & -0.3383 \\ 0 & 0 \end{bmatrix}.$$

It is simple to show that the transformation T_s and T_o,

$$T_s = \begin{bmatrix} 1 & -2 & -3 & -2 & -1 \\ 0 & 1 & 0 & 0 & 0 \\ 0 & 0 & 1 & 0 & 0 \\ 0 & 0 & 0 & 1 & 0 \\ 0 & 0 & 0 & 0 & 1 \end{bmatrix}, \quad T_o = \begin{bmatrix} 1 & 0.1044 \\ 0 & 1 \end{bmatrix},$$

will transform C_1 and D_1 to the following form,

$$T_o^{-1} C_{1\text{P}} T_s = \left[\begin{array}{c|c} 0 & C_{1,02} \\ \hline I_k & 0 \end{array}\right] = \left[\begin{array}{c|cccc} 0 & -2.2811 & -3.2732 & -2.2087 & -1.1044 \\ \hline 1 & 0 & 0 & 0 & 0 \end{array}\right]$$

and

$$T_o^{-1} D_{1\text{PQ}} = \left[\begin{array}{c} D_{1,0} \\ \hline 0 \end{array}\right] = \left[\begin{array}{cc} 0.9409 & -0.3383 \\ \hline 0 & 0 \end{array}\right].$$

Moreover, we have

$$T_s^{-1} A T_s = \left[\begin{array}{c|c} A_{11} & A_{12} \\ \hline A_{21} & A_{22} \end{array}\right]$$

$$= \left[\begin{array}{c|cccc} 5.2714 & -2.4247 & -8.3291 & -7.5428 & 2.7283 \\ \hline -11.8293 & 31.3390 & 21.6962 & 23.6586 & 11.1191 \\ 19.4695 & -48.0062 & -35.5807 & -38.9390 & -15.3720 \\ -17.4591 & 45.0087 & 32.8639 & 35.9182 & 15.3553 \\ 1.2144 & -1.9092 & -2.2257 & -1.4289 & -1.3127 \end{array}\right],$$

$$T_s^{-1} B = \left[\begin{array}{c} B_1 \\ \hline B_2 \end{array}\right] = \left[\begin{array}{ccc} 2.9993 & 1 & 2 \\ \hline -4.4755 & 0 & 0 \\ 7.8569 & 0 & 0 \\ -6.3735 & 0 & 1 \\ 0.1940 & 1 & 0 \end{array}\right],$$

$$T_s^{-1} E = \left[\begin{array}{c} E_1 \\ \hline E_2 \end{array}\right] = \left[\begin{array}{cc} 5.6724 & -13.7425 \\ \hline -62.8474 & -29.3560 \\ 102.9481 & 28.5462 \\ -97.9601 & -22.3008 \\ -0.0958 & 3.1029 \end{array}\right],$$

and $A_{\text{R}} = A_{22}$, $E_{\text{R}} = E_2$,

$$C_{\text{R}} = \begin{bmatrix} -2.2811 & -3.2732 & -2.2087 & -1.1044 \\ -2.4247 & -8.3291 & -7.5428 & 2.7283 \end{bmatrix},$$

and

$$D_{\text{R}} = \begin{bmatrix} 0.9409 & -0.3383 \\ 5.6724 & -13.7425 \end{bmatrix}.$$

Using the algorithm given in Chapter 8, we obtain a gain matrix F,

$$F T_s = \begin{bmatrix} -1.5656 & 4.7579 & 2.1737 & 3.1311 & 1.5656 \\ -299.4859 & 555.2644 & 742.6408 & 597.9718 & 189.8014 \\ 7.4811 & -14.6842 & -19.0100 & -16.9623 & -5.3773 \end{bmatrix},$$

and

$$K_{\text{R}} = \left[\begin{array}{c|c} K_{\text{R}0} & K_{\text{R}1} \end{array}\right] = \left[\begin{array}{c|c} 93.5515 & -4.4388 \\ -143.1777 & 5.6013 \\ 133.7360 & -4.9145 \\ -1.4788 & 0.2622 \end{array}\right].$$

Finally, we obtain a reduced order output feedback controller of the form (7.4.1) with

$$A_{\rm cmp} = 10^3 \cdot \begin{bmatrix} -2.5903 & -3.4139 & -2.7089 & -0.9269 \\ 3.3280 & 4.3868 & 3.4717 & 1.1995 \\ -2.9478 & -3.8917 & -3.0775 & -1.0641 \\ 0.6986 & 0.9299 & 0.7488 & 0.2393 \end{bmatrix},$$

$$C_{\rm cmp} = \begin{bmatrix} 4.7579 & 2.1737 & 3.1311 & 1.5656 \\ 555.2644 & 742.6408 & 597.9718 & 189.8014 \\ -14.6842 & -19.0100 & -16.9623 & -5.3773 \end{bmatrix},$$

and

$$B_{\rm cmp} = 10^3 \cdot \begin{bmatrix} -0.0936 & -4.1798 \\ 0.1432 & 5.3492 \\ -0.1337 & -4.7217 \\ 0.0015 & 1.1362 \end{bmatrix}, \quad D_{\rm cmp} = \begin{bmatrix} 0 & 22.3556 \\ 0 & 894.3952 \\ 0 & -33.1683 \end{bmatrix},$$

which yields the poles of the closed-loop system, when it is applied to the given system, at

$$-97.337, -34.72, -3.591, -1.848, -1.632, -0.248, -1.346, -0.765, -1.$$

Obviously, they are in the stable region. The singular value plots of the resulting closed-loop transfer matrix T_{hw} in Figure 7.4.1 also show that $\|T_{hw}\|_\infty$ is indeed less than 14, the given γ. ▣

7.5. Proofs of Main Results

7.5.A. Proof of Theorem 7.2.1

We need to recall the following two lemmas in order to proceed with our proof of Theorem 7.2.1.

Lemma 7.5.1. Let an auxiliary system $\Sigma_{\rm aux}$ be characterized by

$$\Sigma_{\rm aux} : \begin{cases} \dot{x}_x = A_x \, x_x + B_x \, u_x + E_x \, w_x, \\ h_x = C_x \, x_x + D_x \;\; u_x \,, \end{cases} \tag{7.5.1}$$

where

$$A_x = A_{11\rm P}, \quad B_x = [\, B_{11\rm P} \quad A_{13\rm P} \,], \quad E_x = \begin{bmatrix} E_{a\rm P}^{+} \\ 0 \end{bmatrix},$$

and

$$C_x = \Gamma_{o\rm P} \begin{bmatrix} 0 & 0 \\ 0 & 0 \\ 0 & C_{b\rm P} \end{bmatrix}, \quad D_x = \Gamma_{o\rm P} \begin{bmatrix} I & 0 \\ 0 & C_{d\rm P} C_{d\rm P}' \\ 0 & 0 \end{bmatrix}.$$

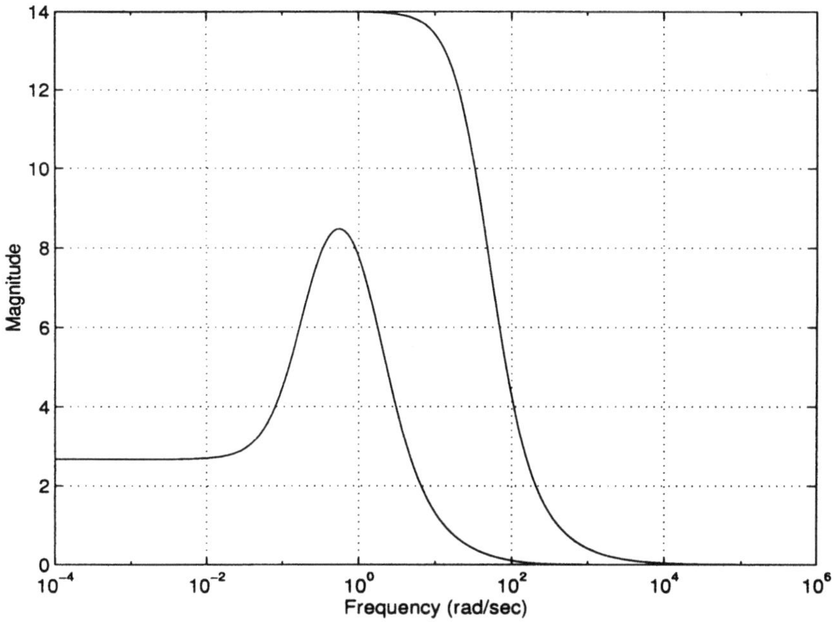

Figure 7.4.1: Max. singular values of T_{hw} under reduced order output feedback.

Then $\Sigma_{\rm aux}$ comprising the state feedback law $u_x = -F_{11}(\gamma)x_x$ is internally stable, i.e.,

$$\lambda(A^{c}_{11P}) = \lambda\{A_{11P} - [B_{11P},\, A_{13P}]F_{11}(\gamma)\} = \lambda\{A_x - B_xF_{11}(\gamma)\} \subset \mathbb{C}^-, \quad (7.5.2)$$

and the resulting closed-loop transfer function from w_x to h_x has H_∞ norm less than γ, i.e.,

$$\|T_{h_x w_x}\|_\infty = \left\| \Gamma_{oP} \begin{bmatrix} -F_{11}(\gamma) \\ [0 \quad C_{bP}] \end{bmatrix} (sI - A^{c}_{11P})^{-1} \begin{bmatrix} E^{+}_{aP} \\ 0 \end{bmatrix} \right\|_\infty < \gamma. \quad (7.5.3)$$

That is $u_x = -F_{11}(\gamma)x_x$ is a γ-suboptimal control law for $\Sigma_{\rm aux}$. ◻

Proof. We first note that Γ_{oP} is nonsingular and $C_{dP}C'_{dP} = I$ which implies that D_x is injective. Furthermore, it is simple to verify that the invariant zeros of (A_x, B_x, C_x, D_x) are given by $\lambda(A^{+}_{aaP})$, and are not on the imaginary axis. Hence $\Sigma_{\rm aux}$ satisfies the assumptions of the *regular* H_∞ control problem. Moreover, it is straightforward to verify that for any $\gamma > \gamma^*$,

$$P_x = (S_{xP} - \gamma^{-2}T_{xP})^{-1} > 0,$$

is the solution of the following well-known H_∞-CARE:

$$P_xA_x + A'_xP_x + \gamma^{-2}P_xE_xE'_xP_x + C'_xC_x$$

$$- [P_x B_x + C_x' D_x](D_x' D_x)^{-1}[B_x' P_x + D_x' C_x] = 0, \tag{7.5.4}$$

with

$$\lambda(A_{xx}^{c}) := \lambda\{A_x + \gamma^{-2} E_x E_x' P_x - B_x (D_x' D_x)^{-1} (B_x' Px) + D_x C_x)\} \in \mathbb{C}^{-}.$$

Then the results of Lemma 7.5.1 follow. ☒

Lemma 7.5.2. Let (A, B, C), where $A \in \mathbb{R}^{n\times n}$, $B \in \mathbb{R}^{n\times m}$ and $C \in \mathbb{R}^{p\times n}$, be right invertible and of minimum phase. Let $F(\varepsilon) \in \mathbb{R}^{m\times n}$ be parameterized in terms of ε and be of the form,

$$F(\varepsilon) = N(\varepsilon)\Gamma(\varepsilon)T(\varepsilon) + R(\varepsilon), \tag{7.5.5}$$

where $N(\varepsilon) \in \mathbb{R}^{m\times p}$, $\Gamma(\varepsilon) \in \mathbb{R}^{p\times p}$, $T(\varepsilon) \in \mathbb{R}^{p\times n}$ and $R(\varepsilon) \in \mathbb{R}^{m\times n}$. Also, $\Gamma(\varepsilon)$ is nonsingular. Moreover, assume that the following conditions hold:

1. $A + BF(\varepsilon)$ is asymptotically stable for all $0 < \varepsilon \le \varepsilon^*$ where $\varepsilon^* > 0$;
2. $T(\varepsilon) \to WC$ as $\varepsilon \to 0$ where W is some $p \times p$ nonsingular matrix;
3. as $\varepsilon \to 0$, $N(\varepsilon)$ tends to some finite matrix N such that $C(sI - A)^{-1}BN$ is invertible;
4. as $\varepsilon \to 0$, $R(\varepsilon)$ tends to some finite matrix R; and
5. $\Gamma^{-1}(\varepsilon) \to 0$ as $\varepsilon \to 0$.

Then as $\varepsilon \to 0$, we have $\left\|C[sI - A - BF(\varepsilon)]^{-1}\right\|_\infty \to 0$. 🄻

Proof. This is a dual version of Lemma 2.2 given by Saberi and Sannuti [113]. The proof of this lemma follows from similar arguments as in [113]. ☒

Now we are ready to proceed with the proof of Theorem 7.2.1. Note that $F(\gamma, \varepsilon, \Lambda_{d\text{P}}, \Delta_{d\text{P}})$ is constructed under the standard ATEA procedure. It can be shown using the techniques of the well-known singular perturbation theory as in Chen [12] that as $\varepsilon \to 0$, the eigenvalues of $A + BF(\gamma, \varepsilon, \Lambda_{d\text{P}}, \Delta_{d\text{P}})$ are given by $\lambda(A_{aa\text{P}}^{-}) \in \mathbb{C}^{-}$, $\lambda(A_{cc\text{P}}^{c}) \in \mathbb{C}^{-}$, $\Lambda_{d\text{P}}/\varepsilon \in \mathbb{C}^{-}$ and $\lambda(A_{11\text{P}}^{c}) \in \mathbb{C}^{-}$ (see Lemma 7.5.1). Hence the closed-loop is internally stable. Moreover, following the results of Chen [12], it can be shown that for any $\lambda_d \in \Lambda_{d\text{P}}/\varepsilon \in \mathbb{C}^{-}$, the corresponding right eigenvector, say $W(\varepsilon)$, satisfies

$$\lim_{\varepsilon\to 0} W(\varepsilon) = \bar{W} \in \mathcal{S}^{+}(\Sigma_{\text{P}}). \tag{7.5.6}$$

In fact, following the same arguments, one can show that as $\varepsilon \to 0$, the eigenvalues of $A+\gamma^{-2}EE'P(\gamma)+BF(\gamma,\varepsilon,\Lambda_{d\mathrm{P}},\Delta_{d\mathrm{P}})$, where $P(\gamma)$ is as defined in (7.3.2), are given by $\lambda(A_{aa\mathrm{P}}^{-}) \in \mathbb{C}^{-}$, $\lambda(A_{cc\mathrm{P}}^{c}) \in \mathbb{C}^{-}$, $\Lambda_{d\mathrm{P}}/\varepsilon \in \mathbb{C}^{-}$ and $\lambda(A_{xx}^{c}) \in \mathbb{C}^{-}$. We will use these properties later on in our proofs of other theorems. This proves the second part of Theorem 7.2.1.

Next, we show that the state feedback law $u = F(\gamma,\varepsilon,\Lambda_{d\mathrm{P}},\Delta_{d\mathrm{P}})x$ yields

$$\|T_{hw}\|_\infty = \left\| [C_2 + D_2F(\gamma,\varepsilon,\Lambda_{d\mathrm{P}},\Delta_{d\mathrm{P}})][sI - A - BF(\gamma,\varepsilon,\Lambda_{d\mathrm{P}},\Delta_{d\mathrm{P}})]^{-1}E \right\|_\infty < \gamma.$$

Without loss of generality but for simplicity of presentation, we assume that the nonsingular transformations $\Gamma_{s\mathrm{P}} = I$ and $\Gamma_{i\mathrm{P}} = I$, i.e., we assume that the system $(A, B, \Gamma_{o\mathrm{P}}^{-1}C_2, \Gamma_{o\mathrm{P}}^{-1}D_2)$ is in the form of the special coordinate basis. In view of (7.2.14), let us partition $F(\gamma,\varepsilon,\Lambda_{d\mathrm{P}},\Delta_{d\mathrm{P}})$ as,

$$F(\gamma,\varepsilon,\Lambda_{d\mathrm{P}},\Delta_{d\mathrm{P}}) = \bar{F}_0(\gamma) + \begin{bmatrix} 0 \\ \bar{F}(\gamma,\varepsilon,\Lambda_{d\mathrm{P}},\Delta_{d\mathrm{P}}) \end{bmatrix},$$

where

$$\bar{F}_0(\gamma) = -\begin{bmatrix} C_{0a\mathrm{P}}^{+} + F_{a0}^{+}(\gamma) & C_{0b\mathrm{P}} + F_{b0}(\gamma) & C_{0a\mathrm{P}}^{-} & C_{0c\mathrm{P}} & C_{0d\mathrm{P}} \\ 0 & 0 & 0 & 0 & 0 \\ 0 & 0 & 0 & 0 & 0 \end{bmatrix},$$

and

$$\bar{F}(\gamma,\varepsilon,\Lambda_{d\mathrm{P}},\Delta_{d\mathrm{P}}) = -\begin{bmatrix} E_{da\mathrm{P}}^{+} + \tilde{F}_{a1}^{+}(\gamma,\varepsilon,\Lambda_{d\mathrm{P}}) & E_{db\mathrm{P}} + \tilde{F}_{b1}(\gamma,\varepsilon,\Lambda_{d\mathrm{P}}) & E_{da\mathrm{P}}^{-} & E_{dc\mathrm{P}} & \tilde{F}_f(\varepsilon,\Lambda_{d\mathrm{P}}) + E_{d\mathrm{P}} \\ E_{ca\mathrm{P}}^{+} & E_{cb\mathrm{P}} & E_{ca\mathrm{P}}^{-} & \Delta_{c\mathrm{P}} & 0 \end{bmatrix}. \quad (7.5.7)$$

Then we have

$$\bar{C} = C_2 + D_2F(\gamma,\varepsilon,\Lambda_{d\mathrm{P}},\Delta_{d\mathrm{P}}) = \Gamma_{o\mathrm{P}}\begin{bmatrix} -F_{a0}^{+}(\gamma) & -F_{b0}(\gamma) & 0 & 0 & 0 \\ 0 & 0 & 0 & 0 & C_{d\mathrm{P}} \\ 0 & C_{b\mathrm{P}} & 0 & 0 & 0 \end{bmatrix},$$

and

$$\bar{A} = A + B\bar{F}_0(\gamma), \quad \bar{B} = \begin{bmatrix} 0 & 0 \\ 0 & 0 \\ 0 & 0 \\ 0 & B_{c\mathrm{P}} \\ B_{d\mathrm{P}} & 0 \end{bmatrix}. \quad (7.5.8)$$

With these definitions, we can write T_{hw} as

$$T_{hw} = \bar{C}\left[sI - \bar{A} - \bar{B}\,\bar{F}(\gamma,\varepsilon,\Lambda_{d\mathrm{P}},\Delta_{d\mathrm{P}})\right]^{-1}E.$$

Then in view of (7.5.7), it can easily be seen that $\bar{F}(\gamma, \varepsilon, \Lambda_{d\mathrm{P}}, \Delta_{d\mathrm{P}})$ has the form,

$$\bar{F}(\gamma, \varepsilon, \Lambda_{d\mathrm{P}}, \Delta_{d\mathrm{P}}) = N\Gamma(\varepsilon)T(\varepsilon) + R,$$

where

$$\Gamma(\varepsilon) = \mathrm{diag}\left\{\frac{1}{\varepsilon^{q_1}}, \frac{1}{\varepsilon^{q_2}}, \cdots, \frac{1}{\varepsilon^{q_{m_{d\mathrm{P}}}}}\right\}, \quad N = -\begin{bmatrix} I_{m_{d\mathrm{P}}} \\ 0 \end{bmatrix},$$

and

$$R = -\begin{bmatrix} E^+_{da\mathrm{P}} & E_{db\mathrm{P}} & E^-_{da\mathrm{P}} & E_{dc\mathrm{P}} & E_{d\mathrm{P}} \\ E^+_{ca\mathrm{P}} & E_{cb\mathrm{P}} & E^-_{ca\mathrm{P}} & \Delta_{c\mathrm{P}} & 0 \end{bmatrix},$$

while $T(\varepsilon)$ satisfies

$$T(\varepsilon) \to TC_m,$$

as $\varepsilon \to 0$, where

$$T = \mathrm{diag}\left[F_{1q_1}, F_{1q_2}, \cdots, F_{m_{d\mathrm{P}}q_{m_{d\mathrm{P}}}}\right],$$

and

$$C_m = [\, F^+_{a1}(\gamma) \quad F_{b1}(\gamma) \quad 0 \quad 0 \quad C_{d\mathrm{P}} \,]. \tag{7.5.9}$$

Using the same arguments as in Chen *et al.* [35], it is straightforward to show that the triple $(\bar{A}, \bar{B}, C_m)$ is right invertible and of minimum phase. Thus, it follows from Lemma 7.5.2 that

$$\left\| C_m \left[sI - \bar{A} - \bar{B}\,\bar{F}(\gamma, \varepsilon, \Lambda_{d\mathrm{P}}, \Delta_{d\mathrm{P}}) \right]^{-1} \right\|_\infty \to 0,$$

as $\varepsilon \to 0$. We should also note that following the same line of reasoning, one can show that the triple $(\bar{A} + \gamma^{-2}EE'P(\gamma), \bar{B}, C_m)$ is right invertible and of minimum phase, and moreover as $\varepsilon \to 0$,

$$\left\| C_m \left[sI - \bar{A} - \gamma^{-2}EE'P(\gamma) - \bar{B}\,\bar{F}(\gamma, \varepsilon, \Lambda_{d\mathrm{P}}, \Delta_{d\mathrm{P}}) \right]^{-1} \right\|_\infty \to 0. \tag{7.5.10}$$

Next, let

$$\bar{C} = \Gamma_{o\mathrm{P}} \begin{bmatrix} 0 \\ C_m \\ 0 \end{bmatrix} + C_e,$$

where

$$C_e = \Gamma_{o\mathrm{P}} \begin{bmatrix} -F^+_{a0}(\gamma) & -F_{b0}(\gamma) & 0 & 0 & 0 \\ -F^+_{a1}(\gamma) & -F_{b1}(\gamma) & 0 & 0 & 0 \\ 0 & C_{b\mathrm{P}} & 0 & 0 & 0 \end{bmatrix}.$$

We have

$$\|T_{hw}\|_\infty \to \left\| C_e \left[sI - \bar{A} - \bar{B}\,\bar{F}(\gamma, \varepsilon, \Lambda_{d\mathrm{P}}, \Delta_{d\mathrm{P}}) \right]^{-1} E \right\|_\infty,$$

as $\varepsilon \to 0$. Following the procedures of Chen [12] or Saberi, Chen and Sannuti [110], it can be shown that

$$C_e\left[sI-\bar{A}-\bar{B}\,\bar{F}(\gamma,\varepsilon,\Lambda_{d\mathrm{P}},\Delta_{d\mathrm{P}})\right]^{-1}E \to \Gamma_{o\mathrm{P}}\begin{bmatrix}-F_{11}(\gamma)\\ [0 \quad C_{b\mathrm{P}}]\end{bmatrix}(sI-A^c_{11\mathrm{P}})^{-1}\begin{bmatrix}E^+_{a\mathrm{P}}\\ 0\end{bmatrix},$$

pointwise in s as $\varepsilon \to 0$. Hence, the results of Theorem 7.2.1 follow readily from Lemma 7.5.1. ⊠

7.5.B. Proof of Theorem 7.2.2

Without loss of generality but for simplicity of presentation, we assume that the nonsingular state and input transformations $\Gamma_{s\mathrm{P}} = I$ and $\Gamma_{i\mathrm{P}} = I$, i.e., the system $(A, B, \Gamma_{o\mathrm{P}}^{-1}C_2, \Gamma_{o\mathrm{P}}^{-1}D_2)$ is in the form of the special coordinate basis. Then it is trivial to show that

$$A + BF(\gamma,\varepsilon,\Lambda_{d\mathrm{P}},\Delta_{c\mathrm{P}}) = \begin{bmatrix}\star & 0 & 0 & \star\\ \star & A^-_{aa\mathrm{P}} & 0 & \star\\ \star & 0 & A^c_{cc\mathrm{P}} & \star\\ \star & 0 & 0 & \star\end{bmatrix},$$

and

$$C_2 + D_2F(\gamma,\varepsilon,\Lambda_{d\mathrm{P}},\Delta_{c\mathrm{P}}) = \Gamma_{o\mathrm{P}}\begin{bmatrix}\star & 0 & 0 & 0\\ 0 & 0 & 0 & \star\\ \star & 0 & 0 & 0\end{bmatrix},$$

where $\star$s represent some sub-matrices which are of no interest to our proof. Hence, for any $\alpha \in \lambda(A^-_{aa\mathrm{P}}) \cup \lambda(A^c_{cc\mathrm{P}})$, the corresponding right eigenvector is in the kernel of $C_2 + D_2F(\gamma,\varepsilon,\Lambda_{d\mathrm{P}},\Delta_{c\mathrm{P}})$. This proves that α is an output decoupling zero of T_{hw}. ⊠

7.5.C. Proof of Theorem 7.3.1

For the sake of simplicity in presentation, we drop in the following proof the arguments of $F(\gamma,\varepsilon,\Lambda_{d\mathrm{P}},\Delta_{c\mathrm{P}})$ and $K(\gamma,\varepsilon,\Lambda_{d\mathrm{Q}},\Delta_{c\mathrm{Q}})$. Also, we assume without loss of generality that $\gamma = 1$. Thus, we will drop the dependency of γ in all the variables.

First, it is simple to verify that the positive semi-definite matrices P of (7.3.2) and Q of (7.3.4) satisfy

$$F_\gamma(P) := \begin{bmatrix}A'P + PA + C_2'C_2 + PEE'P & PB + C_2'D_2\\ B'P + D_2'C_2 & D_2'D_2\end{bmatrix} \ge 0,$$

and

$$G_\gamma(Q) := \begin{bmatrix}AQ + QA' + EE' + QC_2'C_2Q & QC_1' + ED_1'\\ C_1Q + D_1E' & D_1D_1'\end{bmatrix} \ge 0,$$

respectively, i.e., P and Q are the solutions of the quadratic matrix inequalities $F_\gamma(P) \geq 0$ and $G_\gamma(Q) \geq 0$. Moreover, the following auxiliary system,

$$\Sigma_{\mathrm{PQ}} : \begin{cases} \dot{x}_{\mathrm{PQ}} = A_{\mathrm{PQ}}\, x_{\mathrm{PQ}} + B_{\mathrm{PQ}}\, u + E_{\mathrm{PQ}}\, w_{\mathrm{PQ}}, \\ y = C_{1\mathrm{P}}\, x_{\mathrm{PQ}} + D_{1\mathrm{PQ}}\, w_{\mathrm{PQ}}, \\ h_{\mathrm{PQ}} = C_{2\mathrm{P}}\, x_{\mathrm{PQ}} + D_{2\mathrm{P}}\, u, \end{cases} \tag{7.5.11}$$

where

$$F_\gamma(P) = \begin{bmatrix} C'_{2\mathrm{P}} \\ D'_{2\mathrm{P}} \end{bmatrix} [\,C_{2\mathrm{P}} \quad D_{2\mathrm{P}}\,], \quad G_\gamma(Q) = \begin{bmatrix} E_{\mathrm{Q}} \\ D_{1\mathrm{PQ}} \end{bmatrix} [\,E'_{\mathrm{Q}} \quad D'_{1\mathrm{PQ}}\,],$$

and

$$\left. \begin{aligned} A_{\mathrm{PQ}} &:= A + EE'P + (I - QP)^{-1} Q C'_{2\mathrm{P}} C_{2\mathrm{P}}, \\ B_{\mathrm{PQ}} &:= B + (I - QP)^{-1} Q C'_{2\mathrm{P}} D_{2\mathrm{P}}, \\ E_{\mathrm{PQ}} &:= (I - QP)^{-1} E_{\mathrm{Q}}, \\ C_{1\mathrm{P}} &:= C_1 + D_1 E' P, \end{aligned} \right\} \tag{7.5.12}$$

has the following properties: 1) the subsystem $(A_{\mathrm{PQ}}, B_{\mathrm{PQ}}, C_{2\mathrm{P}}, D_{2\mathrm{P}})$ is right invertible and of minimum phase; and 2) the subsystem $(A_{\mathrm{PQ}}, E_{\mathrm{PQ}}, C_{1\mathrm{P}}, D_{1\mathrm{PQ}})$ is left invertible and of minimum phase.

The following lemma is due to [124].

Lemma 7.5.3. For any given compensator Σ_{cmp} of the form

$$\Sigma_{\mathrm{cmp}} : \begin{cases} \dot{v} = A_{\mathrm{cmp}}\, v + B_{\mathrm{cmp}}\, y, \\ u = C_{\mathrm{cmp}}\, v + D_{\mathrm{cmp}}\, y. \end{cases}$$

The following two statements are equivalent:

1. Σ_{cmp} applied to the system Σ defined by (7.3.1) is internally stabilizing and the resulting closed-loop transfer function from w to h has an H_∞ norm less than 1, i.e., $\|T_{hw}\|_\infty < 1$.

2. Σ_{cmp} applied to the new system Σ_{PQ} defined by (7.5.11) is internally stabilizing and the resulting closed loop transfer function from w_{PQ} to h_{PQ} has an H_∞ norm less than 1, i.e., $\|T_{h_{\mathrm{PQ}} w_{\mathrm{PQ}}}\|_\infty < 1$. ◻

Hence, it is sufficient to show Theorem 7.3.1 by showing that Σ_{cmp} of (7.3.5) to (7.3.8) applied to Σ_{PQ} achieves almost disturbance decoupling with internal stability. Observing that

$$C'_{2\mathrm{P}} C_{2\mathrm{P}} = A'P + PA + C'_2 C_2 + PEE'P \quad \text{and} \quad C'_{2\mathrm{P}} D_{2\mathrm{P}} = PB + C'_2 D_2,$$

it is simple to rewrite $A_{\rm cmp}$ of (7.3.6) as

$$A_{\rm cmp} = A_{\rm PQ} + B_{\rm PQ}F + (I - QP)^{-1}KC_{\rm 1P}.$$

Now it is trivial to see that $\Sigma_{\rm cmp}$ of (7.3.5) is simply the well-known full order observer based controller for the system $\Sigma_{\rm PQ}$ with state feedback gain F and observer gain $(I - QP)^{-1}K$. Hence the well-known separation principle holds. Also, noting the facts that $(A_{\rm PQ}, B_{\rm PQ}, C_{\rm 2P}, D_{\rm 2P})$ and $(A_{\rm PQ}, E_{\rm PQ}, C_{\rm 1P}, D_{\rm 1PQ})$ are of minimum phase, and right invertible and left invertible, respectively, it is sufficient to prove Theorem 7.3.1 by showing that as $\varepsilon \to 0$,

1. $A_{\rm PQ} + B_{\rm PQ}F$ is asymptotically stable;
2. $\left\|(C_{\rm 2P} + D_{\rm 2P}F)(sI - A_{\rm PQ} - B_{\rm PQ}F)^{-1}\right\|_\infty \to 0$;
3. $A_{\rm PQ} + (I - QP)^{-1}KC_{\rm 1P}$ is asymptotically stable; and
4. $\left\|[sI - A_{\rm PQ} - (I - QP)^{-1}KC_{\rm 1P}]^{-1}[E_{\rm PQ} + (I - QP)^{-1}KD_{\rm 1PQ}]\right\|_\infty \to 0$.

We shall introduce the following lemma for further development.

Lemma 7.5.4. As $\varepsilon \to 0$, we have

1. $A + EE'P + BF$ is asymptotically stable and

$$\left\|(C_{\rm 2P} + D_{\rm 2P}F)(sI - A - EE'P - BF)^{-1}\right\|_\infty \to 0; \tag{7.5.13}$$

2. $A + QC_2'C_2 + KC_1$ is asymptotically stable and

$$\left\|[sI - A - QC_2'C_2 - KC_1]^{-1}[E_{\rm Q} + KD_{\rm 1PQ}]\right\|_\infty \to 0. \tag{7.5.14}$$

Note that the roles of the above two statements are dual one another. ◻

Proof. It is shown in the proof of Theorem 7.2.1 that for $\varepsilon \to 0$, the matrix $A+EE'P+BF$ is asymptotically stable. In what follows, we will show (7.5.13). By some elementary algebra, it can be shown that

$$C_{\rm 2P} = \Gamma_{o\rm P} \begin{bmatrix} C_{0a\rm P}^{+} + F_{a0}^{+} & C_{0b\rm P} + F_{b0} & C_{0a\rm P}^{-} & C_{0c\rm P} & C_{0d\rm P} \\ F_{a1}^{+} & F_{b1} & 0 & 0 & C_{d\rm P} \\ 0 & 0 & 0 & 0 & 0 \end{bmatrix} \Gamma_{s\rm P}^{-1},$$

and

$$D_{\rm 2P} = D_2 = \Gamma_{o\rm P} \begin{bmatrix} I & 0 & 0 \\ 0 & 0 & 0 \\ 0 & 0 & 0 \end{bmatrix} \Gamma_{i\rm P}^{-1}.$$

Moreover,

$$[C_{2P}+D_{2P}F][sI-A-EE'P-BF]^{-1} = \begin{bmatrix} 0 \\ C_m \\ 0 \end{bmatrix} \left[sI-\bar{A}-EE'P-\bar{B}\,\bar{F}\right]^{-1},$$

where $\bar{A}$ and $\bar{B}$ are as in (7.5.8), $\bar{F}$ is as in (7.5.7) and C_m is given by (7.5.9). In view of (7.5.10), we have the result.

Item 2 of Lemma 7.5.4 is the dual version of Item 1. Hence, the results follow. This completes the proof of Lemma 7.5.4. ☒

Next, we will first show that $A_{PQ}+B_{PQ}F$ is asymptotically stable for some sufficiently small ε and

$$\|[C_{2P}+D_{2P}F][sI-A_{PQ}-B_{PQ}F]^{-1}\|_\infty \to 0,$$

as $\varepsilon \to 0$. In view of Lemma 7.5.4, we have

$$\begin{aligned} sI & - A_{PQ}-B_{PQ}F \\ &= sI-A-EE'P-BF-(I-QP)^{-1}QC_{2P}'[C_{2P}+D_{xP}F] \\ &= \{I-(I-QP)^{-1}QC_{2P}'[C_{2P}+D_{xP}F][sI-A-EE'P-BF]^{-1}\} \\ &\qquad\qquad \cdot[sI-A-EE'P-BF] \\ &\to sI-A-EE'P-BF \quad \text{pointwise in } s \text{ as } \varepsilon\to 0. \end{aligned}$$

This implies that $A_{PQ}+B_{PQ}F$ is asymptotically stable for sufficiently small ε, and

$$\begin{aligned} [C_{2P} & + D_{2P}F][sI-A_{PQ}-B_{PQ}F]^{-1} \\ &= [C_{2P}+D_{2P}F][sI-A-EE'P-BF]^{-1} \\ &\quad \cdot\{I-(I-QP)^{-1}QC_{2P}'[C_{2P}+D_{xP}F][sI-A-EE'P-BF]^{-1}\}^{-1} \\ &\to 0, \quad \text{pointwise in } s \text{ as } \varepsilon\to 0. \end{aligned} \tag{7.5.15}$$

Again, in view of Lemma 7.5.4 and

$$C_{2P}'C_{2P} = A'P+PA+C_2'C_2+PEE'P,$$

$$E_QE_Q' = AQ+QA'+EE'+QC_2'C_2Q,$$

we have the following induction:

$$\begin{aligned} (I & - QP)[sI-A_{PQ}-(I-QP)^{-1}LC_{1P}] \\ &= [(I-QP)(sI-A-EE'P)-QC_{2P}'C_{2P}-LC_1-LD_1E'P] \end{aligned}$$

$$
\begin{aligned}
&= [sI - A - EE'P - QC_{2P}'C_{2P} - LC_1 - LD_1E'P - sQP \\
&\qquad + QPA + QPEE'P] \\
&= [sI - A - EE'P - Q(A'P + PA + C_2'C_2 + PEE'P) \\
&\qquad - LC_1 - LD_1E'P - sQP + QPA + QPEE'P] \\
&= [sI - A - QC_2'C_2 - LC_1 - EE'P - LD_1E'P - QA'P - sQP] \\
&= \left[sI - A - QC_2'C_2 - LC_1 - (E_QE_Q' - AQ - QA' - QC_2'C_2Q)P \right. \\
&\qquad \left. - LD_1E'P - QA'P - sQP\right] \\
&= \left[sI - A - QC_2'C_2 - LC_1 - sQP + AQP + QC_2'C_2QP - E_QE_Q'LD_1E'P\right] \\
&= \left[(sI - A - QC_2'C_2 - LC_1)(I - QP) - (E_Q + LD_{1PQ})E_Q'P\right] \\
&= [sI - A - QC_2'C_2 - LC_1] \\
&\qquad \left[(I - QP) - (sI - A - QC_2'C_2 - LC_1)^{-1}(E_Q + LD_{1PQ})E_Q'P\right] \\
&\to [sI - A - QC_2'C_2 - LC_1](I - QP), \quad \text{pointwise in } s \text{ as } \varepsilon \to 0. \qquad (7.5.16)
\end{aligned}
$$

Hence, $A_{PQ} + (I - QP)^{-1}KC_{1P}$ is asymptotically stable for sufficiently small ε. Now it follows from (7.5.16) that

$$
\begin{aligned}
&[sI - A_{PQ} - (I - QP)^{-1}LC_{1P}]^{-1}[E_{PQ} + (I - QP)^{-1}LD_{1PQ}] \\
&\to (I - QP)^{-1}[sI - A - QC_2'C_2 - LC_1]^{-1}(I - QP)[E_{PQ} + (I - QP)^{-1}LD_{1PQ}] \\
&= (I - QP)^{-1}[sI - A - QC_2'C_2 - LC_1]^{-1}[E_Q + LD_{1PQ}] \\
&\to 0, \quad \text{pointwise in } s \text{ as } \varepsilon \to 0.
\end{aligned}
$$

This completes the proof of Theorem 7.3.1. ⊠

7.5.D. Proof of Theorem 7.3.2

As in the previous proofs, for simplicity, we will assume that $\gamma = 1$ and let $F = F(\gamma, \varepsilon, \Lambda_{dP}, \Delta_{cP})$ and $K = K(\gamma, \varepsilon, \Lambda_{dQ}, \Delta_{cQ})$. Then the closed loop system $T_{hw}(s)$ is given by

$$
[C_2 \quad D_2F]\left(sI - \begin{bmatrix} A & BF \\ -(I - QP)^{-1}KC_1 & A_{\mathrm{cmp}} \end{bmatrix}\right)^{-1} \begin{bmatrix} E \\ -(I - QP)^{-1}KD_1 \end{bmatrix}.
$$

It follows from the proof of Theorem 7.2.2 that for any

$$
\alpha \in \lambda(A_{aaP}^-) \cup \lambda(A_{ccP}^c) \subseteq \lambda(A + BF),
$$

the corresponding right eigenvector, say W, i.e., $(A + BF)W = \alpha W$, satisfies $(C_2 + D_2F)W = 0$. Moreover, it is simple to verify that $(C_{2P} + D_{2P}F)W = 0$ and $PW = 0$.

By duality, one can show that for any $\beta \in \lambda(A_{aaQ}^{-}) \cup \lambda(A_{ccQ}^{c})$, $\beta \in \lambda(A+KC_1)$ and the corresponding left eigenvector, say V, i.e., $V^{H}(A+KC_1) = \beta V^{H}$, satisfies $V^{H}(E+KD_1)=0$ and $V^{H}Q=0$. In view of (7.3.6), we have

$$\begin{aligned} A_{\rm cmp}W &= [A+EE'P+BF+(I-QP)^{-1}QC_{2P}'(C_{2P}+D_{2P}F) \\ &\quad +(I-QP)^{-1}KC_1+(I-QP)^{-1}KD_1E'P]W \\ &= (I-QP)^{-1}KC_1W+(A+BF)W, \end{aligned}$$

and

$$\begin{aligned} V^{H}A_{\rm cmp} &= V^{H}(I-QP)[A+EE'P+BF+(I-QP)^{-1}QC_{2P}'(C_{2P}+D_{2P}F) \\ &\quad +(I-QP)^{-1}KC_1+(I-QP)^{-1}KD_1E'P] \\ &= V^{H}BF+V^{H}(A+KC_1). \end{aligned}$$

Therefore,

$$\begin{bmatrix} A & BF \\ -(I-QP)^{-1}KC_1 & A_{\rm cmp} \end{bmatrix}\begin{bmatrix} W \\ W \end{bmatrix} = \begin{bmatrix} (A+BF)W \\ A_{\rm cmp}W-(I-QP)^{-1}KC_1 \end{bmatrix} = \alpha \begin{bmatrix} W \\ W \end{bmatrix},$$

and

$$[\,C_2 \quad D_2F\,]\begin{bmatrix} W \\ W \end{bmatrix} = (C_2+D_2F)W = 0.$$

This shows that α is an output decoupling zero of $T_{hw}(s)$. Similarly,

$$\begin{aligned} [\,V^{H} \quad -V^{H}\,]&\begin{bmatrix} A & BF \\ -(I-QP)^{-1}KC_1 & A_{\rm cmp} \end{bmatrix} \\ &= [\,V^{H}(I-QP)[A+(I-QP)^{-1}KC_1] \quad V^{H}(BF-A_{\rm cmp})\,] \\ &= \beta\,[\,V^{H} \quad -V^{H}\,], \end{aligned}$$

and

$$[\,V^{H} \quad -V^{H}\,]\begin{bmatrix} E \\ -(I-QP)^{-1}KD_1 \end{bmatrix} = V^{H}(E+KD_1) = 0.$$

This implies that β is an input decoupling zero of $T_{hw}(s)$.

The first part of Item 3 in Theorem 7.3.2 can be verified easily by using (7.5.6) and the fact that

$$\operatorname{Im}(P) = \left[\mathcal{S}^{+}(\Sigma_{P})\right]^{\perp}.$$

The second part is the dual of the first case. This completes the proof of Theorem 7.3.2. ⊠

7.5.E. Proof of Theorem 7.4.1

First, note that the subsystem i) $(A_{PQ}, B_{PQ}, C_{2P}, D_{2P})$ is right invertible and of minimum phase; and ii) the subsystem $(A_{PQ}, E_{PQ}, C_{1P}, D_{1PQ})$ is left invertible and of minimum phase. It follows from Theorem 8.4.2 that there indeed exist gain matrices F and K_R such that the resulting reduced order output feedback control law (7.4.10) internally stabilizes Σ_{PQ} and makes the H_∞ norm of the closed-loop transfer matrix strictly less than any given ε. The second result of Theorem 7.4.1 follows from Lemma 7.5.3. ⊠

Chapter 8

Continuous-time H_∞ Almost Disturbance Decoupling

8.1. Introduction

WE CONSIDER IN this chapter the problem of H_∞ almost disturbance decoupling with measurement feedback and internal stability for continuous-time linear systems. Although in principle it is a special case of the general H_∞ control problem, i.e., the case that $\gamma^* = 0$, the problem of almost disturbance decoupling has a vast history behind it, occupying a central part of classical as well as modern control theory. Several important problems, such as robust control, decentralized control, non-interactive control, model reference or tracking control, H_2 and H_∞ optimal control problems can all be recast into an almost disturbance decoupling problem. Roughly speaking, the basic almost disturbance decoupling problem is to find an output feedback control law such that in the closed-loop system the disturbances are quenched, say in an L_p sense, up to any pre-specified degree of accuracy while maintaining internal stability. Such a problem was originally formulated by Willems ([136] and [137]) and labelled ADDPMS (the almost disturbance decoupling problem with measurement feedback and internal stability). In the case that, instead of a measurement feedback, a state feedback is used, the above problem is termed ADDPS (the almost disturbance decoupling problem with internal stability). The prefix H_∞ in the acronyms H_∞-ADDPMS and H_∞-ADDPS is used to specify that the degree of accuracy in disturbance quenching is measured in L_2-sense.

There is extensive literature on the almost disturbance decoupling problem (See, for example, the recent work [134], [98] and [99] and the references therein). In [134], several variations of the disturbance decoupling problems and their solvability conditions are summarized, and the necessary and sufficient conditions are given, under which the H_∞-ADDPMS and H_∞-ADDPS for continuous-time linear systems are solvable. These conditions are given in terms of geometry subspaces and for strictly proper systems (i.e., without direct feedthrough terms from the control input to the output to be controlled and from the disturbance input to the measurement output). Under these conditions, [98] constructs feedback laws, parameterized explicitly in a single parameter ε, that solve the H_∞-ADDPMS and the H_∞-ADDPS. These results were later extended to proper systems (i.e., with direct feedthrough terms) in [99]. We emphasize that in all the results mentioned above, the internal stability was always with respect to a closed set in the complex plane. Such a closeness restriction, while facilitating the development of the the above results, excludes systems with disturbance affected purely imaginary invariant zero dynamics from consideration. Only recently was this "final" restriction on the internal stability restriction removed by Scherer [119], thus allowing purely imaginary invariant zero dynamics to be affected by the disturbance. More specifically, Scherer [119] gave a set of necessary and sufficient conditions under which the H_∞-ADDPMS and the H_∞-ADDPS, with internal stability being with respect to the open left-half plane, is solvable for general proper linear systems. When the stability is with respect to the open left-half plane, the H_∞-ADDPMS and the H_∞-ADDPS will be referred to as the general H_∞-ADDPMS and the general H_∞-ADDPS, respectively. The explicit construction algorithm for feedback laws that solve these general H_∞-ADDPMS and H_∞-ADDPS under Scherer's necessary and sufficient conditions has only appeared in a very recent paper of Chen, Lin and Hang [24]. The objective of this chapter is to present: 1) easily checkable conditions for the general H_∞-ADDPS and H_∞-ADDPMS; and 2) explicit algorithms to construct solutions that solve these problems. The latter were reported in Chen, Lin and Hang [24].

More specifically, we consider the general H_∞-ADDPMS and the general H_∞-ADDPS, for the following general continuous-time linear system,

$$\Sigma : \begin{cases} \dot{x} = A\ x + B\ u + E\ w, \\ y = C_1\ x + D_1\ w, \\ h = C_2\ x + D_2\ u + D_{22}\ w, \end{cases} \tag{8.1.1}$$

where $x \in \mathbb{R}^n$ is the state, $u \in \mathbb{R}^m$ is the control input, $y \in \mathbb{R}^\ell$ is the measurement, $w \in \mathbb{R}^q$ is the disturbance and $h \in \mathbb{R}^p$ is the output to be controlled. As usual, for convenient reference in future development, throughout this chapter, we define Σ_{P} to be the subsystem characterized by the matrix quadruple (A, B, C_2, D_2) and Σ_{Q} to be the subsystem characterized by the matrix quadruple (A, E, C_1, D_1). The following dynamic feedback control laws are investigated:

$$\Sigma_{\mathrm{cmp}} \ : \begin{cases} \dot{v} = A_{\mathrm{cmp}}\ v + B_{\mathrm{cmp}}\ y, \\ u = C_{\mathrm{cmp}}\ v + D_{\mathrm{cmp}}\ y. \end{cases} \tag{8.1.2}$$

The controller Σ_{cmp} of (8.1.2) is said to be internally stabilizing when applied to the system Σ, if the following matrix is asymptotically stable:

$$A_{\mathrm{cl}} := \begin{bmatrix} A + BD_{\mathrm{cmp}}C_1 & BC_{\mathrm{cmp}} \\ B_{\mathrm{cmp}}C_1 & A_{\mathrm{cmp}} \end{bmatrix}, \tag{8.1.3}$$

i.e., all its eigenvalues lie in the open left-half complex plane. Denote by T_{hw} the corresponding closed-loop transfer matrix from the disturbance w to the output to be controlled h. Then the general H_∞-ADDPMS and the general H_∞-ADDPS can be formally defined as follows.

Definition 8.1.1. The H_∞ almost disturbance decoupling problem with measurement feedback and with internal stability (the H_∞-ADDPMS) for the continuous time system (8.1.1) is said to be solvable if, for any given positive scalar $\gamma > 0$, there exists at least one controller of the form (8.1.2) such that,

1. in the absence of disturbance, the closed-loop system comprising the system (8.1.1) and the controller (8.1.2) is asymptotically stable, i.e., the matrix A_{cl} as given by (8.1.3) is asymptotically stable; and

2. the closed-loop system has an L_2-gain, from the disturbance w to the controlled output h, that is less than or equal to γ, i.e.,

$$\|h\|_2 \le \gamma \|w\|_2, \quad \forall w \in L_2 \text{ and for } (x(0), v(0)) = (0,0). \tag{8.1.4}$$

   Equivalently, the H_∞-norm of the closed-loop transfer matrix from w to h, T_{hw}, is less than or equal to γ, i.e., $\|T_{hw}\|_\infty \le \gamma$.

In the case that $C_1 = I$ and $D_1 = 0$, the general H_∞-ADDPMS as defined above becomes the general H_∞-ADDPS, where only a static state feedback, instead the dynamic output feedback (8.1.2) is necessary. ◻

Clearly, the H_∞-ADDPMS for Σ of (8.1.1) is equivalent to the general H_∞ control problem for Σ with $\gamma^* = 0$. As stated earlier, one of the objectives of this chapter is to construct families of feedback laws of the form (8.1.2), parameterized in a single parameter, say ε, that, under the necessary and sufficient conditions of Scherer [119], solve the above defined general H_∞-ADDPMS and H_∞-ADDPS for general systems whose subsystems Σ_P and Σ_Q may have invariant zeros on the imaginary axis. The feedback laws we are to construct are observer-based. A family of static state feedback laws parameterized in a single parameter is first constructed to solve the general H_∞-ADDPS. A class of observers parameterized in the same parameter ε is then constructed to implement the state feedback laws and thus obtain a family of dynamic measurement feedback laws parameterized in a single parameter ε that solve the general H_∞-ADDPMS. The basic tools we use in the construction of such families of feedback laws are: 1) the special coordinate basis, developed by Sannuti and Saberi [116] and Saberi and Sannuti [111] (see also Chapter 2), in which a linear system is decomposed into several subsystems corresponding to its finite and infinite zero structures as well as its invertibility structures; 2) a block diagonal control canonical form (see also Chapter 2) that puts the dynamics of imaginary invariant zeros into a special canonical form under which the low-gain design technique can be applied; and 3) the H_∞ low-and-high gain design technique. The development of such an H_∞ low-and-high gain design technique was originated in [81] and [83] in the context of H_∞-ADDPMS for special classes of nonlinear systems that specialized to a SISO (and hence square invertible) linear system having no invariant zero in the open right-half plane.

8.2. Solvability Conditions

In this section, we first recall the necessary and sufficient conditions of Scherer [119] under which the general H_∞-ADDPMS and H_∞-ADDPS are solvable. Then we will convert the geometric conditions of Scherer into easily checkable ones using the properties of the special coordinate basis. The following result is a slight generalization of Scherer [119].

Theorem 8.2.1. Consider the general measurement feedback system (8.1.1). Then the general H_∞ almost disturbance decoupling problem for (8.1.1) with internal stability (H_∞-ADDPMS) is solvable, if and only if the following conditions are satisfied:

1. (A, B) is stabilizable;
2. (A, C_1) is detectable;
3. $D_{22} + D_2 S D_1 = 0$, where $S = -(D_2' D_2)^\dagger D_2' D_{22} D_1' (D_1 D_1')^\dagger$;

4. $\mathrm{Im}\,(E + BSD_1) \subset \mathcal{S}^+(\Sigma_P) \cap \{\cap_{\lambda \in \mathbb{C}^0} \mathcal{S}_\lambda(\Sigma_P)\}$;

5. $\mathrm{Ker}\,(C_2 + D_2SC_1) \supset \mathcal{V}^+(\Sigma_Q) \cup \{\cup_{\lambda \in \mathbb{C}^0} \mathcal{V}_\lambda(\Sigma_Q)\}$; and

6. $\mathcal{V}^+(\Sigma_Q) \subset \mathcal{S}^+(\Sigma_P)$. ▣

Remark 8.2.1. Note that if Σ_P is right invertible and of minimum phase, and Σ_Q is left invertible and of minimum phase, then Conditions 4 to 6 of Theorem 8.2.1 are automatically satisfied. Hence, the solvability conditions of the H_∞-ADDPMS for such a case reduce to:

1. (A, B) is stabilizable;

2. (A, C_1) is detectable; and

3. $D_{22} + D_2SD_1 = 0$, where $S = -(D_2'D_2)^\dagger D_2'D_{22}D_1'(D_1D_1')^\dagger$. ▣

Remark 8.2.2. It is simple to verify that for the case when all states of the system (8.1.1) are fully measurable, i.e., $C_1 = I$ and $D_1 = 0$, then the solvability conditions for the general H_∞-ADDPS reduce to the following:

1. (A, B) is stabilizable;

2. $D_{22} = 0$; and

3. $\mathrm{Im}\,(E) \subset \mathcal{S}^+(\Sigma_P) \cap \{\cap_{\lambda \in \mathbb{C}^0} \mathcal{S}_\lambda(\Sigma_P)\}$.

Moreover, in this case, a static state feedback law, i.e., $u = Fx$, where F is a constant matrix and might be parameterized by certain tuning parameters, exists that solves the general H_∞-ADDPS. ▣

Theorem 8.2.1 is quite elegant as it is expressed in terms of the well-known geometric conditions. However, it might be hard to verify these geometric conditions numerically. In what follows, we will present a simple method to check the solvability conditions for the H_∞-ADDPMS for general continuous-time systems.

Step 8.2.0: Let $S = -(D_2'D_2)^\dagger D_2'D_{22}D_1'(D_1D_1')^\dagger$. If $D_{22} + D_2SD_1 \neq 0$, the algorithm stops here. Otherwise, go to Step 8.2.1.

Step 8.2.1: Compute the special coordinate basis of Σ_P, i.e., the quadruple (A, B, C_2, D_2). For easy reference, we append a subscript 'P' to all sub-matrices and transformations in the SCB associated with Σ_P, e.g., Γ_{sP} is the state transformation of the SCB of Σ_P, and A^0_{aaP} is associated with invariant zero dynamics of Σ_P on the imaginary axis.

Step 8.2.2: Next, we denote the set of eigenvalues of A^0_{aaP} with a nonnegative imaginary part as $\{\omega_{P1}, \omega_{P2}, \cdots, \omega_{Pk_P}\}$ and for $i = 1, 2, \cdots, k_P$, choose complex matrices V_{iP}, whose columns form a basis for the eigenspace

$$\left\{ x \in \mathbb{C}^{n^0_{aP}} \mid x^H(\omega_{Pi} I - A^0_{aaP}) = 0 \right\}, \tag{8.2.1}$$

where n^0_{aP} is the dimension of $\mathcal{X}^0_{aP}$. Then, let

$$V_P := [\, V_{1P} \quad V_{2P} \quad \cdots \quad V_{k_P P} \,]. \tag{8.2.2}$$

We also compute $n_{xP} := \dim(\mathcal{X}^+_{aP}) + \dim(\mathcal{X}_{bP})$, and

$$\Gamma^{-1}_{sP}(E + BSD_1) := \begin{bmatrix} E^-_{aP} \\ E^0_{aP} \\ E^+_{aP} \\ E_{bP} \\ E_{cP} \\ E_{dP} \end{bmatrix}. \tag{8.2.3}$$

Step 8.2.3: Let $\Sigma^\star_Q$ be the dual system of Σ_Q and be characterized by a quadruple (A', C_1', E', D_1'). We compute the special coordinate basis of $\Sigma^\star_Q$. Again, for easy reference, we append a subscript 'Q' to all sub-matrices and transformations in the SCB associated with $\Sigma^\star_Q$, e.g., Γ_{sQ} is the state transformation of the SCB of $\Sigma^\star_Q$, and A^0_{aaQ} is associated with invariant zero dynamics of $\Sigma^\star_Q$ on the imaginary axis.

Step 8.2.4: Next, denote the set of eigenvalues of A^0_{aaQ} with a nonnegative imaginary part as $\{\omega_{Q1}, \omega_{Q2}, \cdots, \omega_{Qk_Q}\}$ and for $i = 1, 2, \cdots, k_Q$, choose complex matrices V_{iQ}, whose columns form a basis for the eigenspace

$$\left\{ x \in \mathbb{C}^{n^0_{aQ}} \mid x^H(\omega_{Qi} I - A^0_{aaQ}) = 0 \right\}, \tag{8.2.4}$$

where n^0_{aQ} is the dimension of $\mathcal{X}^0_{aQ}$. Then, let

$$V_Q := [\, V_{1Q} \quad V_{2Q} \quad \cdots \quad V_{k_Q Q} \,]. \tag{8.2.5}$$

We next compute $n_{xQ} := \dim(\mathcal{X}^+_{aQ}) + \dim(\mathcal{X}_{bQ})$, and

$$\Gamma^{-1}_{sQ}(C_2 + D_2 S C_1)' := \begin{bmatrix} E^-_{aQ} \\ E^0_{aQ} \\ E^+_{aQ} \\ E_{bQ} \\ E_{cQ} \\ E_{dQ} \end{bmatrix}. \tag{8.2.6}$$

Step 8.2.5: Finally, compute

$$\Gamma_{sP}^{-1}(\Gamma_{sQ}^{-1})' = \begin{bmatrix} X_a^{-0} & \star & \star \\ \star & \Gamma & \star \\ \star & \star & X_{cd} \end{bmatrix}, \tag{8.2.7}$$

where X_a^{-0} and X_{cd} are of dimensions $(n_{aP}^- + n_{aP}^0) \times (n_{aQ}^- + n_{aQ}^0)$ and $(n_{cP} + n_{dP}) \times (n_{cQ} + n_{dQ})$, respectively, and finally Γ is a sub-matrix of dimension $n_{xP} \times n_{xQ}$. ◻A

We have the following proposition.

Proposition 8.2.1. Consider the general measurement feedback continuous-time system (8.1.1). Then the H_∞ almost disturbance decoupling problem for (8.1.1) with internal stability (H_∞-ADDPMS) is solvable, i.e., $\gamma^* = 0$, if and only if the following conditions are satisfied:

1. (A, B) is stabilizable;
2. (A, C_1) is detectable;
3. $D_{22} + D_2SD_1 = 0$, where $S = -(D_2'D_2)^\dagger D_2'D_{22}D_1'(D_1D_1')^\dagger$;
4. $E_{aP}^+ = 0$, $E_{bP} = 0$ and $V_P^H E_{aP}^0 = 0$;
5. $E_{aQ}^+ = 0$, $E_{bQ} = 0$ and $V_Q^H E_{aQ}^0 = 0$; and
6. $\Gamma = 0$. ◻P

Proof. It is simple to see that the first three conditions are necessary for the H_∞-ADDPMS for (8.1.1) to be solvable. Next, it follows trivially from the properties of the special coordinate basis of Chapter 2 that the geometric condition, $\mathrm{Im}\,(E + BSD_1) \subset \mathcal{S}^+(\Sigma_P) \cap \{\cap_{\lambda \in \mathbb{C}^0} \mathcal{S}_\lambda(\Sigma_P)\}$, is equivalent to the following conditions: $E_{aP}^+ = 0$, $E_{bP} = 0$ and $V_P^H E_{aP}^0 = 0$. Dually, the geometric condition, $\mathrm{Ker}\,(C_2 + D_2SC_1) \supset \mathcal{V}^+(\Sigma_Q) \cup \{\cup_{\lambda \in \mathbb{C}^0} \mathcal{V}_\lambda(\Sigma_Q)\}$, is equivalent to the following conditions: $E_{aQ}^+ = 0$, $E_{bQ} = 0$ and $V_Q^H E_{aQ}^0 = 0$.

Again, following the properties of the special coordinate basis, we have

$$\mathcal{S}^+(\Sigma_P) = \mathrm{Ker}\left\{[0 \quad I_{n_{xP}} \quad 0]\,\Gamma_{sP}^{-1}\right\}, \quad \mathcal{V}^+(\Sigma_Q) = \mathrm{Im}\left\{(\Gamma_{sQ}^{-1})' \begin{bmatrix} 0 \\ I_{n_{xQ}} \\ 0 \end{bmatrix}\right\}.$$

Hence, it is straightforward to verify that $\mathcal{V}^+(\Sigma_Q) \subset \mathcal{S}^+(\Sigma_P)$ is equivalent to

$$[0 \quad I_{n_{xP}} \quad 0]\,\Gamma_{sP}^{-1}\,(\Gamma_{sQ}^{-1})' \begin{bmatrix} 0 \\ I_{n_{xQ}} \\ 0 \end{bmatrix} = \Gamma = 0.$$

Thus, the result follows. ⊠

8.3. Solutions to Full State Feedback Case

In this section, we consider feedback law design for the general H_∞ almost disturbance decoupling problem with internal stability and with full state feedback, where internal stability is with respect to the open left-half plane, i.e., the general H_∞-ADDPS. More specifically, we present a design procedure that constructs a family of parameterized static state feedback laws,

$$u = F(\varepsilon)x, \tag{8.3.1}$$

that solves the general H_∞-ADDPS for the following system,

$$\begin{cases} \dot{x} = A\,x + B\,u + E\,w, \\ y = x \\ h = C_2\,x + D_2\,u + D_{22}\,w. \end{cases} \tag{8.3.2}$$

That is, under this family of state feedback laws, the resulting closed-loop system is asymptotically stable for sufficiently small ε and the H_∞-norm of the closed-loop transfer matrix from w to h, $T_{hw}(s,\varepsilon)$, tends to zero as ε tends to zero, where

$$T_{hw}(s,\varepsilon) = [C_2 + D_2F(\varepsilon)][sI - A - BF(\varepsilon)]^{-1}E + D_{22}. \tag{8.3.3}$$

Clearly, $D_{22} = 0$ is a necessary condition for the solvability of the general H_∞-ADDPS. We present an algorithm for obtaining this $F(\varepsilon)$, following the asymptotic time-scale and eigenstructure assignment (ATEA) procedure. We first use the special coordinate basis of the given system (See Theorem 2.4.1) to decompose the system into several subsystems according to its finite and infinite zero structures as well as its invertibility structures. The new component here is the low-gain design for the part of the zero dynamics corresponding to all purely imaginary invariant zeros. As will be clear shortly, the low-gain component is critical in handling the case when the zero dynamics corresponding to purely imaginary invariant zeros is affected by disturbance. It is well-known that the disturbance affected purely imaginary zero dynamics is difficult to handle and has always been excluded from consideration until recently.

We have in the following a step-by-step algorithm.

Step 8.S.1: (Decomposition of $\Sigma_{\rm P}$). Transform the subsystem $\Sigma_{\rm P}$, i.e., the quadruple (A, B, C_2, D_2), into the special coordinate basis (SCB) as given by Theorem 2.4.1 of Chapter 2. Denote the state, output and input transformation matrices as $\Gamma_{s\rm P}$, $\Gamma_{o\rm P}$ and $\Gamma_{i\rm P}$, respectively.

Step 8.S.2: (Gain matrix for the subsystem associated with $\mathcal{X}_c$). Let F_c be any arbitrary $m_c \times n_c$ matrix subject to the constraint that

$$A_{cc}^c = A_{cc} - B_c F_c, \tag{8.3.4}$$

is a stable matrix. Note that the existence of such an F_c is guaranteed by the property of the special coordinate basis, i.e., (A_{cc}, B_c) is controllable.

Step 8.S.3: (Gain matrix for the subsystems associated with $\mathcal{X}_a^+$ and $\mathcal{X}_b$). Let

$$F_{ab}^+ := \begin{bmatrix} F_{a0}^+ & F_{b0} \\ F_{ad}^+ & F_{bd} \end{bmatrix}, \tag{8.3.5}$$

be any arbitrary $(m_0 + m_d) \times (n_a^+ + n_b)$ matrix subject to the constraint that

$$A_{ab}^{+c} := \begin{bmatrix} A_{aa}^+ & L_{ab}^+ C_b \\ 0 & A_{bb} \end{bmatrix} - \begin{bmatrix} B_{0a}^+ & L_{ad}^+ \\ B_{0b} & L_{bd} \end{bmatrix} F_{ab}^+, \tag{8.3.6}$$

is a stable matrix. Again, note that the existence of such an F_{ab} is guaranteed by the stabilizability of (A, B) and Property 2.4.1 of the special coordinate basis. For future use, let us partition $[\, F_{ad}^+ \quad F_{bd} \,]$ as,

$$[\, F_{ad}^+ \quad F_{bd} \,] = \begin{bmatrix} F_{ad1}^+ & F_{bd1} \\ F_{ad2}^+ & F_{bd2} \\ \vdots & \vdots \\ F_{adm_d}^+ & F_{bdm_d} \end{bmatrix}, \tag{8.3.7}$$

where F_{adi}^+ and F_{bdi} are of dimensions $1 \times n_a^+$ and $1 \times n_b$, respectively.

Step 8.S.4: (Gain matrix for the subsystem associated with $\mathcal{X}_a^0$). The construction of this gain matrix is carried out in the following sub-steps.

Step 8.S.4.1: (Preliminary coordinate transformation). Recalling the definition of $(A_{\rm con}, B_{\rm con})$, i.e., (2.4.27), we have

$$A_{\rm con} - B_{\rm con} [\, 0 \quad 0 \quad F_{ab}^+ \,] = \begin{bmatrix} A_{aa}^- & 0 & A_{aab}^- \\ 0 & A_{aa}^0 & A_{aab}^0 \\ 0 & 0 & A_{ab}^{+c} \end{bmatrix}, \quad B_{\rm con} = \begin{bmatrix} B_{0a}^- & L_{ad}^- \\ B_{0a}^0 & L_{ad}^0 \\ B_{0ab}^+ & L_{abd}^+ \end{bmatrix}, \tag{8.3.8}$$

where

$$B_{0ab}^+ = \begin{bmatrix} B_{0a}^+ \\ B_{0b} \end{bmatrix}, \quad L_{abd}^+ = \begin{bmatrix} L_{ad}^+ \\ L_{bd} \end{bmatrix}, \tag{8.3.9}$$

$$A_{aab}^0 = [\, 0 \quad L_{ab}^0 C_b \,] - [\, B_{0a}^0 \quad L_{ad}^0 \,] F_{ab}^+, \tag{8.3.10}$$

and

$$A_{aab}^- = [\,0 \quad L_{ab}^- C_b\,] - [\,B_{0a}^- \quad L_{ad}^-\,] F_{ab}^+. \tag{8.3.11}$$

Clearly $(A_{\rm con} - B_{\rm con}F_{ab}^+, B_{\rm con})$ remains stabilizable. Construct the following nonsingular transformation matrix,

$$\Gamma_{ab} = \begin{bmatrix} I_{n_a^-} & 0 & 0 \\ 0 & 0 & I_{n_a^+ + n_b} \\ 0 & I_{n_a^0} & T_a^0 \end{bmatrix}^{-1}, \tag{8.3.12}$$

where T_a^0 is the unique solution to the following Lyapunov equation,

$$A_{aa}^0 T_a^0 - T_a^0 A_{ab}^{+c} = A_{aab}^0. \tag{8.3.13}$$

We note here that such a unique solution to the above Lyapunov equation always exists since all the eigenvalues of A_{aa}^0 are on the imaginary axis and all the eigenvalues of A_{ab}^{+c} are in the open left-half plane. It is now easy to verify that

$$\Gamma_{ab}^{-1}(A_{\rm con} - B_{\rm con}F_{ab}^+)\Gamma_{ab} = \begin{bmatrix} A_{aa}^- & A_{aab}^- & 0 \\ 0 & A_{ab}^{+c} & 0 \\ 0 & 0 & A_{aa}^0 \end{bmatrix}, \tag{8.3.14}$$

$$\Gamma_{ab}^{-1} B_{\rm con} = \begin{bmatrix} B_{0a}^- & L_{ad}^- \\ B_{0ab}^+ & L_{abd}^+ \\ B_{0a}^0 + T_a^0 B_{0ab}^+ & L_{ad}^0 + T_a^0 L_{abd}^+ \end{bmatrix}. \tag{8.3.15}$$

Hence, the matrix pair (A_{aa}^0, B_a^0) is controllable, where

$$B_a^0 = [\,B_{0a}^0 + T_a^0 B_{0ab}^+ \quad L_{ad}^0 + T_a^0 L_{abd}^+\,].$$

Step 8.S.4.2: (Further coordinate transformation). Following the proof of Theorem 2.3.2, find nonsingular transformation matrices Γ_{sa}^0 and Γ_{ia}^0 such that (A_{aa}^0, B_a^0) can be transformed into the block diagonal control canonical form,

$$(\Gamma_{sa}^0)^{-1} A_{aa}^0 \Gamma_{sa}^0 = \begin{bmatrix} A_1 & 0 & \cdots & 0 \\ 0 & A_2 & \cdots & 0 \\ \vdots & \vdots & \ddots & \vdots \\ 0 & 0 & \cdots & A_l \end{bmatrix}, \tag{8.3.16}$$

and

$$(\Gamma_{sa}^0)^{-1} B_a^0 \Gamma_{ia}^0 = \begin{bmatrix} B_1 & B_{12} & \cdots & B_{1l} & \star \\ 0 & B_2 & \cdots & B_{2l} & \star \\ \vdots & \vdots & \ddots & \vdots & \vdots \\ 0 & 0 & \cdots & B_l & \star \end{bmatrix}, \tag{8.3.17}$$

where l is an integer and for $i = 1, 2, \cdots, l$,

$$A_i = \begin{bmatrix} 0 & 1 & 0 & \cdots & 0 \\ 0 & 0 & 1 & \cdots & 0 \\ \vdots & \vdots & \vdots & \ddots & \vdots \\ 0 & 0 & 0 & \cdots & 1 \\ -a^i_{n_i} & -a^i_{n_i-1} & -a^i_{n_i-2} & \cdots & -a^i_1 \end{bmatrix}, \quad B_i = \begin{bmatrix} 0 \\ 0 \\ \vdots \\ 0 \\ 1 \end{bmatrix}.$$

We note that all the eigenvalues of A_i are on the imaginary axis. Here the $\star$s represent sub-matrices of less interest.

Step 8.S.4.3: (Subsystem design). For each (A_i, B_i), let $F_i(\varepsilon) \in \mathbb{R}^{1\times n_i}$ be the state feedback gain such that

$$\lambda\{A_i - B_iF_i(\varepsilon)\} = -\varepsilon + \lambda(A_i) \in \mathbb{C}^-. \tag{8.3.18}$$

Note that $F_i(\varepsilon)$ is unique.

Step 8.S.4.4: (Composition of gain matrix for subsystem associated with $\mathcal{X}_a^0$). Let

$$F_a^0(\varepsilon) := \Gamma_{ia}^0 \begin{bmatrix} F_1(\varepsilon) & 0 & \cdots & 0 & 0 \\ 0 & F_2(\varepsilon) & \cdots & 0 & 0 \\ \vdots & \vdots & \ddots & \vdots & \vdots \\ 0 & 0 & \cdots & F_{l-1}(\varepsilon) & 0 \\ 0 & 0 & \cdots & 0 & F_l(\varepsilon) \\ 0 & 0 & \cdots & 0 & 0 \end{bmatrix} (\Gamma_{sa}^0)^{-1}, \tag{8.3.19}$$

where $\varepsilon \in (0, 1]$ is a design parameter whose value is to be specified later.

Clearly, we have

$$|F_a^0(\varepsilon)| \le f_a^0\varepsilon, \quad \varepsilon \in (0, 1], \tag{8.3.20}$$

for some positive constant f_a^0, independent of ε. For future use, we define and partition $F_{ab}(\varepsilon) \in \mathbb{R}^{(m_0+m_d)\times(n_a+n_b)}$ as

$$F_{ab}(\varepsilon) = \begin{bmatrix} F_{ab0}(\varepsilon) \\ F_{abd}(\varepsilon) \end{bmatrix} = \begin{bmatrix} 0_{m_0\times n_a^-} & 0_{m_0\times(n_a^++n_b)} & F_{a0}^0(\varepsilon) \\ 0_{m_d\times n_a^-} & 0_{m_d\times(n_a^++n_b)} & F_{ad}^0(\varepsilon) \end{bmatrix} \Gamma_{ab}^{-1}, \tag{8.3.21}$$

and

$$F_{abd}(\varepsilon) = \begin{bmatrix} F_{abd1}(\varepsilon) \\ F_{abd2}(\varepsilon) \\ \vdots \\ F_{abdm_d}(\varepsilon) \end{bmatrix}, \tag{8.3.22}$$

where $F^0_{a0}(\varepsilon)$ and $F^0_{ad}(\varepsilon)$ are defined as

$$F^0_a(\varepsilon) = \begin{bmatrix} F^0_{a0}(\varepsilon) \\ F^0_{ad}(\varepsilon) \end{bmatrix}. \tag{8.3.23}$$

We also partition $F^0_{ad}(\varepsilon)$ as,

$$F^0_{ad}(\varepsilon) = \begin{bmatrix} F^0_{ad1}(\varepsilon) \\ F^0_{ad2}(\varepsilon) \\ \vdots \\ F^0_{adm_d}(\varepsilon) \end{bmatrix}. \tag{8.3.24}$$

Step 8.S.5: (Gain matrix for the subsystem associated with $\mathcal{X}_d$). This step makes use of subsystems, $i = 1$ to m_d, represented by (2.4.14) of Chapter 2. Let $\Lambda_i = \{\lambda_{i1}, \lambda_{i2}, \cdots, \lambda_{iq_i}\}$, $i = 1$ to m_d, be the sets of q_i elements all in $\mathbb{C}^-$, which are closed under complex conjugation, where q_i and m_d are as defined in Theorem 2.4.1 but associated with the special coordinate basis of $\Sigma_{\rm P}$. Let $\Lambda_d := \Lambda_1 \cup \Lambda_2 \cup \cdots \cup \Lambda_{m_d}$. For $i = 1$ to m_d, we define

$$p_i(s) := \prod_{j=1}^{q_i} (s - \lambda_{ij}) = s^{q_i} + F_{i1} s^{q_i - 1} + \cdots + F_{iq_i - 1} s + F_{iq_i}, \tag{8.3.25}$$

and

$$\tilde{F}_i(\varepsilon) := \frac{1}{\varepsilon^{q_i}} F_i S_i(\varepsilon), \tag{8.3.26}$$

where

$$F_i = [\,F_{iq_i} \quad F_{iq_i-1} \quad \cdots \quad F_{i1}\,], \quad S_i(\varepsilon) = \mathrm{diag}\Big\{1, \varepsilon, \varepsilon^2, \cdots, \varepsilon^{q_i - 1}\Big\}, \tag{8.3.27}$$

Step 8.S.6: (Composition of parameterized gain matrix $F(\varepsilon)$). In this step, various gains calculated in Steps 8.S.2 to 8.S.5 are put together to form a composite state feedback gain matrix $F(\varepsilon)$. Let

$$\tilde{F}_{abd}(\varepsilon) := \begin{bmatrix} F_{abd1}(\varepsilon) F_{1q_1} / \varepsilon^{q_1} \\ F_{abd2}(\varepsilon) F_{2q_2} / \varepsilon^{q_2} \\ \vdots \\ F_{abdm_d}(\varepsilon) F_{m_d q_{m_d}} / \varepsilon^{q_{m_d}} \end{bmatrix}, \tag{8.3.28}$$

$$\tilde{F}^+_{ad}(\varepsilon) := \begin{bmatrix} F^+_{ad1} F_{1q_1} / \varepsilon^{q_1} \\ F^+_{ad2} F_{2q_2} / \varepsilon^{q_2} \\ \vdots \\ F^+_{adm_d} F_{m_d q_{m_d}} / \varepsilon^{q_{m_d}} \end{bmatrix}, \tag{8.3.29}$$

and

$$\tilde{F}_{bd}(\varepsilon) := \begin{bmatrix} F_{bd1}F_{1q_1}/\varepsilon^{q_1} \\ F_{bd2}F_{2q_2}/\varepsilon^{q_2} \\ \vdots \\ F_{bdm_d}F_{m_dq_{m_d}}/\varepsilon^{q_{m_d}} \end{bmatrix}. \tag{8.3.30}$$

Then define the state feedback gain $F(\varepsilon)$ as

$$F(\varepsilon) := -\Gamma_{iP}\left(\tilde{F}^{\star}_{abcd}(\varepsilon) + \tilde{F}_{abcd}(\varepsilon)\right)\Gamma_{sP}^{-1}, \tag{8.3.31}$$

where

$$\tilde{F}^{\star}_{abcd}(\varepsilon) = \begin{bmatrix} C_{0a}^{-} & C_{0a}^{0} & C_{0a}^{+}+F_{a0}^{+} & C_{0b}+F_{b0} & C_{0c} & C_{0d} \\ E_{da}^{-} & E_{da}^{0} & E_{da}^{+}+\tilde{F}_{ad}^{+}(\varepsilon) & E_{db}+\tilde{F}_{bd}(\varepsilon) & E_{dc} & \tilde{F}_{d}(\varepsilon)+E_{d} \\ E_{ca}^{-} & E_{ca}^{0} & E_{ca}^{+} & 0 & F_{c} & 0 \end{bmatrix}, \tag{8.3.32}$$

$$\tilde{F}_{abcd}(\varepsilon) = \begin{bmatrix} F_{ab0}(\varepsilon) & 0 & 0 \\ \tilde{F}_{abd}(\varepsilon) & 0 & 0 \\ 0 & 0 & 0 \end{bmatrix}, \tag{8.3.33}$$

and where

$$E_{d} = \begin{bmatrix} E_{11} & \cdots & E_{1m_d} \\ \vdots & \ddots & \vdots \\ E_{m_d1} & \cdots & E_{m_dm_d} \end{bmatrix}, \tag{8.3.34}$$

$$\tilde{F}_{d}(\varepsilon) = \operatorname{diag}\left\{\tilde{F}_{1}(\varepsilon),\ \tilde{F}_{2}(\varepsilon),\ \cdots,\ \tilde{F}_{m_d}(\varepsilon)\right\}. \tag{8.3.35}$$

⊠A

We have the following theorem.

Theorem 8.3.1. Consider the given system (8.3.2) that satisfies all the conditions in Remark 8.2.2. Then the closed-loop system comprising (8.3.2) and the static state feedback law $u = F(\varepsilon)x$, with $F(\varepsilon)$ given by (8.3.31), has the following properties: For any given $\gamma > 0$, there exists a positive scalar $\varepsilon^* > 0$ such that for all $0 < \varepsilon \le \varepsilon^*$,

1. the closed-loop system is asymptotically stable, i.e., $\lambda\{A+BF(\varepsilon)\} \subset \mathbb{C}^-$;
2. the H_∞-norm of the closed-loop transfer matrix from the disturbance w to the controlled output h is less than γ, i.e., $\|T_{hw}(s,\varepsilon)\|_\infty < \gamma$.

Hence, by Definition 8.1.1, the control law $u = F(\varepsilon)x$ solves the general H_∞-ADDPS for the given system (8.3.2). ⊠T

Proof. See Subsection 8.5.A. ⊠

We illustrate the above result in the following example.

Example 8.3.1. Let us consider a given system of (8.1.1) characterized by $C_1 = I$, $D_1 = 0$ and

$$A = \begin{bmatrix} 0 & 1 & 0 & 0 & 0 & 1 \\ 0 & 0 & 1 & 0 & 0 & 2 \\ 0 & 0 & 0 & 0 & 0 & 3 \\ 0 & 0 & 0 & -1 & 3 & 4 \\ 0 & 0 & 0 & 0 & 4 & 5 \\ 1 & 2 & 3 & 4 & 5 & 6 \end{bmatrix}, \quad B = \begin{bmatrix} 0 & 0 \\ 0 & 0 \\ 0 & 0 \\ 0 & 0 \\ 1 & 0 \\ 0 & 1 \end{bmatrix}, \quad E = \begin{bmatrix} 1 & 0 \\ 2 & 0 \\ 0 & 0 \\ 0 & 0 \\ 0 & 0 \\ 3 & 1 \end{bmatrix}, \tag{8.3.36}$$

$$C_2 = \begin{bmatrix} 0 & 0 & 0 & 0 & 1 & 2 \\ 0 & 0 & 0 & 0 & 0 & 1 \\ 0 & 0 & 0 & 0 & 1 & 0 \end{bmatrix}, \quad D_2 = \begin{bmatrix} 1 & 0 \\ 0 & 0 \\ 0 & 0 \end{bmatrix}, \quad D_{22} = \begin{bmatrix} 0 & 0 \\ 0 & 0 \\ 0 & 0 \end{bmatrix}. \tag{8.3.37}$$

The subsystem $\Sigma_{\rm P}$ is already in the form of the special coordinate basis. It is simple to verify that: i) (A, B) is stabilizable; ii) $\Sigma_{\rm P}$ has three invariant zeros at 0 and one stable invariant zero at -1; iii) $\Sigma_{\rm P}$ has one infinite zero of order zero and one infinite zero of order one; iv) $\Sigma_{\rm P}$ is left invertible; and v)

$$\mathcal{S}^+(\Sigma_{\rm P}) = \operatorname{Im} \left\{ \begin{bmatrix} 1 & 0 & 0 & 0 & 0 \\ 0 & 1 & 0 & 0 & 0 \\ 0 & 0 & 1 & 0 & 0 \\ 0 & 0 & 0 & 1 & 0 \\ 0 & 0 & 0 & 0 & 0 \\ 0 & 0 & 0 & 0 & 1 \end{bmatrix} \right\}, \tag{8.3.38}$$

and

$$\cap_{\lambda \in \mathbb{C}^0} \mathcal{S}_\lambda(\Sigma_{\rm P}) = \operatorname{Im} \left\{ \begin{bmatrix} 1 & 0 & 0 & 0 \\ 0 & 1 & 0 & 0 \\ 0 & 0 & 0 & 0 \\ 0 & 0 & 1 & 0 \\ 0 & 0 & 0 & 0 \\ 0 & 0 & 0 & 1 \end{bmatrix} \right\}. \tag{8.3.39}$$

Hence,

$$\mathcal{S}^+(\Sigma_{\rm P}) \cap \left\{ \cap_{\lambda \in \mathbb{C}^0} \mathcal{S}_\lambda(\Sigma_{\rm P}) \right\} = \operatorname{Im} \left\{ \begin{bmatrix} 1 & 0 & 0 & 0 \\ 0 & 1 & 0 & 0 \\ 0 & 0 & 0 & 0 \\ 0 & 0 & 1 & 0 \\ 0 & 0 & 0 & 0 \\ 0 & 0 & 0 & 1 \end{bmatrix} \right\}. \tag{8.3.40}$$

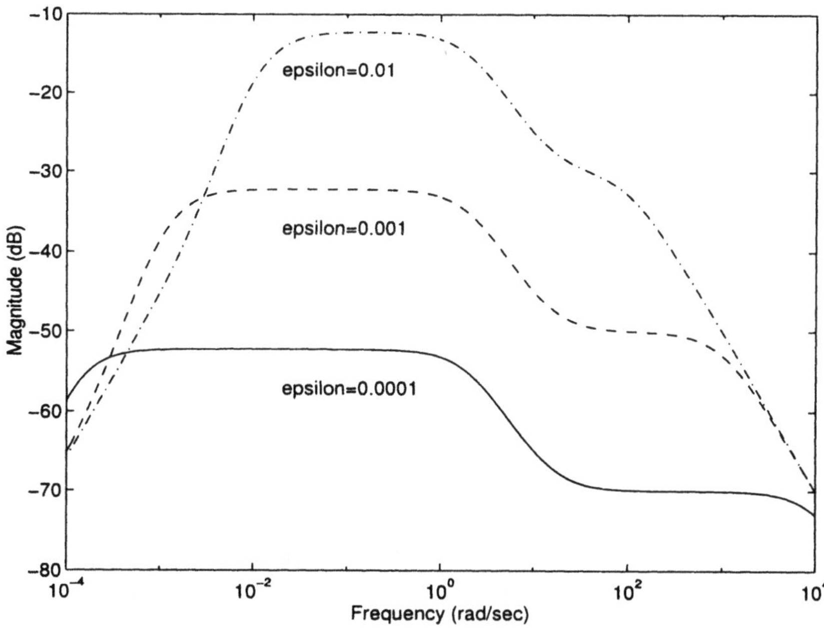

Figure 8.3.1: Max. singular values of T_{hw} — State feedback.

Obviously, $\mathrm{Im}\,(E) \subset \mathcal{S}^+(\Sigma_\mathrm{P}) \cap \{\cap_{\lambda \in \mathbb{C}^0} \mathcal{S}_\lambda(\Sigma_\mathrm{P})\}$ and by Remark 8.2.2, the H_∞-ADDPS is achievable for the given system. Following our algorithm, we obtain a state feedback gain matrix

$$F(\varepsilon) = \begin{bmatrix} 0 & 0 & 0 & 0 & -6 & -2 \\ -\varepsilon^2/3-1 & 2\varepsilon^2/9-\varepsilon-2 & 2\varepsilon/3-\varepsilon^2/27-4 & -4 & -5 & -1/\varepsilon-6 \end{bmatrix}, \tag{8.3.41}$$

which places the closed-loop poles of $A+BF(\varepsilon)$ asymptotically at -1, -2, $-\varepsilon$, $-\varepsilon$, $-\varepsilon$ and $-1/\varepsilon$. The maximum singular value plots of the corresponding closed-loop transfer matrix $T_{hw}(s,\varepsilon)$ in Figure 8.3.1 clearly show that the H_∞-ADDPS is attained as ε tends smaller and smaller. 🄴

8.4. Solutions to Output Feedback Case

We present in this section the designs of both full order and reduced order output feedback controllers that solve the general H_∞-ADDPMS for the given system (8.1.1). Here, by full order controller, we mean that the order of the controller is exactly the same as the given system (8.1.1), i.e, is equal to n. A reduced order controller, on the other hand, refers to a controller whose dynamical order is less than n. We will assume without loss of any generality that $D_{22} = 0$ in the given system (8.1.1) throughout this section.

8.4.1. Full Order Output Feedback

The following is a step-by-step algorithm for constructing a parameterized full order output feedback controller that solves the general H_∞-ADDPMS:

Step 8.F.C.1: (Construction of the gain matrix $F_{\mathrm{P}}(\varepsilon)$). Define an auxiliary system

$$\begin{cases} \dot{x} = A\,x + B\,u + E\,w, \\ y = x \\ h = C_2\,x + D_2\,u + D_{22}\,w, \end{cases} \tag{8.4.1}$$

and then perform Step 8.S.1 to 8.S.6 of the previous section to the above system to obtain a parameterized gain matrix $F(\varepsilon)$. We let $F_{\mathrm{P}}(\varepsilon) = F(\varepsilon)$.

Step 8.F.C.2: (Construction of the gain matrix $K_{\mathrm{Q}}(\varepsilon)$). Define another auxiliary system

$$\begin{cases} \dot{x} = A'\,x + C_1'\,u + C_2'\,w, \\ y = x \\ h = E'\,x + D_1'\,u + D_{22}'\,w, \end{cases} \tag{8.4.2}$$

and then perform Step 8.S.1 to 8.S.6 of the previous section to the above system to get the parameterized gain matrix $F(\varepsilon)$. We let $K_{\mathrm{Q}}(\varepsilon) = F(\varepsilon)'$.

Step 8.F.C.3: (Construction of the full order controller $\Sigma_{\mathrm{FC}}(\varepsilon)$). Finally, the parameterized full order output feedback controller is given by

$$\Sigma_{\mathrm{FC}}(\varepsilon) : \begin{cases} \dot{v} = A_{\mathrm{FC}}(\varepsilon)\,v + B_{\mathrm{FC}}(\varepsilon)\,y, \\ u = C_{\mathrm{FC}}(\varepsilon)\,v + D_{\mathrm{FC}}(\varepsilon)\,y, \end{cases} \tag{8.4.3}$$

where

$$\left.\begin{aligned} A_{\mathrm{FC}}(\varepsilon) &:= A + BF_{\mathrm{P}}(\varepsilon) + K_{\mathrm{Q}}(\varepsilon)C_1, \\ B_{\mathrm{FC}}(\varepsilon) &:= -K_{\mathrm{Q}}(\varepsilon), \\ C_{\mathrm{FC}}(\varepsilon) &:= F_{\mathrm{P}}(\varepsilon), \\ D_{\mathrm{FC}}(\varepsilon) &:= 0. \end{aligned}\right\} \tag{8.4.4}$$

This concludes the algorithm for constructing the full order measurement feedback controller. Ⓐ

We have the following theorem.

Theorem 8.4.1. Consider the given system (8.1.1) with $D_{22} = 0$ satisfying all the conditions in Theorem 8.2.1. Then the closed-loop system comprising (8.1.1) and the full order output feedback controller (8.4.3) has the following properties: For any given $\gamma > 0$, there exists a positive scalar $\varepsilon^* > 0$ such that for all $0 < \varepsilon \leq \varepsilon^*$,

1. the resulting closed-loop system is asymptotically stable; and

2. the H_∞-norm of the resulting closed-loop transfer matrix from the disturbance w to the controlled output h is less than γ, i.e., $\|T_{hw}(s,\varepsilon)\|_\infty < \gamma$.

By Definition 8.1.1, the control law (8.4.3) solves the general H_∞-ADDPMS for the given system (8.1.1). ◻T

Proof. See Subsection 8.5.B. ⊠

We illustrate the above result in the following example.

Example 8.4.1. We reconsider the system (8.1.1) with A, B, E, C_2, D_2 and D_{22} as in Example 8.3.1 but with

$$C_1 = \begin{bmatrix} -1 & -1 & 0 & 1 & 1 & 1 \\ 0 & 0 & 0 & 0 & 1 & 0 \\ 0 & 0 & 0 & 0 & 0 & 1 \end{bmatrix}, \quad D_1 = \begin{bmatrix} 1 & 0 \\ 0 & 0 \\ 0 & 0 \end{bmatrix}. \tag{8.4.5}$$

Using the software toolboxes of Chen [11] and Lin [79], we can easily obtain the following properties of Σ_Q: i) (A, C_1) is detectable; ii) Σ_Q has two stable invariant zeros at -1 and -0.5616, one imaginary axis invariant zero at 0, and one unstable invariant zero at 3.5616; iii) Σ_Q has one infinite zero of order zero and one infinite zero of order one; iv) Σ_Q is left invertible; and v)

$$\mathcal{V}^+(\Sigma_Q) = \operatorname{Im} \left\{ \begin{bmatrix} 1 \\ 1.2808 \\ 0 \\ 0 \\ 0 \\ 0 \end{bmatrix} \right\}, \quad \cup_{\lambda \in \mathbb{C}^0} \mathcal{V}_\lambda(\Sigma_Q) = \operatorname{Im} \left\{ \begin{bmatrix} -2 \\ 1 \\ 2 \\ 0 \\ 0 \\ 0 \end{bmatrix} \right\}. \tag{8.4.6}$$

It is straightforward to see that $\operatorname{Ker}(C_2) \supset \mathcal{V}^+(\Sigma_Q) \cup \{\cup_{\lambda \in \mathbb{C}^0} \mathcal{V}_\lambda(\Sigma_Q)\}$ and $\mathcal{V}^+(\Sigma_Q) \subset \mathcal{S}^+(\Sigma_P)$. By Theorem 8.2.1, the H_∞-ADDPMS is solvable for the

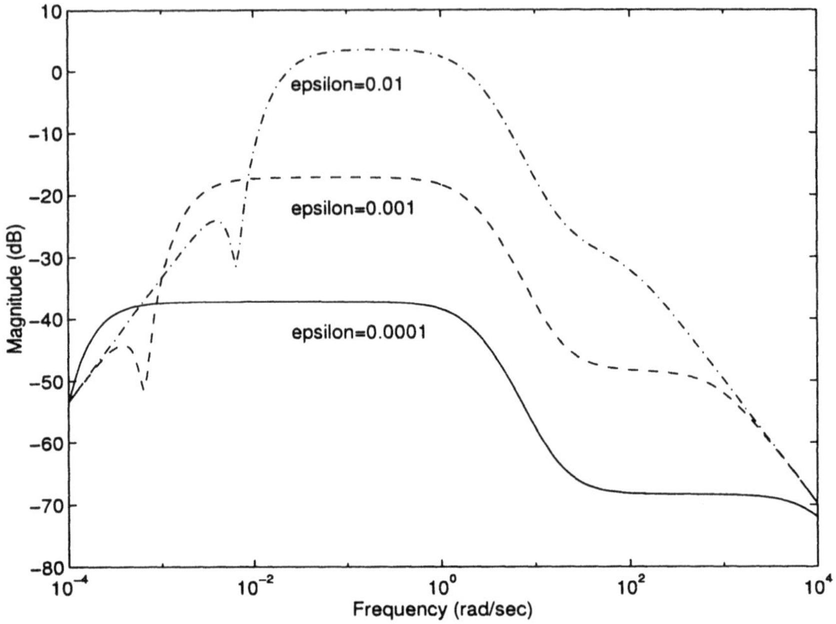

Figure 8.4.1: Max. singular values of T_{hw} — Full order output feedback.

given system. Following our algorithm, we obtain a full order output feedback controller of the form (8.4.3) with $F_\text{P}(\varepsilon)$ as given in (8.3.41) and

$$K_\text{Q}(\varepsilon) = \begin{bmatrix} 2.4375 & 1 & 0.1813 \\ 2.4028 & 2 & -0.0808 \\ 0 & 0 & -3.1758 \\ 0 & -3 & -4 \\ 0 & -8.2462 & -5 \\ -3 & -2 & -1/\varepsilon - 3 \end{bmatrix}, \tag{8.4.7}$$

which places the closed-loop eigenvalues of $A + K_\text{Q}(\varepsilon)C_1$ asymptotically at -0.5616, -1, -4.2462, -4.2787, $-\varepsilon$ and $-1/\varepsilon$. The maximum singular value plots of the corresponding closed-loop transfer matrix $T_{hw}(s,\varepsilon)$ in Figure 8.4.1 show that the H_∞-ADDPMS is attained as ε tends to zero. ▣

8.4.2. Reduced Order Output Feedback

In this subsection, we follow the procedure of Chen *et al.* [33,34] to design a reduced order output feedback controller. We will show that such as a controller structure with appropriately chosen gain matrices also solves the general H_∞-ADDPMS for the system (8.1.1). First, without loss of generality and for simplicity of presentation, we assume that the matrices C_1 and D_1 are already in the form,

$$C_1 = \begin{bmatrix} 0 & C_{1,02} \\ I_k & 0 \end{bmatrix} \quad \text{and} \quad D_1 = \begin{bmatrix} D_{1,0} \\ 0 \end{bmatrix}, \tag{8.4.8}$$

where $k = \ell - \text{rank}(D_1)$ and $D_{1,0}$ is of full rank. Then the given system (8.1.1) can be written as

$$\begin{cases} \begin{pmatrix} \dot{x}_1 \\ \dot{x}_2 \end{pmatrix} = \begin{bmatrix} A_{11} & A_{12} \\ A_{21} & A_{22} \end{bmatrix} \begin{pmatrix} x_1 \\ x_2 \end{pmatrix} + \begin{bmatrix} B_1 \\ B_2 \end{bmatrix} u + \begin{bmatrix} E_1 \\ E_2 \end{bmatrix} w, \\ \begin{pmatrix} y_0 \\ y_1 \end{pmatrix} = \begin{bmatrix} 0 & C_{1,02} \\ I_k & 0 \end{bmatrix} \begin{pmatrix} x_1 \\ x_2 \end{pmatrix} + \begin{bmatrix} D_{1,0} \\ 0 \end{bmatrix} w, \\ h = [\, C_{2,1} \quad C_{2,2} \,] \begin{pmatrix} x_1 \\ x_2 \end{pmatrix} + D_2 \, u + D_{22} \, w, \end{cases} \tag{8.4.9}$$

where the original state x is partitioned into two parts, x_1 and x_2; and y is partitioned into y_0 and y_1 with $y_1 \equiv x_1$. Thus, one needs to estimate only the state x_2 in the reduced order controller design. Next, define an auxiliary subsystem Σ_{QR} characterized by a matrix quadruple $(A_{\text{R}}, E_{\text{R}}, C_{\text{R}}, D_{\text{R}})$, where

$$(A_{\text{R}}, E_{\text{R}}, C_{\text{R}}, D_{\text{R}}) = \left(A_{22}, E_2, \begin{bmatrix} C_{1,02} \\ A_{12} \end{bmatrix}, \begin{bmatrix} D_{1,0} \\ E_1 \end{bmatrix} \right). \tag{8.4.10}$$

The following is a step-by-step algorithm that constructs the reduced order output feedback controller for the general H_∞-ADDPMS.

Step 8.R.C.1: (Construction of the gain matrix $F_{\text{P}}(\varepsilon)$). Define an auxiliary system

$$\begin{cases} \dot{x} = A\, x + B\, u + E\, w, \\ y = x \\ h = C_2\, x + D_2\, u + D_{22}\, w, \end{cases} \tag{8.4.11}$$

and then perform Step 8.S.1 to 8.S.6 of Section 8.3 to the above system to get the parameterized gain matrix $F(\varepsilon)$. We let $F_{\text{P}}(\varepsilon) = F(\varepsilon)$.

Step 8.R.C.2: (Construction of the gain matrix $K_{\text{R}}(\varepsilon)$). Define another auxiliary system

$$\begin{cases} \dot{x} = A_{\text{R}}'\, x + C_{\text{R}}'\, u + C_{2,2}'\, w, \\ y = x \\ h = E_{\text{R}}'\, x + D_{\text{R}}'\, u + D_{22}'\, w, \end{cases} \tag{8.4.12}$$

and then perform Step 8.S.1 to 8.S.6 of Section 8.3 to the above system to get the parameterized gain matrix $F(\varepsilon)$. We let $K_{\text{R}}(\varepsilon) = F(\varepsilon)'$.

Step 8.R.C.3: (Construction of the reduced order controller $\Sigma_{RC}(\varepsilon)$). Let us partition $F_P(\varepsilon)$ and $K_R(\varepsilon)$ as,

$$F_P(\varepsilon) = [\,F_{P1}(\varepsilon) \quad F_{P2}(\varepsilon)\,] \quad \text{and} \quad K_R(\varepsilon) = [\,K_{R0}(\varepsilon) \quad K_{R1}(\varepsilon)\,] \tag{8.4.13}$$

in conformity with the partitions of $x = \begin{pmatrix} x_1 \\ x_2 \end{pmatrix}$ and $y = \begin{pmatrix} y_0 \\ y_1 \end{pmatrix}$, respectively. Then define

$$G_R(\varepsilon) = [\,-K_{R0}(\varepsilon), \;\; A_{21} + K_{R1}(\varepsilon)A_{11} - (A_R + K_R(\varepsilon)C_R)K_{R1}(\varepsilon)\,]. \tag{8.4.14}$$

Finally, the parameterized reduced order output feedback controller is given by

$$\Sigma_{RC}(\varepsilon) \; : \; \begin{cases} \dot{v} = A_{RC}(\varepsilon)\, v + B_{RC}(\varepsilon)\, y, \\ u = C_{RC}(\varepsilon)\, v + D_{RC}(\varepsilon)\, y, \end{cases} \tag{8.4.15}$$

where

$$\left.\begin{aligned} A_{RC}(\varepsilon) &:= A_R + B_2F_{P2}(\varepsilon) + K_R(\varepsilon)C_R + K_{R1}(\varepsilon)B_1F_{P2}(\varepsilon), \\ B_{RC}(\varepsilon) &:= G_R(\varepsilon) + [B_2 + K_{R1}(\varepsilon)B_1]\,[\,0, \;\; F_{P1}(\varepsilon) - F_{P2}(\varepsilon)K_{R1}(\varepsilon)\,], \\ C_{RC}(\varepsilon) &:= F_{P2}(\varepsilon), \\ D_{RC}(\varepsilon) &:= [\,0, \;\; F_{P1}(\varepsilon) - F_{P2}(\varepsilon)K_{R1}(\varepsilon)\,]. \end{aligned}\right\} \tag{8.4.16}$$

This concludes the algorithm for constructing the reduced order measurement feedback controller. ▣

We have the following theorem.

Theorem 8.4.2. Consider the given system (8.1.1) with $D_{22} = 0$ satisfying all the conditions in Theorem 8.2.1. Then the closed-loop system comprising (8.1.1) and the reduced order output feedback controller (8.4.15) has the following properties: For any given $\gamma > 0$, there exists a positive scalar $\varepsilon^* > 0$ such that for all $0 < \varepsilon \le \varepsilon^*$,

1. the resulting closed-loop system is asymptotically stable; and
2. the H_∞-norm of the resulting closed-loop transfer matrix from the disturbance w to the controlled output h is less than γ, i.e., $\|T_{hw}(s,\varepsilon)\|_\infty < \gamma$.

By Definition 8.1.1, the control law (8.4.15) solves the general H_∞-ADDPMS for the given system (8.1.1). ▣

Proof. See Subsection 8.5.C. ▣

We illustrate the above result in the following example.

Example 8.4.2. We again consider the given system as in Examples 8.3.1 and 8.4.1. As all the five conditions of Theorem 8.2.1 are satisfied, the H_∞-ADDPMS for the given system can be solved using a reduced order output feedback controller. We will construct such a controller in the following. First, it is simple to show the transformation T_s and T_o,

$$T_s = \begin{bmatrix} 0 & 0 & 1 & 0 & 0 & 0 \\ 0 & 0 & 0 & 1 & 0 & 0 \\ 0 & 0 & 0 & 0 & 1 & 0 \\ 0 & 0 & 0 & 0 & 0 & 1 \\ 1 & 0 & 0 & 0 & 0 & 0 \\ 0 & 1 & 0 & 0 & 0 & 0 \end{bmatrix}, \quad T_o = \begin{bmatrix} 1 & 1 & 1 \\ 0 & 1 & 0 \\ 0 & 0 & 1 \end{bmatrix}, \tag{8.4.17}$$

will transform C_1 and D_1 to the form of (8.4.8), i.e.,

$$T_o^{-1} C_1 T_s = \left[\begin{array}{c|c} 0 & C_{1,02} \\ \hline I_k & 0 \end{array}\right] = \left[\begin{array}{cc|cccc} 0 & 0 & -1 & -1 & 0 & 1 \\ \hline 1 & 0 & 0 & 0 & 0 & 0 \\ 0 & 1 & 0 & 0 & 0 & 0 \end{array}\right], \tag{8.4.18}$$

and

$$T_o^{-1} D_1 = \left[\begin{array}{c} D_{1,0} \\ \hline 0 \end{array}\right] = \left[\begin{array}{cc} 1 & 0 \\ \hline 0 & 0 \\ 0 & 0 \end{array}\right]. \tag{8.4.19}$$

Moreover, we have

$$T_s^{-1} A T_s = \left[\begin{array}{c|c} A_{11} & A_{12} \\ \hline A_{21} & A_{22} \end{array}\right] = \left[\begin{array}{cc|cccc} 4 & 5 & 0 & 0 & 0 & 0 \\ 5 & 6 & 1 & 2 & 3 & 4 \\ \hline 0 & 1 & 0 & 1 & 0 & 0 \\ 0 & 2 & 0 & 0 & 1 & 0 \\ 0 & 3 & 0 & 0 & 0 & 0 \\ 3 & 4 & 0 & 0 & 0 & -1 \end{array}\right], \tag{8.4.20}$$

$$T_s^{-1} B = \left[\begin{array}{c} B_1 \\ \hline B_2 \end{array}\right] = \left[\begin{array}{cc} 1 & 0 \\ 0 & 1 \\ \hline 0 & 0 \\ 0 & 0 \\ 0 & 0 \\ 0 & 0 \end{array}\right], \quad T_s^{-1} E = \left[\begin{array}{c} E_1 \\ \hline E_2 \end{array}\right] = \left[\begin{array}{cc} 0 & 0 \\ 3 & 1 \\ \hline 1 & 0 \\ 2 & 0 \\ 0 & 0 \\ 0 & 0 \end{array}\right], \tag{8.4.21}$$

and $A_R = A_{22}$, $E_R = E_2$, and

$$C_R = \begin{bmatrix} -1 & -1 & 0 & 1 \\ 0 & 0 & 0 & 0 \\ 1 & 2 & 3 & 4 \end{bmatrix}, \quad D_R = \begin{bmatrix} 1 & 0 \\ 0 & 0 \\ 3 & 1 \end{bmatrix}. \tag{8.4.22}$$

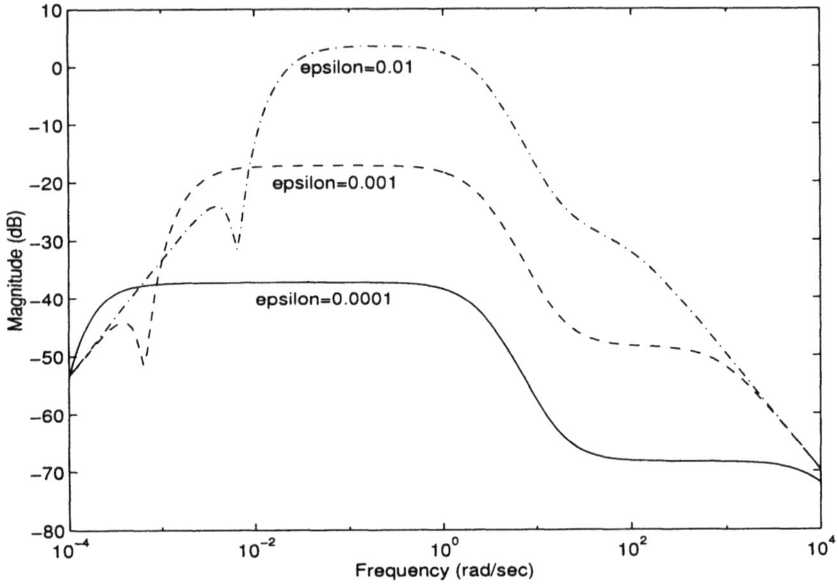

Figure 8.4.2: Max. singular values of T_{hw} — Reduced order output feedback.

Following our algorithm, we obtain

$$F_{\rm P}(\varepsilon)T_s = \left[\begin{array}{c|c} F_{\rm P1}(\varepsilon) & F_{\rm P2}(\varepsilon) \end{array}\right] = \left[\begin{array}{cc|cccccc} -6 & -2 & 0 & 0 & 0 & 0 \\ -5 & -1/\varepsilon-6 & -\varepsilon^2/3-1 & 2\varepsilon^2/9-\varepsilon-2 & 2\varepsilon/3-\varepsilon^2/27-4 & -4 \end{array}\right], \tag{8.4.23}$$

and

$$K_{\rm R}(\varepsilon) = \left[\begin{array}{c|c} K_{\rm R0}(\varepsilon) & K_{\rm R1}(\varepsilon) \end{array}\right] = \left[\begin{array}{r|cr} 1.2000+0.1219\varepsilon & 0 & -0.6663+0.4025\varepsilon \\ 0.8187-0.0609\varepsilon & 0 & -0.8534-0.2012\varepsilon \\ -0.1219\varepsilon & 0 & -0.4025\varepsilon \\ 0 & 0 & 0 \end{array}\right], \tag{8.4.24}$$

which place the eigenvalues of $A_{\rm R} + K_{\rm R}(\varepsilon)C_{\rm R}$ at -0.5616, -1, -3.8303 and $-\varepsilon$. Also, we obtain a reduced order output feedback controller of the form (8.4.15) with all sub-matrices as defined in (8.4.18) to (8.4.24), and with $B_{\rm RC}(\varepsilon)$ and $D_{\rm RC}(\varepsilon)$ being slightly modified to

$$B_{\rm RC}(\varepsilon) = G_{\rm R}(\varepsilon)T_o^{-1} + [B_2 + K_{\rm R1}(\varepsilon)B_1]\,[\,0,\ \ F_{\rm P1}(\varepsilon) - F_{\rm P2}(\varepsilon)K_{\rm R1}(\varepsilon)\,]\,T_o^{-1}, \tag{8.4.25}$$

and

$$D_{\rm RC}(\varepsilon) = [\,0,\ \ F_{\rm P1}(\varepsilon) - F_{\rm P2}(\varepsilon)K_{\rm R1}(\varepsilon)\,]\,T_o^{-1}, \tag{8.4.26}$$

respectively. The maximum singular value plots of the corresponding closed-loop transfer matrix $T_{hw}(s,\varepsilon)$ in Figure 8.4.2 also show that the H_∞-ADDPMS is attained as ε tends to zero. ▣

8.5. Proofs of Main Results

We present the proofs of all the main results of this chapter in this section.

8.5.A. Proof of Theorem 8.3.1

Under the feedback law $u = F(\varepsilon)x$, the closed-loop system on the special coordinate basis can be written as follows,

$$\dot{x}_a^- = A_{aa}^- x_a^- + B_{0a}^- h_0 + L_{ad}^- h_d + L_{ab}^- h_b + E_a^- w, \tag{8.5.1}$$

$$\dot{x}_a^0 = A_{aa}^0 x_a^0 + B_{0a}^0 h_0 + L_{ad}^0 h_d + L_{ab}^0 h_b + E_a^0 w, \tag{8.5.2}$$

$$\dot{x}_{ab}^+ = A_{ab}^{+c} x_{ab}^+ - B_{0ab}^+ F_{a0}^0(\varepsilon)[x_a^0 + T_a^0 x_{ab}^+] + L_{abd}^+[F_{ad}^+,\ F_{bd}]x_{ab}^+ + L_{abd}^+ h_d + E_{ab}^+ w, \tag{8.5.3}$$

$$h_b = [0_{m_b \times n_a^+}, C_b]x_{ab}^+, \tag{8.5.4}$$

$$\dot{x}_c = A_{cc}^c + L_{c0}h_0 + L_{cb}h_b + L_{cd}h_d + E_c w, \tag{8.5.5}$$

$$h_0 = -[F_{a0}^+,\ F_{b0}]x_{ab}^+ - F_{a0}^0(\varepsilon)(x_a^0 + T_a^0 x_{ab}^+), \tag{8.5.6}$$

$$\begin{aligned}\dot{x}_i = A_{q_i} x_i + L_{i0} h_0 + L_{id} h_d - \frac{1}{\varepsilon^{q_i}} B_{q_i} \Big[& F_{adi}^+ F_{iq_i} x_a^+ + F_{bdi} F_{iq_i} x_b \\ & + F_{adi}^0(\varepsilon) F_{iq_i} [x_a^0 + T_a^0 x_{ab}^+] + F_i S_i(\varepsilon) x_i \Big] + E_i w,\end{aligned} \tag{8.5.7}$$

$$h_i = C_{q_i} x_i, \quad i = 1, 2, \cdots, m_d, \tag{8.5.8}$$

where

$$x_{ab}^+ = \begin{pmatrix} x_a^+ \\ x_b \end{pmatrix}, \tag{8.5.9}$$

and B_{0ab}^+ and L_{abd}^+ are as defined in Step 8.S.4.1 of the state feedback design algorithm. We have also used Condition 2 of Remark 8.2.2, i.e., $D_{22} = 0$, and E_a^-, E_a^0, E_{ab}^+, E_b, E_c and E_i, $i = 1, 2, \cdots, m_d$, are defined as follows,

$$\Gamma_{\rm sP}^{-1} E = [\,(E_a^-)' \quad (E_a^0)' \quad (E_{ab}^+)' \quad E_c' \quad E_1' \quad E_2' \quad \cdots \quad E_{m_d}'\,]'. \tag{8.5.10}$$

Condition 4 of the theorem then implies that

$$E_{ab}^+ = 0, \tag{8.5.11}$$

and

$$\operatorname{Im}(E_a^0) \subset \mathcal{S}(A_{aa}^0) := \cap_{w\in\lambda(A_{aa}^0)} \operatorname{Im}\{wI - A_{aa}^0\}. \tag{8.5.12}$$

To complete the proof, we will make two state transformations on the closed-loop system (8.5.1)-(8.5.8). The first state transformation is given as follows,

$$\bar{x}_{ab} = \Gamma_{ab}^{-1} x_{ab}, \quad \bar{x}_c = x_c, \tag{8.5.13}$$

for $i = 1, 2, \cdots, m_d$,

$$\bar{x}_{i1} = x_{i1} + F_{adi}^+ x_a^+ + F_{bdi} x_b + F_{adi}^0(\varepsilon)[x_a^0 + T_a^0 x_{ab}^+], \tag{8.5.14}$$

and for $j = 2, 3, \cdots, q_i$, $i = 1, 2, \cdots, m_d$,

$$\bar{x}_{ij} = x_{ij}, \tag{8.5.15}$$

where

$$x_{ab} = \begin{pmatrix} x_a^- \\ x_a^0 \\ x_{ab}^+ \end{pmatrix} \quad \text{and} \quad \bar{x}_{ab} = \begin{pmatrix} \bar{x}_a^- \\ \bar{x}_{ab}^+ \\ \bar{x}_a^0 \end{pmatrix}. \tag{8.5.16}$$

In the new state variables (8.5.13)-(8.5.15), the closed-loop system becomes,

$$\dot{\bar{x}}_a^- = A_{aa}^- \bar{x}_a^- + A_{aab}^- \bar{x}_{ab}^+ - [B_{0a}^-, L_{ad}^-] F_a^0(\varepsilon) \bar{x}_a^0 + L_{ad}^- \bar{h}_d + E_a^- w, \tag{8.5.17}$$

$$\dot{\bar{x}}_{ab}^+ = A_{ab}^{+c} \bar{x}_{ab}^+ - [B_{0ab}^+, L_{abd}^+] F_a^0(\varepsilon) \bar{x}_a^0 + L_{abd}^+ \bar{h}_d, \tag{8.5.18}$$

$$\dot{\bar{x}}_a^0 = \left[A_{aa}^0 - B_a^0 F_a^0(\varepsilon)\right] \bar{x}_a^0 + (L_{ad}^0 + T_a^0 L_{abd}^+) \bar{h}_d + E_a^0 w, \tag{8.5.19}$$

$$\begin{aligned}\dot{\bar{x}}_c = A_{cc}^c \bar{x}_c + \Big(L_{cb}[0,\ C_b] - [L_{c0},\ L_{cd}] F_{ab}^+\Big) \bar{x}_{ab}^+ \\ - [L_{c0},\ L_{cd}] F_a^0(\varepsilon) \bar{x}_a^0 + L_{cd} \bar{h}_d + E_c w,\end{aligned} \tag{8.5.20}$$

$$h_0 = -[F_{a0}^+,\ F_{b0}] \bar{x}_{ab}^+ - F_{a0}^0(\varepsilon) \bar{x}_a^0, \tag{8.5.21}$$

$$\begin{aligned}\dot{\bar{x}}_i = A_{q_i} \bar{x}_i - \frac{1}{\varepsilon^{q_i}} B_{q_i} F_i S_i(\varepsilon) \bar{x}_i + L_{iab}^+(\varepsilon) \bar{x}_{ab}^+ + L_{ia}^{01}(\varepsilon) F_a^0(\varepsilon) \bar{x}_a^0 \\ + L_{ia}^{02}(\varepsilon) F_a^0(\varepsilon) A_{aa}^0 \bar{x}_a^0 + \bar{L}_{id}(\varepsilon) \bar{h}_d + \bar{E}_i(\varepsilon) w,\end{aligned} \tag{8.5.22}$$

$$\bar{h}_i = h_i + [F_{adi}^+, F_{bdi}] \bar{x}_{ab}^+ + F_{adi}^0 \bar{x}_a^0 = C_{q_i} \bar{x}_i, \quad i = 1, 2, \cdots, m_d, \tag{8.5.23}$$

$$\bar{h}_d = [\bar{h}_1, \bar{h}_2, \cdots, \bar{h}_{m_d}]', \tag{8.5.24}$$

where matrices A_{aab}^-, A_{aab}^0, B_a^0 and L_{abd}^+ are as defined in Step 8.S.4.1 of the state feedback law design algorithm, and $L_{iab}^+(\varepsilon)$, $L_{ia}^{01}(\varepsilon)$, $L_{ia}^{02}(\varepsilon)$, $\bar{L}_{id}(\varepsilon)$ and $\bar{E}_i(\varepsilon)$ are defined in an obvious way and, by (8.3.20), satisfy

$$|L_{iab}^+(\varepsilon)| \le l_{iab}^+, \quad |L_{ia}^{01}(\varepsilon)| \le l_{ia}^{01}, \quad |L_{ia}^{02}(\varepsilon)| \le l_{ia}^{02}, \tag{8.5.25}$$

and

$$|\bar{L}_{id}(\varepsilon)| \le \bar{l}_{id}, \ \ |\bar{E}_i(\varepsilon)| \le \bar{e}_i, \ \ \varepsilon \in (0,1], \tag{8.5.26}$$

for some nonnegative constants l^{+}_{iab}, l^{01}_{ia}, l^{02}_{ia} $\bar{e}_i$ and $\bar{l}_{id}$, independent of ε.

We now proceed to construct the second transformation. We need to recall the following preliminary results from [83].

Lemma 8.5.1. Let the triple $(A_i, B_i, F_i(\varepsilon))$ be as given in Steps 8.S.4.2 and 8.S.4.3 of the state feedback design algorithm. Then, there exists a nonsingular state transformation matrix $Q_i(\varepsilon) \in \mathbb{R}^{n_i \times n_i}$ such that

1. $Q_i(\varepsilon)$ transforms $A_i - B_iF_i(\varepsilon)$ into a real Jordan form, i.e.,

$$\begin{aligned} Q_i^{-1}(\varepsilon)\Big[A_i - B_iF_i(\varepsilon)\Big]Q_i(\varepsilon) &= J_i(\varepsilon) \\ &= \text{blkdiag}\Big\{J_{i0}(\varepsilon), J_{i1}(\varepsilon), J_{i2}(\varepsilon), \cdots, J_{ip_i}(\varepsilon)\Big\}, \end{aligned} \tag{8.5.27}$$

   where

$$J_{i0}(\varepsilon) = \begin{bmatrix} -\varepsilon & 1 & & \\ & \ddots & \ddots & \\ & & -\varepsilon & 1 \\ & & & -\varepsilon \end{bmatrix}_{r_{i0} \times r_{i0}}, \tag{8.5.28}$$

   and for each $j = 1$ to p_i,

$$J_{ij}(\varepsilon) = \begin{bmatrix} J^{\star}_{ij}(\varepsilon) & I_2 & & \\ & \ddots & \ddots & \\ & & J^{\star}_{ij}(\varepsilon) & I_2 \\ & & & J^{\star}_{ij}(\varepsilon) \end{bmatrix}_{2r_{ij} \times 2r_{ij}}, \quad J^{\star}_{ij}(\varepsilon) = \begin{bmatrix} -\varepsilon & \beta_{ij} \\ -\beta_{ij} & -\varepsilon \end{bmatrix}, \tag{8.5.29}$$

   with $\beta_{ij} > 0$ for all $j = 1$ to p_i and $\beta_{ij} \neq \beta_{ik}$ for $j \neq k$.

2. Both $|Q_i(\varepsilon)|$ and $|Q_i^{-1}(\varepsilon)|$ are bounded, i.e.,

$$|Q_i(\varepsilon)| \le \theta_i, \ \ |Q_i^{-1}(\varepsilon)| \le \theta_i, \ \ \varepsilon \in (0,1], \tag{8.5.30}$$

   for some positive constant θ_i, independent of ε.

3. If $E_i \in \mathbb{R}^{n_i \times q}$ is such that

$$\text{Im}(E_i) \subset \cap_{w \in \lambda(A_i)} \text{Im}(wI - A_i), \tag{8.5.31}$$

   then, there exists a $\delta_i \ge 0$, independent of ε, such that

$$|Q_i^{-1}(\varepsilon)E_i| \le \delta_i, \ \ \varepsilon \in (0,1], \tag{8.5.32}$$

and, if we partition $Q_i^{-1}(\varepsilon)E_i$ according to that of $J_i(\varepsilon)$ as,

$$Q_i^{-1}(\varepsilon)E_i = \begin{bmatrix} E_{i0}(\varepsilon) \\ E_{i1}(\varepsilon) \\ \vdots \\ E_{ip_i}(\varepsilon) \end{bmatrix}, \quad E_{i0}(\varepsilon) = \begin{bmatrix} E_{i01}(\varepsilon) \\ E_{i02}(\varepsilon) \\ \vdots \\ E_{i0r_{i0}}(\varepsilon) \end{bmatrix}_{r_{i0}\times 1}, \tag{8.5.33}$$

and

$$E_{ij}(\varepsilon) = \begin{bmatrix} E_{ij1}(\varepsilon) \\ E_{ij2}(\varepsilon) \\ \vdots \\ E_{ijr_{ij}}(\varepsilon) \end{bmatrix}_{2r_{ij}\times 1}, \tag{8.5.34}$$

then, there exists a $\beta_i \geq 0$, independent of ε, such that, for each $j = 0$, to p_i,

$$|E_{ijr_{ij}}(\varepsilon)| \leq \beta_i \varepsilon. \tag{8.5.35}$$

4. If we define a scaling matrix $S_{ai}(\varepsilon)$ as

$$S_{ai}(\varepsilon) = \text{blkdiag}\Big\{S_{ai0}(\varepsilon), S_{ai1}(\varepsilon), S_{ai2}(\varepsilon), \cdots, S_{aip_i}(\varepsilon)\Big\}, \tag{8.5.36}$$

where

$$S_{ai0}(\varepsilon) = \text{diag}\Big\{\varepsilon^{r_{i0}-1}, \varepsilon^{r_{i0}-2}, \cdots, \varepsilon, 1\Big\}, \tag{8.5.37}$$

and for $j = 1$ to p_i,

$$S_{aij}(\varepsilon) = \text{blkdiag}\Big\{\varepsilon^{r_{ij}-1}I_2, \varepsilon^{r_{ij}-2}I_2, \cdots, \varepsilon I_2, I_2\Big\}, \tag{8.5.38}$$

then, there exists a $\kappa_i \geq 0$ independent of ε such that,

$$|F_i(\varepsilon)Q_i(\varepsilon)S_{ai}^{-1}(\varepsilon)| \leq \kappa_i \varepsilon, \quad |F_i(\varepsilon)A_iQ_i(\varepsilon)S_{ai}^{-1}(\varepsilon)| \leq \kappa_i \varepsilon. \tag{8.5.39}$$

🄻

Proof. This is a combination of the results of [83], and (2.2.13) of [80]. ☒

Lemma 8.5.2. Let

$$\tilde{J}_i(\varepsilon) = \text{blkdiag}\Big\{\tilde{J}_{i0}, \tilde{J}_{i1}(\varepsilon), \cdots, \tilde{J}_{ip_i}(\varepsilon)\Big\}, \tag{8.5.40}$$

where

$$\tilde{J}_{i0} = \begin{bmatrix} -1 & 1 & & & \\ & \ddots & \ddots & & \\ & & -1 & 1 \\ & & & -1 \end{bmatrix}_{r_{i0}\times r_{i0}}, \tag{8.5.41}$$

and for each $j = 1$ to p_i,

$$\tilde{J}_{ij}(\varepsilon) = \begin{bmatrix} \tilde{J}^{\star}_{ij}(\varepsilon) & I_2 & & & \\ & \ddots & \ddots & & \\ & & \tilde{J}^{\star}_{ij}(\varepsilon) & I_2 & \\ & & & \tilde{J}^{\star}_{ij}(\varepsilon) \end{bmatrix}_{2r_{ij}\times 2r_{ij}}, \quad \tilde{J}^{\star}_{ij}(\varepsilon) = \begin{bmatrix} -1 & \beta_{ij}/\varepsilon \\ -\beta_{ij}/\varepsilon & -1 \end{bmatrix}, \tag{8.5.42}$$

with $\beta_{ij} > 0$ for all $j = 1$ to p_i and $\beta_j \neq \beta_k$ for $j \neq k$. Then the unique positive definite solution $\tilde{P}_i$ to the Lyapunov equation,

$$\tilde{J}_i(\varepsilon)'\tilde{P}_i + \tilde{P}_i\tilde{J}_i(\varepsilon) = -I, \tag{8.5.43}$$

is independent of ε. ◻

Proof. See [83]. ☒

We now define the following second state transformation on the closed-loop system,

$$\tilde{x}_a^- = \bar{x}_a^-, \quad \tilde{x}_{ab}^+ = \bar{x}_{ab}^+, \tag{8.5.44}$$

$$\tilde{x}_a^0 = [(\tilde{x}_{a1}^0)', (\tilde{x}_{a2}^0)', \cdots (\tilde{x}_{al}^0)']' = S_a(\varepsilon)Q^{-1}(\varepsilon)(\Gamma_{sa}^0)^{-1}\bar{x}_a^0, \tag{8.5.45}$$

with

$$S_a(\varepsilon) = \text{blkdiag}\Big\{S_{a1}(\varepsilon), S_{a2}(\varepsilon), \cdots, S_{al}(\varepsilon)\Big\},$$

$$Q(\varepsilon) = \text{blkdiag}\Big\{Q_1(\varepsilon), Q_2(\varepsilon), \cdots, Q_l(\varepsilon)\Big\},$$

and

$$\tilde{x}_c = \varepsilon\bar{x}_c, \tag{8.5.46}$$

$$\tilde{x}_d = [\tilde{x}_1', \tilde{x}_2', \cdots, \tilde{x}_{m_d}']', \quad \tilde{x}_i = S_i(\varepsilon)\bar{x}_i, \; i = 1, 2, \cdots, m_d, \tag{8.5.47}$$

under which the closed-loop system becomes,

$$\dot{\tilde{x}}_a^- = A_{aa}^-\tilde{x}_a^- + A_{aab}^-(\varepsilon)\tilde{x}_{ab}^+ + A_{aa}^{-0}(\varepsilon)\tilde{x}_a^0 + L_{ad}^-\tilde{h}_d + E_a^-w, \tag{8.5.48}$$

$$\dot{\tilde{x}}_{ab}^+ = A_{ab}^{+c}\tilde{x}_{ab}^+ + A_{aba}^{+0}(\varepsilon)\tilde{x}_a^0 + L_{abd}^+\tilde{h}_d, \tag{8.5.49}$$

$$\dot{\tilde{x}}_a^0 = \tilde{J}(\varepsilon)\tilde{x}_a^0 + \tilde{B}(\varepsilon)\tilde{x}_a^0 + \tilde{L}_{ad}^0(\varepsilon)\tilde{h}_d + \tilde{E}_a^0(\varepsilon)w, \tag{8.5.50}$$

$$\dot{\tilde{x}}_c = A_{cc}^c\tilde{x}_c + \varepsilon[A_{cab}^+\tilde{x}_{ab}^+ + A_{ca}^0(\varepsilon)\tilde{x}_a^0 + L_{cd}\tilde{h}_d + E_cw], \tag{8.5.51}$$

$$h_0 = -[F_{a0}^+, \, F_{b0}]\tilde{x}_{ab}^+ - \tilde{F}_{a0}^0(\varepsilon)\tilde{x}_a^0, \tag{8.5.52}$$

$$\varepsilon\dot{\tilde{x}}_i = (A_{q_i} - B_{q_i}F_i)\tilde{x}_i + \varepsilon\tilde{L}_{iab}^+(\varepsilon)\tilde{x}_{ab}^+ + \varepsilon\tilde{L}_{ia}^0(\varepsilon)\tilde{x}_a^0 + \varepsilon\tilde{L}_{id}(\varepsilon)\tilde{h}_d + \varepsilon\tilde{E}_i(\varepsilon)w, \tag{8.5.53}$$

$$\tilde{h}_i = \bar{h}_i = h_i + [F_{adi}^+, \, F_{bdi}]\tilde{x}_{ab}^+ + \tilde{F}_{adi}^0(\varepsilon)\tilde{x}_a^0 = C_{q_i}\tilde{x}_i, \tag{8.5.54}$$

and

$$\tilde{h}_d = [\tilde{h}_1, \tilde{h}_2, \cdots, \tilde{h}_{m_d}]', \tag{8.5.55}$$

where

$$A_{aa}^{-0}(\varepsilon) = -[B_{0a}^{+},\, L_{ad}^{-}]F_a^0(\varepsilon)\Gamma_{sa}^0 Q(\varepsilon)S_a^{-1}(\varepsilon), \tag{8.5.56}$$

$$A_{aba}^{+0}(\varepsilon) = -[B_{0ab}^{+},\, L_{abd}^{+}]F_a^0(\varepsilon)\Gamma_{sa}^0 Q(\varepsilon)S_a^{-1}(\varepsilon), \tag{8.5.57}$$

$$\tilde{J}(\varepsilon) = \text{blkdiag}\Big\{\varepsilon\tilde{J}_1(\varepsilon), \varepsilon\tilde{J}_2(\varepsilon), \cdots, \varepsilon\tilde{J}_l(\varepsilon)\Big\}, \tag{8.5.58}$$

$$\tilde{B}(\varepsilon) = \begin{bmatrix} 0 & \tilde{B}_{12}(\varepsilon) & \tilde{B}_{13}(\varepsilon) & \cdots & \tilde{B}_{1l}(\varepsilon) \\ 0 & 0 & \tilde{B}_{23}(\varepsilon) & \cdots & \tilde{B}_{2l}(\varepsilon) \\ \vdots & \vdots & \vdots & \ddots & \vdots \\ 0 & 0 & 0 & \cdots & 0 \end{bmatrix}, \tag{8.5.59}$$

with $j = 1, 2, \cdots, l$ and $k = j+1, j+2, \cdots, l$,

$$\tilde{B}_{jk}(\varepsilon) = S_{aj}(\varepsilon)Q_j^{-1}(\varepsilon)B_{jk}F_k(\varepsilon)Q_k(\varepsilon)S_{ak}^{-1}(\varepsilon), \tag{8.5.60}$$

and

$$\tilde{L}_{ad}^0(\varepsilon) = S_a(\varepsilon)Q^{-1}(\varepsilon)(\Gamma_{sa}^0)^{-1}(L_{ad}^0 + T_a^0 L_{abd}^{+}), \tag{8.5.61}$$

$$\tilde{E}_a^0(\varepsilon) = S_a(\varepsilon)Q^{-1}(\varepsilon)(\Gamma_{sa}^0)^{-1}E_a^0 = \begin{bmatrix} \tilde{E}_{a1}^0(\varepsilon) \\ \tilde{E}_{a2}^0(\varepsilon) \\ \vdots \\ \tilde{E}_{al}^0(\varepsilon) \end{bmatrix}, \tag{8.5.62}$$

$$A_{cab}^{+} = L_{cb}[0,\, C_b] - [L_{c0}, L_{cd}]F_{ab}^{+}, \tag{8.5.63}$$

$$A_{ca}^0(\varepsilon) = -[L_{c0},\, L_{cd}]F_a^0(\varepsilon)\Gamma_{sa}^0 Q(\varepsilon)S_a^{-1}(\varepsilon), \tag{8.5.64}$$

$$\tilde{F}_{a0}^0(\varepsilon) = F_{a0}^0(\varepsilon)S_a^{-1}(\varepsilon)Q(\varepsilon)\Gamma_{sa}^0, \tag{8.5.65}$$

$$\tilde{L}_{ia}^0(\varepsilon) = S_i(\varepsilon)[L_{ia}^{01}(\varepsilon)F_a^0(\varepsilon) + L_{ia}^{02}(\varepsilon)F_a^0(\varepsilon)A_{aa}^0]\Gamma_{sa}^0 Q(\varepsilon)S_a^{-1}(\varepsilon), \tag{8.5.66}$$

$$\tilde{L}_{id}(\varepsilon) = S_i(\varepsilon)\bar{L}_{id}(\varepsilon), \quad \tilde{E}_i(\varepsilon) = S_i(\varepsilon)\bar{E}_i(\varepsilon), \tag{8.5.67}$$

$$\tilde{L}_{iab}^{+}(\varepsilon) = S_i(\varepsilon)L_{iab}^{+}(\varepsilon), \quad \tilde{F}_{adi}^0(\varepsilon) = F_{adi}^0(\varepsilon)\Gamma_{sa}^0 Q(\varepsilon)S_a^{-1}(\varepsilon), \tag{8.5.68}$$

and where, for $i = 1$ to l, $\tilde{J}_i(\varepsilon)$ is as defined in Lemma 8.5.2. By (8.3.20), (8.5.25), (8.5.26), and Lemma 8.5.1, we have that, for all $\varepsilon \in (0, 1]$,

$$|A_{aab}^{-}(\varepsilon)| \le a_{aab}^{-}, \quad |\tilde{L}_{ad}^0(\varepsilon)| \le \tilde{l}_{ad}^0, \quad |A_{cab}^{+}| \le a_{cab}^{+}, \tag{8.5.69}$$

$$|A_{aa}^{-0}(\varepsilon)| \le a_{aa}^{-0}\varepsilon, \quad |A_{aba}^{+0}(\varepsilon)| \le a_{aa}^{+0}\varepsilon, \quad |A_{ca}^0(\varepsilon)| \le a_{ca}^0\varepsilon, \quad |\tilde{F}_{a0}^0(\varepsilon)| \le \tilde{f}_{a0}^0\varepsilon, \tag{8.5.70}$$

for $i = 1$ to m_d,

$$|\tilde{L}_{iab}^{+}(\varepsilon)| \le \tilde{l}_{ab}^{+}, \quad |\tilde{L}_{ia}^0(\varepsilon)| \le \tilde{l}_a^0\varepsilon, \tag{8.5.71}$$

and

$$|\tilde{L}_{id}(\varepsilon)| \leq \tilde{l}_d, \ |\tilde{F}^0_{adi}(\varepsilon)| \leq \tilde{f}^0_{ad}\varepsilon, \ |\tilde{E}_i(\varepsilon)| \leq \tilde{e}, \tag{8.5.72}$$

for $i = 1$ to l,

$$|\tilde{E}^0_{ai}(\varepsilon)| \leq \tilde{e}^0_a\varepsilon, \tag{8.5.73}$$

and finally, for $j = 1$ to l, $k = j + 1$ to l,

$$|\tilde{B}_{jk}(\varepsilon)| \leq \tilde{b}_{jk}\varepsilon, \tag{8.5.74}$$

where a^-_{aab}, $\tilde{l}^0_{ad}$, a^+_{cab}, a^{-0}_{aa}, a^{+0}_{aa}, $\tilde{e}^0_a$, a^0_{ca}, $\tilde{f}^0_{a0}$, $\tilde{l}^+_{ab}$, $\tilde{l}^0_a$, $\tilde{l}_d$, $\tilde{f}^0_{ad}$, $\tilde{e}$ and $\tilde{b}_{jk}$ are some positive constants, independent of ε.

We next construct a Lyapunov function for the closed loop system (8.5.48)-(8.5.55). We do this by composing Lyapunov functions for the subsystems. For the subsystem of $\tilde{x}^-_a$, we choose a Lyapunov function,

$$V^-_a(\tilde{x}^-_a) = (\tilde{x}^-_a)' P^-_a \tilde{x}^-_a, \tag{8.5.75}$$

where $P^-_a > 0$ is the unique solution to the Lyapunov equation,

$$(A^-_{aa})' P^-_a + P^-_a A^-_{aa} = -I, \tag{8.5.76}$$

and for the subsystem of $\tilde{x}^+_{ab}$, choose a Lyapunov function,

$$V^+_{ab}(\tilde{x}^+_{ab}) = (\tilde{x}^+_{ab})' P^+_{ab} \tilde{x}^+_{ab}, \tag{8.5.77}$$

where $P^+_{ab} > 0$ is the unique solution to the Lyapunov equation,

$$(A^{+c}_{ab})' P^+_{ab} + P^+_{ab} A^{+c}_{ab} = -I. \tag{8.5.78}$$

The existence of such P^-_{aa} and P^+_{ab} is guaranteed by the fact that both A^-_{aa} and A^{+c}_{ab} are asymptotically stable. For the subsystem of

$$\tilde{x}^0_a = [(\tilde{x}^0_{a1})', (\tilde{x}^0_{a2})', \cdots, (\tilde{x}^0_{al})']', \tag{8.5.79}$$

we choose a Lyapunov function,

$$V^0_a(\tilde{x}^0_a) = \sum_{i=1}^{l} \frac{(\alpha^0_a)^{i-1}}{\varepsilon} (\tilde{x}^0_{ai})' P^0_{ai} \tilde{x}^0_{ai}, \tag{8.5.80}$$

where α^0_a is a positive scalar, whose value is to be determined later, and each P^0_{ai} is the unique solution to the Lyapunov equation,

$$\tilde{J}_i(\varepsilon)' P^0_{ai} + P^0_{ai} \tilde{J}_i(\varepsilon) = -I, \tag{8.5.81}$$

which, by Lemma 8.5.2, is independent of ε. Similarly, for the subsystem $\tilde{x}_c$, choose a Lyapunov function,

$$V_c(\tilde{x}_c) = \tilde{x}_c' P_c \tilde{x}_c, \tag{8.5.82}$$

where $P_c > 0$ is the unique solution to the Lyapunov equation,

$$(A_{cc}^c)' P_c + P_c A_{cc}^c = -I. \tag{8.5.83}$$

The existence of such a P_c is again guaranteed by the fact that A_{cc}^c is asymptotically stable. Finally, for the subsystem of $\tilde{x}_d$, choose a Lyapunov function

$$V_d(\tilde{x}_d) = \sum_{i=1}^{m_d} \tilde{x}_i' P_i \tilde{x}_i, \tag{8.5.84}$$

where each P_i is the unique solution to the Lyapunov equation

$$(A_{q_i} - B_{q_i} F_i)' P_i + P_i (A_{q_i} - B_{q_i} F_i) = -I. \tag{8.5.85}$$

Once again, the existence of such P_i is due to the fact that $A_{q_i} - B_{q_i} F_i$ is asymptotically stable.

We now construct a Lyapunov function for the closed-loop system (8.5.48)-(8.5.55) as follows.

$$V(\tilde{x}_a^-, \tilde{x}_{ab}^+, \tilde{x}_a^0, \tilde{x}_c, \tilde{x}_d) = V_a^-(\tilde{x}_a^-) + \alpha_{ab}^+ V_{ab}^+(\tilde{x}_{ab}^+) + V_a^0(\tilde{x}_a^0) + V_c(\tilde{x}_c) + \alpha_d V_d(\tilde{x}_d), \tag{8.5.86}$$

where $\alpha_{ab}^+ = 2|P_a^-|^2 (a_{aab}^-)^2$ and the value of α_d is to be determined.

Let us first consider the derivative of $V_a^0(\tilde{x}_a^0)$ along the trajectories of the subsystem $\tilde{x}_a^0$ and obtain that,

$$\dot{V}_a^0(\tilde{x}_a^0) = \sum_{i=1}^{l} \left[-(\alpha_a^0)^{i-1} (\tilde{x}_{ai}^0)' \tilde{x}_{ai}^0 + 2 \sum_{j=i+1}^{l} \frac{(\alpha_a^0)^{i-1}}{\varepsilon} (\tilde{x}_{ai}^0)' P_{ai}^0 \tilde{B}_{ij}(\varepsilon) \tilde{x}_{aj}^0 \right]$$
$$+2 \sum_{i=1}^{l} \frac{(\alpha_a^0)^{i-1}}{\varepsilon} \left[(\tilde{x}_{ai}^0)' P_{ai}^0 \tilde{L}_{ad}^0(\varepsilon) \tilde{h}_d + (\tilde{x}_{ai}^0)' P_{ai}^0 \tilde{E}_a^0(\varepsilon) w \right]. \tag{8.5.87}$$

Using (8.5.74), it is straightforward to show that there exists an $\alpha_a^0 > 0$ such that,

$$\dot{V}_a^0(\tilde{x}_a^0) \le -\frac{3}{4} |\tilde{x}_a^0|^2 + \frac{\alpha_1}{\varepsilon} |\tilde{x}_a^0| \cdot |\tilde{h}_d| + \alpha_2 |w|^2, \tag{8.5.88}$$

for some nonnegative constants α_1 and α_2, independent of ε.

In view of (8.5.88), the derivative of V along the trajectory of the closed-loop system (8.5.48)-(8.5.55) can be evaluated as follows,

$$\begin{aligned}\dot{V} \leq & -(\tilde{x}_a^-)'\tilde{x}_a^- + 2(\tilde{x}_a^-)'P_a^- A_{aab}^-(\varepsilon)\tilde{x}_{ab}^+ + 2(\tilde{x}_a^-)'P_a^- A_{aa}^{-0}(\varepsilon)\tilde{x}_a^0 \\ & + 2(\tilde{x}_a^-)'P_a^- L_{ad}^-\tilde{h}_d + 2(\tilde{x}_a^-)'P_a^- E_a^- w - \alpha_{ab}^+(\tilde{x}_{ab}^+)'\tilde{x}_{ab}^+ \\ & + 2\alpha_{ab}^+(x_{ab}^+)'P_{ab}^+ A_{aba}^{+0}(\varepsilon)\tilde{x}_a^0 + 2\alpha_{ab}^+(x_{ab}^+)'P_{ab}^+ L_{abd}^+\tilde{h}_d \\ & -\frac{3}{4}|\tilde{x}_a^0|^2 + \frac{\alpha_1}{\varepsilon}|\tilde{x}_a^0|\cdot|\tilde{h}_d| + \alpha_2|w|^2 - \tilde{x}_c'\tilde{x}_c \\ & +2\varepsilon\tilde{x}_c' P_c[A_{cab}^+\tilde{x}_{ab}^+ + A_{ca}^0(\varepsilon)\tilde{x}_a^0 + L_{cd}\tilde{h}_d + E_c w] \\ & +\alpha_d\sum_{i=1}^{m_d}\Big[-\frac{1}{\varepsilon}\tilde{x}_i'\tilde{x}_i + 2\tilde{x}_i'P_i\tilde{L}_{iab}^+(\varepsilon)\tilde{x}_{ab}^+ \\ & \qquad + 2\tilde{x}_i'P_i\tilde{L}_{ia}^0(\varepsilon)\tilde{x}_a^0 + 2\tilde{x}_i'P_i\tilde{L}_{id}(\varepsilon)\tilde{h}_d + 2\tilde{x}_i'P_i\tilde{E}_i(\varepsilon)w\Big]. \end{aligned} \tag{8.5.89}$$

Using the majorizations (8.5.69)-(8.5.73) and noting the definition of α_{ab}^+ (8.5.86), we can easily verify that, there exist an $\alpha_d > 0$ and an $\varepsilon_1^* \in (0,1]$ such that, for all $\varepsilon \in (0, \varepsilon_1^*]$,

$$\dot{V} \leq -\frac{1}{2}|\tilde{x}_a^-|^2 - \frac{1}{2}|\tilde{x}_{ab}^+|^2 - \frac{1}{2}|\tilde{x}_a^0|^2 - \frac{1}{2\varepsilon}|\tilde{x}_d|^2 + \alpha_3|w|^2, \tag{8.5.90}$$

for some positive constant α_3, independent of ε.

From (8.5.90), it follows that the closed-loop system in the absence of disturbance w is asymptotically stable. It remains to show that, for any given $\gamma > 0$, there exists an $\varepsilon^* \in (0, \varepsilon_1^*]$ such that, for all $\varepsilon \in (0, \varepsilon^*]$,

$$||h||_2 \leq \gamma||w||_2. \tag{8.5.91}$$

To this end, we integrate both sides of (8.5.90) from 0 to ∞. Noting that $V \geq 0$ and $V(t) = 0$ at $t = 0$, we have,

$$||\tilde{h}_d||_2 \leq \left(\sqrt{2\alpha_3\varepsilon}\right)||w||_2, \tag{8.5.92}$$

which, when used in (8.5.88), results in,

$$||\tilde{x}_a^0||_2 \leq \left(\sqrt{\frac{2\alpha_1^2\alpha_3}{\varepsilon}} + \alpha_2\right)||w||_2. \tag{8.5.93}$$

Viewing $\tilde{h}_d$ as disturbance to the dynamics of $\tilde{x}_{ab}^+$ also results in,

$$||\tilde{x}_{ab}^+||_2 \leq \left(\alpha_4\sqrt{\varepsilon}\right)||w||_2, \tag{8.5.94}$$

for some positive constant α_4, independent of ε.

Finally, recalling that

$$h = \Gamma_{\text{OP}} \begin{bmatrix} z_0 \\ \tilde{h}_d - F_{ab}^+ \tilde{x}_{ab}^+ - \tilde{F}_{ad}^0(\varepsilon)\tilde{x}_a^0 \\ z_b \end{bmatrix}, \tag{8.5.95}$$

where

$$\tilde{F}_{ad}^0(\varepsilon) = \begin{bmatrix} \tilde{F}_{ad1}^0(\varepsilon) \\ \tilde{F}_{ad2}^0(\varepsilon) \\ \vdots \\ F_{adm_d}^0(\varepsilon) \end{bmatrix}, \tag{8.5.96}$$

with each $\tilde{F}_{adi}(\varepsilon)$ satisfying (8.5.71) and (8.5.72), we have,

$$\|h\|_2 \le |\Gamma_{\text{OP}}| \left(\sqrt{2\alpha_3\varepsilon} + \alpha_4 |F_{ab}^+| \sqrt{\varepsilon} + \alpha_5 \sqrt{2\alpha_1^2\alpha_3\varepsilon + \alpha_2\varepsilon^2} \right) \|w\|_2, \tag{8.5.97}$$

for some positive constant α_5 independent of ε.

To complete the proof, we choose $\varepsilon^* \in (0, \varepsilon_1^*]$ such that,

$$|\Gamma_{\text{OP}}| \left(\sqrt{2\alpha_3\varepsilon} + \alpha_4 |F_{ab}^+| \sqrt{\varepsilon} + \alpha_5 \sqrt{2\alpha_1^2\alpha_3\varepsilon + \alpha_2\varepsilon^2} \right) \le \gamma. \tag{8.5.98}$$

For use in the proof of measurement feedback results, it is straightforward to verify from the closed-loop equations (8.5.48)-(8.5.55) that the transfer function from $E_a^0 w$ to h is given by

$$T_{ao}^0(s) = T_{ao}(s,\varepsilon) \left[sI - A_{aa}^0 + B_a^0 F_a^0(\varepsilon) \right]^{-1}, \tag{8.5.99}$$

where $T_{ao}(s,\varepsilon) \to 0$ pointwise in s as $\varepsilon \to 0$. ⊠

8.5.B. Proof of Theorem 8.4.1

It is trivial to show the stability of the closed-loop system comprising the given plant (8.1.1) and the full order output feedback controller (8.4.3). The closed-loop poles are given by $\lambda\{A + BF_{\text{P}}(\varepsilon)\}$, which are in $\mathbb{C}^-$ for sufficiently small ε as shown in Theorem 8.3.1, and $\lambda\{A + K_{\text{Q}}(\varepsilon)C_1\}$, which can be dually shown to be in $\mathbb{C}^-$ for sufficiently small ε as well. In what follows, we will show that the full order output feedback controller achieves the H_∞-ADDPMS for (8.1.1), which satisfies all 5 conditions of Theorem 8.2.1. Without loss of any generality but for simplicity of presentation, hereafter we assume throughout the rest of the proof that the subsystem Σ_{P}, i.e., the quadruple (A, B, C_2, D_2), has already been transformed into the special coordinate basis as given in Theorem 2.4.1. To be more specific, we have

$$A = B_0 C_{2,0} + \begin{bmatrix} A_{aa}^- & 0 & 0 & L_{ab}^- C_b & 0 & L_{ad}^- C_d \\ 0 & A_{aa}^0 & 0 & L_{ab}^0 C_b & 0 & L_{ad}^0 C_d \\ 0 & 0 & A_{aa}^+ & L_{ab}^+ C_b & 0 & L_{ad}^+ C_d \\ 0 & 0 & 0 & A_{bb} & 0 & L_{bd} C_d \\ B_c E_{ca}^- & B_c E_{ca}^0 & B_c E_{ca}^+ & L_{cb} C_b & A_{cc} & L_{cd} C_d \\ B_d E_{da}^- & B_d E_{da}^0 & B_d E_{da}^+ & B_d E_{db} & B_d E_{dc} & A_{dd} \end{bmatrix}$$
$$:= B_0 C_{2,0} + \tilde{A}, \tag{8.5.100}$$

$$B = \begin{bmatrix} B_{0a}^- & 0 & 0 \\ B_{0a}^0 & 0 & 0 \\ B_{0a}^+ & 0 & 0 \\ B_{0b} & 0 & 0 \\ B_{0c} & 0 & B_c \\ B_{0d} & B_d & 0 \end{bmatrix}, \quad B_0 = \begin{bmatrix} B_{0a}^- \\ B_{0a}^0 \\ B_{0a}^+ \\ B_{0b} \\ B_{0c} \\ B_{0d} \end{bmatrix}, \tag{8.5.101}$$

$$C_{2,0} = [\, C_{0a}^- \quad C_{0a}^0 \quad C_{0a}^+ \quad C_{0b} \quad C_{0c} \quad C_{0d} \,], \tag{8.5.102}$$

$$C_2 = \begin{bmatrix} C_{0a}^- & C_{0a}^0 & C_{0a}^+ & C_{0b} & C_{0c} & C_{0d} \\ 0 & 0 & 0 & 0 & 0 & C_d \\ 0 & 0 & 0 & C_b & 0 & 0 \end{bmatrix}, \quad D_2 = \begin{bmatrix} I & 0 & 0 \\ 0 & 0 & 0 \\ 0 & 0 & 0 \end{bmatrix}, \tag{8.5.103}$$

$$\mathcal{S}^+(\Sigma_{\rm P}) = \mathrm{Im} \left\{ \begin{bmatrix} I & 0 & 0 & 0 \\ 0 & I & 0 & 0 \\ 0 & 0 & 0 & 0 \\ 0 & 0 & 0 & 0 \\ 0 & 0 & I & 0 \\ 0 & 0 & 0 & I \end{bmatrix} \right\}. \tag{8.5.104}$$

It is simple to note that Condition 4 of Theorem 8.2.1 implies that

$$E = \begin{bmatrix} E_a^- \\ E_a^0 \\ 0 \\ 0 \\ E_c \\ E_d \end{bmatrix}. \tag{8.5.105}$$

Next, for any $\zeta \in \mathcal{V}_\lambda(\Sigma_Q)$ with $\lambda \in \mathbb{C}^0$, we partition ζ as follows,

$$\zeta = \begin{pmatrix} \zeta_a^- \\ \zeta_a^0 \\ \zeta_a^+ \\ \zeta_b \\ \zeta_c \\ \zeta_d \end{pmatrix}. \tag{8.5.106}$$

Then, Condition 5 of Theorem 8.2.1 implies that $C_2\zeta = 0$, or equivalently

$$C_{2,0}\zeta = 0, \quad C_b\zeta_b = 0 \quad \text{and} \quad C_d\zeta_d = 0. \tag{8.5.107}$$

By Definition 2.4.3, we have

$$\begin{bmatrix} A - \lambda I & E \\ C_1 & D_1 \end{bmatrix} \begin{pmatrix} \zeta \\ \eta \end{pmatrix} = 0, \tag{8.5.108}$$

for some appropriate vector η. Clearly, (8.5.108) and (8.5.105) imply that

$$(A - \lambda I)\zeta = -E\eta = \begin{pmatrix} \star \\ \star \\ 0 \\ 0 \\ \star \\ \star \end{pmatrix}, \tag{8.5.109}$$

where $\star$s are some vectors of not much interests. Note that (8.5.107) implies

$$(A - \lambda I)\zeta = (B_0C_{2,0} + \tilde{A} - \lambda I)\zeta = (\tilde{A} - \lambda I)\zeta$$

$$= \begin{bmatrix} \star \\ \star \\ (A_{aa}^+ - \lambda I)\zeta_a^+ + L_{ab}^+C_b\zeta_b + L_{ad}^+C_d\zeta_d \\ (A_{bb} - \lambda I)\zeta_b + L_{bd}C_d\zeta_d \\ \star \\ \star \end{bmatrix}$$

$$= \begin{bmatrix} \star \\ \star \\ (A_{aa}^+ - \lambda I)\zeta_a^+ \\ (A_{bb} - \lambda I)\zeta_b \\ \star \\ \star \end{bmatrix}. \tag{8.5.110}$$

(8.5.109) and (8.5.110) imply

$$(A_{aa}^+ - \lambda I)\zeta_a^+ = 0 \quad \text{and} \quad (A_{bb} - \lambda I)\zeta_b = 0. \tag{8.5.111}$$

Since A_{aa}^+ is unstable, $(A_{aa}^+ - \lambda I)\zeta_a^+ = 0$ implies that $\zeta_a^+ = 0$. Similarly, since (A_{bb}, C_b) is completely observable, $(A_{bb} - \lambda I)\zeta_b = 0$ and $C_b\zeta_b = 0$ imply $\zeta_b = 0$. Thus, ζ has the following property,

$$\zeta = \begin{pmatrix} \zeta_a^- \\ \zeta_a^0 \\ 0 \\ 0 \\ \zeta_c \\ \zeta_d \end{pmatrix} \in \mathcal{S}^+(\Sigma_{\rm P}). \tag{8.5.112}$$

Obviously, (8.5.112) together with Condition 6 of Theorem 8.2.1 imply

$$\mathcal{S}^+(\Sigma_{\rm P}) \supset \mathcal{V}^+(\Sigma_{\rm Q}) \cup \left\{\cup_{\lambda\in\mathbb{C}^0}\mathcal{V}_\lambda(\Sigma_{\rm Q})\right\}. \tag{8.5.113}$$

Next, it is straightforward to verify that $A - sI$ can be partitioned as

$$A - sI = X_1 + X_2C_2 + X_3 + X_4, \tag{8.5.114}$$

where

$$X_1 := \begin{bmatrix} A_{aa}^- - sI & 0 & 0 & L_{ab}^-C_b & 0 & L_{ad}^-C_d \\ 0 & 0 & 0 & 0 & 0 & 0 \\ 0 & 0 & 0 & 0 & 0 & 0 \\ 0 & 0 & 0 & 0 & 0 & 0 \\ B_cE_{ca}^- & B_cE_{ca}^0 & B_cE_{ca}^+ & L_{cb}C_b & A_{cc} - sI & L_{cd}C_d \\ B_dE_{da}^- & B_dE_{da}^0 & B_dE_{da}^+ & B_dE_{db} & B_dE_{dc} & A_{dd} - sI \end{bmatrix}, \tag{8.5.115}$$

$$X_2 = \begin{bmatrix} B_{0a}^- & 0 & 0 \\ B_{0a}^0 & L_{ad}^0 & L_{ab}^0 \\ B_{0a}^+ & L_{ad}^+ & L_{ab}^+ \\ B_{0b} & L_{bd} & 0 \\ B_{0c} & 0 & 0 \\ B_{0d} & 0 & 0 \end{bmatrix}, \tag{8.5.116}$$

$$X_3 = \begin{bmatrix} 0 & 0 & 0 & 0 & 0 & 0 \\ 0 & 0 & 0 & 0 & 0 & 0 \\ 0 & 0 & A_{aa}^+ - sI & 0 & 0 & 0 \\ 0 & 0 & 0 & A_{bb} - sI & 0 & 0 \\ 0 & 0 & 0 & 0 & 0 & 0 \\ 0 & 0 & 0 & 0 & 0 & 0 \end{bmatrix}, \tag{8.5.117}$$

and

$$X_4 = \begin{bmatrix} 0 & 0 & 0 & 0 & 0 & 0 \\ 0 & A_{aa}^0 - sI & 0 & 0 & 0 & 0 \\ 0 & 0 & 0 & 0 & 0 & 0 \\ 0 & 0 & 0 & 0 & 0 & 0 \\ 0 & 0 & 0 & 0 & 0 & 0 \\ 0 & 0 & 0 & 0 & 0 & 0 \end{bmatrix}. \tag{8.5.118}$$

It is simple to see that

$$\operatorname{Im}(X_1) \subset \mathcal{S}^+(\Sigma_P) \cap \{\cap_{\lambda \in \mathbb{C}^0} \mathcal{S}_\lambda(\Sigma_P)\}, \tag{8.5.119}$$

and

$$\operatorname{Ker}(X_3) \supset \mathcal{S}^+(\Sigma_P) \supset \mathcal{V}^+(\Sigma_Q) \cup \{\cup_{\lambda \in \mathbb{C}^0} \mathcal{V}_\lambda(\Sigma_Q)\}. \tag{8.5.120}$$

It follows from the proof of Theorem 8.3.1 that as $\varepsilon \to 0$

$$\left\| [C_2 + D_2 F_P(\varepsilon)][sI - A - BF_P(\varepsilon)]^{-1} \right\|_\infty < \kappa_P, \tag{8.5.121}$$

where κ_P is a finite positive constant and is independent of ε. Moreover, under Condition 4 of Theorem 8.2.1, we have

$$[C_2 + D_2 F_P(\varepsilon)][sI - A - BF_P(\varepsilon)]^{-1} E \to 0, \tag{8.5.122}$$

and

$$[C_2 + D_2 F_P(\varepsilon)][sI - A - BF_P(\varepsilon)]^{-1} X_1 \to 0, \tag{8.5.123}$$

pointwise in s as $\varepsilon \to 0$. By (8.5.99), we have

$$[C_2 + D_2 F_P(\varepsilon)][sI - A - BF_P(\varepsilon)]^{-1} X_4 \to 0, \tag{8.5.124}$$

pointwise in s as $\varepsilon \to 0$. Dually, one can show that

$$\left\| [sI - A - K_Q(\varepsilon)C_1]^{-1}[E + K_Q(\varepsilon)D_1] \right\|_\infty < \kappa_Q, \tag{8.5.125}$$

where κ_Q is a finite positive constant and is independent of ε. If Condition 5 of Theorem 8.2.1 is satisfied, the following results hold,

$$C_2[sI - A - K_Q(\varepsilon)C_1]^{-1}[E + K_Q(\varepsilon)D_1] \to 0, \tag{8.5.126}$$

and

$$X_3[sI - A - K_Q(\varepsilon)C_1]^{-1}[E + K_Q(\varepsilon)D_1] \to 0, \tag{8.5.127}$$

pointwise in s as $\varepsilon \to 0$.

Finally, it is simple to verify that the closed-loop transfer matrix from the disturbance w to the controlled output h under the full order output feedback controller (8.4.3) is given by

$$\begin{aligned}T_{hw}(s,\varepsilon) = {} & [C_2+D_2F_{\rm P}(\varepsilon)][sI-A-BF_{\rm P}(\varepsilon)]^{-1}E \\ & + C_2[sI-A-K_{\rm Q}(\varepsilon)C_1]^{-1}[E+K_{\rm Q}(\varepsilon)D_1]+[C_2+D_2F_{\rm P}(\varepsilon)] \\ & \cdot [sI-A-BF_{\rm P}(\varepsilon)]^{-1}(A-sI)[sI-A-K_{\rm Q}(\varepsilon)C_1]^{-1}[E+K_{\rm Q}(\varepsilon)D_1].\end{aligned}$$

Using (8.5.114), we can rewrite $T_{hw}(s,\varepsilon)$ as

$$\begin{aligned}T_{hw}(s,\varepsilon) = {} & [C_2+D_2F_{\rm P}(\varepsilon)][sI-A-BF_{\rm P}(\varepsilon)]^{-1}E \\ & + C_2[sI-A-K_{\rm Q}(\varepsilon)C_1]^{-1}[E+K_{\rm Q}(\varepsilon)D_1] \\ & + [C_2+D_2F_{\rm P}(\varepsilon)][sI-A-BF_{\rm P}(\varepsilon)]^{-1}(X_1+X_2C_2+X_3+X_4) \\ & \cdot [sI-A-K_{\rm Q}(\varepsilon)C_1]^{-1}[E+K_{\rm Q}(\varepsilon)D_1].\end{aligned}$$

Following (8.5.121) to (8.5.127), and some simple manipulations, it is straightforward to show that as $\varepsilon \to 0$, $T_{hw}(s,\varepsilon) \to 0$, pointwise in s, which is equivalent to $\|T_{hw}\|_\infty \to 0$ as $\varepsilon \to 0$. Hence, the full order output feedback controller (8.4.3) solves the H_∞-ADDPMS for the given plant (8.1.1), provided that all five conditions of Theorem 8.2.1 are satisfied. ⊠

8.5.C. Proof of Theorem 8.4.2

Again, it is trivial to show the stability of the closed-loop system comprising the given plant (8.1.1) and the reduced order measurement feedback controller (8.4.15) as the closed-loop poles are $\lambda\{A+BF_{\rm P}(\varepsilon)\}$ and $\lambda\{A_{\rm R}+K_{\rm R}(\varepsilon)C_{\rm R}\}$, which are asymptotically stable for a sufficiently small ε. Next, it is easy to compute the closed-loop transfer matrix from the disturbance w to the controlled output h under the reduced order output feedback controller,

$$\begin{aligned}T_{hw}(s,\varepsilon) = {} & [C_2 + D_2F_{\rm P}(\varepsilon)][sI - A - BF_{\rm P}(\varepsilon)]^{-1}E \\ & + [C_2 + D_2F_{\rm P}(\varepsilon)][sI - A - BF_{\rm P}(\varepsilon)]^{-1}(A - sI)\begin{pmatrix} 0 \\ I_{n-k}\end{pmatrix} \\ & \cdot [sI - A_{\rm R} - K_{\rm R}(\varepsilon)C_{\rm R}]^{-1}[E_{\rm R} + K_{\rm R}(\varepsilon)D_{\rm R}] \\ & + C_2\begin{pmatrix} 0 \\ I_{n-k}\end{pmatrix}[sI - A_{\rm R} - K_{\rm R}(\varepsilon)C_{\rm R}]^{-1}[E_{\rm R} + K_{\rm R}(\varepsilon)D_{\rm R}].\end{aligned}$$

It was shown in Chen [12] (i.e., Proposition 2.2.1) that

$$\begin{pmatrix} 0 \\ I_{n-k}\end{pmatrix}\mathcal{V}^+(\Sigma_{\rm QR}) = \mathcal{V}^+(\Sigma_{\rm Q}). \tag{8.5.128}$$

Following the same lines of reasoning as in Chen [12], one can also show that

$$\begin{pmatrix} 0 \\ I_{n-k} \end{pmatrix} \cup_{\lambda \in \mathbb{C}^0} \mathcal{V}_\lambda(\Sigma_{\rm QR}) = \cup_{\lambda \in \mathbb{C}^0} \mathcal{V}_\lambda(\Sigma_{\rm Q}). \tag{8.5.129}$$

Hence, we have

$$\begin{pmatrix} 0 \\ I_{n-k} \end{pmatrix} \left(\mathcal{V}^+(\Sigma_{\rm QR}) \cup \left\{\cup_{\lambda \in \mathbb{C}^0} \mathcal{V}_\lambda(\Sigma_{\rm QR})\right\}\right) = \mathcal{V}^+(\Sigma_{\rm Q}) \cup \left\{\cup_{\lambda \in \mathbb{C}^0} \mathcal{V}_\lambda(\Sigma_{\rm Q})\right\}. \tag{8.5.130}$$

The rest of the proof follows from the same lines as those of Theorem 8.4.1. ⊠

Chapter 9

Robust and Perfect Tracking of Continuous-time Systems

9.1. Introduction

IN THIS CHAPTER, we present a so-called robust and perfect tracking (RPT) problem, which was proposed and solved by Liu, Chen and Lin [87]. The development of this chapter follows closely from the results of [87]. The robust and perfect tracking problem is to design a controller such that the resulting closed-loop system is asymptotically stable and the controlled output almost perfectly tracks a given reference signal in the presence of any initial conditions and external disturbances. By almost tracking we mean the ability of a controller to track a given reference signal with arbitrarily fast settling time in the face of external disturbances and initial conditions. More specifically, we consider in this chapter the following multivariable linear time-invariant system,

$$\Sigma : \begin{cases} \dot{x} = A\ x + B\ u + E\ w, & x(0) = x_0, \\ y = C_1\ x + D_1\ w, \\ h = C_2\ x + D_2\ u + D_{22}\ w, \end{cases} \tag{9.1.1}$$

where $x \in \mathbb{R}^n$ is the state, $u \in \mathbb{R}^m$ is the control input, $w \in \mathbb{R}^q$ is the external disturbance, $y \in \mathbb{R}^p$ is the measurement output, and $h \in \mathbb{R}^\ell$ is the output to be controlled. We also assume that the pair (A, B) is stabilizable and (A, C_1) is detectable. For future references, we define Σ_P and Σ_Q to be the subsystems characterized by the matrix quadruples (A, B, C_2, D_2) and (A, E, C_1, D_1), respectively. Given the external disturbance $w \in L_p$, $p \in [1, \infty)$, and any reference signal vector, $r \in \mathbb{R}^\ell$ with r, $\dot{r}$, $\cdots$, $r^{(\kappa-1)}$, $\kappa \geq 1$, being available, and $r^{(\kappa)}$ being either a vector of delta functions or in L_p, the robust and perfect tracking (RPT) problem for the system (9.1.1) is to find a parameterized dynamic measurement control law of the following form

$$\begin{cases} \dot{v} = A_{\rm cmp}(\varepsilon)v + B_{\rm cmp}(\varepsilon)y + G_0(\varepsilon)r + \cdots + G_{\kappa-1}(\varepsilon)r^{(\kappa-1)}, \\ u = C_{\rm cmp}(\varepsilon)v + D_{\rm cmp}(\varepsilon)y + H_0(\varepsilon)r + \cdots + H_{\kappa-1}(\varepsilon)r^{(\kappa-1)}, \end{cases} \tag{9.1.2}$$

such that when (9.1.2) is applied (9.1.1), we have

1. There exists an $\varepsilon^* > 0$ such that the resulting closed-loop system with $r = 0$ and $w = 0$ is asymptotically stable for all $\varepsilon \in (0, \varepsilon^*]$; and

2. Let $h(t, \varepsilon)$ be the closed-loop controlled output response and let $e(t, \varepsilon)$ be the resulting tracking error, i.e., $e(t, \varepsilon) := h(t, \varepsilon) - r(t)$. Then, for any initial condition of the state, $x_0 \in \mathbb{R}^n$,

$$J_p(x_0, w, r, \varepsilon) := \|e\|_p \to 0 \quad \text{as} \quad \varepsilon \to 0. \tag{9.1.3}$$

We introduce in the above formulation some additional information besides the reference signal r, i.e., $\dot{r}, \ddot{r}, \cdots, r^{(\kappa-1)}$, as additional controller inputs. Note that in general, these additional signals can easily be generated without any extra costs. For example, if $r(t) = t^2 \cdot 1(t)$, where $1(t)$ is a unit step function, then one can easily obtain its first order derivative

$$\dot{r}(t) = 2t \cdot 1(t) + t^2 \cdot \delta(t) = 2t \cdot 1(t), \tag{9.1.4}$$

where $\delta(t)$ is a unit impulse function, and the second order derivative

$$\ddot{r}(t) = 2 \cdot 1(t). \tag{9.1.5}$$

These $\dot{r}(t)$ and $\ddot{r}(t)$ can be used to improve the overall tracking performance, while $r^{(3)}(t) = 2 \cdot \delta(t)$ does not exist in the real world and hence cannot be used. We also note that our formulation covers all possible reference signals that have the form, $r(t) = t^k$, $0 \leq k < \infty$. Thus, our method could be applied to approximately track reference signals, which have a Taylor series expansion at $t = 0$. This can be done by truncating the higher order terms of the Taylor series of the given signal. Also, it is simple to see that when $r(t) \equiv 0$, the proposed problem reduces to the well known perfect regulation problem with measurement feedback.

It is appropriate to trace a short history of the literature that dealt with (almost) perfect regulation and (almost) perfect tracking problems. The problem of perfect regulation and its related topics were heavily investigated by many researchers in the 1970's and early 1980's. The perfect regulation problem via state feedback was studied by Kwakernaak and Sivan [70], Francis [53], Kimura [66], and Scherzinger and Davison [120], and was completely solved by Lin *et al.* [86] (see also Lin [82]). The solution to the problem of perfect regulation via measurement output feedback for general linear systems has only been reported recently by Chen *et al.* [26]. There were, however, a couple of different formulations for perfect tracking (see e.g., Lawrence and Rugh [74], Davison and Chow [44], which mainly dealt with state feedback case, and Davison and Scherzinger [45], and the references therein). Their problem formulations are quite different from the RPT problem, as pointed out in details in [87].

In this chapter, we derive a set of necessary and sufficient conditions under which the proposed robust and perfect tracking problem has a solution, and under these conditions, develop algorithms for the construction of parameterized feedback laws that solve the proposed problem. We would like to point out that our problem formulation and design algorithms are capable of tracking any polynomial signals without actually augmenting any additional integrator to the given plant. This is because we have utilized all possible information from the reference $r(t)$. This technique has successfully been used to solve quite a number of practical problems, such as the designs of a hard disk servo system and a gyro mirror targeting system, which will be reported later in Chapters 14 and 16, respectively.

9.2. Solvability Conditions and Solutions

We are now ready to present our main results. We will first derive a set of necessary and sufficient conditions under which the proposed robust and perfect tracking (RPT) problem is solvable for the given plant (9.1.1). In fact, we will show the sufficiency of these conditions by explicitly constructing two types of parameterized control laws: one is of full order, i.e., its dynamical order is equal to n, the order of the plant, and the other is of reduced order, i.e., its dynamical order is less than n.

We have the following theorem.

Theorem 9.2.1. Consider the given system (9.1.1) with its external disturbance $w \in L_p$, $p \in [1, \infty)$, and its initial condition $x(0) = x_0$. Then, for any reference signal $r(t)$, which has all its i-th order derivatives, $i = 0, 1, \cdots, \kappa - 1$, $\kappa \geq 1$, being available and $r^{(\kappa)}(t)$ being either a vector of delta functions or in

L_p, the proposed robust and perfect tracking (RPT) problem is solvable by the control law of (9.1.2) if and only if the following conditions are satisfied:

1. (A, B) is stabilizable and (A, C_1) is detectable;
2. $D_{22} + D_2 S D_1 = 0$, where $S = -(D_2' D_2)^{\dagger} D_2' D_{22} D_1' (D_1 D_1')^{\dagger}$;
3. $\Sigma_{\rm P}$, i.e., (A, B, C_2, D_2), is right invertible and of minimum phase;
4. $\operatorname{Ker}(C_2 + D_2 S C_1) \supset C_1^{-1}\{\operatorname{Im}(D_1)\}$. ⓣ

Proof. We first show that Conditions 1 to 4 in the theorem are necessary. Let us consider the case when $r(t) \equiv 0$, which of course has all its derivatives of any order being available. It is simple to see that the proposed robust and perfect tracking problem then reduces to the perfect regulation problem. Following the results of Chen *et al.* [26], we can reformulate the perfect regulation problem for the given system (9.1.1) as the well studied almost disturbance decoupling problem (see Willems [136,137] for the original formulation of this problem) for the following system,

$$\begin{cases} \dot{x} = A\,x + B\,u + [\,E \quad I\,]\,\tilde{w}, & x(0) = 0, \\ y = C_1\,x \qquad\qquad + [\,D_1 \quad 0\,]\,\tilde{w}, \\ h = C_2\,x + D_2\,u + [\,D_{22} \; 0\,]\,\tilde{w}. \end{cases} \tag{9.2.1}$$

For easy reference, we let $\tilde{\Sigma}_{\rm Q}$ be the subsystem characterized by the matrix quadruple $(A, [\,E \quad I\,], C_1, [\,D_1 \quad 0\,])$. Following the results of the well-known almost disturbance decoupling problem (see e.g., Chapter 8), we can show that if the almost disturbance decoupling problem for the above system is solvable, then the following conditions hold:

1. (A, B) is stabilizable and (A, C_1) is detectable;
2. $D_{22} + D_2 S D_1 = 0$, where $S = -(D_2' D_2)^{\dagger} D_2' D_{22} D_1' (D_1 D_1')^{\dagger}$;
3. $\operatorname{Im}([\,E + BSD_1 \quad I\,]) \subset \mathcal{S}^+(\Sigma_{\rm P})$;
4. $\operatorname{Ker}(C_2 + D_2 S C_1) \supset \mathcal{V}^+(\tilde{\Sigma}_{\rm Q})$.

Clearly, Item 3 above implies that $\mathcal{S}^+(\Sigma_{\rm P}) = \mathbb{R}^n$, which implies that $\Sigma_{\rm P}$ is right invertible without invariant zeros in $\mathbb{C}^+$. Due to the special form of $\tilde{\Sigma}_{\rm Q}$, it is simple to show that $\mathcal{V}^+(\tilde{\Sigma}_{\rm Q}) = C_1^{-1}\{\operatorname{Im}(D_1)\}$. Hence, Items 3 and 4 are respectively equivalent to:

1. $\Sigma_{\rm P}$ is right invertible without invariant zeros in $\mathbb{C}^+$;

2. $\mathrm{Ker}\,(C_2 + D_2 S C_1) \supset C_1^{-1}\{\mathrm{Im}\,(D_1)\}$.

Thus, it remains to show that if the proposed RPT problem is solvable, the subsystem Σ_P must be of minimum phase. In what follows, we proceed to show such a fact.

First, we note that second condition, i.e., $D_{22} + D_2 S D_1 = 0$, implies that if we apply a pre-output feedback law

$$u = Sy, \tag{9.2.2}$$

to the system (9.1.1), the resulting new system will have a direct feedthrough term from w to h equal to 0. Hence, without loss of any generality, we hereafter assume that matrix $D_{22} = 0$ throughout the rest of the proof.

Next, we show that if the robust and perfect tracking problem is solvable for general nonzero reference $r(t)$, Σ_P must be of minimum phase, i.e., Σ_P cannot have any invariant zeros on the imaginary axis. In fact, this condition must hold even for the case when $w = 0$ and $x_0 = 0$, i.e., for the robust and perfect tracking of the following system,

$$\begin{cases} \dot{x} = A\ x + B\ u \\ y = C_1\ x \\ e = C_2\ x + D_2\ u\ - r = h - r. \end{cases} \tag{9.2.3}$$

Now, if we treat r as an external disturbance, then the above problem is again equivalent to the well-known almost disturbance decoupling problem with measurement feedback and with internal stability for the following system,

$$\begin{cases} \dot{x} = A\ x\ +\ B\ u \\ \tilde{y} = \begin{pmatrix} C_1 x \\ r \\ \vdots \\ r^{(\kappa-1)} \end{pmatrix} \\ e = C_2\ x + D_2\ u - r. \end{cases} \tag{9.2.4}$$

Without loss of generality, we assume that the quadruple (A, B, C_2, D_2) has been transformed into the form of the special coordinate basis of Theorem 2.4.1, i.e., we have

$$x = \begin{pmatrix} x_a \\ x_c \\ x_d \end{pmatrix}, \quad x_a = \begin{pmatrix} x_a^- \\ x_a^0 \end{pmatrix}, \quad h = \begin{pmatrix} h_0 \\ h_d \end{pmatrix}, \quad r = \begin{pmatrix} r_0 \\ r_d \end{pmatrix}, \tag{9.2.5}$$

$$e = \begin{pmatrix} e_0 \\ e_d \end{pmatrix} = \begin{pmatrix} h_0 - r_0 \\ h_d - r_d \end{pmatrix}, \quad u = \begin{pmatrix} u_0 \\ u_d \\ u_c \end{pmatrix}, \tag{9.2.6}$$

$$x_d = \begin{pmatrix} x_1 \\ x_2 \\ \vdots \\ x_{m_d} \end{pmatrix}, \quad x_i = \begin{pmatrix} x_{i1} \\ x_{i2} \\ \vdots \\ x_{iq_i} \end{pmatrix}, \quad h_d = \begin{pmatrix} h_1 \\ h_2 \\ \vdots \\ h_{m_d} \end{pmatrix}, \tag{9.2.7}$$

$$r_d = \begin{pmatrix} r_1 \\ r_2 \\ \vdots \\ r_{m_d} \end{pmatrix}, \quad u_d = \begin{pmatrix} u_1 \\ u_2 \\ \vdots \\ u_{m_d} \end{pmatrix}, \tag{9.2.8}$$

and

$$\dot{x}_a^- = A_{aa}^- x_a^- + B_{0a}^- h_0 + L_{ad}^- h_d, \tag{9.2.9}$$

$$\dot{x}_a^0 = A_{aa}^0 x_a^0 + B_{0a}^0 h_0 + L_{ad}^0 h_d, \tag{9.2.10}$$

$$\dot{x}_c = A_{cc} x_c + B_{0c} h_0 + L_{cd} h_d + B_c \left[E_{ca}^- x_a^- + E_{ca}^0 x_a^0 \right] + B_c u_c, \tag{9.2.11}$$

$$e_0 = C_{2,0a}^- x_a^- + C_{2,0a}^0 x_a^0 + C_{2,0c} x_c + C_{2,0d} x_d + u_0 - r_0, \tag{9.2.12}$$

and for each $i = 1, \cdots, m_d$,

$$\dot{x}_i = A_{q_i} x_i + L_{i0} h_0 + L_{id} h_d + B_{q_i} \left[u_i + E_{ia} x_a + E_{ic} x_c + \sum_{j=1}^{m_d} E_{ij} x_j \right], \tag{9.2.13}$$

$$h_i = C_{q_i} x_i = x_{i1}, \quad h_d = C_d x_d, \tag{9.2.14}$$

and finally,

$$e_i = h_i - r_i = C_{q_i} x_i - r_i, \quad e_d = h_d - r_d = C_d x_d - r_d. \tag{9.2.15}$$

Let us define a set of new state variables, i.e., for $i = 1, 2, \cdots, m_d$, we define

$$\bar{x}_i := \begin{pmatrix} x_{i1} \\ x_{i2} \\ \vdots \\ x_{iq_i} \end{pmatrix} - \begin{pmatrix} r_i \\ \dot{r}_i \\ \vdots \\ r_i^{(q_i-1)} \end{pmatrix}, \tag{9.2.16}$$

if $\kappa \geq q_i$, or

$$\bar{x}_i := \begin{pmatrix} x_{i1} \\ \vdots \\ x_{iq} \\ x_{iq+1} \\ \vdots \\ x_{iq_i} \end{pmatrix} - \begin{pmatrix} r_i \\ \vdots \\ r_i^{(\kappa-1)} \\ 0 \\ \vdots \\ 0 \end{pmatrix}, \tag{9.2.17}$$

if $\kappa < q_i$. Then, we have

$$e_i = C_{q_i} \bar{x}_i, \quad e_d = C_d \bar{x}_d, \tag{9.2.18}$$

$$\dot{x}_a^- = A_{aa}^- x_a^- + B_{0a}^- e_0 + L_{ad}^- e_d + [\, B_{0a}^- \quad L_{ad}^- \,]\, r, \tag{9.2.19}$$

$$\dot{x}_a^0 = A_{aa}^0 x_a^0 + B_{0a}^0 e_0 + L_{ad}^0 e_d + [\, B_{0a}^0 \quad L_{ad}^0 \,]\, r, \tag{9.2.20}$$

$$\dot{x}_c = A_{cc} x_c + B_{0c} e_0 + L_{cd} e_d + B_c \left[E_{ca}^- x_a^- + E_{ca}^0 x_a^0\right] + B_c u_c + [\, B_{0c} \quad L_{cd} \,]\, r, \tag{9.2.21}$$

$$e_0 = C_{2,0a}^- x_a^- + C_{2,0a}^0 x_a^0 + C_{2,0c} x_c + C_{2,0d} \bar{x}_d + u_0 - r_0 + C_{2,0d}^* r_d, \tag{9.2.22}$$

for an appropriate dimensional matrix $C_{2,0d}^*$, and for $i = 1, 2, \cdots, m_d$,

$$\dot{\bar{x}}_i = A_{q_i} \bar{x}_i + L_{i0} e_0 + L_{id} e_d + B_{q_i} \left[u_i + E_{ia} x_a + E_{ic} x_c + \sum_{j=1}^{m_d} E_{ij} \bar{x}_j \right] + E_{q_i} \begin{pmatrix} r \\ \vdots \\ r^{(\kappa-1)} \end{pmatrix}, \tag{9.2.23}$$

for an appropriate dimensional matrix E_{q_i}. Note that the disturbances r_0 and r_d in (9.2.22) can be washed out by the following pre-output feedback,

$$u_0 = \tilde{u}_0 + r_0 - C_{2,0d}^* r_d. \tag{9.2.24}$$

Moreover, the subsystem from the controlled input, i.e., $(\, u_0' \;\; u_d' \;\; u_c' \,)'$, to the error output, i.e., $(\, e_0' \;\; e_d \,)'$, is now in the standard form the special coordinate basis of Theorem 2.4.1. It then follows from the result of Chapter 8 (i.e., Proposition 8.2.1) that if the almost disturbance decoupling problem with measurement feedback and with internal stability for the system (9.2.4) is solvable, there must exist a nonzero vector ξ such that

$$\xi^{\mathrm{H}} (\lambda I - A_{aa}^0) = 0 \quad \text{and} \quad \xi^{\mathrm{H}} [\, B_{0a}^0 \quad L_{ad}^0 \,] = 0, \tag{9.2.25}$$

which is implies that $(A_{aa}^0, [\, B_{0a}^0 \;\; L_{ad}^0 \,])$ is not completely controllable. Following Property 2.4.1 of the special coordinate basis of Chapter 2, the uncontrollability of $(A_{aa}^0, [\, B_{0a}^0 \;\; L_{ad}^0 \,])$ implies the unstabilizability of the pair (A, B), which is obviously a contradiction. Hence, x_a^0 must be non-existent. It then follows from Property 2.4.2 of the special coordinate basis that Σ_{P} is of minimum phase. This completes the proof of the necessary part. ⊠

We note that for the case when $D_1 = 0$, then the direct feedthrough term D_{22} must be a zero matrix as well, and the last condition, i.e., Item 4, of Theorem 9.2.1 reduces to $\mathrm{Ker}\,(C_2) \supset \mathrm{Ker}\,(C_1)$.

We will show the sufficiency of those conditions in Theorem 9.2.1 by explicitly constructing parameterized controllers which solve the proposed robust and perfect tracking problem under Conditions 1 to 4 of Theorem 9.2.1. This will be done in the following subsequent subsections. First, we have the following corollary that deals with the state feedback case.

Corollary 9.2.1. Consider the given system (9.1.1) with its external disturbance $w \in L_p$, $p \in [1,\infty)$, its initial condition $x(0) = x_0$. Assume that all its states are measured for feedback, i.e., $C_1 = I$ and $D_1 = 0$. Then, for any reference signal $r(t)$, which has all its i-th order derivatives, $i = 1, 2, \cdots, \kappa - 1$, $\kappa \geq 1$, being available and $r^{(\kappa)}(t)$ being either a vector of delta functions or in L_p, the proposed robust and perfect tracking (RPT) problem is solvable by the control law of (9.1.2) if and only if the following conditions are satisfied:

1. (A, B) is stabilizable;
2. $D_{22} = 0$;
3. $\Sigma_{\rm P}$, i.e., (A, B, C_2, D_2), is right invertible and of minimum phase. ©

9.2.1. Solutions to State Feedback Case

When all states of the plant are measured for feedback, the problem can be solved by a static control law. We construct in this subsection a parameterized state feedback control law,

$$u = F(\varepsilon)x + H_0(\varepsilon)r + \cdots + H_{\kappa-1}(\varepsilon)r^{(\kappa-1)}, \tag{9.2.26}$$

which solves the robust and perfect tracking (RPT) problem for (9.1.1) under the conditions given in Corollary 9.2.1. It is simple to note that we can rewrite the given reference in the following form,

$$\frac{d}{dt}\begin{pmatrix} r \\ \vdots \\ r^{(\kappa-2)} \\ r^{(\kappa-1)} \end{pmatrix} = \begin{bmatrix} 0 & I_\ell & \cdots & 0 \\ \vdots & \vdots & \ddots & \vdots \\ 0 & 0 & \cdots & I_\ell \\ 0 & 0 & \cdots & 0 \end{bmatrix} \begin{pmatrix} r \\ \vdots \\ r^{(\kappa-2)} \\ r^{(\kappa-1)} \end{pmatrix} + \begin{bmatrix} 0 \\ \vdots \\ 0 \\ I_\ell \end{bmatrix} r^{(\kappa)}. \tag{9.2.27}$$

Combining (9.2.27) with the given system, we obtain the following augmented system,

$$\Sigma_{\rm AUG} : \begin{cases} \dot{\boldsymbol{x}} = \boldsymbol{A}\ \boldsymbol{x} + \boldsymbol{B}\ u + \boldsymbol{E}\ \boldsymbol{w} \\ \boldsymbol{y} = \quad \boldsymbol{x} \\ e = \boldsymbol{C}_2\ \boldsymbol{x} + \boldsymbol{D}_2\ u \end{cases} \tag{9.2.28}$$

where

$$\boldsymbol{w} := \begin{pmatrix} w \\ r^{(\kappa)} \end{pmatrix}, \quad \boldsymbol{x} := \begin{pmatrix} r \\ \vdots \\ r^{(\kappa-2)} \\ r^{(\kappa-1)} \\ x \end{pmatrix}, \tag{9.2.29}$$

$$\boldsymbol{A} = \begin{bmatrix} 0 & I_\ell & \cdots & 0 & 0 \\ \vdots & \vdots & \ddots & \vdots & \vdots \\ 0 & 0 & \cdots & I_\ell & 0 \\ 0 & 0 & \cdots & 0 & 0 \\ 0 & 0 & \cdots & 0 & A \end{bmatrix}, \quad \boldsymbol{B} = \begin{bmatrix} 0 \\ \vdots \\ 0 \\ 0 \\ B \end{bmatrix}, \quad \boldsymbol{E} = \begin{bmatrix} 0 & 0 \\ \vdots & \vdots \\ 0 & 0 \\ 0 & I_\ell \\ E & 0 \end{bmatrix}, \tag{9.2.30}$$

and

$$\boldsymbol{C}_2 = [\,-I_\ell \;\; 0 \;\; 0 \;\; \cdots \;\; 0 \;\; C_2\,], \quad \boldsymbol{D}_2 = D_2. \tag{9.2.31}$$

It is then straightforward to show that the subsystem from u to e in the augmented system (9.2.28), i.e., the quadruple $(\boldsymbol{A}, \boldsymbol{B}, \boldsymbol{C}_2, \boldsymbol{D}_2)$, is right invertible and has the same infinite zero structure as that of $\Sigma_{\rm P}$. Furthermore, its invariant zeros contain those of $\Sigma_{\rm P}$ and $\ell \times \kappa$ extra ones at $s = 0$. We are now ready to present a step-by-step algorithm to construct the required control law of the form (9.2.26).

Step 9.S.1. This step is to transform the subsystem from u to e of the augmented system (9.2.28) into the special coordinate basis of Theorem 2.4.1, i.e., to find nonsingular state, input and output transformations Γ_s, Γ_i and Γ_o to put it into the structural form of Theorem 2.4.1 as well as in a small variation of the compact form of (2.4.20) to (2.4.23). It can be shown that the compact form of (2.4.20) to (2.4.23) for the subsystem from u to e of (9.2.28) can be written as,

$$\tilde{A} = \begin{bmatrix} A_{aa}^0 & 0 & 0 & 0 \\ 0 & A_{aa}^- & 0 & L_{ad}^- C_d \\ B_c E_{ca}^0 & B_c E_{ca}^- & A_{cc} & L_{cd} C_d \\ B_d E_{da}^0 & B_d E_{da}^- & B_d E_{dc} & A_{dd} \end{bmatrix}, \tag{9.2.32}$$

$$A_{aa}^0 = \begin{bmatrix} 0 & I_\ell & \cdots & 0 \\ \vdots & \vdots & \ddots & \vdots \\ 0 & 0 & \cdots & I_\ell \\ 0 & 0 & \cdots & 0 \end{bmatrix}, \quad \tilde{B} = \begin{bmatrix} 0 & 0 & 0 \\ B_{0a}^- & 0 & 0 \\ B_{0c} & 0 & B_c \\ B_{0d} & B_d & 0 \end{bmatrix}, \tag{9.2.33}$$

and

$$\tilde{C} = \begin{bmatrix} C_{0a}^0 & C_{0a}^- & C_{0c} & C_{0d} \\ 0 & 0 & 0 & C_d \end{bmatrix}, \quad \tilde{D} = \begin{bmatrix} I_{m_0} & 0 & 0 \\ 0 & 0 & 0 \end{bmatrix}. \tag{9.2.34}$$

Step 9.S.2. Choose an appropriate dimensional matrix F_c such that

$$A_{cc}^c = A_{cc} - B_c F_c \tag{9.2.35}$$

is asymptotically stable. The existence of such an F_c is guaranteed by the property that (A_{cc}, B_c) is completely controllable.

Step 9.S.3. For each x_i of x_d, which is associated with the infinite zero structure of Σ_P or the subsystem from u to e of (9.2.28), we choose an F_i such that

$$p_i(s) = \prod_{j=1}^{q_i} (s - \lambda_{ij}) = s^{q_i} + F_{i1} s^{q_i - 1} + \cdots + F_{iq_{i-1}} s + F_{iq_i} \tag{9.2.36}$$

with all λ_{ij} being in $\mathbb{C}^-$. Let

$$F_i = [\, F_{iq_i} \quad F_{iq_{i-1}} \quad \cdots \quad F_{i1} \,], \quad i = 1, \cdots, m_d. \tag{9.2.37}$$

Step 9.S.4. Next, we construct

$$\boldsymbol{F}(\varepsilon) = -\Gamma_i \begin{bmatrix} C_{0a}^0 & C_{0a}^- & C_{0c} & C_{0d} \\ E_{da}^0 & E_{da}^- & E_{dc} & E_d + F_d(\varepsilon) \\ E_{ca}^0 & E_{ca}^- & F_c & 0 \end{bmatrix} \Gamma_s^{-1}, \tag{9.2.38}$$

where

$$E_d = \begin{bmatrix} E_{11} & \cdots & E_{1m_d} \\ \vdots & \ddots & \vdots \\ E_{m_d 1} & \cdots & E_{m_d m_d} \end{bmatrix}, \tag{9.2.39}$$

$$F_d(\varepsilon) = \text{blkdiag}\left\{ \frac{F_1}{\varepsilon^{q_1}} S_1(\varepsilon), \frac{F_2}{\varepsilon^{q_2}} S_2(\varepsilon), \cdots, \frac{F_{m_d}}{\varepsilon^{q_{m_d}}} S_{m_d}(\varepsilon) \right\}, \tag{9.2.40}$$

and where

$$S_i(\varepsilon) = \text{diag}\left\{1, \varepsilon, \varepsilon^2, \cdots, \varepsilon^{q_i - 1}\right\}. \tag{9.2.41}$$

Step 9.S.5. Finally, we partition

$$\boldsymbol{F}(\varepsilon) = [\, H_0(\varepsilon) \quad \cdots \quad H_{\kappa-1}(\varepsilon) \quad F(\varepsilon) \,], \tag{9.2.42}$$

where $H_i(\varepsilon) \in \mathbb{R}^{m \times \ell}$ and $F(\varepsilon) \in \mathbb{R}^{m \times n}$. This ends the constructive algorithm. ▣

We have the following result.

Theorem 9.2.2. Consider the given system (9.1.1) with its external disturbance $w \in L_p$, $p \in [1, \infty)$, its initial condition $x(0) = x_0$. Assume that all its states are measured for feedback, i.e., $C_1 = I$ and $D_1 = 0$. If Conditions 1 to 3 of Corollary 9.2.1 are satisfied, then, for any reference signal $r(t)$, which has all its i-th order derivatives, $i = 0, 1, \cdots, \kappa - 1$, $\kappa \geq 1$, being available and $r^{(\kappa)}(t)$ being either a vector of delta functions or in L_p, the proposed robust and perfect tracking (RPT) problem is solved by the control law of (9.2.26) with $F(\varepsilon)$ and $H_i(\varepsilon)$, $i = 0, 1, \cdots, \kappa - 1$, as given in (9.2.42). ▣

Proof. See Subsection 9.4.A. ▣

The following remark gives an alternative approach for solving the proposed robust and perfect tracking problem via full state feedback. We leave the proof of this method to readers as an exercise.

Remark 9.2.1. Note that the required gain matrices for the state feedback RPT problem might be computed by solving the following Riccati equation,

$$\boldsymbol{P}\tilde{\boldsymbol{A}}+\tilde{\boldsymbol{A}}'\boldsymbol{P}+\tilde{\boldsymbol{C}}_2'\tilde{\boldsymbol{C}}_2-\left(\boldsymbol{P}\boldsymbol{B}+\tilde{\boldsymbol{C}}_2'\tilde{\boldsymbol{D}}_2\right)\left(\tilde{\boldsymbol{D}}_2'\tilde{\boldsymbol{D}}_2\right)^{-1}\left(\boldsymbol{P}\boldsymbol{B}+\tilde{\boldsymbol{C}}_2'\tilde{\boldsymbol{D}}_2\right)'=0, \quad (9.2.43)$$

for a positive definite solution $\boldsymbol{P} > 0$, where

$$\tilde{\boldsymbol{C}}_2 = \begin{bmatrix} \boldsymbol{C}_2 \\ \varepsilon I_{\kappa\ell+n} \\ 0 \end{bmatrix}, \quad \tilde{\boldsymbol{D}}_2 = \begin{bmatrix} \boldsymbol{D}_2 \\ 0 \\ \varepsilon I_m \end{bmatrix}, \quad (9.2.44)$$

$$\tilde{\boldsymbol{A}} = \begin{bmatrix} \tilde{A}_0 & 0 \\ 0 & A \end{bmatrix}, \quad \tilde{A}_0 = -\varepsilon I_{\kappa\ell} + \begin{bmatrix} 0 & I_\ell & \cdots & 0 \\ \vdots & \vdots & \ddots & \vdots \\ 0 & 0 & \cdots & I_\ell \\ 0 & 0 & \cdots & 0 \end{bmatrix}, \quad (9.2.45)$$

and where $\boldsymbol{B}$, $\boldsymbol{C}_2$ and $\boldsymbol{D}_2$ are as defined in (9.2.30) and (9.2.31). The required gain matrix is then given by

$$\tilde{\boldsymbol{F}}(\varepsilon) = -\left(\tilde{\boldsymbol{D}}_2'\tilde{\boldsymbol{D}}_2\right)^{-1}\left(\boldsymbol{P}\boldsymbol{B} + \tilde{\boldsymbol{C}}_2'\tilde{\boldsymbol{D}}_2\right)' = [\, H_0(\varepsilon) \quad \cdots \quad H_{\kappa-1}(\varepsilon) \quad F(\varepsilon)\,], \quad (9.2.46)$$

where $H_i(\varepsilon) \in \mathbb{R}^{m\times\ell}$ and $F(\varepsilon) \in \mathbb{R}^{m\times n}$. Finally, we note that solutions to the Riccati equation (9.2.43) might have severe numerical problems as ε tends smaller and smaller. Ⓡ

9.2.2. Solutions to Measurement Feedback Case

We will consider two types of measurement feedback control laws, one is of full order controllers whose dynamical order is equal to the order of the given system and the other reduced order controllers with a dynamical order that is less than the order of the given system. Without loss of generality, we assume throughout this subsection that $D_{22} = 0$. If it is nonzero, it can always be washed out by the following pre-output feedback,

$$u = Sy, \quad (9.2.47)$$

with S as given in the second item of Theorem 9.2.1. The following are constructive algorithms for both full and reduced order measurement feedback controllers, which, under the conditions of Theorem 9.2.1, solve the proposed robust and perfect tracking problem.

A. Full Order Measurement Feedback

The following is a step-by-step algorithm for constructing a parameterized full order measurement feedback controller, which solves the robust and perfect tracking problem.

Step 9.F.1. For the given reference $r(t)$ and the given system (9.1.1), we first assume that all the state variables of (9.1.1) are measurable and follow the procedures of the previous subsection to define an auxiliary system,

$$\begin{cases} \dot{\boldsymbol{x}} = \boldsymbol{A}\ \boldsymbol{x} + \boldsymbol{B}\ \boldsymbol{u} + \boldsymbol{E}\ \boldsymbol{w} \\ \boldsymbol{y} = \quad \boldsymbol{x} \\ e = \boldsymbol{C}_2\ \boldsymbol{x} + \boldsymbol{D}_2\ \boldsymbol{u} \end{cases} \tag{9.2.48}$$

Then, we follow Steps 9.S.1 to 9.S.5 of the algorithm of the previous subsection to construct a state feedback gain matrix

$$\boldsymbol{F}(\varepsilon) = [\, H_0(\varepsilon) \quad \cdots \quad H_{\kappa-1}(\varepsilon) \quad F(\varepsilon)\,]\,. \tag{9.2.49}$$

Step 9.F.2. Let $\Sigma_{\mathrm{Q}a}$ be characterized by a matrix quadruple

$$(A_{\mathrm{Q}a}, E_{\mathrm{Q}a}, C_{\mathrm{Q}a}, D_{\mathrm{Q}a}) := (A, [\, E \quad I_n\,], C_1, [\, D_1 \quad 0\,])\,. \tag{9.2.50}$$

This step is to transform this $\Sigma_{\mathrm{Q}a}$ into the special coordinate basis of Theorem 2.4.1. Because of the special structure of the matrix $E_{\mathrm{Q}a}$, it is simple to show that $\Sigma_{\mathrm{Q}a}$ is always right invertible and is free of invariant zeros. Utilize the results of Theorem 2.4.1 to find nonsingular state, input and output transformation $\Gamma_{s\mathrm{Q}}$, $\Gamma_{i\mathrm{Q}}$ and $\Gamma_{o\mathrm{Q}}$ such that

$$\Gamma_{s\mathrm{Q}}^{-1} A \Gamma_{s\mathrm{Q}} = \begin{bmatrix} A_{cc\mathrm{Q}} & L_{cd\mathrm{Q}} \\ E_{dc\mathrm{Q}} & A_{dd\mathrm{Q}} \end{bmatrix} + \begin{bmatrix} B_{0c\mathrm{Q}} \\ B_{0d\mathrm{Q}} \end{bmatrix} [\, C_{0c\mathrm{Q}} \quad 0\,], \tag{9.2.51}$$

$$\Gamma_{s\mathrm{Q}}^{-1} E_{\mathrm{Q}a} \Gamma_{i\mathrm{Q}} = \begin{bmatrix} B_{0c\mathrm{Q}} & 0 & I_{n-k} & 0 \\ B_{0d\mathrm{Q}} & I_k & 0 & 0 \end{bmatrix}, \tag{9.2.52}$$

and

$$\Gamma_{o\mathrm{Q}}^{-1} C_1 \Gamma_{s\mathrm{Q}} = \begin{bmatrix} C_{0c\mathrm{Q}} & 0 \\ 0 & I_k \end{bmatrix}, \quad \Gamma_{o\mathrm{Q}}^{-1} [\, D_1 \quad 0\,] \Gamma_{i\mathrm{Q}} = \begin{bmatrix} I_{\mathrm{p}-k} & 0 & 0 & 0 \\ 0 & 0 & 0 & 0 \end{bmatrix}, \tag{9.2.53}$$

where $k = \mathrm{p} - \mathrm{rank}(D_1)$. It can be verified that the pair (A, C_1) is detectable if and only if the pair

$$\left(A_{cc\mathrm{Q}}, \begin{bmatrix} C_{0c\mathrm{Q}} \\ E_{dc\mathrm{Q}} \end{bmatrix} \right) \tag{9.2.54}$$

is detectable.

Step 9.F.3. Let K_{cQ} be an appropriate dimensional constant matrix such that the eigenvalues of the matrix

$$A_{ccQ}^{c} = A_{ccQ} - K_{cQ} \begin{bmatrix} C_{0cQ} \\ E_{dcQ} \end{bmatrix} = A_{ccQ} - [\, K_{c0Q} \quad K_{cdQ} \,] \begin{bmatrix} C_{0cQ} \\ E_{dcQ} \end{bmatrix} \tag{9.2.55}$$

are all in $\mathbb{C}^-$. Next, we define a parameterized observer gain matrix,

$$K(\varepsilon) = -\Gamma_{sQ} \begin{bmatrix} B_{0cQ} + K_{c0Q} & L_{cdQ} + K_{cdQ}/\varepsilon \\ B_{0dQ} & A_{ddQ} + I_k/\varepsilon \end{bmatrix} \Gamma_{oQ}^{-1}. \tag{9.2.56}$$

Step 9.F.4. Finally, we obtain the following full order measurement feedback control law,

$$\begin{cases} \dot{v} = A_{\rm cmp}(\varepsilon)\, v - K(\varepsilon)\, y + BH_0(\varepsilon)\, r + \cdots + BH_{\kappa-1}(\varepsilon)\, r^{(\kappa-1)}, \\ u = F(\varepsilon)\, v + H_0(\varepsilon)\, r + \cdots + H_{\kappa-1}(\varepsilon)\, r^{(\kappa-1)}, \end{cases} \tag{9.2.57}$$

where $A_{\rm cmp}(\varepsilon) = A + BF(\varepsilon) + K(\varepsilon)C_1$. This completes the construction of the full order measurement feedback controller. 🄰

We have the following theorem.

Theorem 9.2.3. Consider the given system (9.1.1) with its external disturbance $w \in L_p$, $p \in [1, \infty)$, its initial condition $x(0) = x_0$. If Conditions 1 to 4 of Theorem 9.2.1 are satisfied, then, for any reference signal $r(t)$, which has all its i-th order derivatives, $i = 0, 1, \cdots, \kappa-1$, $\kappa \geq 1$, being available and $r^{(\kappa)}(t)$ being either a vector of delta functions or in L_p, then the proposed robust and perfect tracking (RPT) problem is solved by the parameterized full order measurement feedback control laws as given in (9.2.57). 🅃

Proof. See Subsection 9.4.B. ⊠

The following remark yields an alternative way to compute the gain matrix $K(\varepsilon)$ in Step 9.F.3.

Remark 9.2.2. The gain matrix $K(\varepsilon)$ in Step 9.F.3 can also be computed by solving the following Riccati equation,

$$AQ+QA'+(EE'+I)-(QC_1'+ED_1')(D_1D_1'+\varepsilon I)^{-1}(C_1Q+D_1E') = 0, \tag{9.2.58}$$

for a positive definite solution $Q > 0$. The required gain matrix $K(\varepsilon)$ is then given by

$$K(\varepsilon) = -(QC_1' + ED_1')(D_1D_1' + \varepsilon I)^{-1}. \tag{9.2.59}$$

Again, this approach might have some numerical problems when ε is small. 🅁

B. Reduced Order Measurement Feedback

We now present solutions to the robust and perfect tracking problem via reduced order measurement feedback control laws. For simplicity of presentation, we assume that matrices C_1 and D_1 have already been transformed into the following forms,

$$C_1 = \begin{bmatrix} 0 & C_{1,02} \\ I_k & 0 \end{bmatrix} \quad \text{and} \quad D_1 = \begin{bmatrix} D_{1,0} \\ 0 \end{bmatrix}, \tag{9.2.60}$$

where $D_{1,0}$ is of full row rank. Before we present a step-by-step algorithm to to construct a parameterized reduced order measurement feedback controller, we first partition the following system

$$\begin{cases} \dot{x} = A\ x + B\ u + [E \quad I_n]\ \tilde{w}, \\ y = C_1\ x \qquad\qquad + [D_1 \quad 0]\ \tilde{w}, \end{cases} \tag{9.2.61}$$

in conformity with the structures of C_1 and D_1 in (9.2.60), i.e.,

$$\begin{cases} \begin{pmatrix} \dot{x}_1 \\ \dot{x}_2 \end{pmatrix} = \begin{bmatrix} A_{11} & A_{12} \\ A_{21} & A_{22} \end{bmatrix} \begin{pmatrix} x_1 \\ x_2 \end{pmatrix} + \begin{bmatrix} B_1 \\ B_2 \end{bmatrix} u + \begin{bmatrix} E_1 & I_k & 0 \\ E_2 & 0 & I_{n-k} \end{bmatrix} \tilde{w}, \\ \begin{pmatrix} y_0 \\ y_1 \end{pmatrix} = \begin{bmatrix} 0 & C_{1,02} \\ I_k & 0 \end{bmatrix} \begin{pmatrix} x_1 \\ x_2 \end{pmatrix} \qquad\qquad + \begin{bmatrix} D_{1,0} & 0 & 0 \\ 0 & 0 & 0 \end{bmatrix} \tilde{w}, \end{cases} \tag{9.2.62}$$

where

$$\tilde{w} = \begin{pmatrix} w \\ x_0 \cdot \delta(t) \end{pmatrix}. \tag{9.2.63}$$

Obviously, $y_1 = x_1$ is directly available and hence need not to be estimated. Next, we define $\Sigma_{\rm QR}$ to be characterized by

$$(A_{\rm R}, E_{\rm R}, C_{\rm R}, D_{\rm R}) = \left(A_{22}, [E_2 \quad 0 \quad I_{n-k}], \begin{bmatrix} C_{1,02} \\ A_{12} \end{bmatrix}, \begin{bmatrix} D_{1,0} & 0 & 0 \\ E_1 & I_k & 0 \end{bmatrix} \right). \tag{9.2.64}$$

It is again straightforward to verify that $\Sigma_{\rm QR}$ is right invertible with no finite and infinite zeros. Moreover, $(A_{\rm R}, C_{\rm R})$ is detectable if and only if (A, C_1) is detectable. We are ready to present the following algorithm.

Step 9.R.1. For the given reference $r(t)$ and the given system (9.1.1), we again assume that all the state variables of (9.1.1) are measurable and follow the procedures of the previous subsection to define an auxiliary system,

$$\begin{cases} \dot{x} = \boldsymbol{A}\ x + \boldsymbol{B}\ u + \boldsymbol{E}\ \boldsymbol{w} \\ y = \quad x \\ e = \boldsymbol{C}_2\ x + \boldsymbol{D}_2\ u \end{cases} \tag{9.2.65}$$

Then, we follow Steps 9.S.1 to 9.S.5 of the algorithm of the previous subsection to construct a state feedback gain matrix

$$\boldsymbol{F}(\varepsilon) = [\, H_0(\varepsilon) \quad \cdots \quad H_{\kappa-1}(\varepsilon) \quad F(\varepsilon)\,]. \tag{9.2.66}$$

Let us partition $F(\varepsilon)$ in conformity with x_1 and x_2 of (9.2.62) as follows,

$$F(\varepsilon) = [\, F_1(\varepsilon) \quad F_2(\varepsilon)\,]. \tag{9.2.67}$$

Step 9.R.2. Let $K_{\rm R}$ be an appropriate dimensional constant matrix such that the eigenvalues of

$$A_{\rm R} + K_{\rm R} C_{\rm R} = A_{22} + [\, K_{\rm R0} \quad K_{\rm R1}\,] \begin{bmatrix} C_{1,02} \\ A_{12} \end{bmatrix} \tag{9.2.68}$$

are all in $\mathbb{C}^-$. This can be done because $(A_{\rm R}, C_{\rm R})$ is detectable.

Step 9.R.3. Let

$$G_{\rm R} = [\, -K_{\rm R0}, \quad A_{21} + K_{\rm R1} A_{11} - (A_{\rm R} + K_{\rm R} C_{\rm R}) K_{\rm R1}\,], \tag{9.2.69}$$

and

$$\left.\begin{aligned} A_{\rm cmp}(\varepsilon) &= A_{\rm R} + B_2 F_2(\varepsilon) + K_{\rm R} C_{\rm R} + K_{\rm R1} B_1 F_2(\varepsilon), \\ B_{\rm cmp}(\varepsilon) &= G_{\rm R} + (B_2 + K_{\rm R1} B_1)\,[\,0, \quad F_1(\varepsilon) - F_2(\varepsilon) K_{\rm R1}\,], \\ C_{\rm cmp}(\varepsilon) &= F_2(\varepsilon), \\ D_{\rm cmp}(\varepsilon) &= [\,0, \quad F_1(\varepsilon) - F_2(\varepsilon) K_{\rm R1}\,]. \end{aligned}\right\} \tag{9.2.70}$$

Step 9.R.4. Finally, we obtain the following reduced order measurement feedback control law,

$$\begin{cases} \dot v = A_{\rm cmp}(\varepsilon) v + B_{\rm cmp}(\varepsilon) y + G_0(\varepsilon) r + \cdots + G_{\kappa-1}(\varepsilon) r^{(\kappa-1)}, \\ u = C_{\rm cmp}(\varepsilon) v + D_{\rm cmp}(\varepsilon) y + H_0(\varepsilon) r + \cdots + H_{\kappa-1}(\varepsilon) r^{(\kappa-1)}, \end{cases} \tag{9.2.71}$$

where for $i = 0, 1, \cdots, \kappa - 1$,

$$G_i(\varepsilon) = (B_2 + K_{\rm R1} B_1) H_i(\varepsilon). \tag{9.2.72}$$

This completes the construction of the reduced order measurement feedback controller. ◻

Theorem 9.2.4. Consider the given system (9.1.1) with its external disturbance $w \in L_p$, $p \in [1, \infty)$, its initial condition $x(0) = x_0$. If Conditions 1 to 4 of Theorem 9.2.1 are satisfied, then, for any reference signal $r(t)$, which has all its i-th order derivatives, $i = 0, 1, \cdots, \kappa - 1$, $\kappa \geq 1$, being available and $r^{(\kappa)}(t)$ being either a vector of delta functions or in L_p, then the proposed robust and perfect tracking (RPT) problem is solved by the parameterized reduced order measurement feedback control laws of (9.2.71). 🅃

Proof. See Subsection 9.4.C. ⊠

By now, the sufficiency of Theorem 9.2.1 is obvious in view of the results of Theorems 9.2.3 and 9.2.4. The proof of Theorem 9.2.1 is thus completed. ⊠

9.3. Robust and Perfect Tracking for Other References

It is very often in practical control system design to track some references such as sinusoidal functions, which is in L_∞. It is obvious that we could not make the L_∞ norm of the tracking error arbitrarily small if there is a mismatch in the initial value of the output to be controlled and that of the reference signal. Another very common situation could be that the references $r(t)$ might have some entries belonging to one set, say L_{p_1}, and some belonging to another set, say L_{p_2}, for some $p_1 \in [1, \infty]$ and $p_2 \in [1, \infty]$. Thus, for this class of references, we will have to modify our original problem formulation a little bit in order to obtain some meaningful results. Again, we consider a linear system as given in (9.1.1) with an external disturbance

$$w = \begin{pmatrix} w_1 \\ w_2 \\ \vdots \\ w_{\mathrm{q}} \end{pmatrix}, \tag{9.3.1}$$

where $w_i \in L_{p_{w_i}}$, $p_{w_i} \in [1, \infty]$, $i = 1, 2, \cdots, \mathrm{q}$. We also consider a reference

$$r = \begin{pmatrix} r_1 \\ r_2 \\ \vdots \\ r_\ell \end{pmatrix}, \tag{9.3.2}$$

which has the following properties: for $i = 1, 2, \cdots, \ell$, we have r_i, $\dot{r}_i$, $\cdots$, $r_i^{(\kappa_i - 1)}$, $\kappa_i \geq 1$, being available, and $r_i^{(\kappa_i)}$ being a delta function or in $L_{p_{r_i}}$ for some $p_{r_i} \in [1, \infty]$. Then, the general robust and perfect tracking (GRPT) problem for this type of references is to find a parameterized dynamic measurement feedback control law of the form

$$\left\{\begin{aligned} \dot{v} &= A_{\rm cmp}(\varepsilon)v + B_{\rm cmp}(\varepsilon)y + \sum_{i=0}^{\kappa_1-1} G_{1,i}(\varepsilon)r_1^{(i)} + \cdots + \sum_{i=0}^{\kappa_\ell-1} G_{\ell,i}(\varepsilon)r_\ell^{(i)}, \\ u &= C_{\rm cmp}(\varepsilon)v + D_{\rm cmp}(\varepsilon)y + \sum_{i=0}^{\kappa_1-1} H_{1,i}(\varepsilon)r_1^{(i)} + \cdots + \sum_{i=0}^{\kappa_\ell-1} H_{\ell,i}(\varepsilon)r_\ell^{(i)}, \end{aligned}\right. \tag{9.3.3}$$

such that when (9.3.3) is applied to (9.1.1), we have

1. There exists an $\varepsilon^* > 0$ such that the resulting closed-loop system with $r = 0$ and $w = 0$ is asymptotically stable for all $\varepsilon \in (0, \varepsilon^*]$; and

2. The resulting closed-loop error signal e, which is obviously a function of ε, can be decomposed as

$$e = e_{r_1} + \cdots + e_{r_\ell} + e_{w_1} + \cdots + e_{w_q} + e_o, \tag{9.3.4}$$

   and as $\varepsilon \to 0$,

$$\tilde{J}(x_0, w, r, \varepsilon) = \sum_{i=1}^{\ell} \|e_{r_i}\|_{p_{r_i}} + \sum_{i=1}^{q} \|e_{w_i}\|_{p_{w_i}} + \|e_o\|_p \to 0, \tag{9.3.5}$$

   for all $1 \le p < \infty$ and for any $x_0 \in \mathbb{R}^n$. Roughly, e_o is the error due to mismatch in initial conditions of the controlled output and reference, while e_{r_i}, $i = 1, 2, \cdots, \ell$, and e_{w_i}, $i = 1, 2, \cdots, q$, are corresponding to the steady state error.

We have the following result.

Theorem 9.3.1. Consider the given system (9.1.1) with its initial condition $x(0) = x_0$. Also, consider the external disturbance w with its entries $w_i \in L_{p_{w_i}}$, $p_{w_i} \in [1, \infty]$, $i = 1, 2, \cdots, q$. Then, for any reference signal $r(t)$ of the form (9.3.2) with r_i, $\dot{r}_i$, $\cdots$, $r_i^{(\kappa_i-1)}$, $\kappa_i \ge 1$, being available, and $r_i^{(\kappa_i)}$ being a delta function or in $L_{p_{r_i}}$, $p_{r_i} \in [1, \infty]$, $i = 1, 2, \cdots, \ell$, the general robust and perfect tracking (GRPT) problem is solvable by the control law of (9.3.3) if and only if all the same four conditions of Theorem 9.2.1 hold. ⊡

Proof. The proof of this theorem follows from similar lines of reasoning as those of Theorem 9.2.1 with some minor fine tuning. The constructive algorithms of the previous section should be modified as follows:

1. *State Feedback Case.* For the state feedback case, one first needs to obtain an augmented system,

$$\tilde{\Sigma}_{\text{AUG}} : \begin{cases} \dot{\boldsymbol{x}} = \boldsymbol{A}\ \boldsymbol{x} + \boldsymbol{B}\ u + \boldsymbol{E}\ \boldsymbol{w} \\ y = \quad \boldsymbol{x} \\ e = \boldsymbol{C}_2\ \boldsymbol{x} + \boldsymbol{D}_2\ u \end{cases} \tag{9.3.6}$$

with

$$e = h - r, \ . \ \boldsymbol{w} := \begin{pmatrix} w \\ r^{(\kappa)} \end{pmatrix}, \quad \boldsymbol{x} := \begin{pmatrix} \boldsymbol{r}_1 \\ \vdots \\ \boldsymbol{r}_\ell \\ x \end{pmatrix}, \quad \boldsymbol{r}_i = \begin{pmatrix} r_i \\ \vdots \\ r_i^{(\kappa_i - 1)} \end{pmatrix}. \tag{9.3.7}$$

Then, follow the same procedures as in Steps 9.S.1 to 9.S.4 of the previous section to obtain a gain matrix $\boldsymbol{F}(\varepsilon)$, and partition it as follows,

$$\boldsymbol{F}(\varepsilon) = [\, H_{1,0}(\varepsilon) \ \cdots \ H_{1,\kappa_1-1}(\varepsilon) \ \cdots \ H_{\ell,0}(\varepsilon) \ \cdots \ H_{\ell,\kappa_\ell-1}(\varepsilon) \ \ F(\varepsilon)\,]. \tag{9.3.8}$$

The state feedback controller is given by

$$u = F(\varepsilon)x + \sum_{i=0}^{\kappa_1-1} H_{1,i}(\varepsilon) r_1^{(i)} + \cdots + \sum_{i=0}^{\kappa_\ell-1} H_{\ell,i}(\varepsilon) r_\ell^{(i)}. \tag{9.3.9}$$

2. *Full Order Measurement Feedback Case.* One only needs to replace Step 9.F.1 of the algorithm in the previous section with Item 1 above to obtain the desired $\boldsymbol{F}(\varepsilon)$. Steps 9.F.2 and 9.F.3 remain unchanged, and the full order measurement feedback controller is given by,

$$\begin{cases} \dot{v} = A_{\text{cmp}} v - K(\varepsilon) y + \displaystyle\sum_{i=0}^{\kappa_1-1} B H_{1,i}(\varepsilon) r_1^{(i)} + \cdots + \sum_{i=0}^{\kappa_\ell-1} B H_{\ell,i}(\varepsilon) r_\ell^{(i)} \\ u = F(\varepsilon)\, v + \displaystyle\sum_{i=0}^{\kappa_1-1} H_{1,i}(\varepsilon) r_1^{(i)} + \cdots + \sum_{i=0}^{\kappa_\ell-1} H_{\ell,i}(\varepsilon) r_\ell^{(i)}, \end{cases} \tag{9.3.10}$$

where $A_{\text{cmp}} = A + BF(\varepsilon) + K(\varepsilon) C_1$.

3. *Reduced Order Measurement Feedback Case.* Similarly, one again needs only to replace Step 9.R.1 in the algorithm of the previous section with Item 1 above. Steps 9.R.2 and 9.R.3 remain the same, and the reduced order measurement feedback control is given in the form of (9.3.3) with parameterized gain matrices $A_{\text{cmp}}(\varepsilon)$, $B_{\text{cmp}}(\varepsilon)$, $C_{\text{cmp}}(\varepsilon)$, $D_{\text{cmp}}(\varepsilon)$, being

given as in (9.2.70), $H_{j,i}(\varepsilon)$, $j = 1, 2, \cdots, \ell$ and $i = 0, 1, \cdots, \kappa_j - 1$, being given as in (9.3.8), and

$$G_{j,i}(\varepsilon) = (B_2 + K_{\mathrm{R}1} B_1) H_{j,i}(\varepsilon), \tag{9.3.11}$$

$j = 1, 2, \cdots, \ell$ and $i = 0, 1, \cdots, \kappa_j - 1$.

This completes the proof of Theorem 9.3.1. ⊠

Next, we note that in general, it requires infinite gain to achieve the robust and perfect tracking performance. In practical situations, one would have to make some trade-offs between the tracking performance and other requirements in order to design a physically implementable control law. This can be done by adjusting the tuning parameter ε.

Finally, we present a numerical example to illustrate the results of the general robust and perfect tracking design. The plant considered have two controlled outputs. We are going to design a GRPT controller such that when it is applied to the given plant, the first controlled output will robustly and almost perfectly track a ramp signal, while the second one will robustly and almost perfectly track a sinusoidal function.

Example 9.3.1. Consider a linear system given in the form of (9.1.1) with

$$A = \begin{bmatrix} 0 & 1 & 0 \\ 1 & 0 & 1 \\ 0 & 1 & 0 \end{bmatrix}, \quad B = \begin{bmatrix} 1 & 0 \\ 0 & 0 \\ 0 & 1 \end{bmatrix}, \quad E = \begin{bmatrix} 1 & 1 \\ 0 & 1 \\ 1 & 0 \end{bmatrix}, \quad x_0 = \begin{pmatrix} 1 \\ 0 \\ 1 \end{pmatrix}, \tag{9.3.12}$$

and

$$C_2 = \begin{bmatrix} 1 & 1 & 0 \\ 0 & 1 & 0 \end{bmatrix}, \quad D_2 = \begin{bmatrix} 1 & 0 \\ 0 & 0 \end{bmatrix}, \quad D_{22} = 0. \tag{9.3.13}$$

For easy verification, we assume that the external disturbance w is given by

$$w = \begin{bmatrix} 1 \\ \sin(\pi t) \end{bmatrix} \cdot 1(t) \in L_\infty. \tag{9.3.14}$$

Let the reference input be given as,

$$r = \begin{pmatrix} r_1 \\ r_2 \end{pmatrix} = \begin{bmatrix} t \\ \cos(2t) \end{bmatrix} \cdot 1(t). \tag{9.3.15}$$

We note that $\dot{r}_1 = 1(t) \in L_\infty$. Thus, we can achieve the GRPT for the above system and reference without using additional information $\dot{r}_1$.

A. State Feedback Case. We first consider the case when all the state variables of the given system are measurable, i.e., $C_1 = I$ and $D_1 = 0$. It is simple

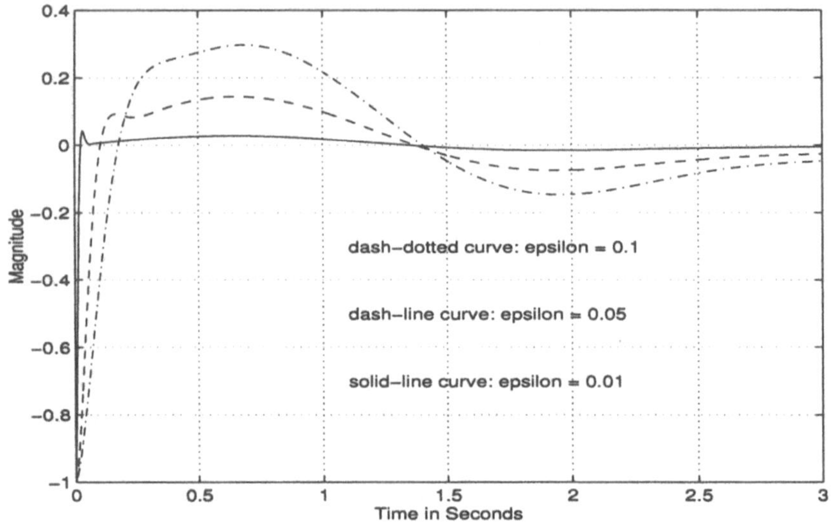

Figure 9.3.1: Tracking error e_2 under state feedback.

to verify that the subsystem Σ_P is invertible and of minimum phase with one invariant zero at $s = -1$ and two infinite zeros of orders 0 and 2, respectively. Hence, the general robust and perfect problem for the system with the given reference is solvable. Following the constructive algorithm for the state feedback case, we obtain a parameterized control law,

$$u = \begin{bmatrix} -1 & -1 & 0 \\ 1-\frac{2}{\varepsilon} & -1-\frac{2}{\varepsilon^2} & -\frac{2}{\varepsilon} \end{bmatrix} x + \begin{bmatrix} 1 & 0 \\ -1 & \frac{2}{\varepsilon^2} \end{bmatrix} r. \tag{9.3.16}$$

The poles of the closed-loop system comprising the given plant and the above control law are located at -1, $-1/\varepsilon \pm j/\varepsilon$. Hence, the closed-loop system is stable for any positive ε. Figure 9.3.1 shows the responses of the error signal $e_2(t) = h_2(t) - r_2(t)$, corresponding to $\varepsilon = 0.1, 0.05$ and 0.01, respectively. Note that $e_1(t) = h_1(t) - r_1(t) \equiv 0$ for all $t \geq 0$. The results clearly show that the general robust and perfect tracking is achieved.

We next consider the robust and perfect tracking with measurement feedback. Let the measurement output $y = C_1 x + D_1 w$ with

$$C_1 = \begin{bmatrix} 1 & 0 & 0 \\ 0 & 1 & 0 \end{bmatrix}, \quad D_1 = \begin{bmatrix} 0 & 0 \\ 0 & 0 \end{bmatrix}. \tag{9.3.17}$$

It is simple to see that $\mathrm{Ker}\,(C_2) = \mathrm{Ker}\,(C_1) = C_1^{-1}\{\mathrm{Im}\,(D_1)\}$ and hence the general robust and perfect tracking problem is solvable via measurement feedback laws. It is interesting to note that the subsystem Σ_Q, i.e., the quadruple

(A, E, C_1, D_1) is of nonminimum phase with an invariant zero at $s = 1$. Thus, the subsystem Σ_Q is not necessary to be of minimum phase and/or left invertible for the solvability of the robust and perfect tracking problem via measurement output feedback, as one would expect from the well-known separation principle arguments.

B.1. Full Order Measurement Feedback Case. Following the constructive algorithm, we obtain the following full order measurement feedback control law,

$$\begin{cases} \dot{v} = \begin{bmatrix} -1-\frac{1}{\varepsilon} & -1 & 0 \\ 0 & -\frac{1}{\varepsilon} & 1 \\ 1-\frac{2}{\varepsilon} & -1-\frac{1}{\varepsilon}-\frac{2}{\varepsilon^2} & -\frac{2}{\varepsilon} \end{bmatrix} v + \begin{bmatrix} 1 & 0 \\ 0 & 0 \\ -1 & \frac{2}{\varepsilon^2} \end{bmatrix} r + \begin{bmatrix} \frac{1}{\varepsilon} & 1 \\ 1 & \frac{1}{\varepsilon} \\ 0 & 1+\frac{1}{\varepsilon} \end{bmatrix} y, \\ u = \begin{bmatrix} -1 & -1 & 0 \\ 1-\frac{2}{\varepsilon} & -1-\frac{2}{\varepsilon^2} & -\frac{2}{\varepsilon} \end{bmatrix} v + \begin{bmatrix} 1 & 0 \\ -1 & \frac{2}{\varepsilon^2} \end{bmatrix} r. \end{cases} \tag{9.3.18}$$

It is straightforward to verify that the closed-loop poles are asymptotically located at -1, $-1/\varepsilon \pm j/\varepsilon$, -1, $-1/\varepsilon$ and $-1/\varepsilon$. Hence, the closed-loop system comprising the given plant and the full order measurement feedback control law is asymptotically stable for all $\varepsilon > 0$. Figure 9.3.2 shows the resulting tracking errors under the full order measurement feedback control law with $\varepsilon = 0.05$, 0.01 and 0.001, respectively. Again, it is clear that the general robust and perfect tracking is achieved.

B.2. Reduced Order Measurement Feedback Case. Again, following our constructive algorithm for the reduced order measurement feedback case, we obtain the following first order dynamic controller,

$$\begin{cases} \dot{v} = \left(-1-\frac{2}{\varepsilon}\right) v + \begin{bmatrix} -\frac{2}{\varepsilon} & -1-\frac{2}{\varepsilon^2}-\frac{2}{\varepsilon} \end{bmatrix} y + \begin{bmatrix} -1 & \frac{2}{\varepsilon^2} \end{bmatrix} r, \\ u = \begin{bmatrix} 0 \\ -\frac{2}{\varepsilon} \end{bmatrix} v + \begin{bmatrix} -1 & -1 \\ 1-\frac{2}{\varepsilon} & -1-\frac{2}{\varepsilon^2}-\frac{2}{\varepsilon} \end{bmatrix} y + \begin{bmatrix} 1 & 0 \\ -1 & \frac{2}{\varepsilon^2} \end{bmatrix} r. \end{cases} \tag{9.3.19}$$

The poles of the closed-loop system comprising the given plant and the above control law are precisely placed at -1, $-1/\varepsilon \pm j/\varepsilon$ and -1. Hence, the closed-loop system is asymptotically stable for all $\varepsilon > 0$. Figure 9.3.3 shows the responses of the error signal $e_2(t)$ with $\varepsilon = 0.1$, 0.05 and 0.01, respectively. As in the state feedback case, the resulting $e_1(t)$ under the reduced order measurement feedback law is identically zero for all $t \geq 0$. The results again clearly show that the general robust and perfect tracking is achieved. ◻E

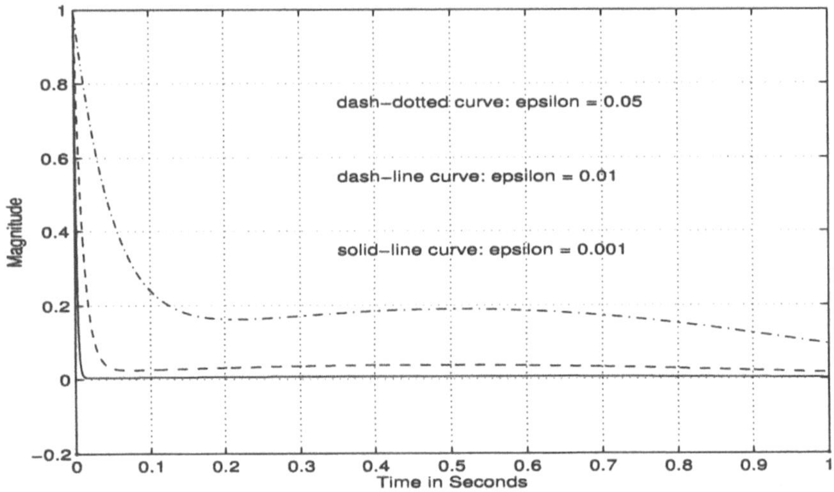

(a) Error signal e_1.

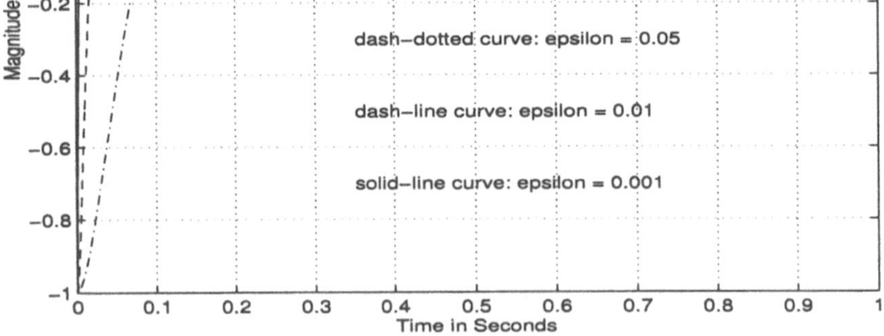

(b) Error signal e_2.

Figure 9.3.2: Tracking errors under full order measurement feedback.

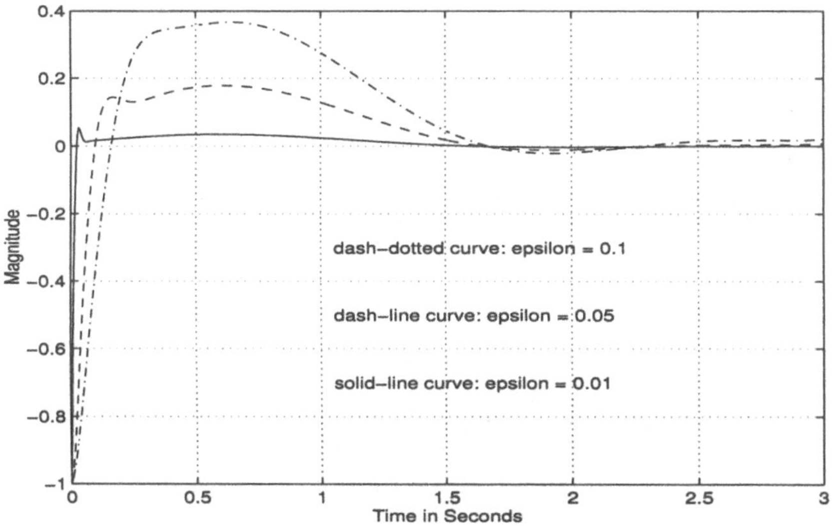

Figure 9.3.3: Tracking error e_2 under reduced order measurement feedback.

9.4. Proofs of Main Results

9.4.A. Proof of Theorem 9.2.2

It was mentioned in the constructive algorithm of Subsection 9.2.1 that, following the structural algorithms of Sannuti and Saberi [116], and Saberi and Sannuti [111], one can transform the system (9.2.28) into the special coordinate basis as given in the compact form of (9.2.32) to (9.2.34). That is there exist nonsingular state, input and output transformation Γ_s, Γ_i and Γ_o such that

$$\begin{pmatrix} \mathbf{r} \\ x \end{pmatrix} = \Gamma_s \begin{pmatrix} \mathbf{r} \\ x_a^- \\ x_c \\ x_d \end{pmatrix} = \begin{bmatrix} I_{\kappa\times\ell} & 0 \\ \star & \tilde{\Gamma}_s \end{bmatrix} \begin{pmatrix} \mathbf{r} \\ x_a^- \\ x_c \\ x_d \end{pmatrix}, \tag{9.4.1}$$

$$e = \Gamma_o \begin{pmatrix} e_0 \\ e_d \end{pmatrix}, \quad u = \Gamma_i \begin{pmatrix} u_0 \\ u_d \\ u_c \end{pmatrix}, \tag{9.4.2}$$

$$\mathbf{r} = \begin{pmatrix} r \\ \dot{r} \\ \vdots \\ r^{(\kappa-1)} \end{pmatrix}, \quad x_d = \begin{pmatrix} x_1 \\ x_2 \\ \vdots \\ x_{m_d} \end{pmatrix}, \quad x_i = \begin{pmatrix} x_{i1} \\ x_{i2} \\ \vdots \\ x_{iq_i} \end{pmatrix}, \tag{9.4.3}$$

$$e_d = \begin{pmatrix} h_1 \\ h_2 \\ \vdots \\ h_{m_d} \end{pmatrix}, \quad u_d = \begin{pmatrix} u_1 \\ u_2 \\ \vdots \\ u_{m_d} \end{pmatrix}, \tag{9.4.4}$$

and

$$\dot{r} = \begin{bmatrix} 0 & I_\ell & \cdots & 0 \\ \vdots & \vdots & \ddots & \vdots \\ 0 & 0 & \cdots & I_\ell \\ 0 & 0 & \cdots & 0 \end{bmatrix} r + \begin{bmatrix} 0 \\ \vdots \\ 0 \\ I_\ell \end{bmatrix} r^{(\kappa)}, \tag{9.4.5}$$

$$\dot{x}_a^- = A_{aa}^- x_a^- + L_{ad}^- e_d + B_{0a}^- e_0 + E_a^- w + G_a^- r^{(\kappa)}, \tag{9.4.6}$$

$$\dot{x}_c = A_{cc} x_c + L_{cd} e_d + B_{0c} e_0 + B_c \Big[u_c + E_{ca}^0 r + E_{ca}^- x_a^- \Big] + E_c w + G_c r^{(\kappa)}, \tag{9.4.7}$$

$$e_0 = C_{0a}^0 r + C_{0a}^- x_a^- + C_{0c} x_c + C_{0d} x_d + u_0, \tag{9.4.8}$$

and for each $i = 1, \cdots, m_d$,

$$\dot{x}_i = A_{q_i} x_i + L_{i0} e_0 + L_{id} e_d + B_{q_i} \left[u_i + E_{ia}^0 r + E_{ia}^- x_a^- + E_{ic} x_c + \sum_{j=1}^{m_d} E_{ij} x_j \right] + E_i w + G_i r^{(\kappa)}, \tag{9.4.9}$$

$$e_i = C_{q_i} x_i = x_{i1}, \quad e_d = C_d x_d. \tag{9.4.10}$$

Now, it is straightforward to see that if $r^{(\kappa)}$ is a vector of delta functions, then the terms $G_a^- r^{(\kappa)}$, $G_c r^{(\kappa)}$ and $G_i r^{(\kappa)}$ can be treated as some additional initial conditions added to the original ones of the states variables, x_a^0, x_c and x_d, respectively. If $r^{(\kappa)}$ is in L_p, $p \in [1, \infty)$, it can be treated as an additional disturbance and can be merged with the original disturbance w. Thus, in both cases, we can write (9.4.6), (9.4.7) and (9.4.9) as

$$\dot{x}_a^- = A_{aa}^- x_a^- + L_{ad}^- e_d + B_{0a}^- e_0 + \bar{E}_a^- \bar{w}, \tag{9.4.11}$$

$$\dot{x}_c = A_{cc} x_c + L_{cd} e_d + B_{0c} e_0 + B_c \Big[u_c + E_{ca}^0 r + E_{ca}^- x_a^- \Big] + \bar{E}_c \bar{w}, \tag{9.4.12}$$

and

$$\dot{x}_i = A_{q_i} x_i + L_{i0} e_0 + L_{id} e_d + B_{q_i} \left[u_i + E_{ia}^0 r + E_{ia}^- x_a^- + E_{ic} x_c + \sum_{j=1}^{m_d} E_{ij} x_j \right] + \bar{E}_i \bar{w}, \tag{9.4.13}$$

with $\bar{w} \in L_p$, $p \in [1, \infty)$, and $\bar{E}_a^-$, $\bar{E}_c$ and $\bar{E}_i$ being some appropriate constant matrices, and with a new but again bounded initial condition, say $\bar{x}_0$.

Next, we note that the control law $u = \boldsymbol{F}\boldsymbol{x}$ with the gain matrix $\boldsymbol{F}$ given in (9.2.38) can be rewritten as,

$$u_0 = -C_{0a}^0 \boldsymbol{r} - C_{0a}^- x_a^- - C_{0c} x_c - C_{0d} x_d, \tag{9.4.14}$$

$$u_c = -F_c x_c - E_{ca}^0 \boldsymbol{r} - E_{ca}^- x_a^-, \tag{9.4.15}$$

and

$$u_i = -E_{ia}^0 \boldsymbol{r} - E_{ia}^- x_a^- - E_{ic} x_c - \sum_{j=1}^{m_d} E_{ij} x_j - \frac{F_i}{\varepsilon^{q_i}} S_i(\varepsilon) \bar{x}_i. \tag{9.4.16}$$

Hence, the closed-loop system comprising the given system and the above control law can be expressed as follows,

$$e_0 = 0, \tag{9.4.17}$$

$$\dot{x}_a^- = A_{aa}^- x_a^- + L_{ad}^- e_d + \bar{E}_a^- \bar{w}, \tag{9.4.18}$$

$$\dot{x}_c = (A_{cc} - B_c F_c) x_c + L_{cd} e_d + \bar{E}_c \bar{w} = A_{cc}^c x_c + L_{cd} e_d + \bar{E}_c \bar{w}, \tag{9.4.19}$$

$$\dot{x}_i = A_{q_i} x_i - B_{qi} \frac{F_i}{\varepsilon^{q_i}} S_i(\varepsilon) x_i + L_{id} e_d + \bar{E}_i \bar{w}, \;\; e_i = C_{q_i} x_i. \tag{9.4.20}$$

Let us define a new state transformation as,

$$\tilde{x}_a^- := x_a^-, \;\; \tilde{x}_c := x_c, \;\; \tilde{x}_d := \begin{pmatrix} \tilde{x}_1 \\ \vdots \\ \tilde{x}_{m_d} \end{pmatrix}, \;\; \tilde{x}_i := S_i(\varepsilon) x_i, \; i = 1, \cdots, m_d. \tag{9.4.21}$$

Then, we have $e_0 = 0$, and

$$\dot{\tilde{x}}_a^- = A_{aa}^- \tilde{x}_a^- + L_{ad}^- \tilde{e}_d + \bar{E}_a^- \bar{w}, \tag{9.4.22}$$

$$\dot{\tilde{x}}_c = A_{cc}^c \tilde{x}_c + L_{cd} \tilde{e}_d + \bar{E}_c \bar{w}, \tag{9.4.23}$$

$$\varepsilon \dot{\tilde{x}}_i = (A_{q_i} - B_{q_i} F_i) \tilde{x}_i + \varepsilon \tilde{L}_{id}(\varepsilon) \tilde{e}_d + \varepsilon \tilde{E}_i(\varepsilon) \bar{w}, \tag{9.4.24}$$

$$\tilde{e}_i = e_i = C_{q_i} \tilde{x}_i, \;\; \tilde{e}_d = e_d = C_d \tilde{x}_d, \tag{9.4.25}$$

$$\tilde{L}_{id}(\varepsilon) = S_i(\varepsilon) L_{id}, \;\; \tilde{E}_i(\varepsilon) = S_i(\varepsilon) \bar{E}_i. \tag{9.4.26}$$

It is simple to show that, for $\varepsilon \in (0, 1]$,

$$|\tilde{L}_{id}(\epsilon)| \le \tilde{l}_d, \;\; |\tilde{E}_i(\epsilon)| \le \theta_i, \; i = 1, \cdots, m_d \tag{9.4.27}$$

for some positive constant $\tilde{l}_d$ and θ_i, which are independent of ε.

We next construct a Lyapunov function for the closed loop system (9.4.22) to (9.4.24). We do this by composing Lyapunov functions for the subsystems. For the subsystem of $\tilde{x}_a^-$, we choose a Lyapunov function,

$$V_a^-(\tilde{x}_a^-) = (\tilde{x}_a^-)' P_a^- \tilde{x}_a^-, \tag{9.4.28}$$

where $P_a^- > 0$ is the unique solution to the Lyapunov equation,

$$(A_{aa}^-)' P_a^- + P_a^- A_{aa}^- = -I, \tag{9.4.29}$$

and for the subsystem of $\tilde{x}_c$, we choose a Lyapunov function,

$$V_c(\tilde{x}_c) = \tilde{x}_c' P_c \tilde{x}_c, \tag{9.4.30}$$

where $P_c > 0$ is the unique solution to the Lyapunov equation,

$$(A_{cc}^c)' P_c + P_c A_{cc}^c = -I. \tag{9.4.31}$$

Finally, for the subsystem of $\tilde{x}_d$, we choose a Lyapunov function

$$V_d(\tilde{x}_d) = \sum_{i=1}^{m_d} \tilde{x}_i' P_i \tilde{x}_i, \tag{9.4.32}$$

where P_i is the unique solution to the Lyapunov equation,

$$(A_{q_i} - B_{q_i} F_i)' P_i + P_i (A_{q_i} - B_{q_i} F_i) = -I. \tag{9.4.33}$$

Since $A_{q_i} - B_{q_i} F_i$ is asymptotically stable, the existence of P_i is guaranteed. We now choose a Lyapunov function for the closed-loop system (9.4.22) to (9.4.24) as follows,

$$V(\tilde{x}_a^-, \tilde{x}_c, \tilde{x}_d) = V_a^-(\tilde{x}_a^-) + V_c(\tilde{x}_c) + \alpha_d V_d(\tilde{x}_d), \tag{9.4.34}$$

where the value of α_d is to be determined. The derivative of V along the trajectory of the closed-loop system (9.4.22) to (9.4.24) can be evaluated as follows,

$$\begin{aligned} \dot{V} = & -(\tilde{x}_a^-)'\tilde{x}_a^- + 2(\tilde{x}_a^-)' P_a^- [L_{ad}^- \tilde{e}_d + \bar{E}_a^- \bar{w}] - \tilde{x}_c' \tilde{x}_c + 2\tilde{x}_c' P_c [L_{cd} \tilde{e}_d + \bar{E}_c \bar{w}] \\ & + \alpha_d \sum_{i=1}^{m_d} \left[-\frac{1}{\varepsilon} \tilde{x}_i' \tilde{x}_i + 2\tilde{x}_i' P_i \tilde{L}_{id}(\varepsilon) \tilde{e}_d + 2\tilde{x}_i' P_i \tilde{E}_i(\varepsilon) \bar{w} \right]. \end{aligned} \tag{9.4.35}$$

It is straightforward to see now that there exist an $\alpha_d > 0$ and $\varepsilon^* \in (0,1]$ such that for all $\varepsilon \in (0, \varepsilon^*]$,

$$\dot{V} \le -\frac{1}{2}|\tilde{x}_a|^2 - \frac{1}{2}|\tilde{x}_c|^2 - \frac{1}{2\varepsilon}|\tilde{x}_d|^2 + \alpha_1 |\bar{w}|^2, \tag{9.4.36}$$

for some positive constant α_1, independent of ε. Thus, the closed-loop system in the absence of disturbance w and reference input r is asymptotically stable.

It remains to show that the resulting tracking error e, which is a function of ε, has the following property,

$$J_p(x_0, w, r, \varepsilon) = \|e\|_p \to 0, \text{ as } \varepsilon \to 0. \tag{9.4.37}$$

We first assume that the disturbance $\bar{w}$ is nonexistent. It follows from (9.4.36) that

$$\dot{V} \leq -\alpha_2 V, \tag{9.4.38}$$

for some positive scalar α_2, independent of ε. Noting the transformation of (9.4.21), we have

$$|\tilde{x}(0)| \leq \alpha_0 |\bar{x}_0|, \tag{9.4.39}$$

for some positive $\alpha_0 > 0$, independent of ε, where $\bar{x}_0$ is the combination of the initial condition of the original system, i.e., x_0, and the additional ones introduced by $r^{(\kappa)}$. Thus,

$$|V(0)| \leq \alpha_3 |\bar{x}_0|^2, \tag{9.4.40}$$

where $\alpha_3 > 0$ and is independent of ε. By the standard comparison theorem, it follows from (9.4.38) that,

$$V \leq V(0) e^{-\alpha_2 t}, \tag{9.4.41}$$

which together with (9.4.40) imply that

$$V \leq \alpha_3 e^{-\alpha_2 t} |\bar{x}_0|^2, \tag{9.4.42}$$

and thus,

$$|\tilde{x}_d(t)| \leq \alpha_4 e^{-\alpha_2 t} |\bar{x}_0| \quad \text{and} \quad |\tilde{e}_d(t)| \leq \alpha_5 e^{-\alpha_2 t} |\bar{x}_0|, \tag{9.4.43}$$

for some positive scalars α_4 and α_5, independent of ε. Now viewing $\tilde{e}_d$ as an input to the subsystem $\tilde{x}_i$ of (9.4.24), one can show that

$$|\tilde{x}_d(t)| \leq \left(\alpha_6 e^{-\alpha_8 t/\varepsilon} + \alpha_7 \varepsilon e^{-\alpha_2 t}\right) |\bar{x}_0|, \tag{9.4.44}$$

and

$$|\tilde{e}_d(t)| \leq \beta_1 \left(\alpha_6 e^{-\alpha_8 t/\varepsilon} + \varepsilon e^{-\alpha_2 t}\right) |\bar{x}_0|, \tag{9.4.45}$$

for some positive scalars α_6, α_7, α_8 and β_1, which are all independent of ε. Noting that

$$e = \Gamma_o \begin{pmatrix} e_0 \\ e_d \end{pmatrix}, \tag{9.4.46}$$

where $e_0 = 0$ and $e_d = \tilde{e}_d$, we then have

$$|e| \le |\Gamma_o|\beta_1(e^{-\alpha_8 t/\varepsilon} + \varepsilon e^{-\alpha_2 t})|\bar{x}_0| = \beta_2(e^{-\alpha_8 t/\varepsilon} + \varepsilon e^{-\alpha_2 t})|\bar{x}_0|. \tag{9.4.47}$$

Thus, for all $1 \le p < \infty$, we have, as $\varepsilon \to 0$,

$$\|e\|_p \le \left(\int_0^\infty \left[\beta_2 m_d \left(e^{-\alpha_8 t/\varepsilon} + \varepsilon e^{-\alpha_2 t}\right)|\bar{x}_0|\right]^p dt\right)^{1/p} \to 0. \tag{9.4.48}$$

Next, we take into consideration the disturbance $\bar{w} \in L_p$, $p \in [1,\infty)$, but with $\bar{x}_0 = 0$. Noting that $\tilde{e}_d$ in (9.4.24) is a part of the state variables of the system and $\varepsilon \tilde{L}_{id}(\varepsilon)$ is negligible compared to $A_{q_i} - B_{q_i}F_i$ for sufficiently small ε, the subsystem (9.4.24) can then be approximated as,

$$\dot{\tilde{x}}_i = \frac{1}{\varepsilon}(A_{q_i} - B_{q_i}F_i)\tilde{x}_i + \tilde{E}_i(\varepsilon)\bar{w}, \tag{9.4.49}$$

where $\bar{w} \in L_p$. Thus, we have

$$\begin{aligned} |e_i| = |\tilde{e}_i| &\le \int_0^t \left| C_{q_i}\exp\left[-\frac{1}{\varepsilon}(A_{q_i} - B_{q_i}F_i)\tau\right] \tilde{E}_i(\varepsilon)\bar{w}(t-\tau)\right| d\tau \\ &\le \beta_3 \int_0^\infty e^{-\beta_4\tau/\varepsilon}|\bar{w}(t-\tau)|d\tau, \end{aligned} \tag{9.4.50}$$

for some positive scalars β_3 and β_4, independent of ε. The result for $p = 1$ is obvious. We proceed to show the case when $1 < p < \infty$. Using the well-known Hölder Inequality, i.e.,

$$\|fg\|_1 \le \|f\|_p \cdot \|g\|_{p^*}, \quad 1/p + 1/p^* = 1, \tag{9.4.51}$$

we have

$$\begin{aligned} |e_i| = |\tilde{e}_i| &\le \beta_3 \int_0^\infty \left[\left(e^{-\beta_4\tau/\varepsilon}\right)^{1/p} |\bar{w}(t-\tau)|\right] \left(e^{-\beta_4\tau/\varepsilon}\right)^{1/p^*} d\tau \\ &\le \beta_3 \left[\int_0^\infty e^{-\beta_4\tau/\varepsilon}|\bar{w}(t-\tau)|^p d\tau\right]^{1/p} \left[\int_0^\infty e^{-\beta_4\tau/\varepsilon} d\tau\right]^{1/p^*} \\ &= \beta_3 \left(\frac{\varepsilon}{\beta_4}\right)^{1/p^*} \left[\int_0^\infty e^{-\beta_4\tau/\varepsilon}|\bar{w}(t-\tau)|^p d\tau\right]^{1/p}. \end{aligned} \tag{9.4.52}$$

Thus,

$$\begin{aligned} \|e_i\|_p^p &\le \beta_3^p \left(\frac{\varepsilon}{\beta_4}\right)^{p/p^*} \int_0^\infty \left[\int_0^\infty e^{-\beta_4\tau/\varepsilon}|\bar{w}(t-\tau)|^p d\tau\right] dt \\ &= \beta_3^p \left(\frac{\varepsilon}{\beta_4}\right)^{p/p^*} \int_0^\infty e^{-\beta_4\tau/\varepsilon} \left[\int_0^\infty |\bar{w}(t-\tau)|^p dt\right] d\tau \end{aligned} \tag{9.4.53}$$

$$= \beta_3^p \left(\frac{\varepsilon}{\beta_4}\right)^{p/p^*} \int_0^\infty e^{-\beta_4 \tau/\varepsilon} \left[\int_0^\infty |\bar{w}(t)|^p dt\right] d\tau \tag{9.4.54}$$

$$= \beta_3^p \left(\frac{\varepsilon}{\beta_4}\right)^{p/p^*} \|\bar{w}\|_p^p \int_0^\infty e^{-\beta_4 \tau/\varepsilon} d\tau$$

$$= \beta_3^p \left(\frac{\varepsilon}{\beta_4}\right)^{1+p/p^*} \|\bar{w}\|_p^p. \tag{9.4.55}$$

Note that we have used the property $\bar{w}(t) = 0$, $t < 0$, to get (9.4.54) from (9.4.53). We would also like to note that the above proof from (9.4.52) to (9.4.55) was inspired by similar arguments reported in Desoer and Vidyasagar [46]. It is now clear,

$$\|e_i\|_p \le \beta_3 \left(\frac{\varepsilon}{\beta_4}\right)^{1/p+1/p^*} \|w\|_p = \left(\frac{\beta_3}{\beta_4}\right) \varepsilon \|\bar{w}\|_p \to 0, \tag{9.4.56}$$

as $\varepsilon \to 0$. In view of (9.4.48) and (9.4.56), the robust and perfect tracking problem is then solved. This completes the proof of Theorem 9.2.2. ⊠

9.4.B. Proof of Theorem 9.2.3

First, let us define a new state variable,

$$x_v = x - v. \tag{9.4.57}$$

Then, it is straightforward to verify that the closed-loop system comprising the given system (9.1.1) and the full order measurement feedback control law of (9.2.57) can be rewritten as follows,

$$\dot{x}_v = \left[A + K(\varepsilon)C_1\right] x_v + \left[E + K(\varepsilon)D_1\right] w, \tag{9.4.58}$$

$$\dot{x} = \left[A+BF(\varepsilon)\right] x - BF(\varepsilon)x_v + BH_0(\varepsilon)r + \cdots + BH_{\kappa-1}(\varepsilon)r^{(\kappa-1)} + Ew, \tag{9.4.59}$$

and

$$h = \left[C_2+D_2F(\varepsilon)\right] x - D_2F(\varepsilon)x_v + D_2H_0(\varepsilon)r + \cdots + D_2H_{\kappa-1}(\varepsilon)r^{(\kappa-1)}. \tag{9.4.60}$$

It is simple to see now the eigenvalues of the closed-loop system are given by $\lambda\{A + BF(\varepsilon)\}$, which have been shown to be in $\mathbb{C}^-$ in Theorem 9.2.2, and $\lambda\{A + K(\varepsilon)C_1\}$, which are equivalent to

$$\lambda\left\{\begin{bmatrix} A_{ccQ} - K_{c0Q}C_{0cQ} & -K_{cdQ}/\varepsilon \\ E_{dcQ} & -I_k/\varepsilon \end{bmatrix}\right\} \to \lambda\left(A_{ccQ}^c\right) \cup \left\{-\frac{1}{\varepsilon}, \cdots, -\frac{1}{\varepsilon}\right\}, \tag{9.4.61}$$

as $\varepsilon \to 0$. Thus, the closed-loop system is asymptotically stable for sufficiently small ε, when the external disturbance $w = 0$ and reference $r = 0$.

Next, we intend to investigate the properties of x_v in the subsystem (9.4.58). Let us transform the subsystem (9.2.50) into the special coordinate basis of Theorem 2.4.1 with nonsingular state, input and output transformations Γ_{sQ}, Γ_{iQ} and Γ_{oQ}, as given in Step 9.F.2 of Subsection 9.2.2. Also, let

$$x_v = \Gamma_{sQ} \begin{pmatrix} x_{cQ} \\ x_{dQ} \end{pmatrix}. \tag{9.4.62}$$

Then, we can rewrite (9.4.58) as,

$$\dot{x}_{cQ} = \Big(A_{ccQ} - K_{c0Q} C_{0cQ}\Big) x_{cQ} - \frac{K_{cdQ}}{\varepsilon} x_{dQ} + E_{cQ} w, \tag{9.4.63}$$

and

$$\dot{x}_{dQ} = -\frac{1}{\varepsilon} x_{dQ} + E_{dcQ} x_{cQ} + E_{dQ} w, \tag{9.4.64}$$

for some appropriate dimensional matrices E_{cQ} and E_{dQ}, independent of ε. Now, let

$$\tilde{x}_{cQ} = x_{cQ} - K_{cdQ} x_{dQ}. \tag{9.4.65}$$

Thus, (9.4.63) and (9.4.64) can be rewritten as,

$$\dot{\tilde{x}}_{cQ} = A^c_{ccQ} \tilde{x}_{cQ} + A^c_{ccQ} K_{cdQ} x_{dQ} + (E_{cQ} - K_{cdQ} E_{dQ}) w, \tag{9.4.66}$$

and

$$\dot{x}_{dQ} = \left(-\frac{1}{\varepsilon} I + E_{dcQ} K_{cdQ}\right) x_{dQ} + E_{dcQ} \tilde{x}_{cQ} + E_{dQ} w, \tag{9.4.67}$$

It is clear to see that as $\varepsilon \to 0$, the poles of the above system are asymptotically given by $\lambda(A^c_{ccQ})$ and k repeated ones at $-1/\varepsilon$. This confirms with what we have claimed earlier in (9.4.61). Following similar arguments as in (9.4.37) to (9.4.56), we can show that for any bounded initial condition and for $w \in L_p$, $p \in [1, \infty)$,

$$\|\tilde{x}_{cQ}\|_p \le \beta_c \|w\|_p \quad \text{and} \quad \|x_{dQ}\|_p \le \beta_d \varepsilon \|w\|_p, \tag{9.4.68}$$

for some positive scalars β_c and β_d, independent of ε. Thus, there exists a scalar β_v, independent of ε, such that

$$\|x_v\|_p \le \beta_v \|w\|_p. \tag{9.4.69}$$

Following (9.2.53), it is simple to verify that

$$C_1^{-1}\{\mathrm{Im}\,(D_1)\} = \mathrm{Ker}\left(\Gamma_{oQ} \begin{bmatrix} 0 & 0 \\ 0 & I_k \end{bmatrix} \Gamma_{sQ}^{-1}\right) = \mathrm{Ker}\left(\begin{bmatrix} 0 & 0 \\ 0 & I_k \end{bmatrix} \Gamma_{sQ}^{-1}\right), \tag{9.4.70}$$

and

$$\left(\begin{bmatrix} 0 & 0 \\ 0 & I_k \end{bmatrix} \Gamma_{sQ}^{-1}\right) x_v = \begin{bmatrix} 0 & 0 \\ 0 & I_k \end{bmatrix} \begin{pmatrix} x_{cQ} \\ x_{dQ} \end{pmatrix} = \begin{pmatrix} 0 \\ x_{dQ} \end{pmatrix}. \tag{9.4.71}$$

Thus, the last condition of Theorem 9.2.1, i.e., $\operatorname{Ker}(C_2) \supset C_1^{-1}\{\operatorname{Im}(D_1)\}$, implies that

$$C_2 x_v = M x_{dQ} \quad \text{and} \quad \|C_2 x_v\|_p \le \beta_m \varepsilon \|w\|_p, \tag{9.4.72}$$

for some appropriate constant matrix M and positive scalar β_m, independent of ε. In fact, for any appropriate matrix N with $\operatorname{Ker}(N) \supset \operatorname{Ker}(C_2)$, we have

$$\|N x_v\|_p \le |N| \cdot \beta_n \cdot \varepsilon \cdot \|w\|_p, \tag{9.4.73}$$

for some positive scalar β_n (independent of ε).

We are now ready to show that the full order measurement feedback control law of (9.2.57) solves the RPT problem. It is straightforward to verify that (9.4.59) and (9.4.60) can be rewritten as

$$\begin{cases} \dot{\boldsymbol{x}} = \ (\boldsymbol{A}+\boldsymbol{B}\boldsymbol{F}) \ \boldsymbol{x} - \ \boldsymbol{B}\boldsymbol{F}(\varepsilon) \ x_v + \boldsymbol{E} \ \boldsymbol{w} \\ e = (\boldsymbol{C}_2+\boldsymbol{D}_2\boldsymbol{F}) \ \boldsymbol{x} - \ \boldsymbol{D}_2\boldsymbol{F}(\varepsilon) \ x_v \end{cases} \tag{9.4.74}$$

where $\boldsymbol{A}$, $\boldsymbol{B}$, $\boldsymbol{E}$, $\boldsymbol{C}_2$ and $\boldsymbol{D}_2$ are as defined in (9.2.30) and (9.2.31). Without loss of any generality, we assume hereafter that the quadruple $(\boldsymbol{A}, \boldsymbol{B}, \boldsymbol{C}_2, \boldsymbol{D}_2)$ is in the form of the special coordinate Following the same procedures as in (9.4.1) to (9.4.20), we can transform (9.4.74) with some appropriate transformations into the following form,

$$\dot{\tilde{x}}_a^- = A_{aa}^- \tilde{x}_a^- + L_{ad}^- \tilde{e}_d + \bar{E}_a^- \bar{w} + N_a^- x_v, \tag{9.4.75}$$

$$\dot{\tilde{x}}_c = A_{cc}^c \tilde{x}_c + L_{cd} \tilde{e}_d + \bar{E}_c \bar{w} + N_c x_v, \tag{9.4.76}$$

$$\dot{x}_i = A_{q_i} x_i - B_{qi} \frac{F_i}{\varepsilon^{q_i}} S_i(\varepsilon) x_i + L_{id} e_d + \bar{E}_i \bar{w} + N_i x_v - \begin{bmatrix} 0 & 0 & B_{q_i} \frac{F_i}{\varepsilon^{q_i}} S_i(\varepsilon) \end{bmatrix} x_v, \tag{9.4.77}$$

$$e_0 = -\,[\, C_{0a}^- \quad C_{0c} \quad C_{0d} \,]\, x_v, \quad e_i = C_{q_i} x_i, \tag{9.4.78}$$

for some appropriate dimensional matrices N_a^-, N_c and N_i, which are all independent of ε. First, it is simple to see that

$$\operatorname{Ker}\Big(-\,[C_{0a} \quad C_{0c} \quad C_{od}]\Big) \supset \operatorname{Ker}(C_2). \tag{9.4.79}$$

In view of (9.4.73), we have

$$\|e_0\|_p \to 0, \quad \text{as } \varepsilon \to 0. \tag{9.4.80}$$

Next, let us define a new state transformation as in (9.4.21), i.e.,

$$\tilde{x}_a^- := x_a^-, \ \ \tilde{x}_c := x_c, \ \ \tilde{x}_d := \begin{pmatrix} \tilde{x}_1 \\ \vdots \\ \tilde{x}_{m_d} \end{pmatrix}, \ \ \tilde{x}_i := S_i(\varepsilon) x_i, \ i = 1, \cdots, m_d. \tag{9.4.81}$$

Then,

$$\dot{\tilde{x}}_a^- = A_{aa}^-\tilde{x}_a^- + L_{ad}^-\tilde{e}_d + [\,\bar{E}_a^- \quad N_a^-\,]\begin{pmatrix}\bar{w}\\ x_v\end{pmatrix}, \tag{9.4.82}$$

$$\dot{\tilde{x}}_c = A_{cc}^c\tilde{x}_c + L_{cd}\tilde{e}_d + [\,\bar{E}_c \quad N_c\,]\begin{pmatrix}\bar{w}\\ x_v\end{pmatrix}, \tag{9.4.83}$$

$$\dot{\tilde{x}}_i = \frac{1}{\varepsilon}(A_{q_i} - B_{q_i}F_i)\tilde{x}_i + \tilde{L}_{id}(\varepsilon)\tilde{e}_d + [\,\tilde{E}_i(\varepsilon) \quad \tilde{N}_i(\varepsilon)\,]\begin{pmatrix}\bar{w}\\ x_v\end{pmatrix} - [\,0 \quad 0 \quad \tilde{M}_i(\varepsilon)\,]x_v, \tag{9.4.84}$$

$$\tilde{e}_i = e_i = C_{q_i}\tilde{x}_i, \quad \tilde{e}_d = e_d = C_d\tilde{x}_d, \tag{9.4.85}$$

where

$$\tilde{M}_i(\varepsilon) = S_i(\varepsilon)B_{q_i}\frac{F_i}{\varepsilon^{q_i}}S_i(\varepsilon), \tag{9.4.86}$$

and

$$\tilde{L}_{id}(\varepsilon) = S_i(\varepsilon)L_{id}, \quad \tilde{N}_i(\varepsilon) = S_i(\varepsilon)N_i, \quad \tilde{E}_i(\varepsilon) = S_i(\varepsilon)\bar{E}_i. \tag{9.4.87}$$

It is clear that the 2-norms of $\tilde{L}_{id}(\varepsilon)$, $\tilde{N}_i(\varepsilon)$ and $\tilde{E}_i(\varepsilon)$ are all bounded, and in view of the special structure of B_{q_i}, $S_i(\varepsilon)$ and F_i of (9.2.37), we have

$$\begin{aligned}\tilde{M}_i(\varepsilon) &= \begin{bmatrix} 0 & 0 & \cdots & 0 \\ \vdots & \vdots & \ddots & \vdots \\ 0 & 0 & \cdots & 0 \\ \dfrac{F_{iq_i}}{\varepsilon} & F_{iq_i-1} & \cdots & \varepsilon^{q_i-2}F_{i1}\end{bmatrix} \\ &= \frac{F_{iq_i}}{\varepsilon}\begin{bmatrix} 0 \\ \vdots \\ 0 \\ C_{q_i}\end{bmatrix} + \hat{M}_i(\varepsilon) = \frac{F_{iq_i}}{\varepsilon}\hat{C}_{q_i} + \hat{M}_i(\varepsilon),\end{aligned} \tag{9.4.88}$$

where $|\hat{M}_i(\varepsilon)| \le \xi_i$ for some positive scalar ξ_i, independent of ε. Thus, (9.4.84) can be rewritten as,

$$\dot{\tilde{x}}_i = \frac{1}{\varepsilon}(A_{q_i} - B_{q_i}F_i)\tilde{x}_i + \tilde{L}_{id}(\varepsilon)\tilde{e}_d + [\,\tilde{E}_i(\varepsilon) \quad \hat{N}_i(\varepsilon)\,]\begin{pmatrix}\bar{w}\\ x_v\end{pmatrix} - \frac{F_{iq_i}}{\varepsilon}[\,0 \quad 0 \quad \hat{C}_{q_i}\,]x_v, \tag{9.4.89}$$

for some bounded $\hat{N}_i(\varepsilon)$. It is clear that

$$\operatorname{Ker}\Big([\,0 \quad 0 \quad \hat{C}_{q_i}\,]\Big) \supset \operatorname{Ker}(C_2). \tag{9.4.90}$$

In view of (9.4.73), we have

$$\left\|\frac{F_{iq_i}}{\varepsilon}[\,0 \quad 0 \quad \hat{C}_{q_i}\,]x_v\right\|_p \le \eta_i\|w\|_p, \tag{9.4.91}$$

for some positive scalar η_i (independent of ε). Hence, we can view

$$\begin{pmatrix} \bar{w} \\ x_v \end{pmatrix} \quad \text{and} \quad \frac{F_{iq_i}}{\varepsilon}[\,0 \quad 0 \quad \hat{C}_{q_i}\,]\,x_v, \tag{9.4.92}$$

as some L_p signals, whose l_p norms are bounded by some ε independent scalars. Then, following the similar procedures as in (9.4.28) to (9.4.56), it is straightforward to show that,

$$\|e_d\|_p \to 0, \ \text{ as } \varepsilon \to 0. \tag{9.4.93}$$

In view of (9.4.80) and (9.4.93), it is clear that the RPT problem is solved by the full order measurement feedback control law (9.2.57). ⊠

9.4.C. Proof of Theorem 9.2.4

We first define a new state variable,

$$x_s = x_2 - v + K_{\mathrm{R}1} x_1. \tag{9.4.94}$$

Again, it is straightforward to verify that the closed-loop system comprising the given system (9.1.1) and the reduced order measurement feedback control law of (9.2.71) can be rewritten as follows,

$$\dot{x}_s = (A_{\mathrm{R}} + K_{\mathrm{R}} C_{\mathrm{R}}) x_s + \left(E_2 + K_{\mathrm{R}} \begin{bmatrix} D_{1,0} \\ E_1 \end{bmatrix} \right) w, \tag{9.4.95}$$

$$\dot{x} = \Big[A + BF(\varepsilon)\Big] x - BF_2(\varepsilon) x_s + BH_0(\varepsilon) r + \cdots + BH_{\kappa-1}(\varepsilon) r^{(\kappa-1)} + Ew, \tag{9.4.96}$$

and

$$h = \Big[C_2 + D_2 F(\varepsilon)\Big] x - D_2 F_2(\varepsilon) x_s + D_2 H_0(\varepsilon) r + \cdots + D_2 H_{\kappa-1}(\varepsilon) r^{(\kappa-1)}. \tag{9.4.97}$$

Thus, it is simple to see that the closed-loop system is asymptotically stable for sufficiently small ε, as the closed-loop poles are given by the eigenvalues of $A + BF(\varepsilon)$ and $A_{\mathrm{R}} + K_{\mathrm{R}} C_{\mathrm{R}}$.

Since $A_{\mathrm{R}} + K_{\mathrm{R}} C_{\mathrm{R}}$ is asymptotically stable, it follows that for any initial condition, $x_s \in L_p$ provided that $w \in L_p$. Next, we rewrite

$$BF_2(\varepsilon) x_s = BF(\varepsilon) \begin{pmatrix} 0 \\ x_s \end{pmatrix} \quad \text{and} \quad D_2 F_2(\varepsilon) x_s = D_2 F(\varepsilon) \begin{pmatrix} 0 \\ x_s \end{pmatrix}. \tag{9.4.98}$$

It follows from (9.2.60) that

$$C_1^{-1}\{\mathrm{Im}\,(D_1)\} = \mathrm{Ker} \left(\begin{bmatrix} 0 & 0 \\ I_k & 0 \end{bmatrix} \right), \tag{9.4.99}$$

and

$$\begin{bmatrix} 0 & 0 \\ I_k & 0 \end{bmatrix} \begin{pmatrix} 0 \\ x_s \end{pmatrix} = 0. \tag{9.4.100}$$

Thus, the last condition of Theorem 9.2.1, i.e., $\mathrm{Ker}\,(C_2) \supset C_1^{-1}\{\mathrm{Im}\,(D_1)\}$, implies that

$$C_2 \begin{pmatrix} 0 \\ x_s \end{pmatrix} = 0 \quad \text{and} \quad N \begin{pmatrix} 0 \\ x_s \end{pmatrix} = 0, \tag{9.4.101}$$

for any appropriate dimensional matrix N with $\mathrm{Ker}\,(N) \supset \mathrm{Ker}\,(C_2)$. Following the same procedures as in (9.4.74) to (9.4.93), we can show that

$$\|e\|_p \to 0, \text{ as } \varepsilon \to 0. \tag{9.4.102}$$

Hence, the RPT problem is solved by the reduced order measurement feedback control law (9.2.71). ⊠

Chapter 10

Infima in Discrete-time H_∞ Optimization

10.1. Introduction

IN THIS CHAPTER, we present computational methods for evaluating the infima of discrete-time H_∞ optimal control problems. The main contributions of this chapter are the non-iterative algorithms that exactly compute the values of infima for systems satisfying certain geometric conditions. If these conditions are not satisfied, one might have to use iterative schemes based on certain reduced order systems for approximating these infima. Most of the results of this chapter were reported earlier in Chen [19], and Chen *et al.* [21].

10.2. Full Information Feedback Case

The main result of this section deals with the non-iterative computation of the infimum for the following full information feedback discrete-time system characterized by:

$$\Sigma : \begin{cases} x(k+1) = A\ x(k) + B\ u(k) + E\ w(k), \\ y(k) = \begin{pmatrix} I \\ 0 \end{pmatrix} x(k) + \begin{pmatrix} 0 \\ I \end{pmatrix} w(k), \\ h(k) = C_2\ x(k) + D_2\ u(k) + D_{22}\ w(k), \end{cases} \tag{10.2.1}$$

where $x \in \mathbb{R}^n$ is the state, $u \in \mathbb{R}^m$ is the control input, $w \in \mathbb{R}^q$ is the external disturbance input, $y \in \mathbb{R}^{n+q}$ is the measurement output, and $h \in \mathbb{R}^\ell$ is the controlled output of Σ. For ease of reference in future development, we define Σ_P to be the subsystem characterized by the matrix quadruple (A, B, C_2, D_2). We first make the following assumptions:

Assumption 10.F.1: (A, B) is stabilizable;

Assumption 10.F.2: Σ_P has no invariant zero on the unit circle;

Assumption 10.F.3: $\mathrm{Im}\,(E) \subset \mathcal{V}^\odot(\Sigma_P) + \mathcal{S}^\odot(\Sigma_P)$; and

Assumption 10.F.4: $D_{22} = 0$. 🄰

In what follows, we state a step-by-step algorithm for the computation of the infimum γ^*.

Step 10.F.1: Without loss of generality but for simplicity of presentation, we assume that the quadruple (A, B, C_2, D_2), i.e., Σ_P, has been partitioned in the form of (2.4.4). Then, transform Σ_P into the special coordinate basis as described in Chapter 2 (see also (2.4.20) to (2.4.23) for the compact form of the special coordinate basis). In this algorithm, for ease of reference in future development, we introduce an additional permutation matrix to the state transformation Γ_s such that the new state variables are ordered as follows:

$$\tilde{x} = \begin{pmatrix} x_c \\ x_a^- \\ x_a^+ \\ x_d \\ x_b \end{pmatrix}. \tag{10.2.2}$$

Next, we compute

$$\Gamma_s^{-1} E = \begin{bmatrix} E_c \\ E_a^- \\ E_a^+ \\ E_d \\ E_b \end{bmatrix}. \tag{10.2.3}$$

Note that Assumption 10.F.3 is equivalent to $E_b = 0$. Also, for economy of notation, we denote n_x the dimension of $\mathbb{R}^n / \mathcal{V}^\odot(\Sigma_P)$, which is equivalent to $n_x = n_a^+ + n_d + n_b$. We note that $n_x = 0$ if and only if the system Σ_P is right invertible and is of minimum phase with no infinite zero of order higher than zero.

Step 10.F.2: Define A_x, B_x, B_{x0}, B_{x1}, E_x, C_x and D_x as follows:

$$A_x := \begin{bmatrix} A_{aa}^+ & L_{ad}^+ C_d & L_{ab}^+ C_b \\ B_d E_{da}^+ & A_{dd} & B_d E_{db} \\ 0 & L_{bd} C_d & A_{bb} \end{bmatrix}, \quad E_x := \begin{bmatrix} E_a^+ \\ E_d \\ E_b \end{bmatrix}, \tag{10.2.4}$$

$$B_x := [\, B_{x0} \;\; B_{x1} \,] := \begin{bmatrix} B_{0a}^+ & 0 \\ B_{0d} & B_d \\ B_{0b} & 0 \end{bmatrix}, \tag{10.2.5}$$

and

$$C_x := \Gamma_o \begin{bmatrix} 0 & 0 & 0 \\ 0 & C_d & 0 \\ 0 & 0 & C_b \end{bmatrix}, \quad D_x = \Gamma_o \begin{bmatrix} I & 0 \\ 0 & 0 \\ 0 & 0 \end{bmatrix}. \tag{10.2.6}$$

It follows from the property of the special coordinate basis that the pair (A_x, B_x) is stabilizable. Next, we find a matrix F_x such that $A_x + B_x F_x$ has no eigenvalue at -1. Then define $\tilde{A}_x$, $\tilde{B}_x$, $\tilde{E}_x$, $\tilde{C}_x$, $\tilde{D}_x$ and $\tilde{D}_{22}$ as:

$$\left.\begin{aligned} \tilde{A}_x &:= (A_x + B_x F_x + I)^{-1}(A_x + B_x F_x - I), \\ \tilde{B}_x &:= 2(A_x + B_x F_x + I)^{-2} B_x, \\ \tilde{E}_x &:= 2(A_x + B_x F_x + I)^{-2} E_x, \\ \tilde{C}_x &:= C_x + D_x F_x, \\ \tilde{D}_x &:= D_x - (C_x + D_x F_x)(A_x + B_x F_x + I)^{-1} B_x, \\ \tilde{D}_{22} &:= D_{22} - (C_x + D_x F_x)(A_x + B_x F_x + I)^{-1} E_x. \end{aligned}\right\} \tag{10.2.7}$$

Step 10.F.3: Solve the following continuous-time algebraic Riccati equation and algebraic Lyapunov equation, both independent of γ:

$$\begin{aligned} 0 = &\left[\tilde{A}_x - \tilde{B}_x(\tilde{D}_x'\tilde{D}_x)^{-1}\tilde{D}_x'\tilde{C}_x\right]\tilde{S}_x + \tilde{S}_x\left[\tilde{A}_x - \tilde{B}_x(\tilde{D}_x'\tilde{D}_x)^{-1}\tilde{D}_x'\tilde{C}_x\right]' \\ &- \tilde{B}_x(\tilde{D}_x'\tilde{D}_x)^{-1}\tilde{B}_x' + \tilde{S}_x\left[\tilde{C}_x'\tilde{C}_x - \tilde{C}_x'\tilde{D}_x(\tilde{D}_x'\tilde{D}_x)^{-1}\tilde{D}_x'\tilde{C}_x\right]\tilde{S}_x, \end{aligned} \tag{10.2.8}$$

$$\begin{aligned} 0 = &\left[\tilde{A}_x - \tilde{B}_x(\tilde{D}_x'\tilde{D}_x)^{-1}\tilde{D}_x'\tilde{C}_x\right]\tilde{T}_x + \tilde{T}_x\left[\tilde{A}_x - \tilde{B}_x(\tilde{D}_x'\tilde{D}_x)^{-1}\tilde{D}_x'\tilde{C}_x\right]' \\ &- \left[\tilde{E}_x - \tilde{B}_x(\tilde{D}_x'\tilde{D}_x)^{-1}\tilde{D}_x'\tilde{D}_{22}\right]\left[\tilde{E}_x - \tilde{B}_x(\tilde{D}_x'\tilde{D}_x)^{-1}\tilde{D}_x'\tilde{D}_{22}\right]', \end{aligned} \tag{10.2.9}$$

for positive definite solution $\tilde{S}_x$ and positive semi-definite solution $\tilde{T}_x$. For future use, we define

$$S_x := (A_x + B_x F_x + I)\tilde{S}_x(A_x' + F_x' B_x' + I)/2, \tag{10.2.10}$$

and

$$T_x := (A_x + B_x F_x + I)\tilde{T}_x(A_x' + F_x' B_x' + I)/2. \tag{10.2.11}$$

Step 10.F.4: The infimum, γ^*, is given by

$$\gamma^* = \sqrt{\lambda_{\max}(\tilde{T}_x \tilde{S}_x^{-1})} = \sqrt{\lambda_{\max}(T_x S_x^{-1})}. \tag{10.2.12}$$

This completes the algorithm for computing γ^* for the full information feedback case. ▣

We have the following theorem.

Theorem 10.2.1. Consider the full information system given by (10.2.1). Then under Assumptions 10.F.1 to 10.F.4,

1. γ^* given by (10.2.12) is indeed its infimum, and

2. for $\gamma > \gamma^*$, the positive semi-definite matrix $P(\gamma)$ given by

$$P(\gamma) = (\Gamma_s^{-1})' \begin{bmatrix} 0 & 0 \\ 0 & (S_x - T_x/\gamma^2)^{-1} \end{bmatrix} \Gamma_s^{-1}, \tag{10.2.13}$$

is the unique solution that satisfies Conditions 2.(a)-2.(c) of Theorem 4.3.1. Moreover, such a solution $P(\gamma)$ does not exist when $\gamma < \gamma^*$. ◻T

Proof. First, we note that it follows from Theorem 2.4.1 and Property 2.4.4 of Chapter 2 that (A_x, B_x, C_x, D_x) is left invertible with no invariant zeros on the unit circle. Following the results of Stoorvogel *et al.* [125] and Lemma 5.3.3, it is straightforward to show that the following three statements are equivalent:

1. There exists a γ suboptimal controller for the full information system (10.2.1).

2. There exists a γ suboptimal controller for the following auxiliary system

$$\begin{cases} x_x(k+1) = A_x \; x_x(k) + B_x \; u_x(k) + E_x \; w_x(k), \\ y_x(k) = \begin{pmatrix} I \\ 0 \end{pmatrix} x_x(k) + \begin{pmatrix} 0 \\ I \end{pmatrix} w_x(k), \\ h_x(k) = C_x \; x_x(k) + D_x \; u_x(k) + D_{22} \; w_x(k), \end{cases} \tag{10.2.14}$$

where A_x, B_x, E_x, C_x and D_x are defined as in (10.2.4) to (10.2.6). Note that $D_{22} = 0$ by the assumption.

3. There exists a γ suboptimal controller for the following auxiliary system

$$\begin{cases} \dot{\tilde{x}}_x = \tilde{A}_x \; \tilde{x}_x + \tilde{B}_x \; \tilde{u}_x + \tilde{E}_x \; \tilde{w}_x, \\ \tilde{y}_x = \begin{pmatrix} I \\ 0 \end{pmatrix} \tilde{x}_x + \begin{pmatrix} 0 \\ I \end{pmatrix} \tilde{w}_x, \\ \tilde{h}_x = \tilde{C}_x \; \tilde{x}_x + \tilde{D}_x \; \tilde{u}_x + \tilde{D}_{22} \; \tilde{w}_x, \end{cases} \tag{10.2.15}$$

where $\tilde{A}_x$, $\tilde{B}_x$, $\tilde{E}_x$, $\tilde{C}_x$, $\tilde{D}_x$ and $\tilde{D}_{22}$ are as defined in (10.2.7).

For future use, we denote Σ_x and $\tilde{\Sigma}_x$ the matrix quadruples (A_x, B_x, C_x, D_x) and $(\tilde{A}_x, \tilde{B}_x, \tilde{C}_x, \tilde{D}_x)$, respectively. Note that by Theorems 4.2.1 and 4.3.1, Items 2 and 3 above are also equivalent to the following:

1. There exists a solution $P_x > 0$ to the following discrete-time algebraic Riccati equation,

$$P_x = A_x' P_x A_x + C_x' C_x - \begin{bmatrix} B_x' P_x A_x + D_x' C_x \\ E_x' P_x A_x \end{bmatrix}' G_x^{-1} \begin{bmatrix} B_x' P_x A_x + D_x' C_x \\ E_x' P_x A_x \end{bmatrix}, \tag{10.2.16}$$

where

$$G_x := \begin{bmatrix} D_x' D_x + B_x' P_x B_x & B_x' P_x E_x \\ E_x' P_x B_x & E_x' P_x E_x - \gamma^2 I \end{bmatrix}, \tag{10.2.17}$$

such that the following conditions are satisfied

$$V_x := B_x' P_x B_x + D_x' D_x > 0, \tag{10.2.18}$$

$$R_x := \gamma^2 I - E_x' P_x E_x + E_x' P_x B_x V_x^{-1} B_x' P_x E_x > 0. \tag{10.2.19}$$

2. There exists a solution $\tilde{P}_x > 0$ to the following continuous-time algebraic Riccati equation,

$$0 = \tilde{P}_x \tilde{A}_x + \tilde{A}_x' \tilde{P}_x + \tilde{C}_x' \tilde{C}_x - \begin{bmatrix} \tilde{B}_x' \tilde{P}_x + \tilde{D}_x' \tilde{C}_x \\ \tilde{E}_x' \tilde{P}_x + \tilde{D}_{22}' \tilde{C}_x \end{bmatrix}' \tilde{G}_x^{-1} \begin{bmatrix} \tilde{B}_x' \tilde{P}_x + \tilde{D}_x' \tilde{C}_x \\ \tilde{E}_x' \tilde{P}_x + \tilde{D}_{22}' \tilde{C}_x \end{bmatrix}, \tag{10.2.20}$$

with

$$\tilde{D}_{22}' [I - \tilde{D}_x (\tilde{D}_x' \tilde{D}_x)^{-1} \tilde{D}_x'] \tilde{D}_{22} < \gamma^2 I, \tag{10.2.21}$$

and

$$\tilde{G}_x := \begin{bmatrix} \tilde{D}_x' \tilde{D}_x & \tilde{D}_x' \tilde{D}_{22} \\ \tilde{D}_{22}' \tilde{D}_x & \tilde{D}_{22}' \tilde{D}_{22} - \gamma^2 I \end{bmatrix}. \tag{10.2.22}$$

Furthermore, the solutions to the above Riccati equations, if they exist, are related by

$$P_x = 2(A_x' + I)^{-1} \tilde{P}_x (A_x + I)^{-1}. \tag{10.2.23}$$

Thus, it is equivalent to show that γ^* given by (10.2.12) is the infimum for the full information system (10.2.1) by showing that it is an infimum for the auxiliary system in (10.2.15). This can be done by first showing the properties of the auxiliary system of (10.2.15) and then applying the results of Chapter 6. We note that the matrix F_x in Step 10.F.2 of the algorithm is a pre-state feedback gain, which is introduced merely to deal with the situation when A_x has eigenvalues at -1 and the inverse of $I + A_x$ does not exist. For the sake of simplicity but without loss of generality, we will hereafter assume that A_x has no eigenvalue at -1 and $F_x = 0$. We will first show the following two facts associated with the auxiliary system (10.2.15): There exists a pre-disturbance feedback to the system in (10.2.15) in the form of,

$$\tilde{u}_x = \tilde{F}_w \tilde{w}_x + \tilde{v}_x, \tag{10.2.24}$$

such that

1. $\tilde{D}_{22} + \tilde{D}_x \tilde{F}_w = 0$, and
2. $\operatorname{Im}(\tilde{E}_x + \tilde{B}_x \tilde{F}_w) \subseteq \mathcal{V}^\odot(\tilde{\Sigma}_x) + \mathcal{S}^\odot(\tilde{\Sigma}_x)$.

In fact, we will show that such an $\tilde{F}_w$ is given by

$$\tilde{F}_w = -(\tilde{D}_x' \tilde{D}_x)^{-1} \tilde{D}_x' \tilde{D}_{22}. \tag{10.2.25}$$

In order to make our proof simpler, we first apply a pre-state feedback law

$$u_x = F_x x_x + v_x = -\begin{bmatrix} 0 & 0 & 0 \\ E_{da}^+ & 0 & E_{db} \end{bmatrix} x_x + v_x, \tag{10.2.26}$$

to the system in (10.2.14) such that the resulting dynamic matrix $A_x + B_x F_x$ has the following format,

$$\begin{bmatrix} A_{aa}^+ & L_{ad}^+ C_d & L_{ab}^+ C_b \\ 0 & A_{dd} & 0 \\ 0 & L_{bd} C_d & A_{bb} \end{bmatrix}, \tag{10.2.27}$$

while the rest of the system matrices in (10.2.14) remain unchanged. Hence, it is without loss of generality that we assume that A_x is already in the form of (10.2.27). Also, we assume that both A_{dd} and A_{bb} have no eigenvalue at -1. Then it is simple to verify that

$$(A_x + I)^{-1} = \begin{bmatrix} (A_{aa}^+ + I)^{-1} & X_1 & X_2 \\ 0 & (A_{dd} + I)^{-1} & 0 \\ 0 & -(A_{bb} + I)^{-1} L_{bd} C_d (A_{dd} + I)^{-1} & (A_{bb} + I)^{-1} \end{bmatrix}, \tag{10.2.28}$$

where

$$X_1 = -(A_{aa}^+ + I)^{-1} \left[L_{ad}^+ - L_{ab}^+ C_b (A_{bb} + I)^{-1} L_{bd} \right] C_d (A_{dd} + I)^{-1}, \tag{10.2.29}$$

$$X_2 = -(A_{aa}^+ + I)^{-1} L_{ab}^+ C_b (A_{bb} + I)^{-1}, \tag{10.2.30}$$

and

$$\begin{aligned} \tilde{D}_x &= D_x - C_x (A_x + I)^{-1} B_x \\ &= \Gamma_o \begin{bmatrix} I & 0 \\ -C_d (A_{dd} + I)^{-1} B_{0d} & -C_d (A_{dd} + I)^{-1} B_d \\ X_3 & C_b (A_{bb} + I)^{-1} L_{bd} C_d (A_{dd} + I)^{-1} B_d \end{bmatrix}, \end{aligned}$$

where

$$X_3 = C_b(A_{bb}+I)^{-1}L_{bd}C_d(A_{dd}+I)^{-1}B_{0d} - C_b(A_{bb}+I)^{-1}B_{0b}. \tag{10.2.31}$$

Define

$$\tilde{\Gamma}_o = \Gamma_o \begin{bmatrix} I & 0 & 0 \\ -C_d(A_{dd}+I)^{-1}B_{0d} & -C_d(A_{dd}+I)^{-1}B_d & 0 \\ X_3 & C_b(A_{bb}+I)^{-1}L_{bd}C_d(A_{dd}+I)^{-1}B_d & I \end{bmatrix}. \tag{10.2.32}$$

We note that $\tilde{\Gamma}_o$ is nonsingular. This follows from the property of the special coordinate basis (see Theorem 2.4.1) that the triple (A_{dd}, B_d, C_d) is square and invertible with no invariant zero, and hence $C_d(A_{dd}+I)^{-1}B_d$ is nonsingular. Then we have

$$\tilde{D}_x = \tilde{\Gamma}_o \begin{bmatrix} I & 0 \\ 0 & I \\ 0 & 0 \end{bmatrix}, \tag{10.2.33}$$

and

$$\tilde{D}_{22} = -C_x(A_x+I)^{-1}E_x = \begin{bmatrix} 0 \\ -C_d(A_{dd}+I)^{-1}E_d \\ C_b(A_{bb}+I)^{-1}L_{bd}C_d(A_{dd}+I)^{-1}E_d \end{bmatrix} = \tilde{\Gamma}_o \begin{bmatrix} 0 \\ X_4 \\ 0 \end{bmatrix}, \tag{10.2.34}$$

where

$$X_4 = [C_d(A_{dd}+I)^{-1}B_d]^{-1}C_d(A_{dd}+I)^{-1}E_d. \tag{10.2.35}$$

It is now obvious to see that the following pre-disturbance feedback law to (10.2.15)

$$\tilde{u}_x = \tilde{F}_w\tilde{w}_x + \tilde{v}_x = -\begin{bmatrix} 0 \\ X_4 \end{bmatrix}\tilde{w}_x + \tilde{v}_x, \tag{10.2.36}$$

guarantees that $\tilde{D}_{22} + \tilde{D}_x\tilde{F}_w = 0$. We also have

$$\tilde{E}_x + \tilde{B}\tilde{F}_w = 2(A_x+I)^{-2}(E_x + B_x\tilde{F}_w) = 2(A_x+I)^{-2}\begin{bmatrix} E_a^+ \\ E_d^* \\ 0 \end{bmatrix}, \tag{10.2.37}$$

where

$$E_d^* = E_d - B_d[C_d(A_{dd}+I)^{-1}B_d]^{-1}C_d(A_{dd}+I)^{-1}E_d. \tag{10.2.38}$$

This shows the first fact. Since $\tilde{D}_x$ is of maximal column rank, it follows that the above $\tilde{F}_w$ is also equivalent to $-(\tilde{D}_x'\tilde{D}_x)^{-1}\tilde{D}_x'\tilde{D}_{22}$. Next, let us proceed to prove the second fact, i.e.,

$$\operatorname{Im}(\tilde{E}_x + \tilde{B}_x\tilde{F}_w) \subseteq \mathcal{V}^{\odot}(\tilde{\Sigma}_x) + \mathcal{S}^{\odot}(\tilde{\Sigma}_x).$$

We will have to apply several nonsingular state transformations to the system

$$\begin{cases} \dot{\tilde{x}}_x = \tilde{A}_x \, \tilde{x}_x + \tilde{B}_x \, \tilde{v}_x + (\tilde{E}_x + \tilde{B}_x \tilde{F}_w) \, \tilde{w}_x, \\ \tilde{h}_x = \tilde{C}_x \, \tilde{x}_x + \tilde{D}_x \, \tilde{v}_x \ , \end{cases} \tag{10.2.39}$$

and transform it into the form of the special coordinate basis as given in Theorem 2.4.1. First let us define a state transformation

$$\tilde{T}_x = (A_x + I)^{-2}. \tag{10.2.40}$$

In view of (10.2.28), it is straightforward, although tedious, to verify that

$$\tilde{T}_x = \begin{bmatrix} (A_{aa}^+ + I)^{-2} & \star & \star \\ 0 & (A_{dd} + I)^{-2} & 0 \\ 0 & X_5 & (A_{bb} + I)^{-2} \end{bmatrix}, \tag{10.2.41}$$

where $\star$s are matrices of not much interest and

$$X_5 = -(A_{bb} + I)^{-1}[L_{bd}C_d(A_{dd} + I)^{-1} + (A_{bb} + I)^{-1}L_{bd}C_d](A_{dd} + I)^{-1}, \tag{10.2.42}$$

and

$$\bar{A}_x := \tilde{T}_x^{-1}\tilde{A}_x\tilde{T}_x = (A_x - I)(A_x + I)^{-1} \tag{10.2.43}$$

$$= \begin{bmatrix} (A_{aa}^+ - I)(A_{aa}^+ + I)^{-1} & \star & 2(A_{aa}^+ + I)^{-1}L_{ab}^+C_b(A_{bb} + I)^{-1} \\ 0 & (A_{dd} - I)(A_{dd} + I)^{-1} & 0 \\ 0 & 2(A_{bb} + I)^{-1}L_{bd}C_d(A_{dd} + I)^{-1} & (A_{bb} - I)(A_{bb} + I)^{-1} \end{bmatrix},$$

$$\bar{B}_x := \tilde{T}_x^{-1}\tilde{B}_x = 2B_x = 2\begin{bmatrix} B_{0a}^+ & 0 \\ B_{0d} & B_d \\ B_{0b} & 0 \end{bmatrix}, \tag{10.2.44}$$

$$\bar{E}_x := \tilde{T}_x^{-1}(\tilde{E}_x + \tilde{B}_x\tilde{F}_w) = 2\begin{bmatrix} E_a^+ \\ E_d^* \\ E_b \end{bmatrix}, \quad \text{where } E_b = 0, \tag{10.2.45}$$

$$\bar{C}_x := \tilde{C}_x\tilde{T}_x$$

$$= \tilde{\Gamma}_o \begin{bmatrix} 0 & 0 & 0 \\ 0 & -[C_d(A_{dd} + I)^{-1}B_d]^{-1}C_d(A_{dd} + I)^{-2} & 0 \\ 0 & -C_b(A_{bb} + I)^{-2}L_{bd}C_d(A_{dd} + I)^{-1} & C_b(A_{bb} + I)^{-2} \end{bmatrix} \tag{10.2.46}$$

$$\bar{D}_x := \tilde{D}_x = \tilde{\Gamma}_o \begin{bmatrix} I & 0 \\ 0 & I \\ 0 & 0 \end{bmatrix}. \tag{10.2.47}$$

In order to bring the system of (10.2.39) into the standard form of the special coordinate basis, we will have to perform another state transformation that will cause the (3, 2) block of $\bar{C}_x$ in the right hand side of (10.2.46) to vanish. The following transformation $\bar{T}_x$ will do the job,

$$\bar{T}_x = \begin{bmatrix} I & 0 & 0 \\ 0 & I & 0 \\ 0 & L_{bd}C_d(A_{dd}+I)^{-1} & (A_{bb}+I)^2 \end{bmatrix}. \tag{10.2.48}$$

It is quite easy to verify this time that

$$\hat{A}_x := \bar{T}_x^{-1}\bar{A}_x\bar{T}_x \tag{10.2.49}$$

$$= \begin{bmatrix} (A_{aa}^+-I)(A_{aa}^++I)^{-1} & \star & 2(A_{aa}^++I)^{-1}L_{ab}^+C_b(A_{bb}+I) \\ 0 & (A_{dd}-I)(A_{dd}+I)^{-1} & 0 \\ 0 & 2(A_{bb}+I)^{-2}L_{bd}C_d(A_{dd}+I)^{-2} & (A_{bb}+I)^{-1}(A_{bb}-I) \end{bmatrix},$$

$$\hat{B}_x := \hat{B}_{x0} := \bar{T}_x^{-1}\bar{B}_x = 2\begin{bmatrix} B_{0a}^+ & 0 \\ B_{0d} & B_d \\ \star & -(A_{bb}+I)^{-2}L_{bd}C_d(A_{dd}+I)^{-1}B_d \end{bmatrix}, \tag{10.2.50}$$

$$\hat{E}_x := \bar{T}_x^{-1}\bar{E}_x$$

$$= 2\begin{bmatrix} E_a^+ \\ E_d^* \\ (A_{bb}+I)^{-2}[E_b - L_{bd}C_d(A_{dd}^++I)^{-1}E_d^*] \end{bmatrix} = 2\begin{bmatrix} E_a^+ \\ E_d^* \\ 0 \end{bmatrix}, \tag{10.2.51}$$

$$\hat{C}_x := \begin{bmatrix} \hat{C}_{x0} \\ \hat{C}_{x1} \end{bmatrix} := \bar{C}_x\bar{T}_x$$

$$= \tilde{\Gamma}_o \left[\begin{array}{ccc} 0 & 0 & 0 \\ 0 & -[C_d(A_{dd}+I)^{-1}B_d]^{-1}C_d(A_{dd}+I)^{-2} & 0 \\ \hline 0 & 0 & C_b \end{array}\right], \tag{10.2.52}$$

$$\hat{D}_x := \bar{D}_x = \tilde{D}_x. \tag{10.2.53}$$

Then we have

$$\hat{A}_x - \hat{B}_{x0}\hat{C}_{x0} = \begin{bmatrix} (A_{aa}^+-I)(A_{aa}^++I)^{-1} & \star & 2(A_{aa}^++I)^{-1}L_{ab}^+C_b(A_{bb}+I) \\ 0 & A_{aa}^* & 0 \\ 0 & 0 & (A_{bb}+I)^{-1}(A_{bb}-I) \end{bmatrix}, \tag{10.2.54}$$

where

$$A_{aa}^{*} = (A_{dd} - I)(A_{dd} + I)^{-1} + 2B_d[C_d(A_{dd} + I)^{-1}B_d]^{-1}C_d(A_{dd} + I)^{-2}. \tag{10.2.55}$$

Define another nonsingular state transformation,

$$\hat{T}_x = \begin{bmatrix} I & 0 & \hat{T}_* \\ 0 & I & 0 \\ 0 & 0 & I \end{bmatrix}, \tag{10.2.56}$$

with $\hat{T}_*$ being a solution to the following general Lyapunov equation

$$(I - A_{aa}^{+})(I + A_{aa}^{+})^{-1}\hat{T}_* + \hat{T}_*(A_{bb} + I)^{-1}(A_{bb} - I) = 2(A_{aa}^{+} - I)^{-1}L_{ab}^{+}C_b(A_{bb} + I).$$

It follows from Kailath [64] that such a solution always exists and is unique if A_{aa}^{+} and A_{bb} have no common eigenvalue. Then it is straightforward to verify that it would transform the (1, 3) block of $\hat{A}_x - \hat{B}_{x0}\hat{C}_{x0}$ in (10.2.54) to 0 while not changing the structures of other blocks. Hence, $\hat{T}_x$ would also transform the system $(\hat{A}_x, \hat{B}_x, \hat{C}_x, \hat{D}_x)$ and $\hat{E}_x$ into the standard form of the special coordinate basis as given in Theorem 2.4.1 since the pair $\{(A_{bb} + I)^{-1}(A_{bb} - I), C_b\}$ is completely observable due to the complete observability of (A_{bb}, C_b). It is now clear from the properties of the special coordinate basis that

$$\operatorname{Im}(\hat{E}_x) \subseteq \mathcal{V}^{\odot}(\hat{\Sigma}_x) + \mathcal{S}^{\odot}(\hat{\Sigma}_x),$$

where $\hat{\Sigma}_x$ is characterized by $(\hat{A}_x, \hat{B}_x, \hat{C}_x, \hat{D}_x)$, which is equivalent to

$$\operatorname{Im}(\tilde{E}_x + \tilde{B}_x\tilde{F}_w) \subseteq \mathcal{V}^{\odot}(\tilde{\Sigma}_x) + \mathcal{S}^{\odot}(\tilde{\Sigma}_x).$$

This proves the second fact.

Next, let us apply a pre-disturbance feedback law,

$$\tilde{u}_x = \tilde{F}_w\tilde{w}_x + \tilde{v}_x = -(\tilde{D}_x'\tilde{D}_x)^{-1}\tilde{D}_x'\tilde{D}_{22}\tilde{w}_x + \tilde{v}_x, \tag{10.2.57}$$

to the auxiliary system (10.2.15). Again, this pre-feedback law will not affect solutions to the H_∞ problem for (10.2.15) or to the solution $\tilde{P}_x$ of (10.2.20)-(10.2.22). After applying this pre-feedback law, we obtain the following new system

$$\begin{cases} \dot{\tilde{x}}_x = \tilde{A}_x\ \tilde{x}_x + \tilde{B}_x\ \tilde{v}_x + \left[\tilde{E}_x - \tilde{B}_x(\tilde{D}_x'\tilde{D}_x)^{-1}\tilde{D}_x'\tilde{D}_{22}\right]\tilde{w}_x, \\ \tilde{y}_x = \begin{pmatrix} I \\ 0 \end{pmatrix}\tilde{x}_x + \begin{pmatrix} 0 \\ I \end{pmatrix}\tilde{w}_x, \\ \tilde{h}_x = \tilde{C}_x\ \tilde{x}_x + \tilde{D}_x\ \tilde{v}_x + 0\ \tilde{w}_x. \end{cases} \tag{10.2.58}$$

Then it follows from Corollary 4.2.1 that the existence condition of a γ suboptimal controller for (10.2.58) is equivalent to the existence of a matrix $\tilde{P}_x > 0$ such that

$$0 = \tilde{P}_x\tilde{A}_x + \tilde{A}_x'\tilde{P}_x + \tilde{C}_x'\tilde{C}_x - (\tilde{P}_x\tilde{B}_x + \tilde{C}_x'\tilde{D}_x)(\tilde{D}_x'\tilde{D}_x)^{-1}(\tilde{P}_x\tilde{B}_x + \tilde{C}_x'\tilde{D}_x)' \\ +\tilde{P}_x\left[\tilde{E}_x - \tilde{B}_x(\tilde{D}_x'\tilde{D}_x)^{-1}\tilde{D}_x'\tilde{D}_{22}\right]\left[\tilde{E}_x - \tilde{B}_x(\tilde{D}_x'\tilde{D}_x)^{-1}\tilde{D}_x'\tilde{D}_{22}\right]'\tilde{P}_x/\gamma^2,$$

is satisfied. Note that the solution $\tilde{P}_x$ to the above Riccati equation is identical to the solution that satisfies (10.2.20)-(10.2.21).

Now, it follows from Theorem 3.3.1 that $(\tilde{A}_x, \tilde{B}_x, \tilde{C}_x, \tilde{D}_x)$ is left invertible, and is free of infinite zeros and stable invariant zeros as well as invariant zeros on the unit circle. Also, in view of the second fact of the auxiliary system of (10.2.58), it satisfies Assumptions 6.F.1 to 6.F.4 of Chapter 6. Following the results of Chapter 6, we can easily show that

$$\gamma^* = \sqrt{\lambda_{\max}(\tilde{T}_x\tilde{S}_x^{-1})}, \tag{10.2.59}$$

and for any $\gamma > \gamma^*$, the positive definite solution $\tilde{P}_x$ of (10.2.20)-(10.2.22) is given by

$$\tilde{P}_x = (\tilde{S}_x - \tilde{T}_x/\gamma^2)^{-1}. \tag{10.2.60}$$

It then follows from (10.2.23) that for any $\gamma > \gamma^*$, the positive definite solution P_x of (10.2.16)-(10.2.19) is given by

$$P_x = 2(A_x' + I)^{-1}(\tilde{S}_x - \tilde{T}_x/\gamma^2)^{-1}(A_x + I)^{-1}, \tag{10.2.61}$$

and hence γ^* can also be obtained from the following expression,

$$\gamma^* = \sqrt{\lambda_{\max}(T_xS_x^{-1})}, \tag{10.2.62}$$

where S_x and T_x are as defined in (10.2.10) and (10.2.11), respectively. Moreover, it is straightforward to verify that

$$P(\gamma) = (\Gamma_s^{-1})'\begin{bmatrix} 0 & 0 \\ 0 & (S_x - T_x/\gamma^2)^{-1} \end{bmatrix}\Gamma_s^{-1},$$

is the unique solution that satisfies Conditions 2.(a)-2.(c) of Theorem 4.3.1.

Finally, note that $(\tilde{A}_x, \tilde{B}_x, \tilde{C}_x, \tilde{D}_x)$ is left invertible, and is free of infinite zeros and stable invariant zeros as well as invariant zeros on the unit circle. It follows from Richardson and Kwong [106] that the solution $\tilde{S}_x$ to the Riccati equation (10.2.8) is positive definite because $(\tilde{A}_x, \tilde{B}_x)$ is controllable, and the solution $\tilde{T}_x$ to the Lyapunov equation (10.2.9) is positive semi-definite. In fact, both of them are unique. This completes the proof of our algorithm. ⊠

The following remarks are in order.

Remark 10.2.1. For the case when $D_{22} \neq 0$, Assumption 10.F.3 should be replaced by the following conditions:

1. $\tilde{D}_{22} := D_{22} - C_x(A_x + I)^{-1}E_x$ is in the range space of $\tilde{D}_x$, and
2. $\text{Im}\Big[\tilde{E}_x - \tilde{B}_x(\tilde{D}_x'\tilde{D}_x)^{-1}\tilde{D}_x'\tilde{D}_{22}\Big] \subseteq \mathcal{V}^{\odot}(\tilde{\Sigma}_x) + \mathcal{S}^{\odot}(\tilde{\Sigma}_x)$.

Then our algorithm would carry through without any problems. We would also like to note that if (A, B, C_2, D_2) is right invertible, then $(\tilde{A}_x, \tilde{B}_x, \tilde{C}_x, \tilde{D}_x)$ is invertible and $\tilde{D}_x$ is square and nonsingular, and $\mathcal{V}^{\odot}(\tilde{\Sigma}_x) + \mathcal{S}^{\odot}(\tilde{\Sigma}_x) = \mathbb{R}^{n_x}$. Hence, the above two conditions will be automatically satisfied. Such a result was first reported in Chen [19]. Ⓡ

Remark 10.2.2. If Assumptions 10.F.3 and 10.F.4 are not satisfied, then one might have to approximate iteratively the infimum γ^* by finding the smallest nonnegative scalar, say $\tilde{\gamma}^* \geq 0$, such that the Riccati equation (10.2.20) and (10.2.21) are satisfied. Ⓡ

We illustrate the above results in the following example.

Example 10.2.1. Consider a full information system (10.2.1) characterized by

$$A = \begin{bmatrix} 1 & 1 & 1 & 1 & 1 \\ 0 & 0 & 0 & 1 & 1 \\ 0 & 0 & 1 & 1 & 1 \\ 1 & 1 & 1 & 1 & 1 \\ 0 & 0 & 0 & 1 & 1 \end{bmatrix}, \quad B = \begin{bmatrix} 0 & 0 & 1 \\ 0 & 0 & 0 \\ 1 & 0 & 0 \\ 0 & 1 & 0 \\ 0 & 0 & 0 \end{bmatrix}, \quad E = \begin{bmatrix} 1 \\ 1 \\ 1 \\ 1 \\ 0 \end{bmatrix}, \tag{10.2.63}$$

and

$$C_2 = \begin{bmatrix} 0 & 0 & -1 & 0 & 0 \\ 0 & 0 & 0 & 1 & 0 \\ 0 & 0 & 0 & 0 & 1 \end{bmatrix}, \quad D_2 = \begin{bmatrix} 1 & 0 & 0 \\ 0 & 0 & 0 \\ 0 & 0 & 0 \end{bmatrix}, \quad D_{22} = \begin{bmatrix} 0 \\ 0 \\ 0 \end{bmatrix}. \tag{10.2.64}$$

It is can be verified that (A, B) is controllable and (A, B, C_2, D_2) is neither right nor left invertible, and is of nonminimum phase with two invariant zeros at 0 and 2, respectively. Moreover, it is already in the form of the special coordinate basis as given in Theorem 2.4.1 and Assumption 10.F.3 is satisfied as $E_b = 0$. Hence, Assumptions 10.F.1 to 10.F.4 are all satisfied. Following the algorithm, we obtain

$$\Gamma_s = I_5, \quad n_x = 3,$$

$$A_x = \begin{bmatrix} 2 & 1 & 1 \\ 1 & 1 & 1 \\ 0 & 1 & 1 \end{bmatrix}, \quad B_x = \begin{bmatrix} 1 & 0 \\ 0 & 1 \\ 0 & 0 \end{bmatrix}, \quad E_x = \begin{bmatrix} 1 \\ 1 \\ 0 \end{bmatrix},$$

$$C_x = \begin{bmatrix} 0 & 0 & 0 \\ 0 & 1 & 0 \\ 0 & 0 & 1 \end{bmatrix}, \quad D_x = \begin{bmatrix} 1 & 0 \\ 0 & 0 \\ 0 & 0 \end{bmatrix},$$

$$\tilde{A}_x = \begin{bmatrix} 0.25 & 0.25 & 0.25 \\ 0.50 & -0.50 & 0.50 \\ -0.25 & 0.75 & -0.25 \end{bmatrix},$$

$$\tilde{B}_x = \begin{bmatrix} 0.3125 & -0.1875 \\ -0.6250 & 1.3750 \\ 0.4375 & -1.0625 \end{bmatrix}, \quad \tilde{E}_x = \begin{bmatrix} 0.125 \\ 0.750 \\ -0.625 \end{bmatrix}$$

and

$$\tilde{C}_x = C_x, \quad \tilde{D}_x = \begin{bmatrix} 1.000 & 0.000 \\ 0.250 & -0.750 \\ -0.125 & 0.375 \end{bmatrix}, \quad \tilde{D}_{22} = \begin{bmatrix} 0.00 \\ -0.50 \\ 0.25 \end{bmatrix}.$$

It is simple to verify that $(\tilde{A}_x, \tilde{B}_x, \tilde{C}_x, \tilde{D}_x)$ is left invertible with two invariant zeros at 1 and 1/3, respectively. Solving Riccati equations (10.2.8) and (10.2.9), we obtain

$$\tilde{S}_x = \begin{bmatrix} 0.227615 & -0.207890 & 0.019725 \\ -0.207890 & 1.202254 & -1.005636 \\ 0.019725 & -1.005636 & 1.014089 \end{bmatrix},$$

and

$$\tilde{T}_x = \begin{bmatrix} 0.09375 & -0.062500 & 0.031250 \\ -0.06250 & 0.041667 & -0.020833 \\ 0.03125 & -0.020833 & 0.010417 \end{bmatrix}.$$

Finally, we get

$$S_x = \begin{bmatrix} 0.562306 & -0.145898 & -0.145898 \\ -0.145898 & 0.618034 & -0.381966 \\ -0.145898 & -0.381966 & 0.618034 \end{bmatrix}, \quad T_x = \begin{bmatrix} 1/3 & 0 & 0 \\ 0 & 0 & 0 \\ 0 & 0 & 0 \end{bmatrix},$$

and the infimum

$$\gamma^* = 0.934173.$$

🅴

10.3. Output Feedback Case

We present in this section a well-conditioned non-iterative algorithm for the exact computation of γ^* of the following measurement feedback discrete-time system Σ,

$$\Sigma : \begin{cases} x(k+1) = A\ x(k) + B\ u(k) + E\ w(k), \\ y(k) \quad = C_1\ x(k) \qquad\qquad\ + D_1\ w(k), \\ h(k) \quad = C_2\ x(k) + D_2\ u(k) + D_{22}\ w(k), \end{cases} \tag{10.3.1}$$

where $x \in \mathbb{R}^n$ is the state, $u \in \mathbb{R}^m$ is the control input, $w \in \mathbb{R}^q$ is the external disturbance input, $y \in \mathbb{R}^p$ is the measurement output, and $h \in \mathbb{R}^\ell$ is the controlled output of Σ. Again, for easy reference, we define Σ_P to be the subsystem characterized by the matrix quadruple (A, B, C_2, D_2) and Σ_Q to be the subsystem characterized by the matrix quadruple (A, E, C_1, D_1). We first make the following assumptions:

Assumption 10.M.1: (A, B) is stabilizable;

Assumption 10.M.2: Σ_P has no invariant zero on the unit circle;

Assumption 10.M.3: $\mathrm{Im}\,(E) \subset \mathcal{V}^{\odot}(\Sigma_P) + \mathcal{S}^{\odot}(\Sigma_P)$;

Assumption 10.M.4: (A, C_1) is detectable;

Assumption 10.M.5: Σ_Q has no invariant zero on the unit circle;

Assumption 10.M.6: $\mathrm{Ker}\,(C_2) \supset \mathcal{V}^{\odot}(\Sigma_Q) \cap \mathcal{S}^{\odot}(\Sigma_Q)$; and

Assumption 10.M.7: $D_{22} = 0$. ▣

As in the previous section, we outline a step-by-step algorithm for the computation of γ^* below:

Step 10.M.1: Define an auxiliary full information problem for

$$\begin{cases} x(k+1) = A\ x(k) + B\ u(k) + E\ w(k), \\ y(k) = \begin{pmatrix} I \\ 0 \end{pmatrix} x(k) + \begin{pmatrix} 0 \\ I \end{pmatrix} w(k), \\ h(k) = C_2\ x(k) + D_2\ u(k) + D_{22}\ w(k), \end{cases} \tag{10.3.2}$$

and perform Steps 10.F.1 to 10.F.3 of the algorithm given in the previous section. For future use and in order to avoid notational confusion, we rename the state transformation of the special coordinate basis for Σ_P as Γ_{sP} and the dimension of A_x as n_{xP}. Also, rename S_x of (10.2.10) and T_x of (10.2.11) as S_{xP} and T_{xP}, respectively.

Step 10.M.2: Define another auxiliary full information problem for

$$\begin{cases} x(k+1) = A'\ x(k) + C_1'\ u(k) + C_2'\ w(k), \\ y(k) = \begin{pmatrix} I \\ 0 \end{pmatrix} x(k) + \begin{pmatrix} 0 \\ I \end{pmatrix} w(k), \\ h(k) = E'\ x(k) + D_1'\ u(k) + D_{22}'\ w(k), \end{cases} \tag{10.3.3}$$

and again perform Steps 10.F.1 to 10.F.3 of the algorithm given in Section 10.2 one more time, but for this auxiliary system. Let $\Sigma_Q^\star$ be the dual system of $\Sigma_{\rm Q}$ and be characterized by (A', C_1', E', D_1'). We rename the state transformation of the special coordinate basis for $\Sigma_{\rm Q}^\star$ as $\Gamma_{s\rm Q}$ and the dimension of A_x as $n_{x\rm Q}$, and S_x of (10.2.10) and T_x of (10.2.11) as $S_{x\rm Q}$ and $T_{x\rm Q}$, respectively.

Step 10.M.3: Partition

$$\Gamma_{s\rm P}^{-1}(\Gamma_{s\rm Q}^{-1})' = \begin{bmatrix} \star & \star \\ \star & \Gamma \end{bmatrix}, \tag{10.3.4}$$

where Γ is a $n_{x\rm P} \times n_{x\rm Q}$ matrix, and define a constant matrix

$$M = \begin{bmatrix} T_{x\rm P}S_{x\rm P}^{-1} + \Gamma S_{x\rm Q}^{-1}\Gamma' S_{x\rm P}^{-1} & -\Gamma S_{x\rm Q}^{-1} \\ -T_{x\rm Q}S_{x\rm Q}^{-1}\Gamma' S_{x\rm P}^{-1} & T_{x\rm Q}S_{x\rm Q}^{-1} \end{bmatrix}. \tag{10.3.5}$$

Step 10.M.4: The infimum γ^* is then given by

$$\gamma^* = \sqrt{\lambda_{\max}(M)}, \tag{10.3.6}$$

where M has only real and nonnegative eigenvalues. ▣

Proof of the Algorithm. Once the result for the full information case is established, the proof of this algorithm is similar to the one given in Section 6.3 of Chapter 6. ▣

The following remarks are in order.

Remark 10.3.1. Consider the given discrete-time system (10.3.1) that satisfies Assumptions 10.M.1 to 10.M.7. Then for any $\gamma > \gamma^*$, where γ^* is given by (10.3.6), the following $P(\gamma)$ and $Q(\gamma)$,

$$P(\gamma) := (\Gamma_{s\rm P}^{-1})' \begin{bmatrix} 0 & 0 \\ 0 & (S_{x\rm P} - T_{x\rm P}/\gamma^2)^{-1} \end{bmatrix} \Gamma_{s\rm P}^{-1}, \tag{10.3.7}$$

and

$$Q(\gamma) := (\Gamma_{s\rm Q}^{-1})' \begin{bmatrix} 0 & 0 \\ 0 & (S_{x\rm Q} - T_{x\rm Q}/\gamma^2)^{-1} \end{bmatrix} \Gamma_{s\rm Q}^{-1}, \tag{10.3.8}$$

satisfy Conditions 2.(a)-2.(g) of Theorem 4.3.1. ▣

Remark 10.3.2. For discrete-time H_∞ control, γ^* for the full information feedback system is in general different from that of the full state feedback system regardless of $D_{22} = 0$ or not. For the state feedback case, i.e., $C_1 = I$ and $D_1 = 0$, we note that the subsystem $\Sigma_{\rm Q}$ is always free of invariant zeros (and hence free of unit circle invariant zeros) and left invertible. Thus, as long as $\Sigma_{\rm P}$ is free of unit circle invariant zeros and satisfies Assumption 10.M.1 to 10.M.3, one can apply the above algorithm to get the infimum, γ^*. For this special case Γ_{sQ}, n_{xQ}, S_{xQ} and T_{xQ} in Step 10.M.2 of the above algorithm can be directly obtained using the following simple procedure: Compute a nonsingular transformation Γ_{sQ} such that

$$\Gamma'_{sQ}E = \begin{bmatrix} 0 \\ \hat{E} \end{bmatrix}, \tag{10.3.9}$$

where $\hat{E}$ is a $n_{xQ} \times n_{xQ}$ nonsingular matrix. Then S_{xQ} and T_{xQ} are respectively given by

$$S_{xQ} = \left(\hat{E}^{-1}\right)' \hat{E}^{-1} \quad \text{and} \quad T_{xQ} = 0, \tag{10.3.10}$$

and hence

$$\gamma^* = \left[\lambda_{\max}(T_{xP}S_{xP}^{-1} + \Gamma S_{xQ}^{-1}\Gamma' S_{xP}^{-1})\right]^{\frac{1}{2}}. \tag{10.3.11}$$

Note that in general, $\gamma^* \geq \{\lambda_{\max}(T_{xP}S_{xP}^{-1})\}^{\frac{1}{2}}$. ®

Remark 10.3.3. For the case when $D_{22} \neq 0$, Assumptions 10.M.3 and 10.M.6 should be replaced by the conditions given in Remark 10.2.1, which is associated with the full information system of (10.3.2), and a set of conditions similar to those in that remark, but for the full information system of (10.3.3). Then our procedure would again carry through and yield the correct result. Note that if Σ_P is right invertible and Σ_Q is left invertible, then all these conditions will be automatically satisfied. The result will then reduce to that of Chen [19]. ®

Remark 10.3.4. If Assumptions 10.M.3 and 10.M.6, i.e., the geometric conditions, and Assumption 10.M.7 are not satisfied, then an iterative scheme might be used to determine the infimum. This can be done by finding the smallest scalar, say $\tilde{\gamma}^*$, such that all the following conditions are satisfied:

1. The Riccati equation

$$0 = \tilde{P}_x\tilde{A}_{xP} + \tilde{A}'_{xP}\tilde{P}_x + \tilde{C}'_{xP}\tilde{C}_{xP} - \begin{bmatrix} \tilde{B}'_{xP}\tilde{P}_x + \tilde{D}'_{xP}\tilde{C}_{xP} \\ \tilde{E}'_{xP}\tilde{P}_x + \tilde{D}'_{22P}\tilde{C}_{xP} \end{bmatrix}'$$
$$\times \begin{bmatrix} \tilde{D}'_{xP}\tilde{D}_{xP} & \tilde{D}'_{xP}\tilde{D}_{22P} \\ \tilde{D}'_{22P}\tilde{D}_{xP} & \tilde{D}'_{22P}\tilde{D}_{22P} - (\tilde{\gamma}^*)^2 I \end{bmatrix}^{-1} \begin{bmatrix} \tilde{B}'_{xP}\tilde{P}_x + \tilde{D}'_{xP}\tilde{C}_{xP} \\ \tilde{E}'_{xP}\tilde{P}_x + \tilde{D}'_{22P}\tilde{C}_{xP} \end{bmatrix},$$

has a positive definite solution $\tilde{P}_x > 0$, which satisfies

$$\tilde{D}'_{22P}[I - \tilde{D}_{xP}(\tilde{D}'_{xP}\tilde{D}_{xP})^{-1}\tilde{D}'_{xP}]\tilde{D}_{22P} < (\tilde{\gamma}^*)^2 I.$$

Here we note that all the sub-matrices in the above Riccati equation are defined as in (10.2.7) but for the auxiliary system (10.3.2) of Step 10.M.1.

2. The Riccati equation

$$0 = \tilde{Q}_x \tilde{A}_{xQ} + \tilde{A}'_{xQ}\tilde{Q}_x + \tilde{C}'_{xQ}\tilde{C}_{xQ} - \begin{bmatrix} \tilde{B}'_{xQ}\tilde{Q}_x + \tilde{D}'_{xQ}\tilde{C}_{xQ} \\ \tilde{E}'_{xQ}\tilde{Q}_x + \tilde{D}'_{22Q}\tilde{C}_{xQ} \end{bmatrix}'$$

$$\times \begin{bmatrix} \tilde{D}'_{xQ}\tilde{D}_{xQ} & \tilde{D}'_{xQ}\tilde{D}_{22Q} \\ \tilde{D}'_{22Q}\tilde{D}_{xQ} & \tilde{D}'_{22Q}\tilde{D}_{22Q} - (\tilde{\gamma}^*)^2 I \end{bmatrix}^{-1} \begin{bmatrix} \tilde{B}'_{xQ}\tilde{Q}_x + \tilde{D}'_{xQ}\tilde{C}_{xQ} \\ \tilde{E}'_{xQ}\tilde{Q}_x + \tilde{D}'_{22Q}\tilde{C}_{xQ} \end{bmatrix},$$

has a positive definite solution $\tilde{Q}_x > 0$, which satisfies

$$\tilde{D}'_{22Q}[I - \tilde{D}_{xQ}(\tilde{D}'_{xQ}\tilde{D}_{xQ})^{-1}\tilde{D}'_{xQ}]\tilde{D}_{22Q} < (\tilde{\gamma}^*)^2 I.$$

Similarly, we note that all the sub-matrices in the above Riccati equation are defined as in (10.2.7) but for the auxiliary system (10.3.3) of Step 10.M.2.

3. Finally, the coupling condition holds, i.e.,

$$\lambda_{\max}\left\{\tilde{P}_x \Gamma \tilde{Q}_x \Gamma'\right\} < (\tilde{\gamma}^*)^2, \tag{10.3.12}$$

where Γ is as defined in (10.3.4). ®

The following example illustrates our computational algorithms.

Example 10.3.1. We consider a discrete-time measurement feedback system (10.3.1) with A, B, E, C_2, D_2 and D_{22} being given as those in Example 10.2.1 of the previous section. We consider the full state feedback case first, i.e., $C_1 = I$ and $D_1 = 0$. Following the algorithm and the simplified procedure in Remark 10.3.2, we obtain those matrices as in the full information case and

$$\Gamma_{sQ} = \begin{bmatrix} 1 & 1 & 1 & 0 & 1 \\ -1 & 0 & 0 & 0 & 0 \\ 0 & -1 & 0 & 0 & 0 \\ 0 & 0 & -1 & 0 & 0 \\ 0 & 0 & 0 & 1 & 0 \end{bmatrix}, \quad n_{xQ} = 1,$$

$$S_{xQ} = 1, \quad T_{xQ} = 0, \quad \Gamma = \begin{bmatrix} 1 \\ 1 \\ 0 \end{bmatrix},$$

and

$$\gamma^* = 3.181043.$$

Now, we consider the computation of γ^* for the given system with an output measurement characterized by

$$C_1 = [\,0 \quad 0 \quad 0 \quad 0 \quad 1\,], \quad D_1 = 0. \tag{10.3.13}$$

It can be shown that (A, C_1) is detectable and (A, E, C_1, D_1) is invertible with three invariant zeros at 0, 0.618 and -1.618, respectively, and one infinite zero of order 2. Hence, Assumption 10.M.6 is automatically satisfied. Following the algorithm, we obtain

$$M = \begin{bmatrix} 52.08746 & 76.55250 & 66.46233 & -0.95905 & 2.61803 & -4.23607 \\ 92.57546 & 138.46401 & 120.13777 & -1.65303 & 5.23607 & -7.85410 \\ 28.03444 & 42.12461 & 36.88854 & -0.69398 & 2.61803 & -2.61803 \\ 19.20270 & 29.28949 & 24.96658 & 0 & 0 & -1.44097 \\ 0 & 0 & 0 & 0 & 0 & 0 \\ -46.97871 & -70.77709 & -61.686918 & 0.95905 & -3.61803 & 4.23607 \end{bmatrix},$$

and

$$\gamma^* = 15.16907.$$

Ⓔ

10.4. Plants with Unit Circle Zeros

We discuss in this section a non-iterative algorithm for computing γ^* of the measurement feedback system (10.3.1) whose subsystems Σ_P and/or Σ_Q have invariant zeros on the unit circle. We assume that (A, B) is stabilizable and (A, C_1) is detectable. Let F and K be matrices of appropriate dimensions such that $A + BF$ and $A + KC_1$ have no eigenvalue at -1 and define

$$\left.\begin{aligned} \tilde{A}_P &:= (A + BF + I)^{-1}(A + BF - I), \\ \tilde{B}_P &:= 2(A + BF + I)^{-1}B, \\ \tilde{E}_P &:= 2(A + BF + I)^{-1}E, \\ \tilde{C}_{2P} &:= (C_2 + D_2F)(A + BF + I)^{-1}, \\ \tilde{D}_{2P} &:= D_2 - (C_2 + D_2F)(A + BF + I)^{-1}B, \\ \tilde{D}_{22P} &:= D_{22} - (C_2 + D_2F)(A + BF + I)^{-1}E, \end{aligned}\right\} \qquad (10.4.1)$$

and

$$\left.\begin{aligned} \tilde{A}_Q &:= (A + KC_1 + I)^{-1}(A + KC_1 - I), \\ \tilde{C}_{1Q} &:= 2C_1(A + KC_1 + I)'^{-1}, \\ \tilde{C}_{2Q} &:= 2C_2(A + KC_1 + I)'^{-1}, \\ \tilde{E}_Q &:= (A + KC_1 + I)^{-1}(E + KD_1), \\ \tilde{D}_{1Q} &:= D_1 - C_1(A + KC_1 + I)^{-1}(E + KD_1), \\ \tilde{D}_{22Q} &:= D_{22} - C_2(A + KC_1 + I)^{-1}(E + KD_1). \end{aligned}\right\} \qquad (10.4.2)$$

Let $\tilde{\Sigma}_P$ denote the system characterized by $(\tilde{A}_P, \tilde{B}_P, \tilde{C}_{2P}, \tilde{D}_{2P})$ and $\tilde{\Sigma}_Q^\star$ denote the system characterized by $(\tilde{A}_Q', \tilde{C}_{1Q}', \tilde{E}_Q', \tilde{D}_{1Q}')$. We also make the following assumptions:

Assumption 10.Z.1: $\mathrm{Im}\,(\tilde{D}_{22\mathrm{P}}) \subset \mathrm{Im}\,(\tilde{D}_{2\mathrm{P}})$;

Assumption 10.Z.2: $\mathrm{Im}\left[\tilde{E}_{\mathrm{P}} - \tilde{B}_{\mathrm{P}}(\tilde{D}_{2\mathrm{P}}'\tilde{D}_{2\mathrm{P}})^{\dagger}\tilde{D}_{2\mathrm{P}}'\tilde{D}_{22\mathrm{P}}\right] \subset \mathcal{V}^-(\tilde{\Sigma}_{\mathrm{P}}) + \mathcal{S}^-(\tilde{\Sigma}_{\mathrm{P}})$;

Assumption 10.Z.3: $\mathrm{Im}\,(\tilde{D}_{22\mathrm{Q}}') \subset \mathrm{Im}\,(\tilde{D}_{1\mathrm{Q}}')$;

Assumption 10.Z.4: $\mathrm{Im}\left[\tilde{C}_{2\mathrm{Q}}' - \tilde{C}_{1\mathrm{Q}}'(\tilde{D}_{1\mathrm{Q}}\tilde{D}_{1\mathrm{Q}}')^{\dagger}\tilde{D}_{1\mathrm{Q}}\tilde{D}_{22\mathrm{Q}}'\right] \subset \mathcal{V}^-(\tilde{\Sigma}_{\mathrm{Q}}^{\star}) + \mathcal{S}^-(\tilde{\Sigma}_{\mathrm{Q}}^{\star})$.

Ⓐ

It can be shown that Assumptions 10.Z.1-10.Z.4 are independent of the choice of F and K in (10.4.1) and (10.4.2). The computation of γ^* for a plant whose subsystems have invariant zeros on the unit circle can be done by slightly modifying the algorithm given in Section 6.4 of Chapter 6. In particular, Σ_{P} in Steps 6.Z.1 and 6.Z.5 should be replaced by $\tilde{\Sigma}_{\mathrm{P}}$ and (6.4.2) should be replaced by the following

$$\Gamma_{s\mathrm{P}}^{-1}\left[\tilde{E}_{\mathrm{P}} - \tilde{B}_{\mathrm{P}}(\tilde{D}_{2\mathrm{P}}'\tilde{D}_{2\mathrm{P}})^{\dagger}\tilde{D}_{2\mathrm{P}}'\tilde{D}_{22\mathrm{P}}\right] = \begin{bmatrix} E_{a\mathrm{P}}^{+} \\ E_{b\mathrm{P}} \\ E_{a\mathrm{P}}^{0} \\ E_{a\mathrm{P}}^{-} \\ E_{c\mathrm{P}} \\ E_{d\mathrm{P}} \end{bmatrix}. \tag{10.4.3}$$

Also, $\Sigma_{\mathrm{Q}}^{\star}$ in Steps 6.Z.2 and 6.Z.5 should be replaced by $\tilde{\Sigma}_{\mathrm{Q}}^{\star}$ and (6.4.19) should be replaced by

$$\Gamma_{s\mathrm{Q}}^{-1}\left[\tilde{C}_{2\mathrm{Q}}' - \tilde{C}_{1\mathrm{Q}}'(\tilde{D}_{1\mathrm{Q}}\tilde{D}_{1\mathrm{Q}}')^{\dagger}\tilde{D}_{1\mathrm{Q}}\tilde{D}_{22\mathrm{Q}}'\right] = \begin{bmatrix} E_{a\mathrm{Q}}^{+} \\ E_{b\mathrm{Q}} \\ E_{a\mathrm{Q}}^{0} \\ E_{a\mathrm{Q}}^{-} \\ E_{c\mathrm{Q}} \\ E_{d\mathrm{Q}} \end{bmatrix}. \tag{10.4.4}$$

The rest of the algorithm remains the same.

Chapter 11

Solutions to Discrete-time H_∞ Problem

11.1. Introduction

THIS CHAPTER IS concerned with the discrete-time H_∞ control problem with full state feedback, full information feedback and general measurement feedback. The objective is to present a solution to the discrete-time H_∞ control problem. One way to approach this problem is to transform the discrete-time H_∞ optimal control problem into an equivalent continuous-time H_∞ control problem via bilinear transformation (see Chapter 3). Then the continuous-time controllers that are solutions to the auxiliary problem can be obtained and transformed back to their discrete-time equivalent using inverse bilinear transformation (see again Chapter 3). Another way is to solve this problem directly in discrete-time setting and in terms of the performance of the original system. This approach leaves the possibility of directly observing the effect of certain physical parameters. Finally, a novel aspect of this chapter is that we show that if certain states or disturbances are observed directly, then this yields the possibility of deriving a reduced order controller. This result corresponds with the continuous-time reduced order controller structure of Chapter 7. In fact, all results presented in this chapter can be regarded as the counterparts of those in Chapter 7.

The main results of this chapter are similar to those in [125], but our presentation is quite different. We arrange them in a way so that it is easier for software implementation.

11.2. Full Information and State Feedbacks

We first consider in this section the following full information feedback system,

$$\Sigma : \begin{cases} x(k+1) = A\ x(k) + B\ u(k) + E\ w(k), \\ y(k) = \begin{pmatrix} I \\ 0 \end{pmatrix} x(k) + \begin{pmatrix} 0 \\ I \end{pmatrix} w(k), \\ h(k) = C_2\ x(k) + D_2\ u(k) + D_{22}\ w(k), \end{cases} \tag{11.2.1}$$

where $x \in \mathbb{R}^n$ is the state, $u \in \mathbb{R}^m$ is the control input, $w \in \mathbb{R}^q$ is the external disturbance input, $y \in \mathbb{R}^{n+q}$ is the measurement output, and $h \in \mathbb{R}^\ell$ is the controlled output of Σ. As usual, we define Σ_P to be the subsystem characterized by the matrix quadruple (A, B, C_2, D_2). We assume that Σ_P has no invariant zero on the unit circle and its infimum is given by γ^*. We are interested in designing a full information feedback control law

$$u(k) = F_1 x(k) + F_2 w(k), \tag{11.2.2}$$

such that when it is applied to the given system (11.2.1), the resulting closed-loop system is asymptotically stable and the resulting closed-loop transfer matrix from w to h has an H_∞-norm less than a given $\gamma > \gamma^*$.

In what follows, we state a step-by-step algorithm for the computation of F_1 and F_2.

Step 11.F.1: Without loss of generality but for simplicity of presentation, we assume that the quadruple (A, B, C_2, D_2), i.e., Σ_P, has been partitioned in the form of (2.4.4). Then, transform Σ_P into the special coordinate basis as described in Chapter 2, i.e., find nonsingular transformations Γ_s, Γ_i and Γ_o such that

$$\Gamma_s^{-1}(A - B_0 C_{2,0})\Gamma_s = \begin{bmatrix} A_{cc} & B_c E_{ca}^- & B_c E_{ca}^+ & L_{cd} C_d & L_{cb} C_b \\ 0 & A_{aa}^- & 0 & L_{ad}^- C_d & L_{ab}^- C_b \\ 0 & 0 & A_{aa}^+ & L_{ad}^+ C_d & L_{ab}^+ C_b \\ B_d E_{dc} & B_d E_{da}^- & B_d E_{da}^+ & A_{dd} & B_d E_{db} \\ 0 & 0 & 0 & L_{bd} C_d & A_{bb} \end{bmatrix},$$

$$\Gamma_o^{-1} \begin{bmatrix} C_{2,0} \\ C_{2,1} \end{bmatrix} \Gamma_s = \begin{bmatrix} C_{0c} & C_{0a}^- & C_{0a}^+ & C_{0b} & C_{0d} \\ 0 & 0 & 0 & C_d & 0 \\ 0 & 0 & 0 & 0 & C_b \end{bmatrix},$$

$$\Gamma_s^{-1} [\, B_0 \quad B_1 \,] \Gamma_i = \begin{bmatrix} B_{0c} & 0 & B_c \\ B_{0a}^- & 0 & 0 \\ B_{0a}^+ & 0 & 0 \\ B_{0d} & B_d & 0 \\ B_{0b} & 0 & 0 \end{bmatrix}, \quad \Gamma_o^{-1} D_2 \Gamma_i = \begin{bmatrix} I_{m_0} & 0 & 0 \\ 0 & 0 & 0 \\ 0 & 0 & 0 \end{bmatrix}.$$

Note that an additional permutation matrix to the state transformation has been introduced here to the original SCB such that the new state variables are ordered as follows:

$$\tilde{x} = \begin{pmatrix} x_c \\ x_a^- \\ x_a^+ \\ x_d \\ x_b \end{pmatrix}. \tag{11.2.3}$$

Next, we compute

$$\Gamma_s^{-1} E/\gamma = \begin{bmatrix} E_c \\ E_a^- \\ E_a^+ \\ E_d \\ E_b \end{bmatrix}. \tag{11.2.4}$$

Step 11.F.2: Let F_c be any appropriate dimensional constant matrix such that all the eigenvalues of $A_{cc} - B_c F_c$ are on the open unit disc. This can be done as (A_{cc}, B_c) is completely controllable.

Step 11.F.3: Define A_x, B_x, E_x, C_x and D_x as follows:

$$A_x := \begin{bmatrix} A_{aa}^+ & L_{ad}^+ C_d & L_{ab}^+ C_b \\ B_d E_{da}^+ & A_{dd} & B_d E_{db} \\ 0 & L_{bd} C_d & A_{bb} \end{bmatrix}, \quad E_x := \begin{bmatrix} E_a^+ \\ E_d \\ E_b \end{bmatrix}, \tag{11.2.5}$$

$$B_x := \begin{bmatrix} B_{0a}^+ & 0 \\ B_{0d} & B_d \\ B_{0b} & 0 \end{bmatrix}, \quad D_{22x} := D_{22}/\gamma, \tag{11.2.6}$$

and

$$C_x := \Gamma_o \begin{bmatrix} 0 & 0 & 0 \\ 0 & C_d & 0 \\ 0 & 0 & C_b \end{bmatrix}, \quad D_x = \Gamma_o \begin{bmatrix} I_{m_0} & 0 \\ 0 & 0 \\ 0 & 0 \end{bmatrix}. \tag{11.2.7}$$

Step 11.F.4: Solve the following discrete-time algebraic Riccati equation:

$$P_x = A_x' P_x A_x + C_x' C_x - \begin{bmatrix} B_x' P_x A_x + D_x' C_x \\ E_x' P_x A_x + D_{22x}' C_x \end{bmatrix}' G_x^{-1} \begin{bmatrix} B_x' P_x A_x + D_x' C_x \\ E_x' P_x A_x + D_{22x}' C_x \end{bmatrix} \tag{11.2.8}$$

where

$$G_x := \begin{bmatrix} D_x' D_x + B_x' P_x B_x & B_x' P_x E_x \\ E_x' P_x B_x & E_x' P_x E_x + D_{22x}' D_{22x} - I \end{bmatrix}, \tag{11.2.9}$$

for $P_x > 0$. Note that because (A_x, B_x, C_x, D_x) is left invertible and only has unstable invariant zeros, such a P_x always exists provided that

$\gamma > \gamma^*$. In fact, one can use the very accurate method given previously in Chapter 3 to obtain this P_x. For future use in the output feedback case, we compute

$$X = (\Gamma_s^{-1})' \begin{bmatrix} 0 & 0 \\ 0 & P_x \end{bmatrix} \Gamma_s^{-1}. \tag{11.2.10}$$

Step 11.F.5: Next, compute

$$F_{1x} = (B_x' P_x B_x + D_x' D_x)^{-1} (B_x' P_x A_x + D_x' C_x), \tag{11.2.11}$$

and

$$F_{2x} = (B_x' P_x B_x + D_x' D_x)^{-1} (B_x' P_x E_x + D_x' D_{22x}). \tag{11.2.12}$$

Then, partition F_{1x} as follows:

$$F_{1x} = \begin{bmatrix} F_{0ax}^+ & F_{0dx} & F_{0bx} \\ F_{dax}^+ & F_{ddx} & F_{dbx} \end{bmatrix}. \tag{11.2.13}$$

Step 11.F.6: Finally, the gain matrices F_1 and F_2 are respectively given by

$$F_1 = -\Gamma_i \begin{bmatrix} C_{0c} & C_{0a}^- & C_{0a}^+ + F_{0ax}^+ & C_{0d} + F_{0dx} & C_{0b} + F_{0bx} \\ E_{dc} & E_{da}^- & F_{dax}^+ & F_{ddx} & F_{dbx} \\ F_c & \star & \star & \star & \star \end{bmatrix} \Gamma_s^{-1}, \tag{11.2.14}$$

and

$$F_2 = -\Gamma_i \begin{bmatrix} F_{2x} \\ \star \end{bmatrix} \gamma, \tag{11.2.15}$$

where $\star$s are some arbitrary matrices with appropriate dimensions. ▣

We have the following theorem.

Theorem 11.2.1. Consider the full information feedback discrete-time system (11.2.1). Then under the full information feedback law,

$$u(k) = F_1 x(k) + F_2 w(k), \tag{11.2.16}$$

with F_1 and F_2 given by (11.2.14) and (11.2.15), respectively, the closed-loop system is asymptotically stable and the H_∞-norm of the closed-loop transfer matrix from the disturbance w to the controlled output h is less than γ. ▣

Proof. It is straightforward to verify that the poles of the closed-loop system comprising the given full information system (11.2.1) with the control law (11.2.16) are given by $A_{cc} - B_c F_c$, A_{aa}^- and $A_x - B_x F_{1x}$. We note that both $A_{cc} - B_c F_c$ and A_{aa}^- are asymptotically stable. Hence, the closed-loop system is stable if and only if $A_x - B_x F_{1x}$ is stable. Moreover, it is also simple to show that its closed-loop transfer matrix from w to h, say T_{hw}, is equal to $\gamma T_{h_x w_x}$, where $T_{h_x w_x}$ is the transfer matrix from w_x to h_x of the closed-loop system comprising the following auxiliary system,

$$\begin{cases} \dot{x}_x = A_x \; x_x + B_x \; u_x + E_x \; w_x, \\ y_x = \begin{pmatrix} I \\ 0 \end{pmatrix} x_x + \begin{pmatrix} 0 \\ I \end{pmatrix} w_x, \\ h_x = C_x \; x_x + D_x \; u_x + D_{22x} \; w_x, \end{cases} \tag{11.2.17}$$

with a full information control law,

$$u_x = -F_{1x} x_x - F_{2x} w_x. \tag{11.2.18}$$

Because (A_x, B_x, C_x, D_x) is left invertible and has only unstable invariant zeros, it follows from the result of [124] that the solution to the Riccati equation (11.2.8) is indeed a positive definite one provided that $\gamma > \gamma^*$. Moreover, we also have $A_x - B_x F_{1x}$ is asymptotically stable and $\|T_{h_x w_x}\|_\infty < 1$. Hence, the result of Theorem 11.2.1 follows. ⊠

We illustrate the above result with a numerical example.

Example 11.2.1. Consider a discrete-time full information system (11.2.1) with matrices A, B, E, C_2, D_2 and D_{22} are as given in Example 10.2.1 of the previous chapter. The infimum for this problem was computed in Example 10.2.1 to be $\gamma^* = 0.934173$. Let us choose a $\gamma = 0.934174$, which is slightly larger than γ^*. Following the above algorithm, we obtain

$$F_1 = \begin{bmatrix} 0 & 0 & -0.745354 & -1.078688 & -1.078688 \\ -1 & -1 & -1.412022 & -1.872678 & -1.872678 \\ -1 & 0 & 0 & 0 & 0 \end{bmatrix},$$

and

$$F_2 = \begin{bmatrix} -0.872677 \\ -1.206011 \\ 0 \end{bmatrix}.$$

The closed-loop poles, i.e., $\lambda(A+BF_1) = \{0, 0, 0, 0, 0.38197\}$. The singular value plot of the closed-loop transfer matrix from w to h in Figure 11.2.1 clearly shows that its H_∞-norm is less than the given $\gamma = 0.934174$. 🄴

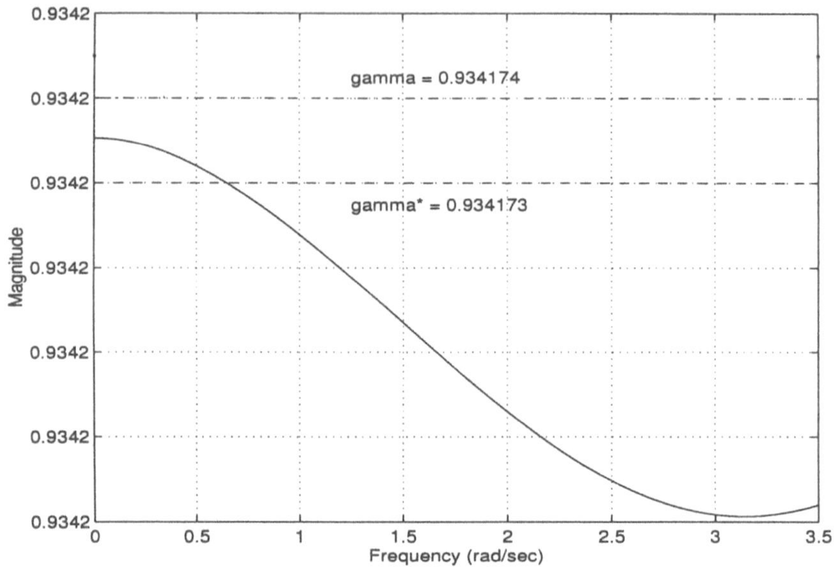

Figure 11.2.1: Singular values of T_{hw} under full information feedback.

As was shown in Chapter 10, for discrete-time systems, the infimum associated with the given full information feedback system is in general different from that associated with its corresponding full state feedback system, i.e.,

$$\begin{cases} x(k+1) = A\ x(k) + B\ u(k) + E\ w(k), \\ y(k) = x(k) \\ h(k) = C_2\ x(k) + D_2\ u(k) + D_{22}\ w(k). \end{cases} \tag{11.2.19}$$

Let γ^* be the infimum associated with the full state feedback problem. Then, for any given $\gamma > \gamma^*$, the following algorithm will produce a static state feedback law that achieves the closed-loop stability as well as the required H_∞-norm bound of the closed-loop transfer matrix from w to h.

Step 11.S.1 to 11.S.4: These steps are identical to Step 11.F.1 to 11.F.4, respectively.

Step 11.S.5: Compute

$$\begin{aligned} H_x := B_x' P_x B_x + D_x' D_x + (B_x' P_x E_x + D_x' D_{22x}) \\ \times (I - D_{22x}' D_{22x} - E_x' P_x E_x)^{-1} (E_x' P_x B_x + D_{22x}' D_x), \end{aligned} \tag{11.2.20}$$

$$\begin{aligned} F_x := H_x^{-1} \Big[B_x' P_x A_x + D_x' C_x + (B_x' P_x E_x + D_x' D_{22x}) \\ \times (I - D_{22x}' D_{22x} - E_x' P_x E_x)^{-1} (E_x' P_x A_x + D_{22x}' C_x) \Big]. \end{aligned} \tag{11.2.21}$$

Then, partition F_x as follows:

$$F_x = \begin{bmatrix} F_{0ax}^+ & F_{0dx} & F_{0bx} \\ F_{dax}^+ & F_{ddx} & F_{dbx} \end{bmatrix}. \tag{11.2.22}$$

Step 11.S.6: The gain matrix F is given by

$$F = -\Gamma_i \begin{bmatrix} C_{0c} & C_{0a}^- & C_{0a}^+ + F_{0ax}^+ & C_{0d} + F_{0dx} & C_{0b} + F_{0bx} \\ E_{dc} & E_{da}^- & F_{dax}^+ & F_{ddx} & F_{dbx} \\ F_c & \star & \star & \star & \star \end{bmatrix} \Gamma_s^{-1}, \tag{11.2.23}$$

where $\star$s are some arbitrary matrices with appropriate dimensions. 🄰

Following the lines of reasoning similar to the proof of Theorem 11.2.1, one can show that the static control law,

$$u(k) = Fx(k), \tag{11.2.24}$$

with F given by (11.2.23), will i) achieve the closed-loop stability, and ii) make the H_∞-norm of the resulting closed-loop transfer matrix from w to h less than the given γ. We illustrate this in the following example.

Example 11.2.2. Let us consider a discrete-time full state feedback system (11.2.19) with matrices A, B, E, C_2, D_2 and D_{22} are as given in Example 10.2.1 of Chapter 10. The infimum for this problem was computed in Example 10.3.1 and is given by $\gamma^* = 3.181043$. Let us choose a $\gamma = 3.181044$, which is slightly larger than γ^*. Following the above algorithm, we obtain

$$F = \begin{bmatrix} 0 & 0 & -0.432563 & -0.885373 & -0.885373 \\ -1 & -1 & -1.479753 & -1.914538 & -1.914538 \\ -1 & 0 & 0 & 0 & 0 \end{bmatrix}.$$

The closed-loop poles, i.e., $\lambda(A + BF) = \{0, 0, 0, 0.27093, 0.38197\}$ and the singular value plot of the closed-loop transfer matrix from w to h in Figure 11.2.2 clearly shows that its H_∞-norm is less that the given $\gamma = 3.181044$. 🄴

11.3. Full Order Output Feedback

We construct solutions to the discrete-time H_∞ control problem for the following measurement feedback discrete-time system,

$$\Sigma : \begin{cases} x(k+1) = A\ x(k) + B\ u(k) + E\ w(k), \\ y(k) \ \ = C_1\ x(k) + D_1\ w(k), \\ h(k) \ \ = C_2\ x(k) + D_2\ u(k) + D_{22}\ w(k), \end{cases} \tag{11.3.1}$$

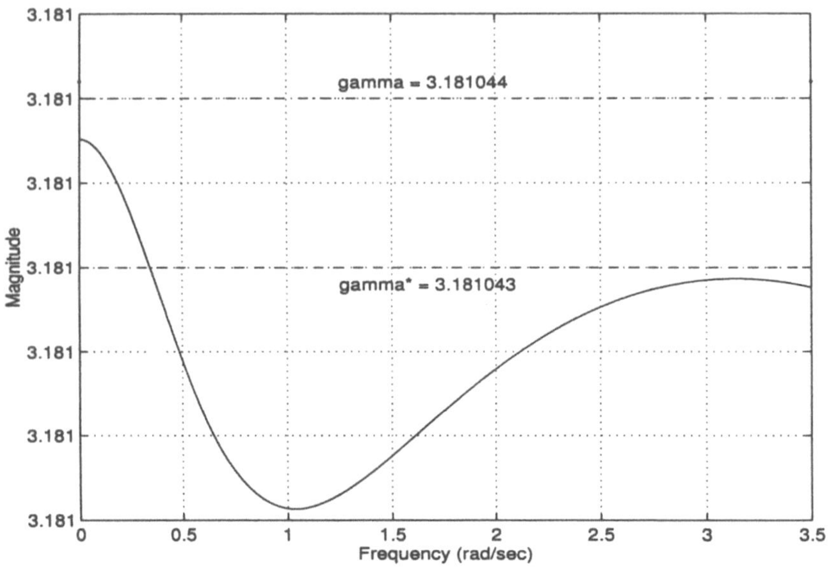

Figure 11.2.2: Singular values of T_{hw} under full state feedback.

where $x \in \mathbf{R}^n$ is the state, $u \in \mathbf{R}^m$ is the control input, $w \in \mathbf{R}^q$ is the external disturbance input, $y \in \mathbf{R}^p$ is the measurement output, and $h \in \mathbf{R}^\ell$ is the controlled output of Σ. Again, for the purpose of easy reference, we define Σ_P to be the subsystem characterized by the matrix quadruple (A, B, C_2, D_2) and Σ_Q to be the subsystem characterized by the matrix quadruple (A, E, C_1, D_1). We assume in this section that both subsystems Σ_P and Σ_Q have no invariant zero on the unit circle.

Let γ^* be the infimum for the given Σ of (11.3.1). Given a positive scalar $\gamma > \gamma^*$, the following algorithm will produce a measurement feedback control law that achieves i) internal stability for the closed-loop system, and ii) the resulting $\|T_{hw}\|_\infty < \gamma$.

Step 11.M.1: Define an auxiliary full information problem for

$$\begin{cases} x(k+1) = A\ x(k) + B\ u(k) + E\ w(k), \\ y(k) = \begin{pmatrix} 0 \\ I \end{pmatrix} x(k) + \begin{pmatrix} I \\ 0 \end{pmatrix} w(k), \\ h(k) = C_2\ x(k) + D_2\ u(k) + D_{22}\ w(k), \end{cases} \tag{11.3.2}$$

and perform Steps 11.F.1 to 11.F.4 of the algorithm given in the previous section to get a positive semi-definite matrix X. Let $P := X$ and compute

$$V := B'PB + D_2'D_2, \tag{11.3.3}$$

and

$$R := \gamma^2 I - D_{22}' D_{22} - E'PE + (E'PB + D_{22}' D_2) V^{\dagger} (B'PE + D_2' D_{22}), \tag{11.3.4}$$

where † denotes the Moore-Penrose (pseudo) inverse. It can be shown that $R > 0$. Next, compute

$$A_s := A - BV^{\dagger}(B'PA + D_2' C_2), \tag{11.3.5}$$

$$C_s := C_2 - D_2 V^{\dagger}(B'PA + D_2' C_2). \tag{11.3.6}$$

and calculate

$$\begin{aligned}
A_{\mathrm{P}} &:= A + ER^{-1}(E'PA_s + D_{22}' C_s),\\
E_{\mathrm{P}} &:= ER^{-\frac{1}{2}},\\
C_{1\mathrm{P}} &:= C_1 + D_1 R^{-1}(E'PA_s + D_{22}' C_s),\\
C_{2\mathrm{P}} &:= (V^{\frac{1}{2}})^{\dagger}\left[B'PA + D_2' C_2 + (B'PE + D_2' D_{22}) R^{-1}(E'PA_s + D_{22}' C_s)\right],\\
D_{1\mathrm{P}} &:= D_1 R^{-\frac{1}{2}},\\
D_{2\mathrm{P}} &:= V^{\frac{1}{2}},\\
D_{22\mathrm{P}} &:= (V^{\frac{1}{2}})^{\dagger}(B'PE + D_2' D_{22}) R^{-\frac{1}{2}}.
\end{aligned}$$

Step 11.M.2: Define another auxiliary full information problem for

$$\begin{cases}
x(k+1) = A' \; x(k) + C_1' \; u(k) + C_2' \; w(k),\\
y(k) = \begin{pmatrix} 0 \\ I \end{pmatrix} x(k) + \begin{pmatrix} I \\ 0 \end{pmatrix} w(k),\\
h(k) = E' \; x(k) + D_1' \; u(k) + D_{22}' \; w(k),
\end{cases} \tag{11.3.7}$$

and again perform Steps 11.F.1 to 11.F.4 of the algorithm in the previous section to get another positive semi-definite matrix X and let $Q := X$. Also, let

$$Y := (\gamma^2 I - QP)^{-1} Q. \tag{11.3.8}$$

Step 11.M.3: Next, compute

$$\left.\begin{aligned}
W_{\mathrm{P}} &:= D_{1\mathrm{P}} D_{1\mathrm{P}}' + C_{1\mathrm{P}} Y C_{1\mathrm{P}}',\\
S_{\mathrm{P}} &:= (C_{2\mathrm{P}} Y C_{1\mathrm{P}} + D_{22\mathrm{P}} D_{1\mathrm{P}}') W_{\mathrm{P}}^{\dagger} (C_{1\mathrm{P}} Y C_{2\mathrm{P}} + D_{1\mathrm{P}} D_{22\mathrm{P}}')\\
&\qquad + \gamma^2 I - D_{22\mathrm{P}} D_{22\mathrm{P}}' - C_{2\mathrm{P}} Y C_{2\mathrm{P}}',\\
A_z &:= A_{\mathrm{P}} - (A_{\mathrm{P}} Y C_{1\mathrm{P}}' + E_{\mathrm{P}} D_{1\mathrm{P}}') W_{\mathrm{P}}^{\dagger} C_{1\mathrm{P}},\\
E_z &:= E_{\mathrm{P}} - (A_{\mathrm{P}} Y C_{1\mathrm{P}}' + E_{\mathrm{P}} D_{1\mathrm{P}}') W_{\mathrm{P}}^{\dagger} D_{1\mathrm{P}},
\end{aligned}\right\} \tag{11.3.9}$$

and

$$\left.\begin{aligned}
A_{\rm PY} &:= A_{\rm P} + (A_z Y C_{2\rm P}' + E_z D_{22\rm P}') S_{\rm P}^{-1} C_{2\rm P},\\
B_{\rm PY} &:= B + (A_z Y C_{2\rm P} + E_z D_{22\rm P}') S_{\rm P}^{-1} D_{2\rm P},\\
E_{\rm PY} &:= \Big[(A_z Y C_{2\rm P} + E_z D_{22\rm P}') S_{\rm P}^{-1} (C_{2\rm P} Y C_{1\rm P} + D_{22\rm P} D_{1\rm P}')\\
&\qquad + A_{\rm P} Y C_{1\rm P}' + E_{\rm P} D_{1\rm P}' \Big] (W_{\rm P}^{\frac{1}{2}})^{\dagger},\\
C_{2\rm PY} &:= S_{\rm P}^{-\frac{1}{2}} C_{2\rm P},\\
D_{1\rm PY} &:= W_{\rm P}^{\frac{1}{2}},\\
D_{2\rm PY} &:= S_{\rm P}^{-\frac{1}{2}} D_{2\rm P},\\
D_{22\rm PY} &:= S_{\rm P}^{-\frac{1}{2}} (C_{2\rm P} Y C_{1\rm P}' + D_{22\rm P} D_{1\rm P}') (W_{\rm P}^{\frac{1}{2}})^{\dagger}.
\end{aligned}\right\} \tag{11.3.10}$$

It can be shown that i) the quadruple $(A_{\rm PY}, B_{\rm PY}, C_{2\rm PY}, D_{2\rm PY})$ is right invertible and of minimum phase with no infinite zero, and ii) the quadruple $(A_{\rm PY}, E_{\rm PY}, C_{1\rm P}, D_{1\rm PY})$ is left invertible and of minimum phase with no infinite zero. Moreover, there exists an appropriate constant matrix $X_{\rm PY}$ such that $D_{2\rm PY} + D_{2\rm PY} X_{\rm PY} D_{1\rm PY} = 0$.

Step 11.M.4: Let

$$F_{1\rm PY} := -D_{2\rm P}^{\dagger} C_{2\rm P} + (I - D_{2\rm P}^{\dagger} D_{2\rm P}) F_0, \tag{11.3.11}$$

$$F_{2\rm PY} := -D_{2\rm PY}^{\dagger} D_{22\rm PY}, \tag{11.3.12}$$

where F_0 is such that $A_{\rm P} + B F_{1\rm PY} = A_{\rm PY} + B_{\rm PY} F_{1\rm PY}$ has all its eigenvalues inside the unit circle. Also, let

$$K_{1\rm PY} := -E_{\rm PY} D_{1\rm PY}^{\dagger} + K_0 (I - D_{1\rm PY} D_{1\rm PY}^{\dagger}), \tag{11.3.13}$$

$$K_{2\rm PY} := -D_{1\rm PY}^{\dagger}, \tag{11.3.14}$$

where K_0 is such that $A_{\rm PY} + K_{1\rm PY} C_{1\rm P}$ is stable. We would like to note that a more systematic procedure to compute the above gain matrices will be given in the next chapter.

Step 11.M.5: Finally, we obtain a measurement output feedback control law,

$$\Sigma_{\rm cmp} \ : \ \begin{cases} v(k+1) = A_{\rm cmp}\, v(k) + B_{\rm cmp}\, y(k), \\ \quad u(k) \ \ = C_{\rm cmp}\, v(k) + D_{\rm cmp}\, y(k), \end{cases} \tag{11.3.15}$$

with

$$\left.\begin{aligned}
D_{\rm cmp} &:= -F_{2\rm PY} K_{2\rm PY},\\
C_{\rm cmp} &:= F_{1\rm PY} - D_{\rm cmp} C_{1\rm P},\\
B_{\rm cmp} &:= B_{\rm PY} D_{\rm cmp} - K_{1\rm PY},\\
A_{\rm cmp} &:= A_{\rm PY} + B_{\rm PY} C_{\rm cmp} + K_{1\rm PY} C_{1\rm P}.
\end{aligned}\right\} \tag{11.3.16}$$

Clearly, $v \in \mathbb{R}^n$, i.e., the obtained controller $\Sigma_{\rm cmp}$ has the same dynamical order as that of the given system Σ. ▣

We have the following theorem.

Theorem 11.3.1. Consider the given discrete-time system Σ of (11.3.1) and the controller $\Sigma_{\rm cmp}$ of (11.3.15) with $A_{\rm cmp}$, $B_{\rm cmp}$, $C_{\rm cmp}$ and $D_{\rm cmp}$ being given by (11.3.16). Also, let $\gamma > \gamma^*$ be given. Then, we have

1. the resulting closed-loop system comprising Σ and $\Sigma_{\rm cmp}$ is asymptotically stable; and

2. the H_∞-norm of the closed-loop transfer matrix from the disturbance w to the controlled output h is less than γ. ▣

Proof. The proof of the above theorem can be carried out in two stages: The first stage involves showing that the following two statements are equivalent:

1. The closed-loop system comprising the given system Σ of (11.3.1) and the controller $\Sigma_{\rm cmp}$ of (11.3.15) is internally stable and its transfer matrix from w to h, $T_{hw}(\Sigma \times \Sigma_{\rm cmp})$, has an H_∞-norm less than γ.

2. The closed-loop system comprising an auxiliary system $\Sigma_{\rm PY}$, where $\Sigma_{\rm PY}$ is given by

$$\begin{cases} x_{\rm PY}(k+1) = A_{\rm PY}\ x_{\rm PY}(k) + B_{\rm PY}\ u(k) + E_{\rm PY}\ w_{\rm PY}(k), \\ y(k) \quad = C_{\rm 1P}\ x_{\rm PY}(k) + D_{\rm 1PY}\ w_{\rm PY}(k), \\ h_{\rm PY}(k) \quad = C_{\rm 2PY}\ x_{\rm PY}(k) + D_{\rm 2PY}\ u(k) + D_{\rm 22PY}\ w_{\rm PY}(k), \end{cases} \tag{11.3.17}$$

and the controller $\Sigma_{\rm cmp}$ of (11.3.15) is internally stable and its transfer matrix from $w_{\rm PY}$ to $h_{\rm PY}$, $T_{h_{\rm PY}w_{\rm PY}}(\Sigma_{\rm PY} \times \Sigma_{\rm cmp})$, has an H_∞-norm less than γ.

The second stage involves showing that the transfer matrix from $w_{\rm PY}$ to $h_{\rm PY}$ of the closed-loop system comprising $\Sigma_{\rm PY}$ and $\Sigma_{\rm cmp}$ is internally stable and is in fact identically zero for all frequencies, i.e., $T_{h_{\rm PY}w_{\rm PY}}(\Sigma_{\rm PY} \times \Sigma_{\rm cmp}) \equiv 0$. It is obvious that $\|T_{h_{\rm PY}w_{\rm PY}}(\Sigma_{\rm PY} \times \Sigma_{\rm cmp})\|_\infty = 0 < \gamma$. Hence, $\Sigma \times \Sigma_{\rm cmp}$ is internally stable and $\|T_{hw}(\Sigma \times \Sigma_{\rm cmp})\|_\infty < \gamma$. We refer interested readers to [125] for more detailed proofs of the above two facts (stages). ▣

Remark 11.3.1. It is clear from the above proof that the design of a γ-suboptimal control law for the original system (11.3.1) is equivalent to finding a control law that solves the H_∞ disturbance decoupling problem with internal stability for the auxiliary system (11.3.17). One can use a more systematic procedure given in Chapter 12 to find such a control law. ▣

The following is an illustrative example.

Example 11.3.1. Let us consider a discrete-time system (11.3.1) with matrices A, B, E, C_2, D_2 and D_{22} are as given in Example 10.2.1 of Chapter 10 and

$$C_1 = [0 \quad 0 \quad 0 \quad 0 \quad 1], \quad D_1 = 0. \tag{11.3.18}$$

The infimum for this problem was computed in Example 10.3.1 and is given by $\gamma^* = 15.16907$. Let us choose a positive scalar $\gamma = 15.17$. Following our algorithm, we obtain a full order output feedback control law (11.3.15) with

$$A_{\rm cmp} = \begin{bmatrix} 0 & 1 & 1.005710 & 1.003529 & -9.516228 \\ 0 & 0 & 0.005710 & 1.003529 & -3.303781 \\ 0 & 0 & 0.691710 & 0.191432 & -1.876073 \\ 0 & 0 & -0.310193 & -0.809744 & 3.217071 \\ 0 & 0 & 0 & 1 & -3.281899 \end{bmatrix},$$

$$B_{\rm cmp} = \begin{bmatrix} 10.519757 \\ 4.307309 \\ 2.067505 \\ -4.026815 \\ 4.281899 \end{bmatrix}, \quad D_{\rm cmp} = \begin{bmatrix} -4.043756 \\ -14.546573 \\ 0 \end{bmatrix},$$

and

$$C_{\rm cmp} = \begin{bmatrix} 0 & 0 & -0.314000 & -0.812097 & 3.231659 \\ -1 & -1 & -1.315903 & -1.813273 & 12.733300 \\ -1 & 0 & 0 & 0 & 0 \end{bmatrix}.$$

The plot of the singular values of the closed-loop transfer matrix from w to h in Figure 11.3.1 shows that

$$\|T_{hw}(\Sigma \times \Sigma_{\rm cmp})\|_\infty < \gamma = 15.17. \tag{11.3.19}$$

The poles of the closed-loop system are given by

$$-0.596025, 0.618045, 0.433068, 0.382376, -0.237186, -0.000212, 0, 0, 0, 0,$$

which are all inside the unit circle. ▣

11.4. Reduced Order Output Feedback

In this section we show that for the singular H_∞ control problem, we can always find a suboptimal solution which has a dynamical order less than that of the plant and is of a reduced order observer-based structure. This result is analogous to that obtained in Chapter 7 for the continuous-time problems.

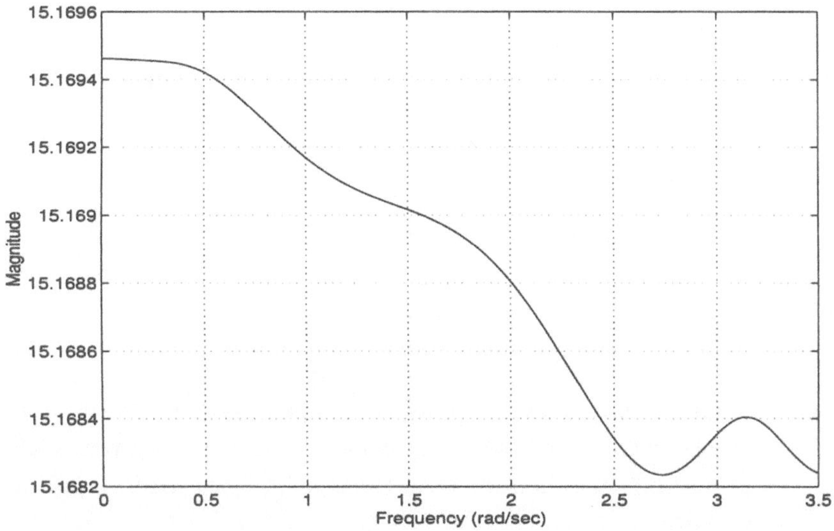

Figure 11.3.1: Singular values of T_{hw} under full order output feedback.

Without loss of generality, we develop such a reduced order observer-based controller for the system $\Sigma_{\rm PY}$ defined in the previous section, i.e.,

$$\begin{cases} x_{\rm PY}(k+1) = A_{\rm PY}\ x_{\rm PY}(k) + B_{\rm PY}\ u(k) + \ E_{\rm PY}\ \ w_{\rm PY}(k), \\ \quad y(k) \quad = C_{\rm 1P}\ x_{\rm PY}(k) \qquad\qquad\quad + D_{\rm 1PY}\ w_{\rm PY}(k), \\ \ h_{\rm PY}(k) \ = C_{\rm 2PY}\ x_{\rm PY}(k) + D_{\rm 2PY}\ u(k) + D_{\rm 22PY}\ w_{\rm PY}(k). \end{cases} \tag{11.4.1}$$

There exists a constant output pre-feedback law $X_{\rm PY}y$ such that after applying this pre-feedback law, namely setting

$$u \longrightarrow X_{\rm PY}y + u, \tag{11.4.2}$$

the direct feed-through term from $w_{\rm PY}$ from $h_{\rm PY}$ disappears. Hence without loss of generality, hereafter we assume that $D_{\rm 22PY} = 0$.

There exists an 'optimal' state feedback gain $F_{\rm PY}$ in the sense that

$$(C_{\rm 2PY} + D_{\rm 2PY}F_{\rm PY})(sI - A_{\rm PY} - B_{\rm PY}F_{\rm PY})^{-1}E_{\rm PY} \equiv 0.$$

with $A_{\rm PY} + B_{\rm PY}F_{\rm PY}$ stable. We need to construct an observer of lower order. Without loss of generality but for simplicity of presentation, we assume that the matrices $C_{\rm 1P}$ and $D_{\rm 1PY}$ are already in the form

$$C_{\rm 1P} = \begin{bmatrix} 0 & C_{1,02} \\ I_{p-m_0} & 0 \end{bmatrix} \quad \text{and} \quad D_{\rm 1PY} = \begin{bmatrix} D_{1,0} \\ 0 \end{bmatrix}, \tag{11.4.3}$$

where m_0 is the rank of D_{1PY} and $D_{1,0}$ is of full rank. Then the given system Σ_{PY} can be written as,

$$\begin{cases} \delta\begin{pmatrix} x_1 \\ x_2 \end{pmatrix} = \begin{bmatrix} A_{11} & A_{12} \\ A_{21} & A_{22} \end{bmatrix} \begin{pmatrix} x_1 \\ x_2 \end{pmatrix} + \begin{bmatrix} E_1 \\ E_2 \end{bmatrix} w_{PY} + \begin{bmatrix} B_1 \\ B_2 \end{bmatrix} u, \\ \begin{pmatrix} y_0 \\ y_1 \end{pmatrix} = \begin{bmatrix} 0 & C_{1,02} \\ I_{p-m_0} & 0 \end{bmatrix} \begin{pmatrix} x_1 \\ x_2 \end{pmatrix} + \begin{bmatrix} D_{1,0} \\ 0 \end{bmatrix} w_{PY} \\ h_{PY} = C_{2PY} \begin{pmatrix} x_1 \\ x_2 \end{pmatrix} + D_{2PY}\, u, \end{cases} \tag{11.4.4}$$

where

$$\begin{pmatrix} x_1 \\ x_2 \end{pmatrix} = x_{PY} \quad \text{and} \quad \begin{pmatrix} y_0 \\ y_1 \end{pmatrix} = y. \tag{11.4.5}$$

We note that $y_1 \equiv x_1$. Thus, one needs to estimate only the state x_2 in the reduced order estimator. Then following closely the procedure given in [29], we first rewrite the state equation for x_1 in terms of the measured output y_1 and state x_2 as follows,

$$y_1(k+1) = A_{11}y_1(k) + A_{12}x_2(k) + E_1 w_{PY}(k) + B_1 u(k), \tag{11.4.6}$$

where y_1 and u are known. Observation of x_2 is made via y_0 and

$$\tilde{y}_1(k) = A_{12}x_2(k) + E_1 w_{PY}(k) = y_1(k+1) - A_{11}y_1(k) - B_1 u(k). \tag{11.4.7}$$

A reduced order system for the estimation of state x_2 is given by

$$\begin{cases} x_2(k+1) = A_R\, x_2(k) + E_R\, w_{PY}(k) + [A_{21} \quad B_2] \begin{pmatrix} y_1(k) \\ u(k) \end{pmatrix}, \\ y_R(k) = C_R\, x_2(k) + D_R\, w_{PY}(k), \end{cases} \tag{11.4.8}$$

where

$$A_R := A_{22}, \quad E_R := E_2, \quad C_R := \begin{bmatrix} C_{1,02} \\ A_{12} \end{bmatrix}, \quad D_R := \begin{bmatrix} D_{1,0} \\ E_1 \end{bmatrix}. \tag{11.4.9}$$

Based on (11.4.8), one can construct a reduced order observer for x_2 as,

$$\hat{x}_2(k+1) = A_R\hat{x}_2(k) + [A_{21} \quad B_2] \begin{pmatrix} y_1(k) \\ u(k) \end{pmatrix} + K_R\Big[C_R\hat{x}_2(k) - y_R(k)\Big], \tag{11.4.10}$$

where K_R is the observer gain matrix which must be chosen such that $A_R + K_R C_R$ is asymptotically stable and

$$(zI - A_R - K_R C_R)^{-1}(E_R - K_R D_R) \equiv 0. \tag{11.4.11}$$

Following the result of Chen [12], i.e., Proposition 2.2.1, one can show that the quadruple (A_R, E_R, C_R, D_R) is left invertible and of minimum phase with no infinite zero, provided that the quadruple $(A_{\rm PY}, E_{\rm PY}, C_{\rm 1PY}, D_{\rm 1PY})$ is left invertible and of minimum phase with no infinite zero. The computation of $K_{\rm R}$ can systematically be done using the procedure given in the next chapter.

At this moment we have a reduced order observer and an optimal state feedback. However, $y_{\rm R}$ contains a future measurement, i.e., the term $y_1(k+1)$ in (11.4.7). We apply a transformation to remove this term. We partition the reduced order observer gain $K_{\rm R} = [K_{\rm R0},\ K_{\rm R1}]$ compatible with the dimensions of the outputs $(y_0',\ \tilde{y}_1')'$, and at the same time define a new variable,

$$v := \hat{x}_2 + K_{\rm R1}\tilde{y}_1.$$

We then obtain the following reduced order estimator based controller,

$$\begin{cases} v(k+1) = (A_{\rm R}+K_{\rm R}C_{\rm R})\ v(k) + (B_2+K_{\rm R1}B_1)\ u(k) + G_{\rm R}\ y(k), \\ \hat{x}_{\rm PY}(k) = \begin{bmatrix} 0 \\ I_{n-p+m_0} \end{bmatrix} v(k) + \begin{bmatrix} 0 & I \\ 0 & -K_{\rm R1} \end{bmatrix} y(k), \\ u(k) = F_{\rm PY}\ \hat{x}_{\rm PY}(k) + X_{\rm PY}\ y(k), \end{cases} \tag{11.4.12}$$

where

$$G_{\rm R} = [-K_{\rm R0},\ A_{21} + K_{\rm R1}A_{11} - (A_{\rm R} + K_{\rm R}C_{\rm R})K_{\rm R1}],$$

and $F_{\rm PY}$ is state feedback gain and $X_{\rm PY}$ is the output pre-feedback gain.

Remark 11.4.1. It is interesting to point out that the state space representation of the reduced order estimator based controller in (11.4.12) might not be minimal and hence the McMillan degree of this controller might be less than the dynamical order of its state space representation (11.4.12). This is mainly due to the stable dynamics which becomes unobservable in the controlled output $h_{\rm PY}$ after the preliminary output feedback law (11.4.2).

A very interesting example is the state feedback case for $C_1 = I$ and $D_1 = 0$. In this case, the preliminary output feedback $X_{\rm PY}$ in (11.4.2) can be chosen such that after this preliminary feedback $C_{\rm 2PY} = 0$ and $A_{\rm PY}$ is stable. Hence we can choose $F_{\rm PY} = 0$ but this implies that the reduced order estimator based controller (11.4.12) has a McMillan degree equal to zero and it reduces to the static state feedback solution, $u = X_{\rm PY}y$. ®

Finally, we note that the reduced order output feedback control law (11.4.12) can be written in the following standard form,

$$\Sigma_{\rm cmp} : \begin{cases} v(k+1) = A_{\rm cmp}\ v(k) + B_{\rm cmp}\ y(k), \\ u(k) = C_{\rm cmp}\ v(k) + D_{\rm cmp}\ y(k), \end{cases} \tag{11.4.13}$$

with

$$\left.\begin{aligned}
A_{\rm cmp} &:= (A_{\rm R} + K_{\rm R} C_{\rm R}) + (B_2 + K_{\rm R1} B_1) F_{\rm PY} \begin{bmatrix} 0 \\ I \end{bmatrix}, \\
B_{\rm cmp} &:= (B_2 + K_{\rm R1} B_1) \left(F_{\rm PY} \begin{bmatrix} 0 & I \\ 0 & -K_{\rm R1} \end{bmatrix} + X_{\rm PY} \right) + G_{\rm R}, \\
C_{\rm cmp} &:= F_{\rm PY} \begin{bmatrix} 0 \\ I \end{bmatrix}, \\
D_{\rm cmp} &:= F_{\rm PY} \begin{bmatrix} 0 & I \\ 0 & -K_{\rm R1} \end{bmatrix} + X_{\rm PY}.
\end{aligned}\right\} \tag{11.4.14}$$

We have the following theorem.

Theorem 11.4.1. Consider the given discrete-time system Σ of (11.3.1). Also, let $\gamma > \gamma^*$ be given. Then, there exist gain matrices $X_{\rm PY}$, $F_{\rm PY}$ and $K_{\rm R}$ such that the resulting controller $\Sigma_{\rm cmp}$ of (11.4.13) with $A_{\rm cmp}$, $B_{\rm cmp}$, $C_{\rm cmp}$ and $D_{\rm cmp}$ being given as in (11.4.14) has the following properties:

1. the resulting closed-loop system comprising Σ and $\Sigma_{\rm cmp}$ is asymptotically stable; and

2. the H_∞-norm of the resulting closed-loop transfer matrix from the disturbance w to the controlled output h is less than γ. ◻

Proof. It is quite obvious because $\Sigma_{\rm PY}$ has the following properties:

1. There exists a constant matrix $X_{\rm PY}$ such that $D_{\rm 2PY} + D_{\rm 2PY} X_{\rm PY} D_{\rm 1PY} = 0$;

2. $(A_{\rm PY}, B_{\rm PY}, C_{\rm 2PY}, D_{\rm 2PY})$ is right invertible and of minimum phase with no infinite zero;

3. $(A_{\rm PY}, E_{\rm PY}, C_{\rm 1P}, D_{\rm 1PY})$ is left invertible and of minimum phase with no infinite zero.

A systematic procedure for computing the gain matrices $X_{\rm PY}$, $F_{\rm PY}$ and $K_{\rm R}$ can be found in Chapter 12. ⊠

The following example illustrates the result of this section.

Example 11.4.1. Consider a discrete-time system of the form (11.3.1) with

$$A = \begin{bmatrix} 1 & 1 \\ 2 & 3 \end{bmatrix}, \quad B = \begin{bmatrix} -2 \\ 1 \end{bmatrix}, \quad E = \begin{bmatrix} 1 \\ 1 \end{bmatrix}, \tag{11.4.15}$$

$$C_1 = \begin{bmatrix} 0 & 1 \\ 1 & 0 \end{bmatrix}, \quad D_1 = \begin{bmatrix} 1 \\ 0 \end{bmatrix}, \tag{11.4.16}$$

and

$$C_2 = \begin{bmatrix} 0.8 & 0.9 \end{bmatrix}, \quad D_2 = 0, \quad D_{22} = 1. \tag{11.4.17}$$

It is simple to verify that the subsystem (A, B, C_2, D_2) is invertible with an unstable invariant zero at 1.5714 and the subsystem (A, E, C_1, D_1) is left invertible with an unstable invariant zero at 2. By utilizing the algorithm for computing γ^* in the previous chapter, we obtain an exact value of the infimum

$$\gamma^* = 3.9631638.$$

In what follows, we will design a γ-suboptimal measurement output control law with $\gamma = 3.963164$. Following the above procedures, we obtain an auxiliary system (11.4.1) with

$$A_{\rm PY} = \begin{bmatrix} 1.14353033 & 1.18520854 \\ 2.34861499 & 3.46328599 \end{bmatrix}, \quad B_{\rm PY} = \begin{bmatrix} -2 \\ 0.96422578 \end{bmatrix},$$

$$E_{\rm PY} = 10^3 \cdot \begin{bmatrix} 1.65382390 \\ 4.83262217 \end{bmatrix}, \quad C_{\rm 1P} = \begin{bmatrix} 0.14353033 & 1.18520854 \\ 1 & 0 \end{bmatrix},$$

$$D_{\rm 1PY} = 10^3 \cdot \begin{bmatrix} 1.65382390 \\ 0 \end{bmatrix}, \quad D_{\rm 22PY} = -3297.4252,$$

$$C_{\rm 2PY} = \begin{bmatrix} -1.74280115 & -2.36309110 \end{bmatrix}, \quad D_{\rm 2PY} = 0.30400789,$$

and finally the controller parameters,

$$A_{\rm cmp} = 0, \quad B_{\rm cmp} = \begin{bmatrix} 0.06254887 & 0.05328328 \end{bmatrix},$$

and

$$C_{\rm cmp} = 0, \quad D_{\rm cmp} = \begin{bmatrix} 6.55844429 & 4.79141397 \end{bmatrix}.$$

The poles of the closed-loop system comprising the given plant and the above controller are given by 0 and $0.4878 \pm j0.1199$. Clearly, they are stable. The plot of the singular values of the closed-loop transfer matrix from w to h in Figure 11.4.1 shows that $\|T_{hw}(\Sigma \times \Sigma_{\rm cmp})\|_\infty$ is indeed less than the given γ. ▣

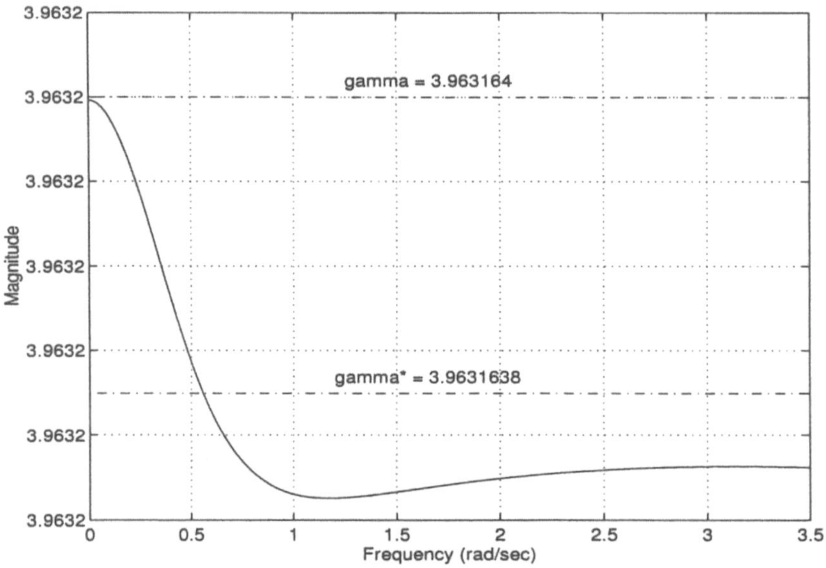

Figure 11.4.1: Singular values of T_{hw} under reduced order output feedback.

Chapter 12

Discrete-time H_∞ Almost Disturbance Decoupling

12.1. Introduction

IN THIS CHAPTER, we consider the problem of H_∞ almost disturbance decoupling for general discrete-time plants whose subsystems are allowed to have invariant zeros on the unit circle of the complex plane. The stability region of a discrete-time system considered in this chapter is defined as usual as the open unit disc. In contrast to the continuous-time case, the problem of almost disturbance decoupling for general discrete-time systems is less studied in the literature. In 1996, Chen, Guo and Lin [21] gave a set of solvability conditions for the H_∞-ADDPMS for the special case when a given plant whose subsystems do not have invariant zeros on the unit circle. Only very recently, has the necessary and sufficient conditions under which the H_∞-ADDPMS for general discrete-time systems been derived by Chen, He and Chen [22]. Solutions to such a general problem have just been reported by Lin and Chen [84]. The results of [22] and [84] form the core of this chapter.

To be more specific, we consider the following standard linear time-invariant discrete-time system Σ characterized by

$$\Sigma : \begin{cases} x(k+1) = A\ x(k) + B\ u(k) + E\ w(k), \\ y(k) = C_1\ x(k) + D_1\ w(k), \\ h(k) = C_2\ x(k) + D_2\ u(k) + D_{22}\ w(k), \end{cases} \tag{12.1.1}$$

where $x \in \mathbb{R}^n$ is the state, $u \in \mathbb{R}^m$ is the control input, $y \in \mathbb{R}^\ell$ is the measurement, $w \in \mathbb{R}^q$ is the disturbance and $h \in \mathbb{R}^p$ is the output to be controlled.

As usual, we denote Σ_P and Σ_Q as the subsystems characterized by matrix quadruples (A, B, C_2, D_2) and (A, E, C_1, D_1), respectively. The following dynamic feedback control laws are investigated:

$$\Sigma_{\rm cmp} : \begin{cases} v(k+1) = A_{\rm cmp}\, v(k) + B_{\rm cmp}\, y(k), \\ u(k) \quad = C_{\rm cmp}\, v(k) + D_{\rm cmp}\, y(k), \end{cases} \tag{12.1.2}$$

The controller $\Sigma_{\rm cmp}$ of (12.1.2) is said to be internally stabilizing when applied to the system Σ, if the following matrix is asymptotically stable:

$$A_{\rm cl} := \begin{bmatrix} A + BD_{\rm cmp}C_1 & BC_{\rm cmp} \\ B_{\rm cmp}C_1 & A_{\rm cmp} \end{bmatrix}, \tag{12.1.3}$$

i.e., all its eigenvalues lie inside the open unit disc of the complex plane. Denote by T_{hw} the corresponding closed-loop transfer matrix from the disturbance w to the controlled output h. Then, the solvability of the H_∞ almost disturbance decoupling problem for general discrete-time systems can be defined as follows.

Definition 12.1.1. The general H_∞ almost disturbance decoupling problem with measurement feedback and with internal stability (H_∞-ADDPMS) for (12.1.1) is said to be solvable if, for any given positive scalar $\gamma > 0$, there exists at least one controller of the form (12.1.2) such that,

1. in the absence of disturbance, the closed-loop system comprising the system (12.1.1) and the controller (12.1.2) is asymptotically stable, i.e., the matrix $A_{\rm cl}$ as given by (12.1.3) is asymptotically stable;
2. the closed-loop system has an L_2-gain, from the disturbance w to the controlled output h, that is less than or equal to γ, i.e.,

$$\|h\|_2 \le \gamma \|w\|_2, \quad \forall w \in L_2 \text{ and for } (x(0), v(0)) = (0, 0). \tag{12.1.4}$$

Equivalently, the H_∞-norm of the closed-loop transfer matrix from w to h, T_{hw}, is less than or equal to γ, i.e., $\|T_{hw}\|_\infty \le \gamma$. ◻

The problem of H_∞ almost disturbance decoupling with state feedback or with full information feedback can be defined in a similar and obvious way. The goal of this chapter is to identify the solvability conditions for these problems and to construct their solutions, if they are existent. The rest of this chapter is organized as follows: In Section 12.2, we give solvability conditions under which the proposed H_∞-ADDPMS for general discrete-time systems is solvable. Sections 12.3 and 12.4 give constructive algorithms that would yield solutions to the general discrete-time H_∞-ADDPMS, if such solutions exist. All proofs of the main results of this chapter are given in Section 12.5 for the sake of clarity in presentation.

12.2. Solvability Conditions

We give in this section the solvability conditions for the general H_∞ almost disturbance decoupling problems with internal stability for the following three cases: the full information feedback, the full state feedback and the measurement feedback. These conditions are characterized in terms of geometric subspaces. We also develop a numerical algorithm that will check these conditions without actually computing any geometric subspaces. The proofs of the main results of this section are given in Section 12.5.

Let us first examine the full information case. We have the following result.

Theorem 12.2.1. Consider the given discrete-time linear time-invariant system Σ of (12.1.1) with the measurement output being

$$y = \begin{pmatrix} x \\ w \end{pmatrix}, \quad \text{or} \quad C_1 = \begin{pmatrix} I \\ 0 \end{pmatrix}, \quad D_1 = \begin{pmatrix} 0 \\ I \end{pmatrix}, \tag{12.2.1}$$

i.e., all the state variables and the disturbances (full information) are measurable and available for feedback. The H_∞ almost disturbance decoupling problem with full information feedback and with internal stability for the given system is solvable if and only if the following conditions are satisfied:

(a) (A, B) is stabilizable;

(b) $\mathrm{Im}\,(D_{22}) \subset \mathrm{Im}\,(D_2)$, i.e., $D_{22} + D_2 S = 0$, where $S = -(D_2' D_2)^\dagger D_2' D_{22}$;

(c) $\mathrm{Im}\,(E + BS) \subset \left\{\mathcal{V}^\odot(\Sigma_\mathrm{P}) + B\mathrm{Ker}\,(D_2)\right\} \cap \left\{\bigcap_{|\lambda|=1} \mathcal{S}_\lambda(\Sigma_\mathrm{P})\right\}$. ▣

Proof. See Subsection 12.5.A. ▣

The result for the general proper measurement feedback case is given in the following theorem.

Theorem 12.2.2. Consider the given discrete-time linear time-invariant system Σ of (12.1.1). The H_∞ almost disturbance decoupling problem with measurement feedback and with internal stability (H_∞-ADDPMS) for (12.1.1) is solvable if and only if the following conditions are satisfied:

(a) (A, B) is stabilizable;

(b) (A, C_1) is detectable;

(c) $D_{22} + D_2 S D_1 = 0$, where $S = -(D_2' D_2)^\dagger D_2' D_{22} D_1' (D_1 D_1')^\dagger$;

(d) $\text{Im}\,(E+BSD_1) \subset \left\{\mathcal{V}^{\odot}(\Sigma_{\rm P}) + B\text{Ker}\,(D_2)\right\} \cap \left\{\bigcap_{|\lambda|=1} \mathcal{S}_\lambda(\Sigma_{\rm P})\right\}$;

(e) $\text{Ker}\,(C_2+D_2SC_1) \supset \left\{\mathcal{S}^{\odot}(\Sigma_{\rm Q}) \cap C_1^{-1}\{\text{Im}\,(D_1)\}\right\} \cup \left\{\bigcup_{|\lambda|=1} \mathcal{V}_\lambda(\Sigma_{\rm Q})\right\}$;

(f) $\mathcal{S}^{\odot}(\Sigma_{\rm Q}) \subset \mathcal{V}^{\odot}(\Sigma_{\rm P})$. ◻T

Proof. See Subsection 12.5.B. ◻

The following theorem deals with the case when the controller structure is restricted to be a strictly proper one, i.e., it is in the form of (12.1.2) with $D_{\rm cmp} \equiv 0$.

Theorem 12.2.3. Consider the given discrete-time linear time-invariant system Σ of (12.1.1). The H_∞ almost disturbance decoupling problem with strictly proper measurement feedback and with internal stability for (12.1.1) is solvable if and only if the following conditions are satisfied:

(a) (A,B) is stabilizable;

(b) (A,C_1) is detectable;

(c) $D_{22}=0$;

(d) $\text{Im}\,(E) \subset \mathcal{V}^{\odot}(\Sigma_{\rm P}) \cap \left\{\bigcap_{|\lambda|=1} \mathcal{S}_\lambda(\Sigma_{\rm P})\right\}$;

(e) $\text{Ker}\,(C_2) \supset \mathcal{S}^{\odot}(\Sigma_{\rm Q}) \cup \left\{\bigcup_{|\lambda|=1} \mathcal{V}_\lambda(\Sigma_{\rm Q})\right\}$;

(f) $\mathcal{S}^{\odot}(\Sigma_{\rm Q}) \subset \mathcal{V}^{\odot}(\Sigma_{\rm P})$;

(g) $A\mathcal{S}^{\odot}(\Sigma_{\rm Q}) \subset \mathcal{V}^{\odot}(\Sigma_{\rm P})$. ◻T

Proof. See Subsection 12.5.E. ◻

The following remarks are in order.

Remark 12.2.1. Note that if $\Sigma_{\rm P}$ is of minimum phase and right invertible with no infinite zeros, and $\Sigma_{\rm Q}$ is of minimum phase and left invertible with no infinite zero, then Conditions (d) to (f) of Theorem 12.2.2 are automatically satisfied. Hence, the solvability conditions of the H_∞-ADDPMS for such a case reduce to:

(a) (A,B) is stabilizable;

(b) (A,C_1) is detectable; and

(c) $D_{22}+D_2SD_1=0$, where $S=-(D_2'D_2)^\dagger D_2'D_{22}D_1'(D_1D_1')^\dagger$. ◻R

Remark 12.2.2. For the special case when all the states of the system (12.1.1) are measurable and available for feedback, i.e., $y = x$, it can be easily derived from Theorem 12.2.2 that the H_∞ almost disturbance decoupling problem with full state feedback and with internal stability for such a system is solvable if and only if the following conditions are satisfied:

(a) (A, B) is stabilizable;

(b) $D_{22} = 0$; and

(c) $\mathrm{Im}\,(E) \subset \mathcal{V}^{\odot}(\Sigma_{\mathrm{P}}) \cap \left\{\bigcap_{|\lambda|=1} \mathcal{S}_\lambda(\Sigma_{\mathrm{P}})\right\}$. Ⓡ

Next, we proceed to develop a numerical algorithm for verification of the solvability conditions of Theorem 12.2.2 without computing any geometric subspaces of Σ_{P} or Σ_{Q}.

Step 12.2.0: Let $S = -(D_2'D_2)^\dagger D_2'D_{22}D_1'(D_1D_1')^\dagger$. If $D_{22} + D_2SD_1 \neq 0$, the H_∞-ADDPMS for (12.1.1) is not solvable and the algorithm stops here. Otherwise, go to the next step.

Step 12.2.1: Compute the special coordinate basis of Σ_{P}, i.e., the quadruple (A, B, C_2, D_2). For easy reference, we append a subscript 'P' to all sub-matrices and transformations in the SCB associated with Σ_{P}, e.g., $\Gamma_{s\mathrm{P}}$ is the state transformation of the SCB of Σ_{P}, $B_{d\mathrm{P}}$ is replacing the sub-matrix B_d, and $A^0_{aa\mathrm{P}}$ is associated with invariant zero dynamics of Σ_{P} on the unit circle.

Step 12.2.2: Next, we denote the set of eigenvalues of $A^0_{aa\mathrm{P}}$ with a nonnegative imaginary part as $\{\omega_{\mathrm{P}1}, \omega_{\mathrm{P}2}, \cdots, \omega_{\mathrm{P}k_{\mathrm{P}}}\}$ and for $i = 1, 2, \cdots, k_{\mathrm{P}}$, choose complex matrices $V_{i\mathrm{P}}$, whose columns form a basis for an appropriate eigenspace $\{x \in \mathbb{C}^{n^0_{a\mathrm{P}}} \mid x^{\mathrm{H}}(\omega_{\mathrm{P}i}I - A^0_{aa\mathrm{P}}) = 0\}$, where $n^0_{a\mathrm{P}}$ is the dimension of $\mathcal{X}^0_{a\mathrm{P}}$. Then, let

$$V_{\mathrm{P}} := [\, V_{1\mathrm{P}} \quad V_{2\mathrm{P}} \quad \cdots \quad V_{k_{\mathrm{P}}\mathrm{P}}\,]. \tag{12.2.2}$$

We also compute $n_{x\mathrm{P}} := \dim(\mathcal{X}^+_{a\mathrm{P}}) + \dim(\mathcal{X}_{b\mathrm{P}}) + \dim(\mathcal{X}_{d\mathrm{P}})$, and

$$\Gamma_{s\mathrm{P}}^{-1}(E + BSD_1) := \begin{bmatrix} E_{c\mathrm{P}} \\ E^-_{a\mathrm{P}} \\ E^0_{a\mathrm{P}} \\ E^+_{a\mathrm{P}} \\ E_{b\mathrm{P}} \\ E_{d\mathrm{P}} \end{bmatrix}. \tag{12.2.3}$$

Step 12.2.3: Let $\Sigma_{\rm Q}^\star$ be the dual system of $\Sigma_{\rm Q}$ and be characterized by a quadruple (A', C_1', E', D_1'). We compute the special coordinate basis of $\Sigma_{\rm Q}^\star$. Again, for easy reference, we append a subscript 'Q' to all sub-matrices and transformations in the SCB associated with $\Sigma_{\rm Q}^\star$, e.g., $\Gamma_{s{\rm Q}}$ is the state transformation of the SCB of $\Sigma_{\rm Q}^\star$, $B_{d{\rm Q}}$ is replacing the sub-matrix B_d, and $A_{aa{\rm Q}}^0$ is associated with invariant zero dynamics of $\Sigma_{\rm Q}^\star$ on the unit circle.

Step 12.2.4: Similarly, we denote the set of eigenvalues of $A_{aa{\rm Q}}^0$ with a non-negative imaginary part as $\{\omega_{{\rm Q}1}, \omega_{{\rm Q}2}, \cdots, \omega_{{\rm Q}k_{\rm Q}}\}$ and for $i = 1, 2, \cdots, k_{\rm Q}$, choose complex matrices $V_{i{\rm Q}}$, whose columns form a basis for an eigenspace $\{x \in \mathbb{C}^{n_{a{\rm Q}}^0} \mid x^{\rm H}(\omega_{{\rm Q}i} I - A_{aa{\rm Q}}^0) = 0\}$, where $n_{a{\rm Q}}^0$ is the dimension of $\mathcal{X}_{a{\rm Q}}^0$. Then, let

$$V_{\rm Q} := [\, V_{1{\rm Q}} \quad V_{2{\rm Q}} \quad \cdots \quad V_{k_{\rm Q}{\rm Q}} \,]. \tag{12.2.4}$$

We next compute $n_{x{\rm Q}} := \dim(\mathcal{X}_{a{\rm Q}}^+) + \dim(\mathcal{X}_{b{\rm Q}}) + \dim(\mathcal{X}_{d{\rm Q}})$, and

$$\Gamma_{s{\rm Q}}^{-1}(C_2 + D_2 S C_1)' := \begin{bmatrix} E_{c{\rm Q}} \\ E_{a{\rm Q}}^- \\ E_{a{\rm Q}}^0 \\ E_{a{\rm Q}}^+ \\ E_{b{\rm Q}} \\ E_{d{\rm Q}} \end{bmatrix}. \tag{12.2.5}$$

Step 12.2.5: Finally, compute

$$\Gamma_{s{\rm P}}^{-1}(\Gamma_{s{\rm Q}}^{-1})' = \begin{bmatrix} \star & \star \\ \star & \Gamma \end{bmatrix}, \tag{12.2.6}$$

where Γ is a $n_{x{\rm P}} \times n_{x{\rm Q}}$ constant matrix. 🄰

The following proposition summaries the result of the above algorithm. It also gives a set of necessary and sufficient conditions, in terms of sub-matrices associated with the SCBs of $\Sigma_{\rm P}$ and $\Sigma_{\rm Q}$, for the solvability of the H_∞-ADDPMS for the general discrete-time system Σ of (12.1.1).

Proposition 12.2.1. Consider the given discrete-time linear time-invariant system Σ of (12.1.1). The H_∞ almost disturbance decoupling problem with measurement feedback and with internal stability (H_∞-ADDPMS) for (12.1.1) is solvable if and only if the following conditions are satisfied:

(a) (A, B) is stabilizable;

(b) (A, C_1) is detectable;

(c) $D_{22} - D_2(D_2'D_2)^\dagger D_2'D_{22}D_1'(D_1D_1')^\dagger D_1 = 0$;

(d) $V_{\rm P}^{\rm H} E_{a\rm P}^0 = 0$, $E_{a\rm P}^+ = 0$, $E_{b\rm P} = 0$, $\operatorname{Im}(E_{d\rm P}) \subset \operatorname{Im}(B_{d\rm P})$;

(e) $V_{\rm Q}^{\rm H} E_{a\rm Q}^0 = 0$, $E_{a\rm Q}^+ = 0$, $E_{b\rm Q} = 0$, $\operatorname{Im}(E_{d\rm Q}) \subset \operatorname{Im}(B_{d\rm Q})$; and

(f) $\Gamma = 0$.

Note that all the matrices in (d)-(f) are well-defined in Steps 12.2.0 to 12.2.5 of the algorithm. ◻P

The above result can be directly verified using the properties of the special coordinate basis of Chapter 2 and the result of Theorem 12.2.2 (see also Chapter 8 for a similar result for continuous-time systems).

12.3. Solutions to State and Full Information Feedback Cases

In this section, we consider feedback control law design for the general H_∞ almost disturbance decoupling problem with internal stability as well as with both full state feedback and full information feedback, where internal stability is with respect to the open unit disc. More specifically, we will first present a design procedure that constructs a family of parameterized static state feedback control laws,

$$u(k) = F(\varepsilon)x(k), \tag{12.3.1}$$

which solves the general H_∞-ADDPMS for the following system,

$$\begin{cases} x(k+1) = A\ x(k) + B\ u(k) + E\ w(k), \\ y(k) = x(k) \\ h(k) = C_2\ x(k) + D_2\ u(k) + D_{22}\ w(k). \end{cases} \tag{12.3.2}$$

That is, under this family of state feedback control laws, the resulting closed-loop system is asymptotically stable for sufficiently small ε and the H_∞-norm of the closed-loop transfer matrix from w to h, $T_{hw}(z,\varepsilon)$, tends to zero as ε tends to zero, where

$$T_{hw}(z,\varepsilon) = [C_2 + D_2F(\varepsilon)][zI - A - BF(\varepsilon)]^{-1}E + D_{22}. \tag{12.3.3}$$

We have the following algorithm for constructing such an $F(\varepsilon)$.

Step 12.S.1: (Decomposition of Σ_P). Transform the subsystem Σ_P, i.e., the matrix quadruple (A, B, C_2, D_2), into the special coordinate basis (SCB) as given by Theorem 2.4.1. Denote the state, output and input transformation matrices as Γ_{sP}, Γ_{oP} and Γ_{iP}, respectively.

Step 12.S.2: (Gain matrix for the subsystem associated with $\mathcal{X}_c$). Let F_c be any constant matrix subject to the constraint that

$$A_{cc}^c = A_{cc} - B_c F_c, \tag{12.3.4}$$

is a stable matrix. Note that the existence of such an F_c is guaranteed by the property of the special coordinate basis, i.e., (A_{cc}, B_c) is controllable.

Step 12.S.3: (Gain matrix for the subsystem associated with $\mathcal{X}_a^+$, $\mathcal{X}_b$ and $\mathcal{X}_d$). Let

$$F_{abd} := \begin{bmatrix} 0 & 0 & F_{a0}^+ & F_{b0} & F_{d0} \\ E_{da}^- & E_{da}^0 & F_{ad}^+ & F_{bd} & F_{dd} \end{bmatrix}, \tag{12.3.5}$$

where

$$F_{abd}^+ := \begin{bmatrix} F_{a0}^+ & F_{b0} & F_{d0} \\ F_{ad}^+ & F_{bd} & F_{dd} \end{bmatrix}, \tag{12.3.6}$$

is any constant matrix subject to the constraint that

$$A_{abd}^{+c} := \begin{bmatrix} A_{aa}^+ & L_{ab}^+ C_b & L_{ad}^+ C_d \\ 0 & A_{bb} & L_{bd} C_d \\ B_d E_{da}^+ & B_d E_{db} & A_{dd} \end{bmatrix} - \begin{bmatrix} B_{0a}^+ & 0 \\ B_{0b} & 0 \\ B_{0d} & B_d \end{bmatrix} F_{abd}^+ \tag{12.3.7}$$

is an asymptotically stable matrix. Again, the existence of such an F_{abd}^+ is guaranteed by the property of the special coordinate basis.

Step 12.S.4: (Gain matrix for the subsystem associated with A_{aa}^0). The construction of this gain matrix is carried out in the following sub-steps.

Step 12.S.4.1: (Preliminary coordinate transformation). Noting that

$$A_{\rm con} := \begin{bmatrix} A_{aa} & L_{ab} C_b & L_{ad} C_d \\ 0 & A_{bb} & L_{bd} C_d \\ B_d E_{da} & B_d E_{db} & A_{dd} \end{bmatrix}, \quad B_{\rm con} := \begin{bmatrix} B_{0a} & 0 \\ B_{0b} & 0 \\ B_{0d} & B_d \end{bmatrix},$$

we have

$$A_{\rm con} - B_{\rm con} F_{abd} = \begin{bmatrix} A_{aa}^- & 0 & A_{abd}^- \\ 0 & A_{aa}^0 & A_{abd}^0 \\ 0 & 0 & A_{abd}^{+c} \end{bmatrix}, \quad B_{\rm con} = \begin{bmatrix} B_{0a}^- & 0 \\ B_{0a}^0 & 0 \\ B_{0abd}^+ & \tilde{B}_d \end{bmatrix}, \tag{12.3.8}$$

where

$$B_{0abd}^{+} = \begin{bmatrix} B_{0a}^{+} \\ B_{0b} \\ B_{0d} \end{bmatrix}, \quad \tilde{B}_d = \begin{bmatrix} 0 \\ 0 \\ B_d \end{bmatrix}, \tag{12.3.9}$$

$$A_{abd}^{0} = [\,0 \quad L_{ab}^{0}C_b \quad L_{ad}^{0}C_d\,] - [\,B_{0a}^{0} \quad 0\,]\,F_{abd}^{+}, \tag{12.3.10}$$

and

$$A_{abd}^{-} = [\,0 \quad L_{ab}^{-}C_b \quad L_{ad}^{-}C_d\,] - [\,B_{0a}^{-} \quad 0\,]\,F_{abd}^{+}. \tag{12.3.11}$$

Clearly, the pair $(A_{\rm con} - B_{\rm con}F_{abd}, B_{\rm con})$ remains stabilizable. Construct the following nonsingular transformation matrix,

$$\Gamma_{abd} = \begin{bmatrix} I_{n_a^-} & 0 & 0 \\ 0 & 0 & I_{n_a^+ + n_b + n_d} \\ 0 & I_{n_a^0} & T_a^0 \end{bmatrix}^{-1}, \tag{12.3.12}$$

where T_a^0 is the unique solution to the following Lyapunov equation,

$$A_{aa}^{0} T_a^0 - T_a^0 A_{abd}^{+c} = A_{abd}^{0}. \tag{12.3.13}$$

We note here that such a unique solution to the above Lyapunov equation always exists since all the eigenvalues of A_{aa}^0 are on the unit circle and all the eigenvalues of A_{abd}^{+c} are on the open unit disc. It is now easy to verify that

$$\Gamma_{abd}^{-1}(A_{\rm con} - B_{\rm con}F_{abd})\Gamma_{abd} = \begin{bmatrix} A_{aa}^{-} & A_{aab}^{-} & 0 \\ 0 & A_{abd}^{+c} & 0 \\ 0 & 0 & A_{aa}^{0} \end{bmatrix}, \tag{12.3.14}$$

and

$$\Gamma_{abd}^{-1}B_{\rm con} = \begin{bmatrix} B_{0a}^{-} & 0 \\ B_{0abd}^{+} & \tilde{B}_d \\ B_{0a}^{0} + T_a^0 B_{0abd}^{+} & T_a^0 \tilde{B}_d \end{bmatrix}. \tag{12.3.15}$$

Hence, the matrix pair (A_{aa}^0, B_a^0) is controllable, where

$$B_a^0 = [\,B_{0a}^{0} + T_a^0 B_{0abd}^{+} \quad T_a^0 \tilde{B}_d\,]. \tag{12.3.16}$$

Step 12.S.4.2: (Further coordinate transformation). Use the results of Chapter 2 to find nonsingular transformation matrices Γ_{sa}^0 and Γ_{ia}^0 such that (A_{aa}^0, B_a^0) can be transformed into the block diagonal control canonical form,

$$(\Gamma_{sa}^0)^{-1} A_{aa}^0 \Gamma_{sa}^0 = \begin{bmatrix} A_1 & 0 & \cdots & 0 \\ 0 & A_2 & \cdots & 0 \\ \vdots & \vdots & \ddots & \vdots \\ 0 & 0 & \cdots & A_l \end{bmatrix}, \tag{12.3.17}$$

and

$$(\Gamma_{sa}^0)^{-1}B_a^0\Gamma_{ia}^0 = \begin{bmatrix} B_1 & B_{12} & \cdots & B_{1l} & \star \\ 0 & B_2 & \cdots & B_{2l} & \star \\ \vdots & \vdots & \ddots & \vdots & \vdots \\ 0 & 0 & \cdots & B_l & \star \end{bmatrix}, \tag{12.3.18}$$

where l is an integer and for $i = 1, 2, \cdots, l$,

$$A_i = \begin{bmatrix} 0 & 1 & 0 & \cdots & 0 \\ 0 & 0 & 1 & \cdots & 0 \\ \vdots & \vdots & \vdots & \ddots & \vdots \\ 0 & 0 & 0 & \cdots & 1 \\ -a_{n_i}^i & -a_{n_i-1}^i & -a_{n_i-2}^i & \cdots & -a_1^i \end{bmatrix}, \quad B_i = \begin{bmatrix} 0 \\ 0 \\ \vdots \\ 0 \\ 1 \end{bmatrix}. \tag{12.3.19}$$

We note that all the eigenvalues of A_i are on the unit circle. Here, the $\star$s represent sub-matrices of less interest.

Step 12.S.4.3: (Subsystem design). For each (A_i, B_i), let $F_i(\varepsilon) \in \mathbb{R}^{1\times n_i}$ be the state feedback gain such that

$$\lambda\{A_i - B_iF_i(\varepsilon)\} = \{(1-\varepsilon)e^{j\theta_{i1}}, \cdots, (1-\varepsilon)e^{j\theta_{in_i}}\},$$

where $e^{j\theta_{i\ell}}$, $\ell = 1, 2, \cdots, n_i$, are the eigenvalues of A_i. Clearly, all the eigenvalues of $A_i + B_iF_i(\varepsilon)$ are on the open unit disc and $F_i(\varepsilon)$ is unique.

Step 12.S.4.4: (Composition of gain matrix for subsystem associated with $\mathcal{X}_a^0$). Let

$$F_a^0(\varepsilon) := \Gamma_{ia}^0 \begin{bmatrix} F_1(\varepsilon) & 0 & \cdots & 0 & 0 \\ 0 & F_2(\varepsilon) & \cdots & 0 & 0 \\ \vdots & \vdots & \ddots & \vdots & \vdots \\ 0 & 0 & \cdots & F_{l-1}(\varepsilon) & 0 \\ 0 & 0 & \cdots & 0 & F_l(\varepsilon) \\ 0 & 0 & \cdots & 0 & 0 \end{bmatrix} (\Gamma_{sa}^0)^{-1}, \tag{12.3.20}$$

where $\varepsilon \in (0, 1]$ is a design parameter whose value is to be specified later. For future use, we partition

$$F_a^0(\varepsilon) = \begin{bmatrix} F_{a0}^0(\varepsilon) \\ F_{ad}^0(\varepsilon) \end{bmatrix}, \tag{12.3.21}$$

and

$$F_a^0(\varepsilon)T_a^0 = \begin{bmatrix} F_{a0+}^0(\varepsilon) & F_{a0b}^0(\varepsilon) & F_{a0d}^0(\varepsilon) \\ F_{ad+}^0(\varepsilon) & F_{adb}^0(\varepsilon) & F_{add}^0(\varepsilon) \end{bmatrix}. \tag{12.3.22}$$

Step 12.S.5: (Composition of parameterized gain matrix $F(\varepsilon)$). In this step, various gains calculated in Steps 12.S.3 to 12.S.5 are put together to form a composite state feedback gain matrix $F(\varepsilon)$. It is given by

$$F(\varepsilon) := -\Gamma_{iP}\big[F_0 + F_\star(\varepsilon)\big]\Gamma_{sP}^{-1}, \tag{12.3.23}$$

where

$$F_0 = \begin{bmatrix} C_{0a}^- & C_{0a}^0 & C_{0a}^+ + F_{a0}^+ & C_{0b} + F_{b0} & C_{0c} & C_{0d} + F_{d0} \\ E_{da}^- & E_{da}^0 & F_{ad}^+ & F_{bd} & E_{dc} & F_{dd} \\ E_{ca}^- & E_{ca}^0 & E_{ca}^+ & 0 & F_c & 0 \end{bmatrix}, \tag{12.3.24}$$

and

$$F_\star(\varepsilon) = \begin{bmatrix} 0 & F_{a0}^0(\varepsilon) & F_{a0+}^0(\varepsilon) & F_{a0b}^0(\varepsilon) & 0 & F_{a0d}^0(\varepsilon) \\ 0 & F_{ad}^0(\varepsilon) & F_{ad+}^0(\varepsilon) & F_{adb}^0(\varepsilon) & 0 & F_{add}^0(\varepsilon) \\ 0 & 0 & 0 & 0 & 0 & 0 \end{bmatrix}. \tag{12.3.25}$$

This completes the construction of the parameterized state feedback gain matrix $F(\varepsilon)$. ▣

We have the following theorem.

Theorem 12.3.1. Consider the given system (12.3.2) in which all the states are available for feedback. Assume that the problem of H_∞ almost disturbance decoupling with internal stability for (12.3.2) is solvable, i.e., the solvability conditions of Remark 12.2.2 are satisfied. Then, the closed-loop system comprising (12.3.2) and the full state feedback control law,

$$u(k) = F(\varepsilon)x(k), \tag{12.3.26}$$

with $F(\varepsilon)$ given by (12.3.23), has the following properties: For any given $\gamma > 0$, there exists a positive scalar $\varepsilon^* > 0$ such that for all $0 < \varepsilon \le \varepsilon^*$,

1. the closed-loop system is asymptotically stable, i.e., $\lambda\{A+BF(\varepsilon)\}$ are on the open unit disc; and
2. the H_∞-norm of the closed-loop transfer matrix from the disturbance w to the controlled output h is less than γ, i.e., $\|T_{hw}(z,\varepsilon)\|_\infty < \gamma$.

Hence, by Definition 12.1.1, the control law of (12.3.26) solves the H_∞-ADDPMS for (12.3.2). ▣

Proof. See Subsection 12.5.C. ▣

Next, we proceed to design a parameterized control law,

$$u(k) = F_x(\varepsilon)x(k) + F_w w(k), \tag{12.3.27}$$

which solves the H_∞ almost disturbance decoupling problem with internal stability for the following full information system,

$$\begin{cases} x(k+1) = A\ x(k) + B\ u(k) + E\ w(k), \\ y(k) = \begin{pmatrix} I \\ 0 \end{pmatrix} x(k) + \begin{pmatrix} 0 \\ I \end{pmatrix} w(k), \\ h(k) = C_2\ x(k) + D_2\ u(k) + D_{22}\ w(k). \end{cases} \tag{12.3.28}$$

That is, under the above full information feedback control law, the resulting closed-loop system is asymptotically stable for sufficiently small ε and the H_∞-norm of the closed-loop transfer matrix from w to h, $T_{hw}(z,\varepsilon)$, tends to zero as ε tends to zero, where

$$T_{hw}(z,\varepsilon) = [C_2 + D_2F_x(\varepsilon)][zI - A - BF_x(\varepsilon)]^{-1}(E + BF_w) + (D_{22} + D_2F_w).$$

The following is a step-by-step algorithm for constructing $F_x(\varepsilon)$ and F_w.

Step 12.F.1: (Computation of S). Compute

$$S = -(D_2'D_2)^\dagger D_2'D_{22}. \tag{12.3.29}$$

Step 12.F.2: (Computation of $F_x(\varepsilon)$). Follow Steps 12.S.1 to 12.S.5 of the previous algorithm to yield a gain matrix $F(\varepsilon)$. Then, let

$$F_x(\varepsilon) = F(\varepsilon). \tag{12.3.30}$$

Also, we need to retain the transformation matrices Γ_{sP} and Γ_{iP}, as well as the sub-matrix B_d of the SCB of Σ_P in order to compute F_w in the next step.

Step 12.F.3: (Construction of gain matrix F_w). Let

$$\Gamma_{sP}^{-1}(E + BS) = \begin{bmatrix} E_a^- \\ E_a^0 \\ E_a^+ \\ E_b \\ E_c \\ E_d \end{bmatrix}. \tag{12.3.31}$$

Then, the gain matrix F_w is given by

$$F_w = -\Gamma_{iP} \begin{bmatrix} 0 \\ (B_d' B_d)^{-1} B_d' E_d \\ 0 \end{bmatrix} + S. \qquad (12.3.32)$$

It is interesting to note that the first portion of matrix F_w is used to clean up the disturbance associated with E_d and in the range space of B_d, while the second portion is used to reject disturbance entering into the system through D_{22}. ▣

We have the following result.

Theorem 12.3.2. Consider the given system (12.3.28) in which all the states and the disturbances are available for feedback. Assume that the problem of H_∞ almost disturbance decoupling with internal stability for (12.3.28) is solvable, i.e., the solvability conditions of Theorem 12.2.1 are satisfied. Then, the closed-loop system comprising (12.3.28) and the full information feedback control law,

$$u(k) = F_x(\varepsilon)x(k) + F_w w(k), \qquad (12.3.33)$$

with $F_x(\varepsilon)$ and F_w being given by (12.3.30) and (12.3.32), respectively, has the following properties: For any given $\gamma > 0$, there exists a positive scalar $\varepsilon^* > 0$ such that for all $0 < \varepsilon \leq \varepsilon^*$,

1. the closed-loop system is asymptotically stable, i.e., $\lambda\{A + BF_x(\varepsilon)\}$ are on the open unit disc; and

2. the H_∞-norm of the closed-loop transfer matrix from the disturbance w to the controlled output h is less than γ, i.e., $\|T_{hw}(z,\varepsilon)\|_\infty < \gamma$.

Hence, by Definition 12.1.1, the control law of (12.3.33) solves the H_∞-ADDPMS for (12.3.28). ▣

Proof. See Subsection 12.5.D. ▣

We illustrate the results of this section with the following example.

Example 12.3.1. Consider a discrete-time system characterized by (12.1.1) with

$$A = \begin{bmatrix} 1 & 1 & 1 & 1 & 0 \\ 0 & 1 & 1 & 1 & 0 \\ 0 & 0 & 0.1 & 1 & 0 \\ 0 & 0 & 0 & 0 & 1 \\ 0.1 & 0.1 & 0.1 & 0.1 & 0.1 \end{bmatrix}, \quad B = \begin{bmatrix} 0 \\ 0 \\ 0 \\ 0 \\ 1 \end{bmatrix}, \quad E = \begin{bmatrix} 0 & 1 \\ 0 & 0 \\ 0 & 0 \\ 0 & 0 \\ \alpha_e & 0 \end{bmatrix}, \qquad (12.3.34)$$

where α_e is a scalar, and

$$C_2 = \begin{bmatrix} 0 & 0 & 0 & 1 & 0 \\ 0 & 0 & 1 & 0 & 0 \end{bmatrix}, \quad D_2 = \begin{bmatrix} 0 \\ 0 \end{bmatrix}, \quad D_{22} = \begin{bmatrix} 0 & 0 \\ 0 & 0 \end{bmatrix}. \tag{12.3.35}$$

We will consider both the state feedback case and the full information feedback case in this example. Using the toolbox of Chen [14], we can verify that (A, B) is controllable and Σ_P, i.e., (A, B, C_2, D_2), is left invertible with two invariant zeros at $z = 1$ and one infinite zero of order 2. Moreover,

$$\mathcal{V}^{\odot}(\Sigma_P) = \text{Im}\left\{\begin{bmatrix} 1 & 0 \\ 0 & 1 \\ 0 & 0 \\ 0 & 0 \\ 0 & 0 \end{bmatrix}\right\}, \quad B\text{Ker}\,(D_2) = \text{Im}\left\{\begin{bmatrix} 0 \\ 0 \\ 0 \\ 0 \\ 1 \end{bmatrix}\right\}, \tag{12.3.36}$$

and

$$\bigcap_{|\lambda|=1} \mathcal{S}_\lambda(\Sigma_P) = \text{Im}\left\{\begin{bmatrix} 1 & 0 & 0 \\ 0 & 0 & 0 \\ 0 & 0 & 0 \\ 0 & 1 & 0 \\ 0 & 0 & 1 \end{bmatrix}\right\}. \tag{12.3.37}$$

Also, we have

$$\mathcal{V}^{\odot}(\Sigma_P) \cap \left\{\bigcap_{|\lambda|=1} \mathcal{S}_\lambda(\Sigma_P)\right\} = \text{Im}\left\{\begin{bmatrix} 1 \\ 0 \\ 0 \\ 0 \\ 0 \end{bmatrix}\right\},$$

and

$$\left\{\mathcal{V}^{\odot}(\Sigma_P) + B\text{Ker}\,(D_2)\right\} \cap \left\{\bigcap_{|\lambda|=1} \mathcal{S}_\lambda(\Sigma_P)\right\} = \text{Im}\left\{\begin{bmatrix} 1 & 0 \\ 0 & 0 \\ 0 & 0 \\ 0 & 0 \\ 0 & 1 \end{bmatrix}\right\}.$$

It is clear to see now that the H_∞ almost disturbance decoupling problem with internal stability (H_∞-ADDPS) using state feedback for the given system is solvable if and only if $\alpha_e = 0$ and the H_∞-ADDPS using full information feedback for the given system is always solvable. Following the algorithms of this section, we obtain the following parameterized gain matrices,

$$F_x(\varepsilon) = \begin{bmatrix} -0.526316(\varepsilon-1)^2 - 1.052632(\varepsilon-1) - 0.626316 \\ -0.775623(\varepsilon-1)^2 - 2.603878(\varepsilon-1) - 1.928255 \\ -0.798061(\varepsilon-1)^2 - 2.763490(\varepsilon-1) - 2.066429 \\ -(\varepsilon-1)^2 - 4.2(\varepsilon-1) - 3.31 \\ -2(\varepsilon-1) - 2.2 \end{bmatrix}', \tag{12.3.38}$$

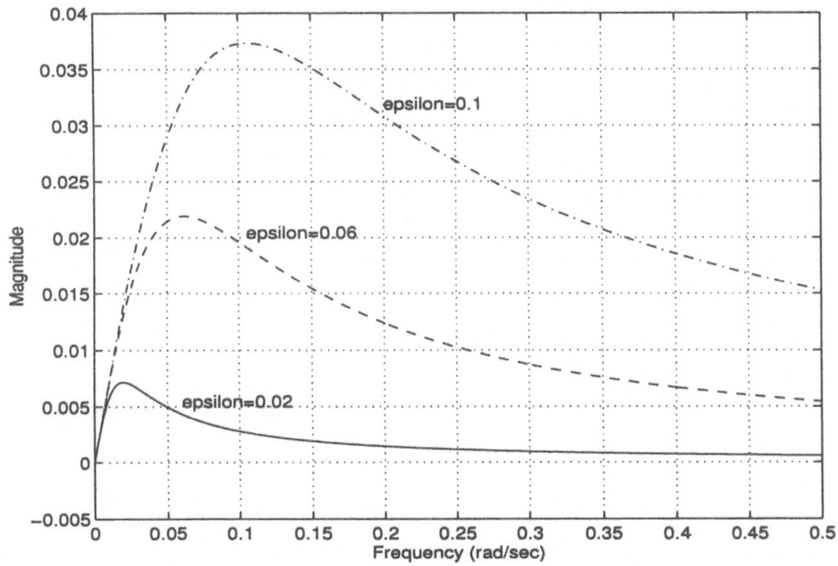

Figure 12.3.1: Max. singular values of T_{hw} — Full information case.

which places the eigenvalues of $A + BF_x(\varepsilon)$ around at 0, 0, 0, $1 - \varepsilon$ and $1 - \varepsilon$, and

$$F_w = [\,-\alpha_e \quad 0\,]. \tag{12.3.39}$$

The maximum singular value plots of the corresponding closed-loop transfer matrix $T_{hw}(z, \varepsilon)$ in Figure 12.3.1 clearly show that the H_∞-ADDPS using full information feedback (or state feedback when $\alpha_e = 0$) is attained as ε tends smaller and smaller. ⓔ

12.4. Solutions to Measurement Feedback Case

We present in this section the design of both full order and reduced order output feedback controllers that solve the general H_∞-ADDPMS for the given system (12.1.1). Here, by a full order controller, we mean that the order of the controller is exactly the same as the given system (12.1.1), i.e, is equal to n. A reduced order controller, on the other hand, refers to a controller whose dynamical order is less than n.

12.4.1. Full Order Output Feedback

In this subsection, we focus on the design of a full order proper measurement feedback control law, which solves the H_∞-ADDPMS for the given system

(12.1.1) under the solvability conditions of Theorem 12.2.2. For the case when the given system satisfies the conditions listed in Theorem 12.2.3, a slight modification on the algorithm below, i.e., letting $N = 0$ in Step 12.F.C.1, would yield a strictly proper solution. The following is a step-by-step algorithm for constructing a parameterized full order output feedback controller that solves the H_∞-ADDPMS for (12.1.1).

Step 12.F.C.1: (Computation of N). Utilize the properties of the special coordinate basis to compute two constant matrices X and Y such that $\mathcal{V}^{\odot}(\Sigma_{\mathrm{P}}) = \operatorname{Ker}(X)$ and $\mathcal{S}^{\odot}(\Sigma_{\mathrm{Q}}) = \operatorname{Im}(Y)$. Then, compute

$$N = -(B'X'XB + D_2'D_2)^{\dagger} \begin{bmatrix} B'X' & D_2' \end{bmatrix} \begin{bmatrix} XAY & XE \\ C_2Y & D_{22} \end{bmatrix} \times \begin{bmatrix} Y'C_1' \\ D_1' \end{bmatrix} (C_1YY'C_1' + D_1D_1')^{\dagger}. \quad (12.4.1)$$

Step 12.F.C.2: (Construction of the gain matrix $F_{\mathrm{P}}(\varepsilon)$). Define an auxiliary system

$$\begin{cases} x(k+1) = \tilde{A}\ x(k) + B\ u(k) + \tilde{E}\ w(k), \\ y(k) = x(k) \\ h(k) = \tilde{C}_2\ x(k) + D_2\ u(k) + 0\ w(k), \end{cases} \quad (12.4.2)$$

where

$$\tilde{A} := A + BNC_1, \quad (12.4.3)$$

$$\tilde{E} := E + BND_1, \quad (12.4.4)$$

$$\tilde{C}_2 := C_2 + D_2NC_1, \quad (12.4.5)$$

and then perform Steps 12.S.1 to 12.S.5 of the previous section to the above system (12.4.2) to obtain a parameterized gain matrix $F(\varepsilon)$. We let $F_{\mathrm{P}}(\varepsilon) = F(\varepsilon)$.

Step 12.F.C.3: (Construction of the gain matrix $K_{\mathrm{Q}}(\varepsilon)$). Define another auxiliary system

$$\begin{cases} x(k+1) = \tilde{A}'\ x(k) + C_1'\ u(k) + \tilde{C}_2'\ w(k), \\ y(k) = x(k) \\ h(k) = \tilde{E}'\ x(k) + D_1'\ u(k) + 0\ w(k), \end{cases} \quad (12.4.6)$$

and then perform Steps 12.S.1 to 12.S.6 of the previous section to the above system to get the parameterized gain matrix $F(\varepsilon)$. Similarly, we let $K_{\mathrm{Q}}(\varepsilon) = F(\varepsilon)'$.

Step 12.F.C.4: (Construction of the full order controller $\Sigma_{FC}(\varepsilon)$). Finally, the parameterized full order output feedback controller is given by

$$\Sigma_{FC}(\varepsilon) : \begin{cases} v(k+1) = A_{FC}(\varepsilon)\, v(k) + B_{FC}(\varepsilon)\, y(k), \\ u(k) \quad = C_{FC}(\varepsilon)\, v(k) + D_{FC}(\varepsilon)\, y(k), \end{cases} \tag{12.4.7}$$

where

$$\left.\begin{aligned} A_{FC}(\varepsilon) &:= A + BNC_1 + BF_P(\varepsilon) + K_Q(\varepsilon)C_1, \\ B_{FC}(\varepsilon) &:= -K_Q(\varepsilon), \\ C_{FC}(\varepsilon) &:= F_P(\varepsilon), \\ D_{FC}(\varepsilon) &:= N. \end{aligned}\right\} \tag{12.4.8}$$

Note that if the given system satisfies all conditions of Theorem 12.2.3, then one can choose matrix $D_{FC}(\varepsilon) = N = 0$ and obtain a strictly proper control law. ◻A

We have the following theorem.

Theorem 12.4.1. Consider the given system Σ of (12.1.1). Assume that the problem of H_∞ almost disturbance decoupling with internal stability for (12.1.1) is solvable, i.e., the solvability conditions of Theorem 12.2.2 are satisfied. Then, the closed-loop system comprising (12.1.1) and the full order measurement feedback controller (12.4.7) has the following properties: For any given $\gamma > 0$, there exists a positive scalar $\varepsilon^* > 0$ such that for all $0 < \varepsilon \le \varepsilon^*$,

1. the closed-loop system is asymptotically stable; and
2. the H_∞-norm of the closed-loop transfer matrix from the disturbance w to the controlled output h is less than γ, i.e., $\|T_{hw}(z,\varepsilon)\|_\infty < \gamma$.

Hence, by Definition 12.1.1, the control law of (12.4.7) solves the H_∞-ADDPMS for (12.1.1). ◻T

Proof. See Subsection 12.5.E. ⊠

We illustrate the above result in the following example.

Example 12.4.1. We now consider a discrete-time system characterized by (12.1.1) with A, B, E, C_2, D_2 and D_{22} being given as in Example 12.3.1, and

$$C_1 = \begin{bmatrix} 0.5 & 0.1 & 0.5 & 0.2 & 0.1 \\ 1 & 0 & 0 & 0 & 0 \end{bmatrix}, \quad D_1 = \begin{bmatrix} 1 & 0 \\ 0 & 0 \end{bmatrix}. \tag{12.4.9}$$

For simplicity, we let $\alpha_e = 1$ in matrix E. Using the toolbox of Chen [14] again, one can verify that (A, C_1) is observable and Σ_Q, i.e., (A, E, C_1, D_1), is invertible with one infinite zero of order one and four invariant zeros at -0.6554, $0.3777 \pm j0.6726$, and 1. Moreover,

$$\mathcal{S}^{\odot}(\Sigma_Q) = \operatorname{Im}\left\{\begin{bmatrix}1\\0\\0\\0\\0\end{bmatrix}\right\}, \quad C_1^{-1}\{\operatorname{Im}(D_1)\} = \operatorname{Im}\left\{\begin{bmatrix}1\\0\\0\\0\\0\end{bmatrix}\right\}, \tag{12.4.10}$$

and

$$\bigcup_{|\lambda|=1} \mathcal{V}_\lambda(\Sigma_Q) = \operatorname{Im}\left\{\begin{bmatrix}0\\1\\0\\0\\0\end{bmatrix}\right\}. \tag{12.4.11}$$

Hence,

$$\left\{\mathcal{S}^{\odot}(\Sigma_Q) \cap C_1^{-1}\{\operatorname{Im}(D_1)\}\right\} \cup \left\{\bigcup_{|\lambda|=1} \mathcal{V}_\lambda(\Sigma_Q)\right\} = \operatorname{Im}\left\{\begin{bmatrix}1&0\\0&1\\0&0\\0&0\\0&0\end{bmatrix}\right\}. \tag{12.4.12}$$

It is ready to see now that all conditions in Theorem 12.2.2 are satisfied. Hence, the H_∞-ADDPMS for the given system is solvable. Following the algorithm of this subsection, we obtain a full order output feedback controller of the form (12.4.7) with

$$N = [\,-1 \quad 0.4\,], \tag{12.4.13}$$

$$F_P(\varepsilon) = \begin{bmatrix} -0.526316(\varepsilon-1)^2 - 1.052632(\varepsilon-1) - 0.526316 \\ -0.775623(\varepsilon-1)^2 - 2.603878(\varepsilon-1) - 1.828255 \\ -0.798061(\varepsilon-1)^2 - 2.763490(\varepsilon-1) - 1.566429 \\ -(\varepsilon-1)^2 - 4.2(\varepsilon-1) - 3.11 \\ -2(\varepsilon-1) - 2.1 \end{bmatrix}', \tag{12.4.14}$$

which places the eigenvalues of $\tilde{A} + BF_P(\varepsilon)$ around at 0, 0, 0, $1-\varepsilon$ and $1-\varepsilon$, and

$$K_Q(\varepsilon) = \begin{bmatrix} -10 & 4 \\ -10\varepsilon & 5\varepsilon \\ 0 & 0 \\ 0 & 0 \\ 0 & 0 \end{bmatrix}, \tag{12.4.15}$$

which places the eigenvalues of $\tilde{A} + K_Q(\varepsilon)C_1$ at -0.6554, $0.3777 \pm j0.6726$, 0 and $1-\varepsilon$. The maximum singular value plots of the corresponding closed-

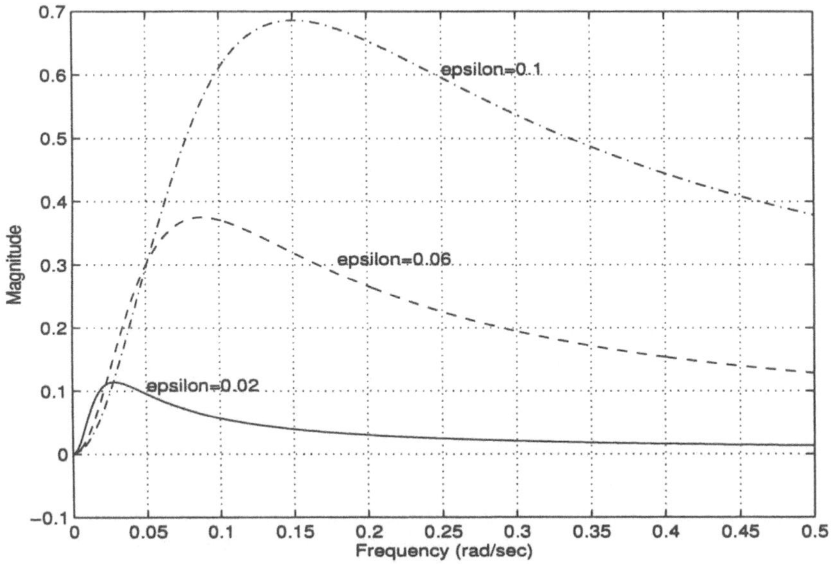

Figure 12.4.1: Max. singular values of T_{hw} — Full order output feedback.

loop transfer matrix $T_{hw}(z,\varepsilon)$ in Figure 12.4.1 show that the H_∞-ADDPMS is attained as ε tends to zero. 🅴

12.4.2. Reduced Order Output Feedback

In this subsection, we follow the procedure of Chapter 8 to design a reduced order output feedback controller. We will show that such a controller structure with appropriately chosen gain matrices also solves the general H_∞-ADDPMS for the discrete-time system (12.1.1). First of all, without loss of generality but for simplicity of presentation, we assume that the matrices C_1 and D_1 are already in the form,

$$C_1 = \begin{bmatrix} 0 & C_{1,02} \\ I_k & 0 \end{bmatrix} \quad \text{and} \quad D_1 = \begin{bmatrix} D_{1,0} \\ 0 \end{bmatrix}, \tag{12.4.16}$$

where $k = \ell - \text{rank}(D_1)$ and $D_{1,0}$ is of full rank. Next, we follow Steps 12.F.C.1 and 12.F.C.2 of the previous subsection to compute the constant matrix N, and form the following system,

$$\begin{cases} x(k+1) = \tilde{A}\ x(k) + B\ u(k) + \tilde{E}\ w(k), \\ y(k) \ = C_1\ x(k) \qquad\qquad\quad + D_1\ w(k), \\ h(k) \ = \tilde{C}_2\ x(k) + D_2\ u(k) + 0\ w(k), \end{cases} \tag{12.4.17}$$

where $\tilde{A}$, $\tilde{E}$ and $\tilde{C}_2$ are defined as in (12.4.3)-(12.4.5). Then, partition (12.4.17) as follows,

$$\begin{cases} \begin{pmatrix} x_1(k+1) \\ x_2(k+1) \end{pmatrix} = \begin{bmatrix} A_{11} & A_{12} \\ A_{21} & A_{22} \end{bmatrix} \begin{pmatrix} x_1(k) \\ x_2(k) \end{pmatrix} + \begin{bmatrix} B_1 \\ B_2 \end{bmatrix} u(k) + \begin{bmatrix} E_1 \\ E_2 \end{bmatrix} w(k), \\ \begin{pmatrix} y_0(k) \\ y_1(k) \end{pmatrix} = \begin{bmatrix} 0 & C_{1,02} \\ I_k & 0 \end{bmatrix} \begin{pmatrix} x_1(k) \\ x_2(k) \end{pmatrix} + \begin{bmatrix} D_{1,0} \\ 0 \end{bmatrix} w(k), \\ h(k) = [C_{2,1} \quad C_{2,2}] \begin{pmatrix} x_1(k) \\ x_2(k) \end{pmatrix} + D_2 \, u(k) + 0 \, w(k), \end{cases}$$

where the state x of (12.4.17) is partitioned to two parts, x_1 and x_2; and y is partitioned to y_0 and y_1 with $y_1 \equiv x_1$. Thus, one needs to estimate only the state x_2 in the reduced order controller design. Next, define an auxiliary subsystem $\Sigma_{\rm QR}$ characterized by a matrix quadruple $(A_{\rm R}, E_{\rm R}, C_{\rm R}, D_{\rm R})$, where

$$(A_{\rm R}, E_{\rm R}, C_{\rm R}, D_{\rm R}) = \left(A_{22}, E_2, \begin{bmatrix} C_{1,02} \\ A_{12} \end{bmatrix}, \begin{bmatrix} D_{1,0} \\ E_1 \end{bmatrix} \right). \tag{12.4.18}$$

The following is a step-by-step algorithm that constructs the reduced order output feedback controller for the general discrete-time H_∞-ADDPMS.

Step 12.R.C.1: (Construction of the gain matrix $F_{\rm P}(\varepsilon)$). Define an auxiliary system

$$\begin{cases} x(k+1) = \tilde{A} \, x(k) + B \, u(k) + \tilde{E} \, w(k), \\ y(k) = x(k) \\ h(k) = \tilde{C}_2 \, x(k) + D_2 \, u(k) + 0 \, w(k), \end{cases} \tag{12.4.19}$$

and then perform Steps 12.S.1 to 12.S.5 of the previous section to the above system to obtain a parameterized gain matrix $F(\varepsilon)$. Furthermore, we let $F_{\rm P}(\varepsilon) = F(\varepsilon)$.

Step 12.R.C.2: (Construction of the gain matrix $K_{\rm R}(\varepsilon)$). Define another auxiliary system

$$\begin{cases} x(k+1) = A_{\rm R}' \, x(k) + C_{\rm R}' \, u(k) + C_{2,2}' \, w(k), \\ y(k) = x(k) \\ h(k) = E_{\rm R}' \, x(k) + D_{\rm R}' \, u(k) + 0 \, w(k), \end{cases} \tag{12.4.20}$$

and then perform Steps 12.S.1 to 12.S.5 of the previous section to the above system to obtain a parameterized gain matrix $F(\varepsilon)$. Similarly, we let $K_{\rm R}(\varepsilon) = F(\varepsilon)'$.

Step 12.R.C.3: (Construction of the reduced order controller $\Sigma_{RC}(\varepsilon)$). Let us partition $F_P(\varepsilon)$ and $K_R(\varepsilon)$ as,

$$F_P(\varepsilon) = [\, F_{P1}(\varepsilon) \quad F_{P2}(\varepsilon)\,] \quad \text{and} \quad K_R(\varepsilon) = [\, K_{R0}(\varepsilon) \quad K_{R1}(\varepsilon)\,] \tag{12.4.21}$$

in conformity with the partitions of $x = \begin{pmatrix} x_1 \\ x_2 \end{pmatrix}$ and $y = \begin{pmatrix} y_0 \\ y_1 \end{pmatrix}$, respectively. Then define

$$G_R(\varepsilon) = [\, -K_{R0}(\varepsilon), \;\; A_{21} + K_{R1}(\varepsilon)A_{11} - (A_R + K_R(\varepsilon)C_R)K_{R1}(\varepsilon)\,]. \tag{12.4.22}$$

Finally, the parameterized reduced order output feedback controller is given by

$$\Sigma_{RC}(\varepsilon) \; : \; \begin{cases} v(k+1) = A_{RC}(\varepsilon)\; v(k) \;+\; B_{RC}(\varepsilon)\; y(k), \\ \quad u(k) \;\;\; = C_{RC}(\varepsilon)\; v(k) \;+\; D_{RC}(\varepsilon)\; y(k), \end{cases} \tag{12.4.23}$$

where

$$\left.\begin{aligned} A_{RC}(\varepsilon) &:= A_R + B_2 F_{P2}(\varepsilon) + K_R(\varepsilon)C_R + K_{R1}(\varepsilon)B_1 F_{P2}(\varepsilon), \\ B_{RC}(\varepsilon) &:= G_R(\varepsilon) + [B_2 + K_{R1}(\varepsilon)B_1]\,[\,0, \;\; F_{P1}(\varepsilon) - F_{P2}(\varepsilon)K_{R1}(\varepsilon)\,], \\ C_{RC}(\varepsilon) &:= F_{P2}(\varepsilon), \\ D_{RC}(\varepsilon) &:= [\,0, \;\; F_{P1}(\varepsilon) - F_{P2}(\varepsilon)K_{R1}(\varepsilon)\,] + N. \end{aligned}\right\} \tag{12.4.24}$$

Note that the reduced order control law in general has a nonzero direct feedthrough term from y to u. ▣

We have the following theorem.

Theorem 12.4.2. Consider the given system Σ of (12.1.1). Assume that the problem of H_∞ almost disturbance decoupling with internal stability for (12.1.1) is solvable, i.e., the solvability conditions of Theorem 12.2.2 are satisfied. Then, the closed-loop system comprising (12.1.1) and the reduced order measurement feedback controller (12.4.23) has the following properties: For any given $\gamma > 0$, there exists a positive scalar $\varepsilon^* > 0$ such that for all $0 < \varepsilon \le \varepsilon^*$,

1. the closed-loop system is asymptotically stable; and
2. the H_∞-norm of the closed-loop transfer matrix from the disturbance w to the controlled output h is less than γ, i.e., $\|T_{hw}(z,\varepsilon)\|_\infty < \gamma$.

Hence, by Definition 12.1.1, the control law of (12.4.23) solves the H_∞-ADDPMS for (12.1.1). ▣

Proof. See Subsection 12.5.F. ▣

We illustrate the above result in the following example.

Example 12.4.2. We again consider the system as given in Example 12.4.1. In what follows, we will construct a reduced order output feedback controller. We first partition

$$\tilde{A} = \left[\begin{array}{c|c} A_{11} & A_{12} \\ \hline A_{21} & A_{22} \end{array}\right] = \left[\begin{array}{c|cccc} 1 & 1 & 1 & 1 & 0 \\ \hline 0 & 1 & 1 & 1 & 0 \\ 0 & 0 & 0.1 & 1 & 0 \\ 0 & 0 & 0 & 0 & 1 \\ 0 & 0 & -0.4 & -0.1 & 0 \end{array}\right], \tag{12.4.25}$$

$$B = \left[\begin{array}{c} B_1 \\ \hline B_2 \end{array}\right] = \left[\begin{array}{c} 0 \\ \hline 0 \\ 0 \\ 0 \\ 1 \end{array}\right], \quad \tilde{E} = \left[\begin{array}{c} E_1 \\ \hline E_2 \end{array}\right] = \left[\begin{array}{cc} 0 & 1 \\ \hline 0 & 0 \\ 0 & 0 \\ 0 & 0 \\ 0 & 0 \end{array}\right], \tag{12.4.26}$$

and $A_R = A_{22}$, $E_R = E_2$, and

$$C_R = \begin{bmatrix} 0.1 & 0.5 & 0.2 & 0.1 \\ 1 & 1 & 1 & 0 \end{bmatrix}, \quad D_R = \begin{bmatrix} 1 & 0 \\ 0 & 1 \end{bmatrix}. \tag{12.4.27}$$

Following our algorithm, we obtain

$$F_P(\varepsilon)' = \left[\begin{array}{c} F_{P1}(\varepsilon)' \\ \hline F_{P2}(\varepsilon)' \end{array}\right] = \left[\begin{array}{c} -0.526316(\varepsilon-1)^2 - 1.052632(\varepsilon-1) - 0.526316 \\ \hline -0.775623(\varepsilon-1)^2 - 2.603878(\varepsilon-1) - 1.828255 \\ -0.798061(\varepsilon-1)^2 - 2.763490(\varepsilon-1) - 1.566429 \\ -(\varepsilon-1)^2 - 4.2(\varepsilon-1) - 3.11 \\ -2(\varepsilon-1) - 2.1 \end{array}\right], \tag{12.4.28}$$

and

$$K_R(\varepsilon) = \left[\begin{array}{c|c} K_{R0}(\varepsilon) & K_{R1}(\varepsilon) \end{array}\right] = \left[\begin{array}{c|c} 0 & -\varepsilon \\ 0 & 0 \\ 0 & 0 \\ 0 & 0 \end{array}\right], \tag{12.4.29}$$

which places the eigenvalues of $A_R + K_R(\varepsilon)C_R$ at -0.6554, $0.3777 \pm j0.6726$, and $1-\varepsilon$. Also, we obtain a reduced order output feedback controller of the form (12.4.23) with all sub-matrices as defined in (12.4.24). The maximum singular value plots of the corresponding closed-loop transfer matrix $T_{hw}(z, \varepsilon)$ in Figure 12.4.2 show that the H_∞-ADDPMS is attained as ε tends to zero. 🄴

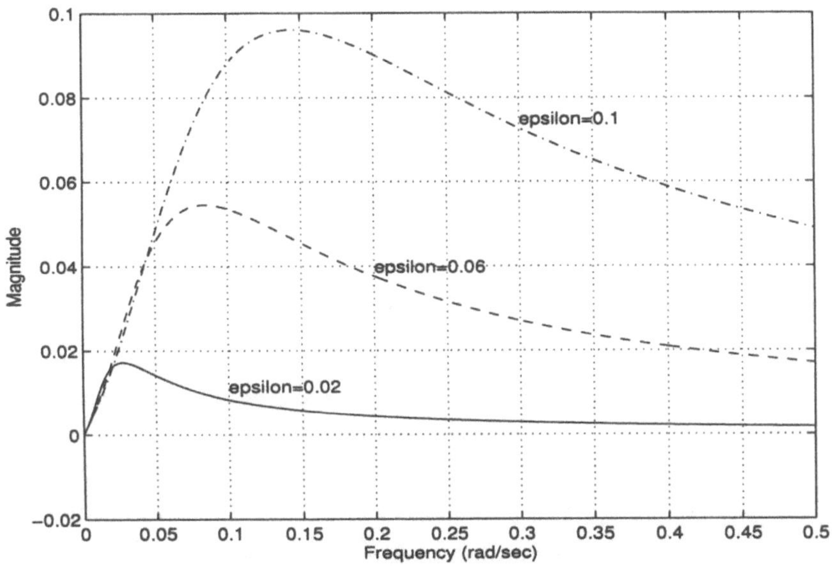

Figure 12.4.2: Max. singular values of T_{hw} — Reduced order output feedback.

12.5. Proofs of Main Results

12.5.A. Proof of Theorem 12.2.1

We show the result of Theorem 12.2.1, i.e., the solvability conditions of the H_∞-ADDPMS for the following full information system,

$$\Sigma_{\rm FI} \ : \ \begin{cases} x(k+1) = A\ x(k) + B\ u(k) + E\ w(k), \\ y(k) = \begin{pmatrix} I \\ 0 \end{pmatrix} x(k) + \begin{pmatrix} 0 \\ I \end{pmatrix} w(k), \\ h(k) \ \cdot = C_2\ x(k) + D_2\ u(k) + D_{22}\ w(k). \end{cases} \tag{12.5.1}$$

We first define the following auxiliary continuous-time system,

$$\check{\Sigma}_{\rm FI} \ : \ \begin{cases} \dot{\check{x}} = \check{A}\ \check{x} + \check{B}\ \check{u} + \check{E}\ \check{w}, \\ \check{y} = \begin{pmatrix} I \\ 0 \end{pmatrix} \check{x} + \begin{pmatrix} 0 \\ I \end{pmatrix} \check{w}, \\ \check{z} = \check{C}_2\ \check{x} + \check{D}_2\ \check{u} + \check{D}_{22}\ \check{w}, \end{cases} \tag{12.5.2}$$

where $\breve{A}$, $\breve{B}$, $\breve{E}$, $\breve{C}_2$, $\breve{D}_2$ and $\breve{D}_{22}$ are defined as

$$\left.\begin{aligned}
\breve{A} &= (A + BF_0 + I)^{-1}(A + BF_0 - I),\\
\breve{B} &= \sqrt{2}(A + BF_0 + I)^{-1}B,\\
\breve{E} &= \sqrt{2}(A + BF_0 + I)^{-1}E,\\
\breve{C}_2 &= \sqrt{2}(C_2 + D_2F_0)(A + BF_0 + I)^{-1},\\
\breve{D}_2 &= D_2 - (C_2 + D_2F_0)(A + BF_0 + I)^{-1}B,\\
\breve{D}_{22} &= D_{22} - (C_2 + D_2F_0)(A + BF_0 + I)^{-1}E,
\end{aligned}\right\} \tag{12.5.3}$$

and where F_0 is chosen such that $A + BF_0$ has no eigenvalue at -1. This can always be done provided that (A, B) is stabilizable. For future use, we denote $\breve{\Sigma}_{\rm P}$ as the subsystem characterized by $(\breve{A}, \breve{B}, \breve{C}_2, \breve{D}_2)$. It was shown in Glover [57] (see also Chapter 3) that the infimum of H_∞ optimization for the discrete-time system (12.5.1) is equivalent to that of H_∞ optimization for the auxiliary continuous-time system (12.5.2). Thus, as a direct consequence, the H_∞-ADDPMS for the discrete-time system (12.5.1) is solvable if and only if the H_∞-ADDPMS for the continuous-time system (12.5.2) is solvable. Following the results of Scherer [118,119], one can show that the H_∞-ADDPMS for (12.5.2) is solvable if and only if the following conditions are satisfied:

(a) $(\breve{A}, \breve{B})$ is stabilizable;

(b) there exists a matrix $\breve{S}$ such that $\breve{D}_{22} + \breve{D}_2\breve{S} = 0$; and

(c) $\operatorname{Im}(\breve{E} + \breve{B}\breve{S}) \subset \mathcal{S}^+(\breve{\Sigma}_{\rm P}) \cap \left\{\bigcap_{\lambda \in \mathbf{C}^0} \mathcal{S}_\lambda(\breve{\Sigma}_{\rm P})\right\}$.

It is simple to show that (A, B) is stabilizable if and only if $(\breve{A}, \breve{B})$ is stabilizable. Hence, it is sufficient to show Theorem 12.2.1 by showing that the following two statements are equivalent:

1. The first statement:

   (a) There exists a S such that $D_{22} + D_2S = 0$;

   (b) $\operatorname{Im}(E + BS) \subset \left\{\mathcal{V}^\circ(\Sigma_{\rm P}) + B\operatorname{Ker}(D_2)\right\} \cap \left\{\bigcap_{|\lambda|=1} \mathcal{S}_\lambda(\Sigma_{\rm P})\right\}$.

2. The second statement:

   (a) There exists a $\breve{S}$ such that $\breve{D}_{22} + \breve{D}_2\breve{S} = 0$;

   (b) $\operatorname{Im}(\breve{E} + \breve{B}\breve{S}) \subset \mathcal{S}^+(\breve{\Sigma}_{\rm P}) \cap \left\{\bigcap_{\lambda \in \mathbf{C}^0} \mathcal{S}_\lambda(\breve{\Sigma}_{\rm P})\right\}$.

Statement 1 ⇒ Statement 2: It is without loss of any generality to assume that matrix D_{22} in (12.5.1) is equal to 0. Also, by the definitions of the geometric subspaces $\mathcal{V}^{\times}$, $\mathcal{S}^{\times}$, $\mathcal{V}_{\lambda}$ and $\mathcal{S}_{\lambda}$, it is simple to verify that they are all invariant under any state feedback, output injection laws, and nonsingular input as well as nonsingular output transformations. Hereafter, we will assume that the subsystem $\Sigma_{\rm P}$, i.e., the quadruple (A, B, C_2, D_2), is in the form of the special coordinate basis of Theorem 2.4.1. For easy reference in future development, we further assume that the state space of $\Sigma_{\rm P}$ has been decomposed as follows:

$$\mathcal{X} = \mathcal{X}_a^{0*} \oplus \mathcal{X}_a^{-} \oplus \mathcal{X}_c \oplus \mathcal{X}_a^{+} \oplus \mathcal{X}_b \oplus \mathcal{X}_d \oplus \mathcal{X}_a^{01}, \tag{12.5.4}$$

where $\mathcal{X}_a^{01}$ is corresponding the zero dynamics of $\Sigma_{\rm P}$ associated with the invariant zero at $z = -1$ and $\mathcal{X}_a^{0*}$ is corresponding to the zero dynamics of $\Sigma_{\rm P}$ associated with the rest invariant zeros on the unit circle. More specifically, we let

$$A = \begin{bmatrix} A_{aa}^{0*} & 0 & 0 & 0 & L_{ab}^{0*}C_b & L_{ad}^{0*}C_d & 0 \\ 0 & A_{aa}^{-} & 0 & 0 & L_{ab}^{-}C_b & L_{ad}^{-}C_d & 0 \\ B_cE_{ca}^{0*} & B_cE_{ca}^{-} & A_{cc} & B_cE_{ca}^{+} & L_{cb}C_b & L_{cd}C_d & B_cE_{ca}^{01} \\ 0 & 0 & 0 & A_{aa}^{+} & L_{ab}^{+}C_b & L_{ad}^{+}C_d & 0 \\ 0 & 0 & 0 & 0 & A_{bb} & L_{bd}C_d & 0 \\ B_dE_{da}^{0*} & B_dE_{da}^{-} & B_dE_{dc} & B_dE_{da}^{+} & B_dE_{db} & A_{dd} & B_dE_{da}^{01} \\ 0 & 0 & 0 & 0 & L_{ab}^{01}C_b & L_{ad}^{01}C_d & A_{aa}^{01} \end{bmatrix} + B_0C_{2,0}, \tag{12.5.5}$$

$$B = [\,B_0 \quad B_1\,] = \begin{bmatrix} B_{0a}^{0*} & 0 & 0 \\ B_{0a}^{-} & 0 & 0 \\ B_{0c} & 0 & B_c \\ B_{0a}^{+} & 0 & 0 \\ B_{0b} & 0 & 0 \\ B_{0d} & B_d & 0 \\ B_{0a}^{01} & 0 & 0 \end{bmatrix}, \quad E = \begin{bmatrix} E_a^{0*} \\ E_a^{-} \\ E_c \\ E_a^{+} \\ E_b \\ E_d \\ E_a^{01} \end{bmatrix}, \tag{12.5.6}$$

$$D_2 = \begin{bmatrix} I & 0 & 0 \\ 0 & 0 & 0 \\ 0 & 0 & 0 \end{bmatrix}, \tag{12.5.7}$$

and

$$C_2 = \begin{bmatrix} C_{2,0} \\ C_{2,1} \end{bmatrix} = \begin{bmatrix} C_{0a}^{0*} & C_{0a}^{-} & C_{0c} & C_{0a}^{+} & C_{0b} & C_{0d} & C_{0a}^{01} \\ 0 & 0 & 0 & 0 & 0 & C_d & 0 \\ 0 & 0 & 0 & 0 & C_b & 0 & 0 \end{bmatrix}, \tag{12.5.8}$$

where A_{aa}^{01} has all its eigenvalues at -1 and A_{aa}^{0*} has all its eigenvalues on the unit circle, but excluding the point -1. Then, the condition in Statement 1(b) is equivalent to that

$$E_a^+ = 0, \quad E_b = 0, \quad E_a^{01} = (I + A_{aa}^{01})X_a^{01}, \quad E_d = B_d X_d, \tag{12.5.9}$$

for some appropriately dimensional X_a^{01} and X_d, and

$$E_a^{0*} = Y_{aa}^{0*} X_a^{0*}, \tag{12.5.10}$$

where Y_{aa}^{0*} is a matrix whose columns span $\cap_{\alpha \in \lambda(A_{aa}^{0*})} \mathrm{Im}\,(\alpha I - A_{aa}^{0*})$ and X_a^{0*} is an appropriately dimensional matrix.

Let us now choose F_0 as,

$$F_0 = -\begin{bmatrix} C_{0a}^{0*} & C_{0a}^{-} & C_{0c} & C_{0a}^{+} & C_{0b} & C_{0d} & C_{0a}^{01} \\ E_{da}^{0*} & E_{da}^{-} & E_{dc} & E_{da}^{+} & E_{db} & 0 & E_{da}^{01} - \hat{E}_{da}^{01} \\ E_{ca}^{0*} & E_{ca}^{-} & 0 & E_{ca}^{+} & 0 & 0 & 0 \end{bmatrix}. \tag{12.5.11}$$

Then, we have

$$\hat{A} = A + BF_0 = \begin{bmatrix} A_{aa}^{0*} & 0 & 0 & 0 & L_{ab}^{0*}C_b & L_{ad}^{0*}C_d & 0 \\ 0 & A_{aa}^{-} & 0 & 0 & L_{ab}^{-}C_b & L_{ad}^{-}C_d & 0 \\ 0 & 0 & A_{cc} & 0 & L_{cb}C_b & L_{cd}C_d & 0 \\ 0 & 0 & 0 & A_{aa}^{+} & L_{ab}^{+}C_b & L_{ad}^{+}C_d & 0 \\ 0 & 0 & 0 & 0 & A_{bb} & L_{bd}C_d & 0 \\ 0 & 0 & 0 & 0 & 0 & A_{dd} & B_d\hat{E}_{da}^{01} \\ 0 & 0 & 0 & 0 & L_{ab}^{01}C_b & L_{ad}^{01}C_d & A_{aa}^{01} \end{bmatrix}, \tag{12.5.12}$$

and

$$\hat{C}_2 = C_2 + D_2F_0 = \begin{bmatrix} 0 & 0 & 0 & 0 & 0 & 0 & 0 \\ 0 & 0 & 0 & 0 & 0 & C_d & 0 \\ 0 & 0 & 0 & 0 & C_b & 0 & 0 \end{bmatrix}. \tag{12.5.13}$$

For simplicity, we further assume that A_{cc}, A_{bb} and A_{dd} have no eigenvalue at -1. Otherwise, some additional pre-state feedback will relocate them to somewhere else. Also, $\hat{E}_{da}^{01}$ is chosen such that $\hat{A}$ has no eigenvalue at -1. Next, it can be computed that

$$(A+BF_0+I)^{-1}=\begin{bmatrix} (I+A_{aa}^{0*})^{-1} & 0 & 0 & 0 & X_{15} & X_{16} & X_{17} \\ 0 & (I+A_{aa}^{-})^{-1} & 0 & 0 & X_{25} & X_{26} & X_{27} \\ 0 & 0 & (I+A_{cc})^{-1} & 0 & X_{35} & X_{36} & X_{37} \\ 0 & 0 & 0 & (I+A_{aa}^{+})^{-1} & X_{45} & X_{46} & X_{47} \\ 0 & 0 & 0 & 0 & X_{55} & X_{56} & X_{57} \\ 0 & 0 & 0 & 0 & X_{65} & X_{66} & X_{67} \\ 0 & 0 & 0 & 0 & X_{75} & X_{76} & X_{77} \end{bmatrix}, \tag{12.5.14}$$

where

$$X_{55}=(I+A_{bb})^{-1}\Big\{I-L_{bd}C_d(I+A_{dd})^{-1}B_d\hat{E}_{da}^{01}\Delta^{-1}L_{ab}^{01}C_b(I+A_{bb})^{-1}\Big\}, \tag{12.5.15}$$

$$X_{56}=-(I+A_{bb})^{-1}L_{bd}\Big\{I+L_{bd}C_d(I+A_{dd})^{-1}B_d\hat{E}_{da}^{01}\Delta^{-1} \times\big[L_{ad}^{01}-L_{ab}^{01}C_b(I+A_{bb})^{-1}L_{bd}\big]\Big\}C_d(I+A_{dd})^{-1}, \tag{12.5.16}$$

$$X_{66}=(I+A_{dd})^{-1}\times \Big\{B_d\hat{E}_{da}^{01}\Delta^{-1}\big[L_{ad}^{01}-L_{ab}^{01}C_b(I+A_{bb})^{-1}L_{bd}\big]C_d(I+A_{dd})^{-1}+I\Big\}, \tag{12.5.17}$$

and

$$X_{57}=(I+A_{bb})^{-1}L_{bd}C_d(I+A_{dd})^{-1}B_d\hat{E}_{da}^{01}\Delta^{-1}, \tag{12.5.18}$$

$$X_{65}=(I+A_{dd})^{-1}B_d\hat{E}_{da}^{01}\Delta^{-1}L_{ab}^{01}C_b(I+A_{bb})^{-1}, \tag{12.5.19}$$

$$X_{67}=-(I+A_{dd})^{-1}B_d\hat{E}_{da}^{01}\Delta^{-1}, \tag{12.5.20}$$

$$X_{75}=-\Delta^{-1}L_{ab}^{01}C_b(I+A_{bb})^{-1}, \tag{12.5.21}$$

$$X_{76}=\Delta^{-1}\big[L_{ab}^{01}C_b(I+A_{bb})^{-1}L_{bd}-L_{ad}^{01}\big]C_d(I+A_{dd})^{-1}, \tag{12.5.22}$$

$$X_{77}=\Delta^{-1}, \tag{12.5.23}$$

$$X_{15}=-(I+A_{aa}^{0*})^{-1}\big(L_{ab}^{0*}C_bX_{55}+L_{ad}^{0*}C_dX_{65}\big), \tag{12.5.24}$$

$$X_{16}=-(I+A_{aa}^{0*})^{-1}\big(L_{ab}^{0*}C_bX_{56}+L_{ad}^{0*}C_dX_{66}\big), \tag{12.5.25}$$

$$X_{17}=-(I+A_{aa}^{0*})^{-1}\big(L_{ab}^{0*}C_bX_{57}+L_{ad}^{0*}C_dX_{67}\big), \tag{12.5.26}$$

$$X_{25}=-(I+A_{aa}^{-})^{-1}\big(L_{ab}^{-}C_bX_{55}+L_{ad}^{-}C_dX_{65}\big), \tag{12.5.27}$$

$$X_{26} = -(I + A_{aa}^-)^{-1}\big(L_{ab}^- C_b X_{56} + L_{ad}^- C_d X_{66}\big), \tag{12.5.28}$$

$$X_{27} = -(I + A_{aa}^-)^{-1}\big(L_{ab}^- C_b X_{57} + L_{ad}^- C_d X_{67}\big), \tag{12.5.29}$$

$$X_{35} = -(I + A_{cc})^{-1}\big(L_{cb} C_b X_{55} + L_{cd} C_d X_{65}\big), \tag{12.5.30}$$

$$X_{36} = -(I + A_{cc})^{-1}\big(L_{cb} C_b X_{56} + L_{cd} C_d X_{66}\big), \tag{12.5.31}$$

$$X_{37} = -(I + A_{cc})^{-1}\big(L_{cb} C_b X_{57} + L_{cd} C_d X_{67}\big), \tag{12.5.32}$$

$$X_{45} = -(I + A_{aa}^+)^{-1}\big(L_{ab}^+ C_b X_{55} + L_{ad}^+ C_d X_{65}\big), \tag{12.5.33}$$

$$X_{46} = -(I + A_{aa}^+)^{-1}\big(L_{ab}^+ C_b X_{56} + L_{ad}^+ C_d X_{66}\big), \tag{12.5.34}$$

$$X_{47} = -(I + A_{aa}^+)^{-1}\big(L_{ab}^+ C_b X_{57} + L_{ad}^+ C_d X_{67}\big), \tag{12.5.35}$$

and where

$$\Delta = I + A_{aa}^{01} + \big[L_{ab}^{01} C_b (I + A_{bb})^{-1} L_{bd} - L_{ad}^{01}\big] C_d (I + A_{dd})^{-1} B_d \hat{E}_{da}^{01}. \tag{12.5.36}$$

Furthermore, we have

$$\breve{B} = \sqrt{2}\begin{bmatrix} (I+A_{aa}^{0*})^{-1}B_{0a}^{0*}+X_{15}B_{0b}+X_{16}B_{0d}+X_{17}B_{0a}^{01} & X_{16}B_d & 0 \\ (I+A_{aa}^{-})^{-1}B_{0a}^{-}+X_{25}B_{0b}+X_{26}B_{0d}+X_{27}B_{0a}^{01} & X_{26}B_d & 0 \\ (I+A_{cc})^{-1}B_{0c}+X_{35}B_{0b}+X_{36}B_{0d}+X_{37}B_{0a}^{01} & X_{36}B_d & X_{cc} \\ (I+A_{aa}^{+})^{-1}B_{0a}^{+}+X_{45}B_{0b}+X_{46}B_{0d}+X_{47}B_{0a}^{01} & X_{46}B_d & 0 \\ X_{55}B_{0b}+X_{56}B_{0d}+X_{57}B_{0a}^{01} & X_{56}B_d & 0 \\ X_{65}B_{0b}+X_{66}B_{0d}+X_{67}B_{0a}^{01} & X_{66}B_d & 0 \\ X_{75}B_{0b}+X_{76}B_{0d}+X_{77}B_{0a}^{01} & X_{76}B_d & 0 \end{bmatrix}, \tag{12.5.37}$$

where $X_{cc} = (I + A_{cc})^{-1} B_c$,

$$\breve{E} = \sqrt{2}\begin{bmatrix} (I+A_{aa}^{0*})^{-1}Y_{aa}^{0*}X_a^{0*}+X_{16}B_dX_d+X_{17}(I+A_{aa}^{01})X_a^{01} \\ (I+A_{aa}^{-})^{-1}E_a^{-}+X_{26}B_dX_d+X_{27}(I+A_{aa}^{01})X_a^{01} \\ (I+A_{cc})^{-1}E_c+X_{36}B_dX_d+X_{37}(I+A_{aa}^{01})X_a^{01} \\ X_{46}B_dX_d+X_{47}(I+A_{aa}^{01})X_a^{01} \\ X_{56}B_dX_d+X_{57}(I+A_{aa}^{01})X_a^{01} \\ X_{66}B_dX_d+X_{67}(I+A_{aa}^{01})X_a^{01} \\ X_{76}B_dX_d+X_{77}(I+A_{aa}^{01})X_a^{01} \end{bmatrix}, \tag{12.5.38}$$

$$\check{D}_2 = \begin{bmatrix} I & 0 & 0 \\ -C_d(X_{65}B_{0b}+X_{66}B_{0d}+X_{67}B_{0a}^{01}) & -C_dX_{66}B_d & 0 \\ -C_b(X_{55}B_{0b}+X_{56}B_{0d}+X_{57}B_{0a}^{01}) & -C_bX_{56}B_d & 0 \end{bmatrix}, \tag{12.5.39}$$

and

$$\check{D}_{22} = \begin{bmatrix} 0 \\ -C_d[X_{66}B_dX_d + X_{67}(I + A_{aa}^{01})X_a^{01}] \\ -C_b[X_{56}B_dX_d + X_{57}(I + A_{aa}^{01})X_a^{01}] \end{bmatrix}. \tag{12.5.40}$$

Next, let us define

$$\check{S} := \begin{bmatrix} 0 \\ -X_d + \hat{E}_{da}^{01}X_a^{01} \\ 0 \end{bmatrix}. \tag{12.5.41}$$

Noting that

$$I + A_{aa}^{01} = \Delta - [L_{ab}^{01}C_b(I + A_{bb})^{-1}L_{bd} - L_{ad}^{01}]C_d(I + A_{dd})^{-1}B_d\hat{E}_{da}^{01}, \tag{12.5.42}$$

it is straightforward to verify that

$$\check{D}_{22} + \check{D}_2\check{S} = \begin{bmatrix} 0 \\ -C_d[X_{67}(I + A_{aa}^{01})X_a^{01} + X_{66}B_d\hat{E}_{da}^{01}X_a^{01}] \\ -C_b[X_{57}(I + A_{aa}^{01})X_a^{01} + X_{56}B_d\hat{E}_{da}^{01}X_a^{01}] \end{bmatrix} = 0, \tag{12.5.43}$$

which shows that Statement 2(a) holds, and

$$\begin{aligned} \check{E} + \check{B}\check{S} &= \sqrt{2}\begin{bmatrix} (I+A_{aa}^{0*})^{-1}Y_{aa}^{0*}X_a^{0*} + X_{16}B_d\hat{E}_{da}^{01}X_a^{01} + X_{17}(I+A_{aa}^{01})X_a^{01} \\ (I+A_{aa}^{-})^{-1}E_a^{-} + X_{26}B_d\hat{E}_{da}^{01}X_a^{01} + X_{27}(I+A_{aa}^{01})X_a^{01} \\ (I+A_{cc})^{-1}E_c + X_{36}B_d\hat{E}_{da}^{01}X_a^{01} + X_{37}(I+A_{aa}^{01})X_a^{01} \\ X_{46}B_d\hat{E}_{da}^{01}X_a^{01} + X_{47}(I+A_{aa}^{01})X_a^{01} \\ X_{56}B_d\hat{E}_{da}^{01}X_a^{01} + X_{57}(I+A_{aa}^{01})X_a^{01} \\ X_{66}B_d\hat{E}_{da}^{01}X_a^{01} + X_{67}(I+A_{aa}^{01})X_a^{01} \\ X_{76}B_d\hat{E}_{da}^{01}X_a^{01} + X_{77}(I+A_{aa}^{01})X_a^{01} \end{bmatrix} \\ &= \sqrt{2}\begin{bmatrix} (I+A_{aa}^{0*})^{-1}Y_{aa}^{0*}X_a^{0*} \\ \star \\ \star \\ 0 \\ 0 \\ 0 \\ \star \end{bmatrix}, \end{aligned} \tag{12.5.44}$$

where $\star$s are matrices of not much interest. Let the state space of $\check{\Sigma}_{\rm P}$, i.e., the matrix quadruple $(\check{A}, \check{B}, \check{C}_2, \check{D}_2)$, be decomposed as follows:

$$\check{\mathcal{X}} = \check{\mathcal{X}}_a^0 \oplus \check{\mathcal{X}}_a^- \oplus \check{\mathcal{X}}_c \oplus \check{\mathcal{X}}_a^{+*} \oplus \check{\mathcal{X}}_b \oplus \check{\mathcal{X}}_a^{+1} \oplus \check{\mathcal{X}}_d, \tag{12.5.45}$$

where $\check{\mathcal{X}}_a^0$, $\check{\mathcal{X}}_a^-$, $\check{\mathcal{X}}_c$, $\check{\mathcal{X}}_b$ and $\check{\mathcal{X}}_d$ are the usual subspaces defined in the special coordinate basis of $\check{\Sigma}_{\rm P}$, while $\check{\mathcal{X}}_a^{+1}$ is corresponding to the zero dynamics of $\check{\Sigma}_{\rm P}$ associated with the invariant zero at $s = 1$, and $\check{\mathcal{X}}_a^{+*}$ is corresponding to the zero dynamics of $\check{\Sigma}_{\rm P}$ associated with the rest unstable invariant zeros (excluding the point $s = 1$). It was shown in Chapter 3, i.e., (3.4.55), that $\check{\mathcal{X}}$ of $\check{\Sigma}_{\rm P}$ and $\mathcal{X}$ of $\Sigma_{\rm P}$ are related by

$$\check{\mathcal{X}}_a^0 = \mathcal{X}_a^{0*}, \quad \check{\mathcal{X}}_a^- = \mathcal{X}_a^-, \quad \check{\mathcal{X}}_c = \mathcal{X}_c, \quad \check{\mathcal{X}}_a^{+*} = \mathcal{X}_a^+, \tag{12.5.46}$$

and

$$\check{\mathcal{X}}_b = \mathcal{X}_b, \quad \check{\mathcal{X}}_a^{+1} = \mathcal{X}_d, \quad \check{\mathcal{X}}_d = \mathcal{X}_a^{01}. \tag{12.5.47}$$

Moreover, the zero dynamics of $\check{\Sigma}_{\rm P}$ corresponding to the imaginary axis invariant zeros are fully characterized by the eigenstructure of the following matrix,

$$\check{A}_{aa}^0 := (A_{aa}^{0*} + I)^{-1}(A_{aa}^{0*} - I). \tag{12.5.48}$$

Noting (12.5.10), it is ready to verify that

$$\operatorname{Im}\left\{(I + A_{aa}^{0*})^{-1} Y_{aa}^{0*}\right\} = \bigcap_{\beta \in \lambda(\check{A}_{aa}^0)} \operatorname{Im}\left\{\beta I - \check{A}_{aa}^0\right\}. \tag{12.5.49}$$

It is now straightforward to see from (12.5.44) and the properties of the special coordinate basis that

$$\operatorname{Im}(\check{E} + \check{B}\check{S}) \subset \mathcal{S}^+(\check{\Sigma}_{\rm P}) \cap \left\{\cap_{\lambda \in \mathbb{C}^0} \mathcal{S}_\lambda(\check{\Sigma}_{\rm P})\right\}, \tag{12.5.50}$$

i.e., Statement 2(b) holds.

Statement 2 $\Rightarrow$ Statement 1: It follows by reversing the above arguments using the well-known bilinear transformation and the results of Chapter 3. Thus, it is omitted. This completes the proof of Theorem 12.2.1. ⊠

12.5.B. Proof of Theorem 12.2.2

For simplicity of presentation, we assume throughout this proof that matrix A has no eigenvalue at -1. Then, we define the following auxiliary continuous-time system,

$$\check{\Sigma} : \begin{cases} \dot{\check{x}} = \check{A}\,\check{x} + \check{B}\,\check{u} + \check{E}\,\check{w}, \\ \check{y} = \check{C}_1\,\check{x} + \check{D}_1\,\check{w}, \\ \check{z} = \check{C}_2\,\check{x} + \check{D}_2\,\check{u} + \check{D}_{22}\,\check{w}, \end{cases} \tag{12.5.51}$$

where $\check{A}$, $\check{B}$, $\check{E}$, $\check{C}_1$, $\check{D}_1$, $\check{C}_2$, $\check{D}_2$ and $\check{D}_{22}$ are defined as

$$\left.\begin{aligned}
\check{A} &= (A+I)^{-1}(A-I), \\
\check{B} &= \sqrt{2}(A+I)^{-1}B, \\
\check{E} &= \sqrt{2}(A+I)^{-1}E, \\
\check{C}_1 &= \sqrt{2}C_1(A+I)^{-1}, \\
\check{D}_1 &= D_1 - C_1(A+I)^{-1}E, \\
\check{C}_2 &= \sqrt{2}C_2(A+I)^{-1}, \\
\check{D}_2 &= D_2 - C_2(A+I)^{-1}B, \\
\check{D}_{22} &= D_{22} - C_2(A+I)^{-1}E.
\end{aligned}\right\} \tag{12.5.52}$$

For easy reference later on, we let $\check{\Sigma}_P$ denote the subsystem characterized by $(\check{A}, \check{B}, \check{C}_2, \check{D}_2)$ and $\check{\Sigma}_Q$ denote the subsystem characterized by characterized by $(\check{A}, \check{E}, \check{C}_1, \check{D}_1)$, respectively. Following the result of Glover [57], one can show that the following two statements are equivalent:

1. The H_∞-ADDPMS for the originally given discrete-time system Σ of (12.1.1) is solvable.

2. The H_∞-ADDPMS for the auxiliary continuous-time system $\check{\Sigma}$ of (12.5.51) is solvable.

It follows from Theorem 8.2.1 (see also Scherer [118,119]) that the second statement above is also equivalent to the following conditions:

(a) $(\check{A}, \check{B})$ is stabilizable.

(b) $(\check{A}, \check{C}_1)$ is detectable.

(c) $\check{D}_{22} + \check{D}_2 \check{S} \check{D}_1 = 0$, where $\check{S} = -(\check{D}_2' \check{D}_2)^{\dagger} \check{D}_2' \check{D}_{22} \check{D}_1' (\check{D}_1 \check{D}_1')^{\dagger}$.

(d) $\operatorname{Im}(\check{E} + \check{B}\check{S}\check{D}_1) \subset \mathcal{S}^+(\check{\Sigma}_P) \cap \left\{\bigcap_{\lambda \in \mathbf{C}^0} \mathcal{S}_\lambda(\check{\Sigma}_P)\right\}$.

(e) $\operatorname{Ker}(\check{C}_2 + \check{D}_2\check{S}\check{C}_1) \supset \mathcal{V}^+(\check{\Sigma}_Q) \cup \left\{\bigcup_{\lambda \in \mathbf{C}^0} \mathcal{V}_\lambda(\check{\Sigma}_Q)\right\}$.

(f) $\mathcal{V}^+(\check{\Sigma}_Q) \subset \mathcal{S}^+(\check{\Sigma}_P)$.

First, it is simple to check that the triple $(\check{A}, \check{B}, \check{C}_1)$ is stabilizable and detectable if and only if the triple (A, B, C) is stabilizable and detectable. Next, following the proof in Subsection 12.5.A, we have the following equivalent statements:

1. Statement I:

   (a) $D_{22} + D_2 S D_1 = 0$, where $S = -(D_2' D_2)^\dagger D_2' D_{22} D_1' (D_1 D_1')^\dagger$;

   (b) $\mathrm{Im}\,(E + BS) \subset \left\{\mathcal{V}^\odot(\Sigma_P) + B\mathrm{Ker}\,(D_2)\right\} \cap \left\{\bigcap_{|\lambda|=1} \mathcal{S}_\lambda(\Sigma_P)\right\}$.

2. Statement II:

   (a) $\check{D}_{22} + \check{D}_2 \check{S} \check{D}_1 = 0$, where $\check{S} = -(\check{D}_2' \check{D}_2)^\dagger \check{D}_2' \check{D}_{22} \check{D}_1' (\check{D}_1 \check{D}_1')^\dagger$;

   (b) $\mathrm{Im}\,(\check{E} + \check{B}\check{S}\check{D}_1) \subset \mathcal{S}^+(\check{\Sigma}_P) \cap \left\{\bigcap_{\lambda \in \mathbb{C}^0} \mathcal{S}_\lambda(\check{\Sigma}_P)\right\}$.

Dualizing the arguments of Subsection 12.5.A, we can show that the following two statements are also equivalent:

1. Statement A:

   (a) $D_{22} + D_2 S D_1 = 0$, where $S = -(D_2' D_2)^\dagger D_2' D_{22} D_1' (D_1 D_1')^\dagger$;

   (b) $\mathrm{Ker}\,(C_2 + D_2 S C_1) \supset \left\{\mathcal{S}^\odot(\Sigma_Q) \cap C_1^{-1}\{\mathrm{Im}\,(D_1)\}\right\} \cup \left\{\bigcup_{|\lambda|=1} \mathcal{V}_\lambda(\Sigma_Q)\right\}$.

2. Statement B:

   (a) $\check{D}_{22} + \check{D}_2 \check{S} \check{D}_1 = 0$, where $\check{S} = -(\check{D}_2' \check{D}_2)^\dagger \check{D}_2' \check{D}_{22} \check{D}_1' (\check{D}_1 \check{D}_1')^\dagger$;

   (b) $\mathrm{Ker}\,(\check{C}_2 + \check{D}_2 \check{S} \check{C}_1) \supset \mathcal{V}^+(\check{\Sigma}_Q) \cup \left\{\bigcup_{\lambda \in \mathbb{C}^0} \mathcal{V}_\lambda(\check{\Sigma}_Q)\right\}$.

Finally, it was shown in Chapter 3 that

$$\mathcal{V}^\odot(\Sigma_P) = \mathcal{S}^+(\check{\Sigma}_P), \quad \mathcal{S}^\odot(\Sigma_P) = \mathcal{V}^+(\check{\Sigma}_P), \tag{12.5.53}$$

and

$$\mathcal{V}^\odot(\Sigma_Q) = \mathcal{S}^+(\check{\Sigma}_Q), \quad \mathcal{S}^\odot(\Sigma_Q) = \mathcal{V}^+(\check{\Sigma}_Q). \tag{12.5.54}$$

Hence, the following two statements are equivalent:

1. $\mathcal{S}^\odot(\Sigma_Q) \subset \mathcal{V}^\odot(\Sigma_P)$.

2. $\mathcal{V}^+(\check{\Sigma}_Q) \subset \mathcal{S}^+(\check{\Sigma}_P)$.

Thus, the result of Theorem 12.2.2 follows. ⊠

12.5.C. Proof of Theorem 12.3.1

Under the feedback control law $u = F(\varepsilon)x$, the closed-loop system on the special coordinate basis can be written as follows,

$$\delta(x_a^-) = A_{aa}^- x_a^- + B_{0a}^- h_0 + L_{ad}^- h_d + L_{ab}^- h_b + E_a^- w, \tag{12.5.55}$$

$$\delta(x_a^0) = A_{aa}^0 x_a^0 + B_{0a}^0 h_0 + L_{ad}^0 h_d + L_{ab}^0 h_b + E_a^0 w, \tag{12.5.56}$$

$$\delta(x_{abd}^+) = A_{abd}^{+c} x_{abd}^+ + [B_{0abd}^+, B_{abd}^+] F_a^0(\varepsilon)[x_a^0 + T_a^0 x_{abd}^+] + E_{abd}^+ w, \tag{12.5.57}$$

$$x_c^+ = A_{cc}^c x_c + B_{0c} h_0 + L_{cb} h_b + L_{cd} h_d + E_c w, \tag{12.5.58}$$

$$h_0 = \left[F_{a0}^+,\ F_{b0},\ F_{d0}\right] x_{abd}^+ + F_{a0}^0(\varepsilon)(x_a^0 + T_a^0 x_{abd}^+), \tag{12.5.59}$$

$$h_b = \left[0_{m_b\times n_a^+},\ C_b,\ 0_{m_b\times n_d}\right] x_{abd}^+,\quad h_d = \left[0_{m_b\times n_a^+},\ 0_{m_b\times n_b},\ C_d\right] x_{abd}^+ \tag{12.5.60}$$

where

$$x_{abd}^+ = \begin{pmatrix} x_a^+ \\ x_b \\ x_d \end{pmatrix}, \tag{12.5.61}$$

and B_{0abd}^+ is as defined in Step 12.S.4.1 of the state feedback design algorithm. We have also used Condition (b) of Remark 12.2.2, i.e., $D_{22} = 0$, and E_a^-, E_a^0, E_{abd}^+, E_b and E_c are defined as follows,

$$\Gamma_{sP}^{-1} E = \left[(E_a^-)'\ (E_a^0)'\ (E_{ab}^+)'\ E_c'\ E_d'\right]',\quad E_{abd}^+ = \left[(E_{ab}^+)'\ E_d'\right]'. \tag{12.5.62}$$

Condition (c) of Remark 12.2.2 then implies that

$$E_{abd}^+ = 0, \tag{12.5.63}$$

and

$$\mathrm{Im}(E_a^0) \subset \mathcal{S}(A_{aa}^0) := \cap_{\omega\in\lambda(A_{aa}^0)} \mathrm{Im}\{\omega I - A_{aa}^0\}. \tag{12.5.64}$$

To complete the proof, we will make two state transformations on the closed-loop system (12.5.55)-(12.5.60). The first state transformation is given as follows,

$$\bar{x}_{abd} = \Gamma_{abd}^{-1} x_{abd},\quad \bar{x}_c = x_c, \tag{12.5.65}$$

where

$$x_{abd} = \begin{pmatrix} x_a^- \\ x_a^0 \\ x_{abd}^+ \end{pmatrix} \quad \text{and} \quad \bar{x}_{abd} = \begin{pmatrix} \bar{x}_a^- \\ \bar{x}_{abd}^+ \\ \bar{x}_a^0 \end{pmatrix}. \tag{12.5.66}$$

In the new state variables (12.5.65), the closed-loop system becomes,

$$\delta(\bar{x}_a^-) = A_{aa}^- \bar{x}_a^- + A_{aabd+}^- \bar{x}_{abd}^+ + B_{0a}^- F_{a0}^0(\varepsilon)\bar{x}_a^0 + E_a^- w, \tag{12.5.67}$$

$$\delta(\bar{x}_{abd}^+) = A_{abd}^{+c}\bar{x}_{abd}^+ + [B_{0abd}^+, B_{abd}^+]F_a^0(\varepsilon)\bar{x}_a^0, \tag{12.5.68}$$

$$\delta(\bar{x}_a^0) = (A_{aa}^0 + B_a^0 F_a^0(\varepsilon))\bar{x}_a^0 + E_a^0 w, \tag{12.5.69}$$

$$\delta(\bar{x}_c) = A_{cc}^c\bar{x}_c + A_{cabd+}\bar{x}_{abd}^+ + B_{0c}F_{a0}^0(\varepsilon)\bar{x}_a^0 + E_c w, \tag{12.5.70}$$

$$h_0 = \begin{bmatrix} F_{a0}^+, \ F_{b0}, \ F_{d0} \end{bmatrix} x_{abd}^+ + F_{a0}^0(\varepsilon)\bar{x}_a^0, \tag{12.5.71}$$

$$h_b = \begin{bmatrix} 0_{m_b \times n_a^+}, \ C_b, \ 0_{m_b \times n_d} \end{bmatrix} x_{abd}^+, \quad h_d = \begin{bmatrix} 0_{m_b \times n_a^+}, \ 0_{m_b \times n_b}, C_d \end{bmatrix} x_{abd}^+, \tag{12.5.72}$$

where

$$A_{aabd+}^- = B_{0a}^- \begin{bmatrix} F_{a0}^+ & F_{b0} & F_{d0} \end{bmatrix} + L_{ad}^- \begin{bmatrix} 0 & 0 & C_d \end{bmatrix} + L_{ab}^- \begin{bmatrix} 0 & C_b & 0 \end{bmatrix}, \tag{12.5.73}$$

and

$$A_{cabd+} = B_{0c} \begin{bmatrix} F_{a0}^+ & F_{b0} & F_{d0} \end{bmatrix} + L_{cb} \begin{bmatrix} 0 & C_b & 0 \end{bmatrix} + L_{cd} \begin{bmatrix} 0 & 0 & C_d \end{bmatrix}. \tag{12.5.74}$$

We now proceed to construct the second transformation. We need to recall the following preliminary results from Lin [82].

Lemma 12.5.1. Consider a single input pair (A, B) in the form of (12.3.19) with all eigenvalues of A on the unit circle. Let $F(\varepsilon) \in \mathbb{R}^{1\times n}$ be the unique matrix such that $\lambda\{A + BF(\varepsilon)\} = (1-\varepsilon)\lambda(A)$, $\varepsilon \in (0, 1]$. Then, there exists a nonsingular transformation matrix $Q(\varepsilon) \in \mathbb{R}^{n\times n}$ such that

1. $Q(\varepsilon)$ transforms $A + BF(\varepsilon)$ into a real Jordan form, i.e.,

$$\begin{aligned} Q^{-1}(\varepsilon)[A + BF(\varepsilon)]Q(\varepsilon) &= J(\varepsilon) \\ &:= \text{blkdiag}\left\{J_{-1}(\varepsilon), J_{+1}(\varepsilon), J_1(\varepsilon), \cdots, J_l(\varepsilon)\right\}, \end{aligned} \tag{12.5.75}$$

where

$$J_{-1}(\varepsilon) = \begin{bmatrix} -(1-\varepsilon) & 1 & & & \\ & \ddots & \ddots & & \\ & & -(1-\varepsilon) & 1 \\ & & & -(1-\varepsilon) \end{bmatrix}_{n_{-1}\times n_{-1}}, \tag{12.5.76}$$

$$J_{+1}(\varepsilon) = \begin{bmatrix} 1-\varepsilon & 1 & & \\ & \ddots & \ddots & \\ & & 1-\varepsilon & 1 \\ & & & 1-\varepsilon \end{bmatrix}_{n_{+1}\times n_{+1}}, \tag{12.5.77}$$

and for each $i = 1$ to l,

$$J_i(\varepsilon) = \begin{bmatrix} J_i^\star(\varepsilon) & I_2 & & & \\ & \ddots & \ddots & & \\ & & J_i^\star(\varepsilon) & I_2 & \\ & & & J_i^\star(\varepsilon) \end{bmatrix}_{2n_i \times 2n_i}, \tag{12.5.78}$$

$$J_i^\star(\varepsilon) = (1-\varepsilon)\begin{bmatrix} \alpha_i & \beta_i \\ -\beta_i & \alpha_i \end{bmatrix}, \tag{12.5.79}$$

with $\alpha_i^2 + \beta_i^2 = 1$ for all $i = 1$ to l and $\alpha_i \neq \alpha_j$ for $i \neq j$.

2. Both $|Q(\varepsilon)|$ and $|Q^{-1}(\varepsilon)|$ are bounded, i.e.,

$$|Q(\varepsilon)| \leq \theta, \quad |Q^{-1}(\varepsilon)| \leq \theta, \quad \varepsilon \in (0,1] \tag{12.5.80}$$

for some positive constant θ, independent of ε.

3. Let $E \in \mathbb{R}^{n \times q}$ is such that

$$\mathrm{Im}(E) \subset \cap_{w \in \lambda(A)} \mathrm{Im}(wI - A), \tag{12.5.81}$$

where q is any integer. Then, there exists a $\sigma \geq 0$, independent of ε, such that

$$|Q^{-1}(\varepsilon)E| \leq \sigma, \quad \varepsilon \in (0,1], \tag{12.5.82}$$

and, if we partition $Q^{-1}(\varepsilon)E$ according to that of $J(\varepsilon)$ as,

$$Q^{-1}(\varepsilon)E = \begin{bmatrix} E_0(\varepsilon) \\ E_1(\varepsilon) \\ \vdots \\ E_l(\varepsilon) \end{bmatrix}, \tag{12.5.83}$$

$$E_0(\varepsilon) = \begin{bmatrix} E_{01}(\varepsilon) \\ E_{02}(\varepsilon) \\ \cdot \ \vdots \\ E_{0n_0}(\varepsilon) \end{bmatrix}_{n_0 \times q}, \quad E_i(\varepsilon) = \begin{bmatrix} E_{i1}(\varepsilon) \\ E_{i2}(\varepsilon) \\ \vdots \\ E_{in_i}(\varepsilon) \end{bmatrix}_{2n_i \times q}, \tag{12.5.84}$$

then, there exists a $\beta \geq 0$, independent of ε, such that, for each $i = 0$, to l,

$$|E_{in_i}(\varepsilon)| \leq \beta\varepsilon. \tag{12.5.85}$$

4. Let

$$S(\varepsilon) = \mathrm{blkdiag}\{S_{-1}(\varepsilon), S_{+1}(\varepsilon), S_1(\varepsilon), S_2(\varepsilon), \cdots, S_l(\varepsilon)\} \tag{12.5.86}$$

where

$$S_{-1}(\varepsilon) = \text{diag}\{\varepsilon^{n_{-1}-1}, \varepsilon^{n_{-1}-2}, \cdots, \varepsilon, 1\}, \tag{12.5.87}$$

$$S_{+1}(\varepsilon) = \text{diag}\{\varepsilon^{n_{+1}-1}, \varepsilon^{n_{+1}-2}, \cdots, \varepsilon, 1\}, \tag{12.5.88}$$

and for each $i = 1$ to l,

$$S_i(\varepsilon) = \text{blkdiag}\{\varepsilon^{n_i-1} I_2, \varepsilon^{n_i-2} I_2, \cdots, \varepsilon I_2, I_2\}. \tag{12.5.89}$$

Then,

(a)

$$S(\varepsilon)J(\varepsilon)S^{-1}(\varepsilon) = \tilde{J}(\varepsilon) := \text{blkdiag}\left\{\tilde{J}_{-1}(\varepsilon), \tilde{J}_{+1}(\varepsilon), \tilde{J}_1(\varepsilon), \cdots, \tilde{J}_l(\varepsilon)\right\} \tag{12.5.90}$$

where

$$\tilde{J}_{-1}(\varepsilon) = \begin{bmatrix} -(1-\varepsilon) & \varepsilon & & \\ & \ddots & \ddots & \\ & & -(1-\varepsilon) & \varepsilon \\ & & & -(1-\varepsilon) \end{bmatrix}_{n_{-1}\times n_{-1}}, \tag{12.5.91}$$

$$\tilde{J}_{+1}(\varepsilon) = \begin{bmatrix} (1-\varepsilon) & \varepsilon & & \\ & \ddots & \ddots & \\ & & (1-\varepsilon) & \varepsilon \\ & & & (1-\varepsilon) \end{bmatrix}_{n_{+1}\times n_{+1}}, \tag{12.5.92}$$

and for each $i = 1$ to l,

$$\tilde{J}_i(\varepsilon) = \begin{bmatrix} J_i^\star(\varepsilon) & \varepsilon I_2 & & \\ & \ddots & \ddots & \\ & & J_i^\star(\varepsilon) & \varepsilon I_2 \\ & & & J_i^\star(\varepsilon) \end{bmatrix}_{2n_i\times 2n_i}, \tag{12.5.93}$$

$$J_i^\star(\varepsilon) = (1-\varepsilon)\begin{bmatrix} \alpha_i & \beta_i \\ -\beta_i & \alpha_i \end{bmatrix}, \tag{12.5.94}$$

with $\beta_i > 0$ for all $i = 1$ to l and $\beta_i \neq \beta_j$ for $i \neq j$.

(b) The unique positive definite solution $\tilde{P}(\varepsilon)$ to the Lyapunov equation

$$\tilde{J}'(\varepsilon)\tilde{P}\tilde{J}(\varepsilon) - \tilde{P} = -\varepsilon I, \tag{12.5.95}$$

is bounded, i.e., there exist positive definite matrices $\tilde{P}_1$ and $\tilde{P}_2$, independent of ε, such that

$$\tilde{P}_1 \leq \tilde{P}(\varepsilon) \leq \tilde{P}_2, \ \ \forall \varepsilon \in (0, \varepsilon^*], \tag{12.5.96}$$

for some $\varepsilon^* \in (0, 1]$.

5. There exist $\alpha, \beta \geq 0$, independent of ε, such that

$$|F(\varepsilon)Q(\varepsilon)S^{-1}(\varepsilon)| \leq \alpha\varepsilon, \tag{12.5.97}$$

and

$$|F(\varepsilon)AQ(\varepsilon)S^{-1}(\varepsilon)| \leq \beta\varepsilon, \tag{12.5.98}$$

for all $\varepsilon \in (0,1]$. ◻

We now define the following second state transformation on the closed-loop system,

$$\tilde{x}_a^- = \bar{x}_a^-, \quad \tilde{x}_{abd}^+ = \bar{x}_{abd}^+, \quad \tilde{x}_c = \varepsilon\bar{x}_c, \tag{12.5.99}$$

and

$$\tilde{x}_a^0 = [(\tilde{x}_{a1}^0)', (\tilde{x}_{a2}^0)', \cdots (\tilde{x}_{al}^0)']' = S_a(\varepsilon)Q_a^{-1}(\varepsilon)(\Gamma_{sa}^0)^{-1}\bar{x}_a^0, \tag{12.5.100}$$

with

$$S_a(\varepsilon) = \text{blkdiag}\{S_{a1}(\varepsilon), S_{a2}(\varepsilon), \cdots, S_{al}(\varepsilon)\}, \tag{12.5.101}$$

and

$$Q_a(\varepsilon) = \text{blkdiag}\{Q_{a1}(\varepsilon), Q_{a2}(\varepsilon), \cdots, Q_{al}(\varepsilon)\}, \tag{12.5.102}$$

where $Q_{ai}(\varepsilon)$ and $S_{ai}(\varepsilon)$ are the $Q(\varepsilon)$ and $S(\varepsilon)$ of Lemmas 12.5.1 for the triple (A_i, B_i, F_i). Hence, the properties of Lemma 12.5.1 all apply. In these new state variables, the closed-loop system becomes,

$$\delta(\tilde{x}_a^-) = A_{aa}^-\tilde{x}_a^- + A_{aabd+}^-\tilde{x}_{abd}^+ + B_{0a}^-F_{a0}^0(\varepsilon)\Gamma_{sa}^0Q_a(\varepsilon)S_a^{-1}(\varepsilon)\tilde{x}_a^0 + E_a^-w, \tag{12.5.103}$$

$$\delta(\tilde{x}_{abd}^+) = A_{abd}^{+c}\tilde{x}_{abd}^+ + [B_{0abd}^+, B_{abd}^+]F_a^0(\varepsilon)\Gamma_{sa}^0Q_a(\varepsilon)S_a^{-1}(\varepsilon)\tilde{x}_a^0, \tag{12.5.104}$$

$$\delta(\tilde{x}_a^0) = \tilde{J}_a(\varepsilon)\tilde{x}_a^0 + \tilde{B}(\varepsilon)\tilde{x}_a^0 + \tilde{E}_a^0(\varepsilon)w, \tag{12.5.105}$$

$$\delta(\tilde{x}_c) = A_{cc}^c\tilde{x}_c + \varepsilon\Big[A_{cabd+}\tilde{x}_{abd}^+ + B_{0c}F_{a0}^0(\varepsilon)\Gamma_{sa}^0Q_a(\varepsilon)S_a^{-1}(\varepsilon)\tilde{x}_a^0 + E_cw\Big], \tag{12.5.106}$$

$$h_0 = \Big[F_{a0}^+, F_{b0}, F_{d0}\Big]x_{abd}^+ + F_{a0}^0(\varepsilon)\Gamma_{sa}^0Q_a(\varepsilon)S_a^{-1}(\varepsilon)\tilde{x}_a^0, \tag{12.5.107}$$

and

$$h_b = \Big[0_{m_b\times n_a^+}, C_b, 0_{m_b\times n_d}\Big]x_{abd}^+, \quad h_d = \Big[0_{m_b\times n_a^+}, 0_{m_b\times n_b}, C_d\Big]x_{abd}^+, \tag{12.5.108}$$

where

$$\tilde{J}_a(\varepsilon) = \text{blkdiag}\{\varepsilon\tilde{J}_{a1}(\varepsilon), \varepsilon\tilde{J}_{a2}(\varepsilon), \cdots, \varepsilon\tilde{J}_{al}(\varepsilon)\}, \tag{12.5.109}$$

$$\tilde{B}(\varepsilon) = \begin{bmatrix} 0 & \tilde{B}_{12}(\varepsilon) & \tilde{B}_{13}(\varepsilon) & \cdots & \tilde{B}_{1l}(\varepsilon) \\ 0 & 0 & \tilde{B}_{23}(\varepsilon) & \cdots & \tilde{B}_{2l}(\varepsilon) \\ \vdots & \vdots & \vdots & \ddots & \vdots \\ 0 & 0 & 0 & \cdots & 0 \end{bmatrix}, \tag{12.5.110}$$

with for $i = 1, 2, \cdots, l$ and $j = i+1, i+2, \cdots, l$,

$$\tilde{B}_{ij}(\varepsilon) = S_{ai}(\varepsilon)Q_{ai}^{-1}(\varepsilon)B_{ij}F_j(\varepsilon)Q_{aj}(\varepsilon)S_{aj}^{-1}(\varepsilon), \tag{12.5.111}$$

and finally

$$\tilde{E}_a^0(\varepsilon) = S_a(\varepsilon)Q_a^{-1}(\varepsilon)(\Gamma_{sa}^0)^{-1}E_a^0 = \begin{bmatrix} \tilde{E}_{a1}^0(\varepsilon) \\ \tilde{E}_{a2}^0(\varepsilon) \\ \cdots \\ \tilde{E}_{al}^0(\varepsilon) \end{bmatrix}, \tag{12.5.112}$$

and where, for $i = 1$ to l, $\tilde{J}_{ai}(\varepsilon)$ is the $\tilde{J}(\varepsilon)$ of Lemma 12.5.1 for the matrix triple (A_i, B_i, F_i).

By Lemma 12.5.1, we have that, for all $\varepsilon \in (0, 1]$,

$$|F_a^0(\varepsilon)\Gamma_{sa}^0 Q_a(\varepsilon)S_a^{-1}(\varepsilon)| \le \tilde{f}_{a0}^0 \tag{12.5.113}$$

for $i = 1$ to l,

$$|\tilde{E}_{ai}^0(\varepsilon)| \le \tilde{e}_a^0\varepsilon, \tag{12.5.114}$$

and finally, for $i = 1$ to l, $j = i+1$ to l,

$$|\tilde{B}_{ij}(\varepsilon)| \le \tilde{b}_{ij}\varepsilon, \tag{12.5.115}$$

where $\tilde{f}_{a0}^0$, $\tilde{e}_a^0$ and $\tilde{b}_{ij}$ are some positive constants, independent of ε.

We next construct a Lyapunov function for the closed loop system (12.5.103)-(12.5.108). We do this by composing Lyapunov functions for the subsystems. For the subsystem of $\tilde{x}_a^-$, we choose a Lyapunov function,

$$V_a^-(\tilde{x}_a^-) = (\tilde{x}_a^-)'P_a^-\tilde{x}_a^-, \tag{12.5.116}$$

where $P_a^- > 0$ is the unique solution to the Lyapunov equation,

$$(A_{aa}^-)'P_a^-A_{aa}^- - P_a^- = -I, \tag{12.5.117}$$

and for the subsystem of $\tilde{x}_{abd}^+$, choose a Lyapunov function,

$$V_{abd}^+(\tilde{x}_{abd}^+) = (\tilde{x}_{abd}^+)'P_{abd}^+\tilde{x}_{abd}^+, \tag{12.5.118}$$

where $P_{abd}^+ > 0$ is the unique solution to the Lyapunov equation,

$$(A_{abd}^{+c})'P_{abd}^+A_{abd}^{+c} - P_{abd}^+ = -I. \tag{12.5.119}$$

The existence of such P_a^- and P_{ab}^+ is guaranteed by the fact that both A_{aa}^- and A_{abd}^{+c} are asymptotically stable. For the subsystem of

$$\tilde{x}_a^0 = [(\tilde{x}_{a1}^0)', (\tilde{x}_{a2}^0)', \cdots, (\tilde{x}_{al}^0)']', \tag{12.5.120}$$

we choose a Lyapunov function,

$$V_a^0(\tilde{x}_a^0) = \sum_{i=1}^{l} \frac{(\alpha_a^0)^{i-1}}{\varepsilon} (\tilde{x}_{ai}^0)' P_{ai}^0(\varepsilon) \tilde{x}_{ai}^0, \tag{12.5.121}$$

where α_a^0 is a positive scalar, whose value is to be determined later, and each $P_{ai}^0(\varepsilon)$ is the unique solution to the Lyapunov equation,

$$\tilde{J}_{ai}(\varepsilon)' P_{ai}^0 \tilde{J}_{ai}(\varepsilon) - P_{ai}^0 = -\varepsilon I, \tag{12.5.122}$$

which, by Lemma 12.5.1, satisfies,

$$P_{ai}(\varepsilon) \le \bar{P}_{ai} \tag{12.5.123}$$

for some $\bar{P}_{ai}$ independent of ε. Similarly, for the subsystem $\tilde{x}_c$, choose a Lyapunov function,

$$V_c(\tilde{x}_c) = \tilde{x}_c' P_c \tilde{x}_c, \tag{12.5.124}$$

where $P_c > 0$ is the unique solution to the Lyapunov equation,

$$(A_{cc}^c)' P_c A_{cc}^c - P_c = -I. \tag{12.5.125}$$

The existence of such a P_c is again guaranteed by the fact that A_{cc}^c is asymptotically stable.

We now construct a Lyapunov function for the closed-loop system (12.5.103)-(12.5.108) as follows,

$$V(\tilde{x}_a^-, \tilde{x}_{abd}^+, \tilde{x}_a^0, \tilde{x}_c) = V_a^-(\tilde{x}_a^-) + \alpha_{abd}^+ V_{abd}^+(\tilde{x}_{abd}^+) + V_a^0(\tilde{x}_a^0) + V_c(\tilde{x}_c), \tag{12.5.126}$$

where $\alpha_{abd}^+ = 2|P_a^{-1}|^2 |A_{aa}^-|^2 + 1$.

Let us first consider the difference of $V_a^0(\tilde{x}_a^0)$ along the trajectories of the subsystem $\tilde{x}_a^0$ and obtain that,

$$\begin{aligned} \Delta V_a^0 = \sum_{i=1}^{l} \Bigg[& -(\alpha_a^0)^{i-1} (\tilde{x}_{ai}^0)' \tilde{x}_{ai}^0 + 2 \sum_{j=i+1}^{l} \frac{(\alpha_a^0)^{i-1}}{\varepsilon} (\tilde{x}_{ai}^0)' \tilde{J}_{ai}'(\varepsilon) P_{ai}^0(\varepsilon) \\ & \times \left[\tilde{B}_{ij}(\varepsilon) \tilde{x}_{aj}^0 + \tilde{E}_{ai}^0(\varepsilon) w \right] \\ & + \frac{(\alpha_a^0)^{i-1}}{\varepsilon} \left(\sum_{j=i+1}^{l} \tilde{B}_{ij}(\varepsilon) \tilde{x}_{aj}^0(\varepsilon) + \tilde{E}_{ai}^0(\varepsilon) w \right)' P_{ai}^0(\varepsilon) \\ & \times \left(\sum_{j=i+1}^{l} \tilde{B}_{ij}(\varepsilon) \tilde{x}_{aj}^0(\varepsilon) + \tilde{E}_{ai}^0(\varepsilon) w \right) \Bigg] \end{aligned} \tag{12.5.127}$$

Using (12.5.114), (12.5.115) and Lemma 12.5.1, it is straightforward to show that, there exists an $\alpha_a^0 > 0$ such that,

$$\Delta V_a^0 \le -\frac{3}{4}|\tilde{x}_a^0|^2 + \alpha_1 |w|^2, \tag{12.5.128}$$

for some nonnegative constants α_1, independent of ε.

In view of (12.5.128), the difference of V along the trajectory of the closed-loop system (12.5.103)-(12.5.108) can be evaluated as follows,

$$\begin{aligned}\Delta V \le & -(\tilde{x}_a^-)'\tilde{x}_a^- + 2(\tilde{x}_a^-)'(A_{aa}^-)'P_a^- \\ & \times[A_{aabd}^-(\varepsilon)\tilde{x}_{abd}^+ + B_{0a}^- F_{a0}^0(\varepsilon)\Gamma_{sa}^0 Q_a(\varepsilon)S_a^{-1}(\varepsilon)\tilde{x}_a^0 + E_a^- w] \\ & -\alpha_{abd}^+(\tilde{x}_{abd}^+)'\tilde{x}_{abd}^+ + 2\alpha_{abd}^+(\tilde{x}_{abd}^+)'(A_{abd}^{+c})'P_{abd}^+ \\ & \times[B_{0abd}^+, B_{abd}^+]F_a^0(\varepsilon)\Gamma_{sa}^0 Q_a(\varepsilon)S_a^{-1}(\varepsilon)\tilde{x}_a^0 \\ & -\frac{3}{4}\left|\tilde{x}_a^0\right|^2 + \alpha_1|w|^2 - \tilde{x}_c'\tilde{x}_c + 2\varepsilon\tilde{x}_c'(A_{cc}^{+c})'P_c[A_{cabd+}\tilde{x}_{abd}^+ \\ & +B_{0c}F_{a0}^0(\varepsilon)\Gamma_{sa}^0 Q_a(\varepsilon)S_a^{-1}(\varepsilon)\tilde{x}_a^0 + E_c w]\end{aligned} \tag{12.5.129}$$

Using (12.5.113) and noting the definition of α_{ab}^+ (12.5.126), we can easily verify that, there exists an $\varepsilon_1^* \in (0,1]$ such that, for all $\varepsilon \in (0, \varepsilon_1^*]$,

$$\Delta V \le -\frac{1}{2}|\tilde{x}_a^-|^2 - \frac{1}{2}|\tilde{x}_{ab}^+|^2 - \frac{1}{2}|\tilde{x}_a^0|^2 - \frac{1}{2}|\tilde{x}_c|^2 + \alpha_2|w|^2, \tag{12.5.130}$$

for some positive constant α_2, independent of ε.

From (12.5.130), it follows that the closed-loop system in the absence of disturbance w is asymptotically stable. It remains to show that, for any given $\gamma > 0$, there exists an $\varepsilon^* \in (0, \varepsilon_1^*]$ such that, for all $\varepsilon \in (0, \varepsilon^*]$,

$$\|h\|_2 \le \gamma\|w\|_2. \tag{12.5.131}$$

To this end, we sum both sides of (12.5.130) from 0 to ∞. Noting that $V \ge 0$ and $V(k) = 0$ at $k = 0$, we have,

$$\|\tilde{x}_a^0\|_2 \le \left(\sqrt{2\alpha_3}\right)\|w\|_2, \tag{12.5.132}$$

which, when used together with (12.5.113) in (12.5.104), results in,

$$\|\tilde{x}_{abd}^+\|_2 \le \alpha_3\varepsilon\|w\|_2. \tag{12.5.133}$$

for some positive constant α_3, independent of ε.

Finally, recalling that

$$h = \Gamma_{\text{oP}}\begin{pmatrix} h_0 \\ h_d \\ h_b \end{pmatrix}, \tag{12.5.134}$$

where h_0, h_d, h_b are as defined in the closed-loop system (12.5.103)-(12.5.108), we have,

$$\|h\|_2 \le \alpha_4 |\Gamma_{\rm OP}| \varepsilon \|w\|_2, \tag{12.5.135}$$

for some positive constant α_4 independent of ε.

To complete the proof, we choose $\varepsilon^* \in (0, \varepsilon_1^*]$ such that,

$$\alpha_4 |\Gamma_{\rm OP}| \varepsilon \le \gamma. \tag{12.5.136}$$

Finally, for the use in the proof of Theorem 12.4.1, it is straightforward to verify from the closed-loop system equations (12.5.103)-(12.5.108) that the transfer function from $E_a^0 w$ to h is given by

$$T_{a0}^0(z,\varepsilon) = T_{a0}(z,\varepsilon)\left[sI - A_{aa}^0 - B_a^0 F_a^0(\varepsilon)\right]^{-1}, \tag{12.5.137}$$

where $T_{a0}(z,\varepsilon) \to 0$ pointwise in z as $\varepsilon \to 0$. ⊠

12.5.D. Proof of Theorem 12.3.2

Without loss of any generality, but for simplicity of presentation, we assume that the matrix quadruple (A, B, C_2, D_2) is in the form of the special coordinate basis of Theorem 2.4.1. It is simple to verify that if Condition (b) of Theorem 12.2.1 holds, we have

$$D_{22} + D_2 F_w = D_{22} + D_2 S - \begin{bmatrix} I & 0 & 0 \\ 0 & 0 & 0 \\ 0 & 0 & 0 \end{bmatrix} \begin{bmatrix} 0 \\ (B_d' B_d)^{-1} B_d' E_d \\ 0 \end{bmatrix} = 0. \tag{12.5.138}$$

Also, Condition (c) of Theorem 12.2.1 implies that

$$E + BS = \begin{bmatrix} E_a^- \\ E_a^0 \\ 0 \\ 0 \\ E_c \\ B_d X_d \end{bmatrix}, \tag{12.5.139}$$

with an appropriately dimensional X_d, and

$$E_a^0 = Y_a^0 X_a^0, \tag{12.5.140}$$

where Y_a^0 is a matrix whose columns span $\cap_{\alpha\in\lambda(A_{aa}^0)}\mathrm{Im}\,(\alpha I - A_{aa}^0)$ and X_a^0 is an appropriately dimensional matrix. Next, it is simple to verify that

$$E + BF_w = \begin{bmatrix} E_a^- \\ E_a^0 \\ 0 \\ 0 \\ E_c \\ B_dX_d - B_d(B_d'B_d)^{-1}B_d'B_dX_d \end{bmatrix} = \begin{bmatrix} E_a^- \\ E_a^0 \\ 0 \\ 0 \\ E_c \\ 0 \end{bmatrix}. \tag{12.5.141}$$

Hence, we have

$$\mathrm{Im}\,(E + BF_w) \subset \mathcal{V}^{\odot}(\Sigma_{\mathrm{P}}) \cap \left\{\cap_{|\lambda|=1}\mathcal{S}_\lambda(\Sigma_{\mathrm{P}})\right\}, \tag{12.5.142}$$

and the result follows from Theorem 12.3.1. ⊠

12.5.E. Proof of Theorems 12.2.3 and 12.4.1

In what follows, we are mainly to examine the result of Theorem 12.4.1. However, the result of Theorem 12.2.3 will become very obvious as we proceed.

Let us first apply a pre-output feedback control law,

$$u = Sy + \hat{u}, \tag{12.5.143}$$

with $S = -(D_2'D_2)^\dagger D_2'D_{22}D_1'(D_1D_1')^\dagger$, to the given system Σ of (12.1.1). Under Condition (c) of Theorem 12.2.2, we have $D_{22} + D_2SD_1 = 0$. We also have a new system,

$$\begin{cases} x(k+1) = (A+BSC_1)\; x(k) + B\; \hat{u}(k) + (E+BSD_1)\; w(k), \\ \quad y(k) \;\;= \quad C_1 \quad x(k) \quad + \quad D_1 \quad w(k), \\ \quad h(k) \;\;= (C_2+D_2SD_1)\; x(k) + D_2\; \hat{u}(k) + \quad 0 \quad w(k). \end{cases} \tag{12.5.144}$$

We denote $\hat{\Sigma}_{\mathrm{P}}$ and $\hat{\Sigma}_{\mathrm{Q}}$ the subsystems characterized by the matrix quadruples $(A + BSC_1, B, C_2 + D_2SC_1, D_2)$ and $(A + BSC_1, E + BSD_1, C_1, D_1)$, respectively. Recalling the definitions of $\mathcal{V}^{\odot}$ and $\mathcal{S}^{\odot}$, which are invariant under any state feedback and output injection laws, we have $\mathcal{V}^{\odot}(\Sigma_{\mathrm{P}}) = \mathcal{V}^{\odot}(\hat{\Sigma}_{\mathrm{P}})$, $\mathcal{S}^{\odot}(\Sigma_{\mathrm{Q}}) = \mathcal{V}^{\odot}(\hat{\Sigma}_{\mathrm{Q}})$, and

$$\begin{bmatrix} A+BSC_1 \\ C_2+D_2SC_1 \end{bmatrix} \mathcal{V}^{\odot}(\Sigma_{\mathrm{P}}) \subset \left(\mathcal{V}^{\odot}(\Sigma_{\mathrm{P}}) \oplus \{0\}\right) + \mathrm{Im}\left\{\begin{bmatrix} B \\ D_2 \end{bmatrix}\right\}, \tag{12.5.145}$$

as well as

$$[\,A+BSC_1 \;\; E+BSD_1\,]\left\{\left(\mathcal{S}^{\odot}(\Sigma_{\mathrm{Q}}) \oplus \mathbf{R}^q\right) \cap \mathrm{Ker}\,\{[\,C_1 \;\; D_1\,]\}\right\} \subset \mathcal{S}^{\odot}(\Sigma_{\mathrm{Q}}). \tag{12.5.146}$$

Furthermore, it can be easily verified that Condition (d) of Theorem 12.2.2 implies

$$\operatorname{Im}\left\{\begin{bmatrix} E+BSD_1 \\ 0 \end{bmatrix}\right\} \subset \left(\mathcal{V}^{\odot}(\Sigma_{\mathrm{P}}) \oplus \{0\}\right) + \operatorname{Im}\left\{\begin{bmatrix} B \\ D_2 \end{bmatrix}\right\}, \tag{12.5.147}$$

and that Condition (e) of Theorem 12.2.2 implies

$$\left(\mathcal{S}^{\odot}(\Sigma_{\mathrm{Q}}) \oplus \mathbf{R}^q\right) \cap \operatorname{Ker}\{[C_1 \;\; D_1]\} \subset \operatorname{Ker}\{[C_2 \;\; 0]\}. \tag{12.5.148}$$

Next, it is ready to show that (12.5.145) and (12.5.147) together with Condition (f) of Theorem 12.2.2 imply that

$$\begin{bmatrix} A+BSC_1 & E+BSD_1 \\ C_2+D_2SC_1 & 0 \end{bmatrix}\left(\mathcal{S}^{\odot}(\Sigma_{\mathrm{Q}}) \oplus \mathbf{R}^q\right) \subset \left(\mathcal{V}^{\odot}(\Sigma_{\mathrm{P}}) \oplus \{0\}\right) + \operatorname{Im}\left\{\begin{bmatrix} B \\ D_2 \end{bmatrix}\right\}, \tag{12.5.149}$$

and that (12.5.146) and (12.5.148) together with Condition (f) of Theorem 12.2.2 imply that

$$\begin{bmatrix} A+BSC_1 & E+BSD_1 \\ C_2+D_2SC_1 & 0 \end{bmatrix}\left\{\left(\mathcal{S}^{\odot}(\Sigma_{\mathrm{Q}}) \oplus \mathbf{R}^q\right) \cap \operatorname{Ker}\{[C_1 \;\; D_1]\}\right\} \subset \left(\mathcal{V}^{\odot}(\Sigma_{\mathrm{P}}) \oplus \{0\}\right). \tag{12.5.150}$$

Finally, (12.5.149) and (12.5.150) imply that there exists a matrix $\tilde{N}$, which satisfies the following condition,

$$\left(\begin{bmatrix} A+BSC_1 & E+BSD_1 \\ C_2+D_2SD_1 & 0 \end{bmatrix} + \begin{bmatrix} B \\ D_2 \end{bmatrix}\tilde{N}[C_1 \;\; D_1]\right)\left(\mathcal{S}^{\odot}(\Sigma_{\mathrm{Q}}) \oplus \mathbf{R}^q\right) \subset \left(\mathcal{V}^{\odot}(\Sigma_{\mathrm{P}}) \oplus \{0\}\right). \tag{12.5.151}$$

It is simple to verify that matrix $\tilde{N} := N - S$, where N is given as in (12.4.1), is one of the solutions to (12.5.151). Following the result of [128], one can show that matrix N of (12.4.1) or $N = \tilde{N} + S$ with $\tilde{N}$ being any solution of (12.5.151) has the following properties:

$$\operatorname{Im}(E + BND_1) \subset \mathcal{V}^{\odot}(\Sigma_{\mathrm{P}}), \tag{12.5.152}$$

$$\operatorname{Ker}(C_2 + D_2NC_1) \supset \mathcal{S}^{\odot}(\Sigma_{\mathrm{Q}}), \tag{12.5.153}$$

and

$$(A + BNC_1)\mathcal{S}^{\odot}(\Sigma_{\mathrm{Q}}) \subset \mathcal{V}^{\odot}(\Sigma_{\mathrm{P}}), \quad D_2ND_1 = 0. \tag{12.5.154}$$

Noting that $D_2ND_1 = 0$, it can be further showed using the compact form of the special coordinate basis that

$$\operatorname{Im}(E + BND_1) \subset \mathcal{V}^{\odot}(\Sigma_{\mathrm{P}}) \cap \left\{\cap_{|\lambda|=1}\mathcal{S}_{\lambda}(\Sigma_{\mathrm{P}})\right\}, \tag{12.5.155}$$

and

$$\text{Ker}\,(C_2 + D_2NC_1) \supset \mathcal{S}^\odot(\Sigma_Q) \cup \left\{\cup_{|\lambda|=1}\mathcal{V}_\lambda(\Sigma_Q)\right\}. \tag{12.5.156}$$

Now, let us apply the following pre-output feedback law, $u = Ny + \tilde{u}$, to the system (12.1.1). We obtain

$$\begin{cases} x(k+1) = \tilde{A}\ x(k) + B\ \tilde{u}(k) + \tilde{E}\ w(k), \\ y(k) \quad = C_1\ x(k) \qquad\qquad + D_1\ w(k), \\ h(k) \quad = \tilde{C}_2\ x(k) + D_2\ \tilde{u}(k) + 0\ \ w(k), \end{cases} \tag{12.5.157}$$

where $\tilde{A}$, $\tilde{E}$ and $\tilde{C}_2$ are as defined in (12.4.3) to (12.4.5). Clearly, it is sufficient to prove Theorem 12.4.1 by showing the following controller

$$\tilde{\Sigma}_{FC}(\varepsilon) : \begin{cases} v(k+1) = A_{FC}(\varepsilon)\ v(k) + B_{FC}(\varepsilon)\ y(k), \\ \tilde{u}(k) \quad = C_{FC}(\varepsilon)\ v(k) + \quad 0 \quad\ y(k), \end{cases} \tag{12.5.158}$$

with $A_{FC}(\varepsilon)$, $B_{FC}(\varepsilon)$ and $C_{FC}(\varepsilon)$ being given as in (12.4.8), solves the H_∞-ADDPMS for (12.5.157). For simplicity of presentation, we denote $\tilde{\Sigma}_P$ the subsystem,

$$\left(\tilde{A}, B, \tilde{C}_2, D_2\right) := \left(A + BNC_1, B, C_2 + D_2NC_1, D_2\right), \tag{12.5.159}$$

and denote $\tilde{\Sigma}_Q$ the subsystem,

$$\left(\tilde{A}, \tilde{E}, C_1, D_1\right) := \left(A + BNC_1, E + BND_1, C_1, D_1\right). \tag{12.5.160}$$

It is simple to see that $(\tilde{A}, B, C_1)$ remains stabilizable and detectable. Also, it is trivial to show the stability of the closed-loop system comprising the given plant (12.5.157) and the controller (12.5.158). The closed-loop eigenvalues are given by $\lambda\{\tilde{A} + BF_P(\varepsilon)\}$, which are in $\mathbb{C}^\odot$ for sufficiently small ε as shown in Theorem 12.3.1, and $\lambda\{\tilde{A} + K_Q(\varepsilon)C_1\}$, which can be dually shown to be in $\mathbb{C}^\odot$ for sufficiently small ε as well. In what follows, we will show that the controller (12.5.158) achieves the H_∞-ADDPMS for (12.5.157), under all the conditions of Theorem 12.2.2. By (12.5.154)-(12.5.156), and the fact that $\mathcal{V}^\odot$, $\mathcal{S}^\odot$, $\mathcal{V}_\lambda$ as well as $\mathcal{S}_\lambda$ are all invariant under any state feedback and output injection laws, we have that Conditions (d) to (f) of Theorem 12.2.2 are equivalent to the following conditions:

($\tilde{d}$). $\text{Im}\,(\tilde{E}) \subset \mathcal{V}^\odot(\tilde{\Sigma}_P) \cap \left\{\bigcap_{|\lambda|=1}\mathcal{S}_\lambda(\tilde{\Sigma}_P)\right\}$;

($\tilde{e}$). $\text{Ker}\,(\tilde{C}_2) \supset \mathcal{S}^\odot(\tilde{\Sigma}_Q) \cup \left\{\bigcup_{|\lambda|=1}\mathcal{V}_\lambda(\tilde{\Sigma}_Q)\right\}$;

($\tilde{f}$). $\mathcal{S}^\odot(\tilde{\Sigma}_Q) \subset \mathcal{V}^\odot(\tilde{\Sigma}_P)$; and

$(\tilde{g})$. $\tilde{A}\mathcal{S}^{\odot}(\tilde{\Sigma}_{\mathrm{Q}}) \subset \mathcal{V}^{\odot}(\tilde{\Sigma}_{\mathrm{P}})$.

In view of the above, the result of Theorem 12.2.3 is obvious. We will continue on the proof of Theorem 12.4.1.

Next, without of loss any generality but for simplicity of presentation, we assume throughout the rest of the proof that the subsystem $\tilde{\Sigma}_{\mathrm{P}}$, i.e., the quadruple $(\tilde{A}, B, \tilde{C}_2, D_2)$, has already been transformed into the special coordinate basis as given in Theorem 2.4.1. To be more specific, we have

$$\tilde{A} = B_0 C_{2,0} + \begin{bmatrix} A_{aa}^- & 0 & 0 & L_{ab}^- C_b & 0 & L_{ad}^- C_d \\ 0 & A_{aa}^0 & 0 & L_{ab}^0 C_b & 0 & L_{ad}^0 C_d \\ 0 & 0 & A_{aa}^+ & L_{ab}^+ C_b & 0 & L_{ad}^+ C_d \\ 0 & 0 & 0 & A_{bb} & 0 & L_{bd} C_d \\ B_c E_{ca}^- & B_c E_{ca}^0 & B_c E_{ca}^+ & L_{cb} C_b & A_{cc} & L_{cd} C_d \\ B_d E_{da}^- & B_d E_{da}^0 & B_d E_{da}^+ & B_d E_{db} & B_d E_{dc} & A_{dd} \end{bmatrix}$$
$$:= B_0 C_{2,0} + \bar{A}, \tag{12.5.161}$$

$$B = \begin{bmatrix} B_{0a}^- & 0 & 0 \\ B_{0a}^0 & 0 & 0 \\ B_{0a}^+ & 0 & 0 \\ B_{0b} & 0 & 0 \\ B_{0c} & 0 & B_c \\ B_{0d} & B_d & 0 \end{bmatrix}, \quad B_0 = \begin{bmatrix} B_{0a}^- \\ B_{0a}^0 \\ B_{0a}^+ \\ B_{0b} \\ B_{0c} \\ B_{0d} \end{bmatrix}, \tag{12.5.162}$$

$$\tilde{C}_2 = \begin{bmatrix} C_{0a}^- & C_{0a}^0 & C_{0a}^+ & C_{0b} & C_{0c} & C_{0d} \\ 0 & 0 & 0 & 0 & 0 & C_d \\ 0 & 0 & 0 & C_b & 0 & 0 \end{bmatrix}, \tag{12.5.163}$$

$$C_{2,0} = [\, C_{0a}^- \quad C_{0a}^0 \quad C_{0a}^+. \quad C_{0b} \quad C_{0c} \quad C_{0d} \,], \quad D_2 = \begin{bmatrix} I & 0 & 0 \\ 0 & 0 & 0 \\ 0 & 0 & 0 \end{bmatrix}, \tag{12.5.164}$$

and

$$\mathcal{V}^{\odot}(\tilde{\Sigma}_{\mathrm{P}}) = \mathrm{Im} \left\{ \begin{bmatrix} I & 0 & 0 \\ 0 & I & 0 \\ 0 & 0 & 0 \\ 0 & 0 & 0 \\ 0 & 0 & I \\ 0 & 0 & 0 \end{bmatrix} \right\}. \tag{12.5.165}$$

It is simple to note that Condition $(\tilde{d})$ implies that

$$\tilde{E} = \begin{bmatrix} E_a^- \\ E_a^0 \\ 0 \\ 0 \\ E_c \\ 0 \end{bmatrix}. \tag{12.5.166}$$

Next, for any $\zeta \in \mathcal{V}_\lambda(\tilde{\Sigma}_Q)$ with $\lambda \in \mathbb{C}^\circ$, we partition ζ as follows,

$$\zeta = \begin{pmatrix} \zeta_a^- \\ \zeta_a^0 \\ \zeta_a^+ \\ \zeta_b \\ \zeta_c \\ \zeta_d \end{pmatrix}. \tag{12.5.167}$$

Then, Condition $(\tilde{e})$ implies that $\tilde{C}_2\zeta = 0$, or equivalently

$$C_{2,0}\zeta = 0, \quad C_b\zeta_b = 0 \quad \text{and} \quad C_d\zeta_d = 0. \tag{12.5.168}$$

By Definition 2.4.3, we have

$$\begin{bmatrix} \tilde{A} - \lambda I & \tilde{E} \\ C_1 & D_1 \end{bmatrix} \begin{pmatrix} \zeta \\ \eta \end{pmatrix} = 0, \tag{12.5.169}$$

for some appropriate vector η. Clearly, (12.5.169) and (12.5.166) imply that

$$(\tilde{A} - \lambda I)\zeta = -\tilde{E}\eta = \begin{pmatrix} \star \\ \star \\ 0 \\ 0 \\ \star \\ 0 \end{pmatrix}, \tag{12.5.170}$$

where $\star$s are some vectors of not much interests. Note that (12.5.168) implies

$$\begin{aligned} (\tilde{A} - \lambda I)\zeta &= (B_0 C_{2,0} + \bar{A} - \lambda I)\zeta = (\bar{A} - \lambda I)\zeta \\ &= \begin{bmatrix} \star \\ \star \\ (A_{aa}^+ - \lambda I)\zeta_a^+ + L_{ab}^+ C_b \zeta_b + L_{ad}^+ C_d \zeta_d \\ (A_{bb} - \lambda I)\zeta_b + L_{bd} C_d \zeta_d \\ \star \\ (A_{dd} - \lambda I)\zeta_d + B_d \zeta_x \end{bmatrix} \end{aligned}$$

$$= \begin{bmatrix} \star \\ \star \\ (A_{aa}^{+} - \lambda I)\zeta_a^{+} \\ (A_{bb} - \lambda I)\zeta_b \\ \star \\ (A_{dd} - \lambda I)\zeta_d + B_d\zeta_x \end{bmatrix}, \tag{12.5.171}$$

where

$$\zeta_x = E_{da}^{-}\zeta_a^{-} + E_{da}^{0}\zeta_a^{0} + E_{da}^{+}\zeta_a^{+} + E_{db}\zeta_b + E_{dc}\zeta_c. \tag{12.5.172}$$

(12.5.170) and (12.5.171) imply

$$(A_{aa}^{+} - \lambda I)\zeta_a^{+} = 0, \quad (A_{bb} - \lambda I)\zeta_b = 0, \tag{12.5.173}$$

and

$$(A_{dd} - \lambda I)\zeta_d + B_d\zeta_x = 0. \tag{12.5.174}$$

Since A_{aa}^{+} has all its eigenvalues in $\mathbb{C}^{\otimes}$, $(A_{aa}^{+} - \lambda I)\zeta_a^{+} = 0$ implies that $\zeta_a^{+} = 0$. Similarly, since (A_{bb}, C_b) is completely observable, $(A_{bb}-\lambda I)\zeta_b = 0$ and $C_b\zeta_b = 0$ imply $\zeta_b = 0$. Also, (12.5.174) and $C_d\zeta_d = 0$ imply that

$$\begin{bmatrix} A_{dd} - \lambda I & B_d \\ C_d & 0 \end{bmatrix} \begin{pmatrix} \zeta_d \\ \zeta_x \end{pmatrix} = 0. \tag{12.5.175}$$

Because (A_{dd}, B_d, C_d) is invertible and is free of invariant zeros, (12.5.175) implies that $\zeta_d = 0$ and $\zeta_x = 0$. Thus, we have

$$\zeta \in \operatorname{Ker}\left\{B_d\,[\,E_{da}^{-} \quad E_{da}^{0} \quad E_{da}^{+} \quad E_{db} \quad E_{dc} \quad 0\,]\right\}, \tag{12.5.176}$$

and hence

$$\mathcal{V}_\lambda(\tilde{\Sigma}_{\mathrm{Q}}) \subset \operatorname{Ker}\left\{B_d\,[\,E_{da}^{-} \quad E_{da}^{0} \quad E_{da}^{+} \quad E_{db} \quad E_{dc} \quad 0\,]\right\}. \tag{12.5.177}$$

Moreover, ζ has the following property,

$$\zeta = \begin{pmatrix} \zeta_a^{-} \\ \zeta_a^{0} \\ 0 \\ 0 \\ \zeta_c \\ 0 \end{pmatrix} \in \mathcal{V}^{\odot}(\tilde{\Sigma}_{\mathrm{P}}). \tag{12.5.178}$$

Obviously, (12.5.178) together with Condition $(\tilde{f})$ imply

$$\mathcal{V}^{\odot}(\tilde{\Sigma}_{\mathrm{P}}) \supset \mathcal{S}^{\odot}(\tilde{\Sigma}_{\mathrm{Q}}) \cup \left\{\cup_{\lambda\in\mathbb{C}^{\mathrm{o}}}\mathcal{V}_\lambda(\tilde{\Sigma}_{\mathrm{Q}})\right\}. \tag{12.5.179}$$

Similarly, for any $\xi \in \mathcal{S}^{\odot}(\tilde{\Sigma}_{\rm Q})$, Conditions ($\tilde{e}$) and ($\tilde{g}$) imply that $\tilde{C}_2\xi = 0$ and

$$\tilde{A}\xi = \begin{pmatrix} \star \\ \star \\ 0 \\ 0 \\ \star \\ 0 \end{pmatrix}. \tag{12.5.180}$$

Now, it is straightforward to show that

$$\xi \in \operatorname{Ker}\left\{B_d\,[\,E_{da}^- \;\; E_{da}^0 \;\; E_{da}^+ \;\; E_{db} \;\; E_{dc} \;\; 0\,]\right\}, \tag{12.5.181}$$

and hence

$$\mathcal{S}^{\odot}(\tilde{\Sigma}_{\rm Q}) \subset \operatorname{Ker}\left\{B_d\,[\,E_{da}^- \;\; E_{da}^0 \;\; E_{da}^+ \;\; E_{db} \;\; E_{dc} \;\; 0\,]\right\}. \tag{12.5.182}$$

(12.5.177) and (12.5.182) imply that

$$\operatorname{Ker}\left\{B_d\,[\,E_{da}^- \;\; E_{da}^0 \;\; E_{da}^+ \;\; E_{db} \;\; E_{dc} \;\; 0\,]\right\} \supset \mathcal{S}^{\odot}(\tilde{\Sigma}_{\rm Q}) \cup \left\{\cup_{\lambda \in \mathbf{C}^{\rm o}} \mathcal{V}_\lambda(\tilde{\Sigma}_{\rm Q})\right\}. \tag{12.5.183}$$

Next, we partition $\tilde{A} - zI$ as follows,

$$\tilde{A} - zI = X_1 + X_2C_2 + X_3 + X_4 + X_5, \tag{12.5.184}$$

where

$$X_1 := \begin{bmatrix} A_{aa}^- - zI & 0 & 0 & L_{ab}^- C_b & 0 & L_{ad}^- C_d \\ 0 & 0 & 0 & 0 & 0 & 0 \\ 0 & 0 & 0 & 0 & 0 & 0 \\ 0 & 0 & 0 & 0 & 0 & 0 \\ B_c E_{ca}^- & B_c E_{ca}^0 & B_c E_{ca}^+ & L_{cb} C_b & A_{cc} - zI & L_{cd} C_d \\ 0 & 0 & 0 & 0 & 0 & 0 \end{bmatrix}, \tag{12.5.185}$$

$$X_2 = \begin{bmatrix} B_{0a}^- & 0 & 0 \\ B_{0a}^0 & L_{ad}^0 & L_{ab}^0 \\ B_{0a}^+ & L_{ad}^+ & L_{ab}^+ \\ B_{0b} & L_{bd} & 0 \\ B_{0c} & 0 & 0 \\ B_{0d} & 0 & 0 \end{bmatrix}, \tag{12.5.186}$$

$$X_3 = \begin{bmatrix} 0 & 0 & 0 & 0 & 0 & 0 \\ 0 & 0 & 0 & 0 & 0 & 0 \\ 0 & 0 & A_{aa}^+ - zI & 0 & 0 & 0 \\ 0 & 0 & 0 & A_{bb} - zI & 0 & 0 \\ 0 & 0 & 0 & 0 & 0 & 0 \\ 0 & 0 & 0 & 0 & 0 & A_{dd} - zI \end{bmatrix}, \tag{12.5.187}$$

$$X_4 = \begin{bmatrix} 0 & 0 & 0 & 0 & 0 & 0 \\ 0 & A_{aa}^0 - zI & 0 & 0 & 0 & 0 \\ 0 & 0 & 0 & 0 & 0 & 0 \\ 0 & 0 & 0 & 0 & 0 & 0 \\ 0 & 0 & 0 & 0 & 0 & 0 \\ 0 & 0 & 0 & 0 & 0 & 0 \end{bmatrix}, \tag{12.5.188}$$

and

$$X_5 = \begin{bmatrix} 0 & 0 & 0 & 0 & 0 & 0 \\ 0 & 0 & 0 & 0 & 0 & 0 \\ 0 & 0 & 0 & 0 & 0 & 0 \\ 0 & 0 & 0 & 0 & 0 & 0 \\ 0 & 0 & 0 & 0 & 0 & 0 \\ B_d E_{da}^- & B_d E_{da}^0 & B_d E_{da}^+ & B_d E_{db} & B_d E_{dc} & 0 \end{bmatrix}. \tag{12.5.189}$$

It is simple to see that

$$\mathrm{Im}\,(X_1) \subset \mathcal{V}^{\odot}(\tilde{\Sigma}_{\mathrm{P}}) \cap \left\{ \cap_{|\lambda|=1} \mathcal{S}_\lambda(\tilde{\Sigma}_{\mathrm{P}}) \right\}, \tag{12.5.190}$$

$$\mathrm{Ker}\,(X_3) \supset \mathcal{V}^{\odot}(\tilde{\Sigma}_{\mathrm{P}}) \supset \mathcal{S}^{\odot}(\tilde{\Sigma}_{\mathrm{Q}}) \cup \left\{ \cup_{|\lambda|=1} \mathcal{V}_\lambda(\tilde{\Sigma}_{\mathrm{Q}}) \right\}. \tag{12.5.191}$$

Also, (12.5.183) implies that

$$\mathrm{Ker}\,(X_5) \supset \mathcal{S}^{\odot}(\tilde{\Sigma}_{\mathrm{Q}}) \cup \left\{ \cup_{|\lambda|=1} \mathcal{V}_\lambda(\tilde{\Sigma}_{\mathrm{Q}}) \right\}. \tag{12.5.192}$$

It follows from the proof of Theorem 12.3.1 that as $\varepsilon \to 0$

$$\left\| [\tilde{C}_2 + D_2 F_{\mathrm{P}}(\varepsilon)][zI - \tilde{A} - BF_{\mathrm{P}}(\varepsilon)]^{-1} \right\|_\infty < \kappa_{\mathrm{P}}, \tag{12.5.193}$$

where κ_{P} is a finite positive constant and is independent of ε. Moreover, under Condition $(\tilde{d})$, we have

$$[\tilde{C}_2 + D_2 F_{\mathrm{P}}(\varepsilon)][zI - \tilde{A} - BF_{\mathrm{P}}(\varepsilon)]^{-1} \tilde{E} \to 0, \tag{12.5.194}$$

and

$$[\tilde{C}_2 + D_2 F_{\mathrm{P}}(\varepsilon)][zI - \tilde{A} - BF_{\mathrm{P}}(\varepsilon)]^{-1} X_1 \to 0, \tag{12.5.195}$$

pointwise in z as $\varepsilon \to 0$. Following (12.5.137), we can show that

$$[\tilde{C}_2 + D_2 F_{\mathrm{P}}(\varepsilon)][zI - \tilde{A} - BF_{\mathrm{P}}(\varepsilon)]^{-1} X_4 \to 0, \tag{12.5.196}$$

pointwise in z as $\varepsilon \to 0$. Dually, one can show that

$$\left\| [zI - \tilde{A} - K_{\mathrm{Q}}(\varepsilon) C_1]^{-1} [\tilde{E} + K_{\mathrm{Q}}(\varepsilon) D_1] \right\|_\infty < \kappa_{\mathrm{Q}}, \tag{12.5.197}$$

where $\kappa_{\rm Q}$ is a finite positive constant and is independent of ε. If Condition ($\tilde{e}$) is satisfied, the following results hold,

$$\tilde{C}_2[zI - \tilde{A} - K_{\rm Q}(\varepsilon)C_1]^{-1}[\tilde{E} + K_{\rm Q}(\varepsilon)D_1] \to 0, \tag{12.5.198}$$

$$X_3[zI - \tilde{A} - K_{\rm Q}(\varepsilon)C_1]^{-1}[\tilde{E} + K_{\rm Q}(\varepsilon)D_1] \to 0, \tag{12.5.199}$$

and

$$X_5[zI - \tilde{A} - K_{\rm Q}(\varepsilon)C_1]^{-1}[\tilde{E} + K_{\rm Q}(\varepsilon)D_1] \to 0, \tag{12.5.200}$$

pointwise in z as $\varepsilon \to 0$.

Finally, it is simple to verify that the closed-loop transfer matrix from the disturbance w to the controlled output h of the closed-loop system comprising the system (12.5.157) and the controller (12.5.158) is given by

$$\begin{aligned} T_{hw}(z,\varepsilon) = {} & [\tilde{C}_2+D_2F_{\rm P}(\varepsilon)][zI-\tilde{A}-BF_{\rm P}(\varepsilon)]^{-1}\tilde{E} \\ & +\tilde{C}_2[zI-\tilde{A}-K_{\rm Q}(\varepsilon)C_1]^{-1}[\tilde{E}+K_{\rm Q}(\varepsilon)D_1]+[\tilde{C}_2+D_2F_{\rm P}(\varepsilon)] \\ & \cdot[zI-\tilde{A}-BF_{\rm P}(\varepsilon)]^{-1}(\tilde{A}-zI)[zI-\tilde{A}-K_{\rm Q}(\varepsilon)C_1]^{-1}[\tilde{E}+K_{\rm Q}(\varepsilon)D_1]. \end{aligned}$$

Using (12.5.184), we can rewrite $T_{hw}(z,\varepsilon)$ as

$$\begin{aligned} T_{hw}(z,\varepsilon) = {} & [\tilde{C}_2+D_2F_{\rm P}(\varepsilon)][zI-\tilde{A}-BF_{\rm P}(\varepsilon)]^{-1}\tilde{E} \\ & +\tilde{C}_2[zI-\tilde{A}-K_{\rm Q}(\varepsilon)C_1]^{-1}[\tilde{E}+K_{\rm Q}(\varepsilon)D_1] \\ & +[\tilde{C}_2+D_2F_{\rm P}(\varepsilon)][zI-\tilde{A}-BF_{\rm P}(\varepsilon)]^{-1}(X_1+X_2\tilde{C}_2+X_3+X_4+X_5) \\ & \cdot[zI-\tilde{A}-K_{\rm Q}(\varepsilon)C_1]^{-1}[\tilde{E}+K_{\rm Q}(\varepsilon)D_1]. \end{aligned}$$

Following (12.5.193) to (12.5.200), and some simple manipulations, it is straightforward to show that as $\varepsilon \to 0$, $T_{hw}(z,\varepsilon) \to 0$, pointwise in z, which is equivalent to $\|T_{hw}\|_\infty \to 0$ as $\varepsilon \to 0$. Hence, the full order output feedback controller (12.4.7) solves the H_∞-ADDPMS for the given plant (12.1.1), provided that all the conditions of Theorem 12.2.2 are satisfied. ⊠

12.5.F. Proof of Theorem 12.4.2

It is sufficient to show Theorem 12.4.2 by showing that the following controller,

$$\tilde{\Sigma}_{\rm RC}(\varepsilon) : \begin{cases} v(k+1) = A_{\rm RC}(\varepsilon)\, v(k) + B_{\rm RC}(\varepsilon)\, y(k), \\ \tilde{u}(k) \quad = C_{\rm RC}(\varepsilon)\, v(k) + \tilde{D}_{\rm RC}(\varepsilon)\, y(k), \end{cases} \tag{12.5.201}$$

with $A_{\rm RC}(\varepsilon)$, $B_{\rm RC}(\varepsilon)$, $C_{\rm RC}(\varepsilon)$ being given as in (12.4.24), and

$$\tilde{D}_{\rm RC}(\varepsilon) = [0,\ \ F_{\rm P1}(\varepsilon) - F_{\rm P2}(\varepsilon)K_{\rm R1}(\varepsilon)\,], \tag{12.5.202}$$

solves the H_∞-ADDPMS for (12.5.157). Again, it is trivial to show the stability of the closed-loop system comprising with (12.5.157) and the controller (12.5.201) as the closed-loop poles are $\lambda\{\tilde{A}+BF_{\rm P}(\varepsilon)\}$ and $\lambda\{A_{\rm R}+K_{\rm R}(\varepsilon)C_{\rm R}\}$, which are asymptotically stable for sufficiently small ε. Next, it is easy to compute the corresponding closed-loop transfer matrix from the disturbance w to the controlled output h,

$$
\begin{aligned}
T_{hw}(z,\varepsilon) &= [\tilde{C}_2 + D_2F_{\rm P}(\varepsilon)][zI-\tilde{A}-BF_{\rm P}(\varepsilon)]^{-1}\tilde{E} \\
&\quad + [\tilde{C}_2 + D_2F_{\rm P}(\varepsilon)][zI-\tilde{A}-BF_{\rm P}(\varepsilon)]^{-1}(\tilde{A}-zI)\begin{pmatrix} 0 \\ I_{n-k}\end{pmatrix} \\
&\quad \cdot [zI-\tilde{A}_{\rm R}-K_{\rm R}(\varepsilon)C_{\rm R}]^{-1}[E_{\rm R}+K_{\rm R}(\varepsilon)D_{\rm R}] \\
&\quad + \tilde{C}_2\begin{pmatrix} 0 \\ I_{n-k}\end{pmatrix}[zI-\tilde{A}_{\rm R}-K_{\rm R}(\varepsilon)C_{\rm R}]^{-1}[E_{\rm R}+K_{\rm R}(\varepsilon)D_{\rm R}].
\end{aligned}
$$

Following the result of Chen [12] (i.e., Proposition 2.2.1), one can show that

$$
\begin{pmatrix} 0 \\ I_{n-k}\end{pmatrix} \mathcal{S}^{\odot}(\Sigma_{\rm QR}) = \mathcal{S}^{\odot}(\tilde{\Sigma}_{\rm Q}) \cap C_1^{-1}\{{\rm Im}\,(D_1)\}, \tag{12.5.203}
$$

and

$$
\begin{pmatrix} 0 \\ I_{n-k}\end{pmatrix} \cup_{|\lambda|=1} \mathcal{V}_\lambda(\Sigma_{\rm QR}) = \cup_{|\lambda|=1}\mathcal{V}_\lambda(\tilde{\Sigma}_{\rm Q}). \tag{12.5.204}
$$

Hence, we have

$$
\begin{aligned}
&\begin{pmatrix} 0 \\ I_{n-k}\end{pmatrix} \left(\mathcal{S}^{\odot}(\Sigma_{\rm QR}) \cup \left\{\cup_{|\lambda|=1}\mathcal{V}_\lambda(\Sigma_{\rm QR})\right\}\right) \\
&\quad = \left\{\mathcal{S}^{\odot}(\tilde{\Sigma}_{\rm Q}) \cap C_1^{-1}\{{\rm Im}\,(D_1)\}\right\} \cup \left\{\cup_{|\lambda|=1}\mathcal{V}_\lambda(\tilde{\Sigma}_{\rm Q})\right\} \\
&\quad \subset \mathcal{S}^{\odot}(\tilde{\Sigma}_{\rm Q}) \cup \left\{\cup_{|\lambda|=1}\mathcal{V}_\lambda(\tilde{\Sigma}_{\rm Q})\right\}.
\end{aligned} \tag{12.5.205}
$$

The rest of the proof follows from the same lines as those of Theorem 12.4.1. ⊠

Chapter 13

Robust and Perfect Tracking of Discrete-time Systems

13.1. Introduction

THIS CHAPTER IS a counterpart of Chapter 9. We present in this chapter (see also [25]) the robust and perfect tracking problem for the following discrete-time system,

$$\Sigma : \begin{cases} x(k+1) = A\ x(k) + B\ u(k) + E\ w(k), & x(0) = x_0, \\ y(k) \ \ = C_1\ x(k) \qquad\qquad\ \ + D_1\ w(k), \\ h(k) \ \ = C_2\ x(k) + D_2\ u(k) + D_{22}\ w(k), \end{cases} \tag{13.1.1}$$

where $x \in \mathbb{R}^n$ is the state, $u \in \mathbb{R}^m$ is the control input, $w \in \mathbb{R}^q$ is the external disturbance, $y \in \mathbb{R}^p$ is the measurement output, and $h \in \mathbb{R}^\ell$ is the output to be controlled. We also assume that the pair (A, B) is stabilizable and (A, C_1) is detectable. For future references, we define Σ_P and Σ_Q to be the subsystems characterized by the matrix quadruples (A, B, C_2, D_2) and (A, E, C_1, D_1), respectively. Given the external disturbance $w \in L_p$, $p \in [1, \infty]$, and any reference signal vector $r \in \mathbb{R}^\ell$, the robust and perfect tracking (RPT) problem for the discrete-time system (9.1.1) is to find a parameterized dynamic measurement feedback control law of the following form

$$\begin{cases} v(k+1) = A_{\rm cmp}(\varepsilon)v(k) + B_{\rm cmp}(\varepsilon)y(k) + G(\varepsilon)r(k), \\ \ \ u(k) \ \ = C_{\rm cmp}(\varepsilon)v(k) + D_{\rm cmp}(\varepsilon)y(k) + H(\varepsilon)r(k), \end{cases} \tag{13.1.2}$$

such that when (13.1.2) is applied to (13.1.1),

1. There exists an $\varepsilon^* > 0$ such that the resulting closed-loop system with $r = 0$ and $w = 0$ is asymptotically stable for all $\varepsilon \in (0, \varepsilon^*]$; and

2. Let $h(k, \varepsilon)$ be the closed-loop controlled output response and let $e(k, \varepsilon)$ be the resulting tracking error, i.e., $e(k, \varepsilon) := h(k, \varepsilon) - r(k)$. Then, for any initial condition of the state, $x_0 \in \mathbf{R}^n$,

$$J_p(x_0, w, r, \varepsilon) := \|e\|_p \to 0 \quad \text{as} \quad \varepsilon \to 0. \tag{13.1.3}$$

In this chapter, we will derive a set of necessary and sufficient conditions under which the proposed robust and perfect tracking problem has a solution, and under these conditions, develop algorithms for the construction of feedback laws that solve the proposed problem. It turns out that the solvability condition for the above proposed RPT problem is quite restrictive compared to its counterpart in the continuous-time case (see Chapter 9). We will later introduce a modified problem, which can be solved for a much larger class of discrete-time systems. This modified formulation will yield internally stabilizing control laws that are capable of tracking reference signals with some delays. If we know the reference signal few steps ahead, the modified tracking control scheme will then track the reference precisely after certain steps, provided that the given plant satisfies a new set of more relaxed conditions.

13.2. Solvability Conditions and Solutions

The following theorem gives a set of necessary and sufficient conditions under which the proposed robust and perfect tracking (RPT) problem is solvable for the given plant (13.1.1). As in Chapter 9, we will show the sufficiency of these conditions by explicitly constructing the required control laws. It turns out that for the discrete-time robust and perfect tracking problem, the required control laws can be chosen to be independent of any tuning parameter.

Theorem 13.2.1. Consider the given system (13.1.1) with its external disturbance $w \in L_p$, $p \in [1, \infty]$, and its initial condition $x(0) = x_0$. Then, for any reference signal $r(k)$, the proposed robust and perfect tracking (RPT) problem is solvable by the control law of (13.1.2) if and only if the following conditions are satisfied:

1. (A, B) is stabilizable and (A, C_1) is detectable;
2. $D_{22} + D_2 S D_1 = 0$, where $S = -(D_2' D_2)^\dagger D_2' D_{22} D_1' (D_1 D_1')^\dagger$;
3. Σ_P is right invertible and of minimum phase with no infinite zeros;
4. $\operatorname{Ker}(C_2 + D_2 S C_1) \supset C_1^{-1}\{\operatorname{Im}(D_1)\}$. ◻

Proof. We first show that Conditions 1 to 4 in the theorem are necessary. Let us consider the case when $r(k) \equiv 0$. Then, the proposed robust and perfect tracking problem reduces to the perfect regulation problem. Following along the same idea as in Chapter 9, we can reformulate the perfect regulation problem for the given system (13.1.1) as the well studied almost disturbance decoupling problem for the following system,

$$\begin{cases} x(k+1) = A\ x(k) + B\ u(k) + [E \quad I]\,\tilde{w}(k), & x(0) = 0, \\ y(k) \ = C_1\ x(k) + [D_1 \ \ 0]\,\tilde{w}(k), \\ h(k) \ = C_2\ x(k) + D_2\ u(k) + [D_{22}\ 0]\,\tilde{w}(k). \end{cases} \tag{13.2.1}$$

For easy reference, we let $\tilde{\Sigma}_{\rm Q}$ be the subsystem characterized by the matrix quadruple $(A, [E \quad I], C_1, [D_1 \quad 0])$. Following the results of the discrete-time almost disturbance decoupling problem of Chapter 12, we can show that if the almost disturbance decoupling problem for the above system is solvable, then the following conditions hold:

1. (A, B) is stabilizable and (A, C_1) is detectable;
2. $D_{22} + D_2 S D_1 = 0$, where $S = -(D_2' D_2)^\dagger D_2' D_{22} D_1' (D_1 D_1')^\dagger$;
3. $\mathrm{Im}\,([E + BSD_1 \quad I]) \subset \mathcal{V}^{\odot}(\Sigma_{\rm P}) + B\mathrm{Ker}\,(D_2)$;
4. $\mathrm{Ker}\,(C_2 + D_2 S C_1) \supset \mathcal{S}^{\odot}(\tilde{\Sigma}_{\rm Q}) \cap C_1^{-1}\{\mathrm{Im}\,(D_1)\}$;
5. $\mathcal{S}^{\odot}(\tilde{\Sigma}_{\rm Q}) \subset \mathcal{V}^{\odot}(\Sigma_{\rm P})$.

Next, it is simple to see that $\mathcal{S}^{\odot}(\tilde{\Sigma}_{\rm Q}) = \mathbb{R}^n$ and hence Item 5 implies that $\mathcal{V}^{\odot}(\Sigma_{\rm P}) = \mathbb{R}^n$, or equivalently, $\Sigma_{\rm P}$ is right invertible with on infinite zeros and no invariant zeros in $\mathbb{C}^{\otimes}$. Furthermore, Item 4 reduces to the condition $\mathrm{Ker}\,(C_2 + D_2 S C_1) \supset C_1^{-1}\{\mathrm{Im}\,(D_1)\}$. Thus, it remains to show that if the proposed RPT problem is solvable, the subsystem $\Sigma_{\rm P}$ must be of minimum phase. In what follows, we proceed to show such a fact.

First, we note that the second condition, i.e., $D_{22} + D_2 S D_1 = 0$, implies that if we apply a pre-output feedback law

$$u(k) = Sy(k), \tag{13.2.2}$$

to the system (13.1.1), the resulting new system will have a zero direct feedthrough term from w to h. Hence, without loss of any generality, we hereafter assume that $D_{22} = 0$ throughout the rest of the proof.

Next, we show that if the robust and perfect tracking problem is solvable for general nonzero reference $r(k)$, $\Sigma_{\rm P}$ must be of minimum phase, i.e., $\Sigma_{\rm P}$ cannot have any invariant zeros on the unit circle. In fact, this condition must hold even for the case when $w = 0$ and $x_0 = 0$, i.e., for the robust and perfect tracking of the following system,

$$\begin{cases} x(k+1) = A\ x(k) + B\ u(k) \\ y(k) \ = C_1\ x(k) \\ e(k) \ = C_2\ x(k) + D_2\ u(k) - r(k) = h(k) - r(k). \end{cases} \tag{13.2.3}$$

Now, if we treat r as an external disturbance, then the above problem is again equivalent to the well-known almost disturbance decoupling problem with measurement feedback and with internal stability for the following system,

$$\begin{cases} x(k+1) = A\ x(k) + B\ u(k) \\ \tilde{y}(k) \ = \begin{pmatrix} C_1 x(k) \\ r(k) \end{pmatrix} \\ e(k) \ = C_2\ x(k) + D_2\ u(k) - r(k). \end{cases} \tag{13.2.4}$$

Without loss of generality, we assume that the quadruple (A, B, C_2, D_2) has been transformed into the form of the special coordinate basis of Theorem 2.4.1, i.e., we have

$$x = \begin{pmatrix} x_a^- \\ x_a^0 \\ x_c \end{pmatrix}, \quad h = h_0, \quad r = r_0, \quad e = e_0 = h_0 - r_0, \quad u = \begin{pmatrix} u_0 \\ u_c \end{pmatrix}, \tag{13.2.5}$$

and

$$x_a^-(k+1) = A_{aa}^- x_a^-(k) + B_{0a}^- h_0(k), \tag{13.2.6}$$

$$x_a^0(k+1) = A_{aa}^0 x_a^0(k) + B_{0a}^0 h_0(k), \tag{13.2.7}$$

$$x_c(k+1) = A_{cc}x_c(k) + B_{0c}h_0(k) + B_c\left[E_{ca}^- x_a^-(k) + E_{ca}^0 x_a^0(k)\right] + B_c u_c(k), \tag{13.2.8}$$

$$e_0(k) = C_{2,0a}^- x_a^-(k) + C_{2,0a}^0 x_a^0(k) + C_{2,0c}x_c(k) + u_0(k) - r_0(k). \tag{13.2.9}$$

In order to bring the subsystem from u to e into the standard form of the special coordinate basis, we need to change h_0 in (13.2.6) to (13.2.8) to $e_0 + r_0$, i.e.,

$$x_a^-(k+1) = A_{aa}^- x_a^-(k) + B_{0a}^- e_0(k) + B_{0a}^- r_0(k), \tag{13.2.10}$$

$$x_a^0(k+1) = A_{aa}^0 x_a^0(k) + B_{0a}^0 e_0(k) + B_{0a}^0 r_0(k), \tag{13.2.11}$$

$$x_c(k+1) = A_{cc}x_c(k) + B_{0c}e_0(k) + B_c\left[E_{ca}^- x_a^-(k) + E_{ca}^0 x_a^0(k)\right] + B_c u_c(k) + B_{0c}r_0(k). \tag{13.2.12}$$

It is simple to see that the subsystem from the controlled input, i.e., $(u_0'\ \ u_c')'$, to the error output, i.e., e_0, is now in the standard form of the special coordinate basis of Theorem 2.4.1. It then follows from the result of Chapter 12 (i.e., Proposition 12.2.1) that if the almost disturbance decoupling problem with measurement feedback and with internal stability for the system (13.2.4) is solvable, there must exist a nonzero vector ξ such that

$$\xi^{\mathrm{H}}(\lambda I - A^0_{aa}) = 0 \quad \text{and} \quad \xi^{\mathrm{H}} B^0_{0a} = 0, \tag{13.2.13}$$

which implies that the matrix pair (A^0_{aa}, B^0_{0a}) is not completely controllable. Following Property 2.4.1 of the special coordinate basis of Chapter 2, the uncontrollability of (A^0_{aa}, B^0_{0a}) implies the unstabilizability of the pair (A, B), which is obviously a contradiction. Hence, x^0_a must be non-existent. It then follows from Property 2.4.2 of the special coordinate basis that Σ_{P} is of minimum phase. This completes the proof of the necessary part. ☒

We note that for the case when $D_1 = 0$, then the direct feedthrough term D_{22} must be a zero matrix as well, and the last condition, i.e., Item 4, of Theorem 13.2.1 reduces to $\mathrm{Ker}\,(C_2) \supset \mathrm{Ker}\,(C_1)$.

We will show the sufficiency of those conditions in Theorem 13.2.1 by explicitly constructing controllers which solve the proposed robust and perfect tracking problem under Conditions 1 to 4 of Theorem 13.2.1. This will be done in the following subsequent subsections. It turns out that the control laws, which solve the robust and perfect tracking for the given plant (13.1.1) under the solvability of Theorem 13.2.1, need not be parameterized by any tuning parameter. Thus, (13.1.2) can be replaced by

$$\begin{cases} v(k+1) = A_{\mathrm{cmp}} v(k) + B_{\mathrm{cmp}} y(k) + G r(k), \\ \quad u(k) \;\; = C_{\mathrm{cmp}} v(k) + D_{\mathrm{cmp}} y(k) + H r(k), \end{cases} \tag{13.2.14}$$

and furthermore, the resulting tracking error $e(k)$ can be made identically zero for all $k \geq 0$. We first have the following corollary that deals with the state feedback case.

Corollary 13.2.1. Consider the given system (13.1.1) with its external disturbance $w \in L_p$, $p \in [1, \infty]$, its initial condition $x(0) = x_0$. Assume that all its states are measured for feedback, i.e., $C_1 = I$ and $D_1 = 0$. Then, for any reference signal $r(k)$, the proposed robust and perfect tracking (RPT) problem is solvable by the control law of (13.2.14) if and only if the following conditions are satisfied:

1. (A, B) is stabilizable;
2. $D_{22} = 0$;
3. Σ_{P} is right invertible and of minimum phase with no infinite zeros. ©

13.2.1. Solutions to State Feedback Case

When all states of the plant are measured for feedback, the problem can be solved by a static control law. We construct in this subsection a state feedback control law,

$$u = Fx + Hr, \tag{13.2.15}$$

which solves the robust and perfect tracking (RPT) problem for (13.1.1) under the conditions given in Corollary 13.2.1. We have the following algorithm.

Step 13.S.1. This step is to transform the subsystem from u to h of the given system (13.1.1) into the special coordinate basis of Theorem 2.4.1, i.e., to find nonsingular state, input and output transformations Γ_s, Γ_i and Γ_o to put it into the structural form of Theorem 2.4.1 as well as in the compact form of (2.4.20) to (2.4.23), i.e.,

$$\tilde{A} = \Gamma_s^{-1}(A - B_0 C_{2,0})\Gamma_s = \begin{bmatrix} A_{aa}^- & 0 \\ B_c E_{ca}^- & A_{cc} \end{bmatrix}, \tag{13.2.16}$$

$$\tilde{B} = \Gamma_s^{-1} B \Gamma_i = \Gamma_s^{-1} [\, B_0 \quad B_1 \,] \Gamma_i = \begin{bmatrix} B_{0a}^- & 0 \\ B_{0c} & B_c \end{bmatrix}, \tag{13.2.17}$$

$$\tilde{D}_2 = \Gamma_o^{-1} D_2 \Gamma_i = [\, I_{m_0} \quad 0 \,], \tag{13.2.18}$$

and

$$\tilde{C}_2 = \Gamma_o^{-1} C_2 \Gamma_s = \Gamma_o^{-1} C_{2,0} \Gamma_s = [\, C_{2,0a}^- \quad C_{2,0c} \,]. \tag{13.2.19}$$

Step 13.S.2. Choose an appropriate dimensional matrix F_c such that

$$A_{cc}^c = A_{cc} - B_c F_c \tag{13.2.20}$$

is asymptotically stable. The existence of such an F_c is guaranteed by the property that (A_{cc}, B_c) is completely controllable.

Step 13.S.3. Finally, we let

$$F = -\Gamma_i \begin{bmatrix} C_{2,0a}^- & C_{2,0c} \\ E_{ca}^- & F_c \end{bmatrix} \Gamma_s^{-1} \quad \text{and} \quad H = \Gamma_i \begin{bmatrix} I \\ 0 \end{bmatrix} \Gamma_o^{-1}. \tag{13.2.21}$$

This ends the constructive algorithm. 🄰

We have the following result.

Theorem 13.2.2. Consider the given discrete-time system (13.1.1) with any external disturbance $w(k)$ and any initial condition $x(0)$. Assume that all its states are measured for feedback, i.e., $C_1 = I$ and $D_1 = 0$. If Conditions 1 to 3 of Corollary 13.2.1 are satisfied, then, for any reference signal $r(k)$, the proposed robust and perfect tracking (RPT) problem is solved by the control law of (13.2.15) with F and H as given in (13.2.21). ⊡

Proof. It is straightforward to verify that the closed-loop system comprising the given plant (13.1.1) and the control law (13.2.15) with F and H being given as in (13.2.21) can be written as,

$$x(k+1) = \Gamma_s \begin{bmatrix} A_{aa}^- & 0 \\ 0 & A_{cc}^c \end{bmatrix} \Gamma_s^{-1} x(k) + \Gamma_s \begin{bmatrix} B_{0a}^- \\ B_{0c} \end{bmatrix} \Gamma_o^{-1} r(k) + Ew(k), \quad (13.2.22)$$

and

$$h(k) = r(k). \quad (13.2.23)$$

Clearly, the resulting closed-loop system is asymptotically stable and $e(k) \equiv 0$ for all $k \geq 0$. Thus, the robust and perfect tracking problem is solved. ⊠

13.2.2. Solutions to Measurement Feedback Case

Without loss of generality, we assume throughout this subsection that matrix $D_{22} = 0$. If it is nonzero, it can always be washed out by the following pre-output feedback,

$$u(k) = Sy(k), \quad (13.2.24)$$

with S as given in the second item of Theorem 13.2.1. It turns out that for discrete-time systems, the full order observer based control law is not capable of achieving the robust and perfect tracking performance, because there is a delay of one step in the observer itself. Thus, we will focus on the construction of a reduced order measurement feedback control law to solve the RPT problem. For simplicity of presentation, we assume that matrices C_1 and D_1 have already been transformed into the following forms,

$$C_1 = \begin{bmatrix} 0 & C_{1,02} \\ I_\kappa & 0 \end{bmatrix} \quad \text{and} \quad D_1 = \begin{bmatrix} D_{1,0} \\ 0 \end{bmatrix}, \quad (13.2.25)$$

where $D_{1,0}$ is of full row rank. Before we present a step-by-step algorithm to construct a reduced order measurement feedback controller, we first partition the following system

$$\begin{cases} x(k+1) = A\ x(k) + B\ u(k) + [E \quad I_n]\ \tilde{w}(k), \\ \ \ y(k) \ = C_1\ x(k) \qquad\qquad\quad + [D_1 \quad 0]\ \tilde{w}(k), \end{cases} \quad (13.2.26)$$

in conformity with the structures of C_1 and D_1 in (13.2.25), i.e.,

$$\left\{\begin{aligned} \begin{pmatrix}\delta(x_1)\\ \delta(x_2)\end{pmatrix} &= \begin{bmatrix}A_{11} & A_{12}\\ A_{21} & A_{22}\end{bmatrix}\begin{pmatrix}x_1\\ x_2\end{pmatrix} + \begin{bmatrix}B_1\\ B_2\end{bmatrix}u + \begin{bmatrix}E_1 & I_\kappa & 0\\ E_2 & 0 & I_{n-\kappa}\end{bmatrix}\tilde{w},\\ \begin{pmatrix}y_0\\ y_1\end{pmatrix} &= \begin{bmatrix}0 & C_{1,02}\\ I_\kappa & 0\end{bmatrix}\begin{pmatrix}x_1\\ x_2\end{pmatrix} + \begin{bmatrix}D_{1,0} & 0 & 0\\ 0 & 0 & 0\end{bmatrix}\tilde{w},\end{aligned}\right. \tag{13.2.27}$$

where $\delta(x_1) = x_1(k+1)$ and $\delta(x_2) = x_2(k+1)$. Obviously, $y_1 = x_1$ is directly available and hence need not be estimated. Next, let $\Sigma_{\rm QR}$ to be characterized by

$$(A_{\rm R}, E_{\rm R}, C_{\rm R}, D_{\rm R}) = \left(A_{22}, [\,E_2 \quad 0 \quad I_{n-\kappa}\,], \begin{bmatrix}C_{1,02}\\ A_{12}\end{bmatrix}, \begin{bmatrix}D_{1,0} & 0 & 0\\ E_1 & I_\kappa & 0\end{bmatrix}\right). \tag{13.2.28}$$

It is straightforward to verify that $\Sigma_{\rm QR}$ is right invertible with no finite and infinite zeros. Moreover, $(A_{\rm R}, C_{\rm R})$ is detectable if and only if (A, C_1) is detectable. We are ready to present the following algorithm.

Step 13.R.1. For the given system (13.1.1), we again assume that all the state variables of (13.1.1) are measurable and then follow Steps 13.S.1 to 13.S.3 of the algorithm of the previous subsection to construct gain matrices F and H. We also partition F in conformity with x_1 and x_2 of (13.2.27) as follows,

$$F = [\,F_1 \quad F_2\,]. \tag{13.2.29}$$

Step 13.R.2. Let $K_{\rm R}$ be an appropriate dimensional constant matrix such that the eigenvalues of

$$A_{\rm R} + K_{\rm R} C_{\rm R} = A_{22} + [\,K_{\rm R0} \quad K_{\rm R1}\,]\begin{bmatrix}C_{1,02}\\ A_{12}\end{bmatrix} \tag{13.2.30}$$

are all in $\mathbb{C}^{\odot}$. This can be done because $(A_{\rm R}, C_{\rm R})$ is detectable.

Step 13.R.3. Let

$$G_{\rm R} = [\,-K_{\rm R0}, \quad A_{21} + K_{\rm R1}A_{11} - (A_{\rm R} + K_{\rm R}C_{\rm R})K_{\rm R1}\,], \tag{13.2.31}$$

$$\left.\begin{aligned} A_{\rm cmp} &= A_{\rm R} + B_2F_2 + K_{\rm R}C_{\rm R} + K_{\rm R1}B_1F_2,\\ B_{\rm cmp} &= G_{\rm R} + (B_2 + K_{\rm R1}B_1)\,[\,0, \quad F_1 - F_2K_{\rm R1}\,],\\ C_{\rm cmp} &= F_2,\\ D_{\rm cmp} &= [\,0, \quad F_1 - F_2K_{\rm R1}\,],\end{aligned}\right\} \tag{13.2.32}$$

and

$$G = (B_2 + K_{\rm R1}B_1)H. \tag{13.2.33}$$

Step 13.R.4. Finally, we obtain the following reduced order measurement feedback control law,

$$\begin{cases} v(k+1) = A_{\rm cmp}\ v(k) + B_{\rm cmp}\ y(k) + G\ r(k), \\ \ u(k) \ = C_{\rm cmp}\ v(k) + D_{\rm cmp}\ y(k) + H\ r(k). \end{cases} \tag{13.2.34}$$

This completes the construction of the reduced order measurement feedback controller. Ⓐ

Theorem 13.2.3. Consider the given system (13.1.1) with any external disturbance $w(k)$ and any initial condition $x(0)$. If Conditions 1 to 4 of Theorem 13.2.1 are satisfied, then, for any reference signal $r(k)$, the proposed robust and perfect tracking (RPT) problem is solved by the reduced order measurement feedback control laws of (13.2.34). Ⓣ

Proof. We first define a new state variable,

$$x_s(k) = x_2(k) - v(k) + K_{\rm R1}x_1(k). \tag{13.2.35}$$

It is straightforward to verify that the closed-loop system comprising the given system (13.1.1) and the reduced order measurement feedback control law of (13.2.34) can be rewritten as follows,

$$x_s(k+1) = (A_{\rm R} + K_{\rm R}C_{\rm R})x_s(k) + \left(E_2 + K_{\rm R}\begin{bmatrix} D_{1,0} \\ E_1 \end{bmatrix}\right) w(k), \tag{13.2.36}$$

$$x(k+1) = (A + BF)x(k) - BF_2x_s(k) + BHr(k) + Ew(k), \tag{13.2.37}$$

and

$$h(k) = (C_2 + D_2F)x(k) - D_2F_2x_s(k) + D_2Hr(k). \tag{13.2.38}$$

Thus, it is simple to see that the closed-loop system is asymptotically stable, as the closed-loop poles are given by the eigenvalues of $A + BF$ and $A_{\rm R} + K_{\rm R}C_{\rm R}$.

Next, if follows from (13.2.18), (13.2.19), (13.2.21) and (13.2.29) that

$$D_2F_2x_s = D_2F\begin{pmatrix} 0 \\ x_s \end{pmatrix} = -C_2\begin{pmatrix} 0 \\ x_s \end{pmatrix}. \tag{13.2.39}$$

Also, it follows from (13.2.25) that

$$C_1^{-1}\{{\rm Im}\,(D_1)\} = {\rm Ker}\left(\begin{bmatrix} 0 & 0 \\ I_K & 0 \end{bmatrix}\right) \quad \text{and} \quad \begin{bmatrix} 0 & 0 \\ I_K & 0 \end{bmatrix}\begin{pmatrix} 0 \\ x_s \end{pmatrix} = 0. \tag{13.2.40}$$

Therefore, the last condition of Theorem 13.2.1, i.e., ${\rm Ker}\,(C_2) \supset C_1^{-1}\{{\rm Im}\,(D_1)\}$, implies that

$$C_2\begin{pmatrix} 0 \\ x_s \end{pmatrix} = 0, \tag{13.2.41}$$

and $D_2F_2x_s(k) = 0$ for all $k \geq 0$. It is now simple to show that (13.2.38) reduces to

$$h(k) = r(k). \tag{13.2.42}$$

Hence, the RPT problem is solved by the reduced order measurement feedback control law (13.2.34). ⊠

The sufficiency of Theorem 13.2.1 is obvious now in view of the result of Theorem 13.2.3. The proof of Theorem 13.2.1 is thus completed. ⊠

We illustrate the results of this section in the following examples.

Example 13.2.1. Consider a discrete-time system characterized by (13.1.1) with

$$A = \begin{bmatrix} 1 & 1 & 0 \\ 0 & 0 & 1 \\ 1 & -0.5 & 1 \end{bmatrix}, \quad B = \begin{bmatrix} 1 & 0 \\ 0 & 0 \\ 1 & -1 \end{bmatrix}, \quad E = \begin{bmatrix} 1 \\ 2 \\ 3 \end{bmatrix}, \tag{13.2.43}$$

$$C_2 = \begin{bmatrix} 1 & 0 & 1 \\ 0 & 1 & 0 \end{bmatrix}, \quad D_2 = \begin{bmatrix} 1 & 0 \\ 0 & 1 \end{bmatrix}, \quad D_{22} = \begin{bmatrix} 0 \\ 0 \end{bmatrix}, \tag{13.2.44}$$

and any disturbance $w(k)$ and any initial condition $x(0)$. It can readily be verified that (A, B) is completely controllable and the subsystem Σ_P is invertible and of minimum phase with three invariant zeros at 0, $-\sqrt{2}/2$, $\sqrt{2}/2$, and with no infinite zero.

A. State Feedback Case. Let us first assume that all the state variables of the given plant are available for feedback, i.e., $C_1 = I$ and $D_1 = 0$. Clearly, all conditions in Corollary 13.2.1 are satisfied and hence the RPT problem is solvable. Following the constructive algorithm for the state feedback case, we obtain

$$u(k) = Fx(k) + Gr(k) = -\begin{bmatrix} 1 & 0 & 1 \\ 0 & 1 & 0 \end{bmatrix} x(k) + \begin{bmatrix} 1 & 0 \\ 0 & 1 \end{bmatrix} r(k). \tag{13.2.45}$$

It is straightforward to verify that the closed-loop system comprising the given plant and the above control law is internally stable and the resulting output to be controlled $h(k) \equiv r(k)$ for all $k \geq 0$. Thus, the robust and perfect tracking performance is achieved.

B. Measurement Feedback Case. Let the measurement output be given as

$$y(k) = C_1x(k) + D_1w(k) = \begin{bmatrix} 1 & 1 & 1 \\ 0 & 2 & 0 \end{bmatrix} x(k) + \begin{bmatrix} 0 \\ 0 \end{bmatrix} w(k). \tag{13.2.46}$$

It is again simple to verify that (A, C_1) is observable and $\text{Ker}(C_2) = \text{Ker}(C_1)$. Thus, all conditions of Theorem 13.2.1 hold and the RPT problem for the given plant is solvable via a measurement feedback controller. Let us first perform some appropriate state and output transformations such that C_1 can be converted into the form of (13.2.25). This can be done by the following transformations,

$$x = \Gamma_s \tilde{x} = \begin{bmatrix} 1 & -1 & -1 \\ 0 & 1 & 0 \\ 0 & 0 & 1 \end{bmatrix} \tilde{x} \quad \text{and} \quad y = \Gamma_o \tilde{y} = \begin{bmatrix} 1 & 0 \\ 0 & 2 \end{bmatrix} \tilde{y}. \tag{13.2.47}$$

Then, we have

$$\tilde{x}(k+1) = \begin{bmatrix} 2 & -1.5 & 0 \\ 0 & 0 & 1 \\ 1 & -1.5 & 0 \end{bmatrix} \tilde{x}(k) + \begin{bmatrix} 2 & -1 \\ 0 & 0 \\ 1 & -1 \end{bmatrix} u(k) + \begin{bmatrix} 6 \\ 2 \\ 3 \end{bmatrix} w(k), \tag{13.2.48}$$

and

$$\tilde{y}(k) = \begin{bmatrix} 1 & 0 & 0 \\ 0 & 1 & 0 \end{bmatrix} \tilde{x}(k) + \begin{bmatrix} 0 \\ 0 \end{bmatrix} w(k). \tag{13.2.49}$$

Following the constructive algorithm of the measurement feedback case, we obtain a reduced order measurement feedback law of the form (13.2.34) with

$$A_{\rm cmp} = 0, \quad B_{\rm cmp} = [0 \quad 0.25], \quad G = [1 \quad -1], \tag{13.2.50}$$

and

$$C_{\rm cmp} = \begin{bmatrix} 0 \\ 0 \end{bmatrix}, \quad D_{\rm cmp} = \begin{bmatrix} -1 & 0.5 \\ 0 & -0.5 \end{bmatrix}, \quad H = \begin{bmatrix} 1 & 0 \\ 0 & 1 \end{bmatrix}. \tag{13.2.51}$$

It happens for this particular example that the obtained measurement feedback control law is equivalent to the following static measurement feedback law,

$$u(k) = \begin{bmatrix} -1 & 0.5 \\ 0 & -0.5 \end{bmatrix} y(k) + \begin{bmatrix} 1 & 0 \\ 0 & 1 \end{bmatrix} r(k). \tag{13.2.52}$$

It is straightforward to show that the closed-loop comprising the given plant and the above static feedback law is asymptotically stable with its closed-loop poles being placed at 0, $\sqrt{2}/2$ and $-\sqrt{2}/2$, and the resulting output to be controlled $h(k) = r(k)$ for all $k \geq 0$. Hence, the RPT problem for the given plant is solved by the control law (13.2.52). ▣

13.3. An Almost Perfect Tracking Problem

As it has been seen in the previous section, the solvability conditions for the robust and perfect tracking problem, especially the restriction on the infinite zeros of the given system, are too strong to be satisfied in most practical situations. We introduce in this section a modified problem, the so-called almost perfect tracking problem, which can be solved for a much larger class of discrete-time systems with any infinite zero structure. This modified formulation will yield internally stabilizing control laws that are capable of tracking reference signal $r(k)$ with some delays. If we know the reference signal few steps ahead, the modified tracking control scheme will then track the reference precisely after certain steps.

For simplicity, we consider in this section the discrete-time system (13.1.1) without external disturbances, i.e.,

$$\Sigma : \begin{cases} x(k+1) = A\ x(k) + B\ u(k), & x(0) = x_0, \\ y(k) = C_1\ x(k) \\ h(k) = C_2\ x(k) + D_2\ u(k). \end{cases} \tag{13.3.1}$$

Let us first consider the reference $r(k) \in \mathbb{R}^\ell$ to be tracked is a known vector sequence, which implies that $r(k+d)$, $0 \le d \le \kappa_d$, is known for some integer $\kappa_d \ge 0$. This is a quite reasonable assumption in most practical situations when one wants to track references such as step functions, ramp functions and sinusoidal functions. We will later deal with the case when $r(k+d)$, $d > 0$, is unknown. We are ready to formally define the almost perfect tracking problem. Given the discrete-time system (13.3.1) with initial condition $x(0) = x_0$ and the reference $r(k)$ with $r(k+d)$, $0 \le d \le \kappa_d$, being known for a nonnegative integer κ_d, the (κ_d, κ_0) almost perfect tracking problem, where κ_0 is another nonnegative integer, is to find a dynamic measurement feedback control law of the following form,

$$\begin{cases} v(k+1) = A_{\rm cmp}v(k) + B_{\rm cmp}y(k) + G_0 r(k) + \cdots + G_{\kappa_d} r(k+\kappa_d), \\ u(k) = C_{\rm cmp}v(k) + D_{\rm cmp}y(k) + H_0 r(k) + \cdots + H_{\kappa_d} r(k+\kappa_d), \end{cases} \tag{13.3.2}$$

such that when (13.3.2) is applied to (13.3.1),

1. The resulting closed-loop system is internally stable; and

2. For any initial condition $x_0 \in \mathbb{R}^n$, the resulting tracking error satisfies:

$$J(x_0, u, \kappa_d, \kappa_0) := \sum_{k=\kappa_0}^{\infty} |e(k)| = 0, \tag{13.3.3}$$

i.e., $e(k) = 0$, or $h(k) = r(k)$, for all $k \ge \kappa_0$.

We have the following theorem.

Theorem 13.3.1. Consider the discrete-time plant (13.3.1) with $x(0) = x_0$, and with i) (A, B) being stabilizable and (A, C_1) being observable; and ii) Σ_P being right invertible and of minimum phase. Let the infinite zero structure (see Chapter 2 for its definition) of Σ_P be given as $S_\infty^\star(\Sigma_P) = \{q_1, \cdots, q_{m_d}\}$, with $q_1 \leq \cdots \leq q_{m_d}$, and let the controllability index of (A', C_1') be $\mathcal{C} = \{k_1, \cdots, k_p\}$, with $k_1 \leq \cdots \leq k_p$. Then, the (κ_d, κ_0) almost perfect tracking problem is solvable for any reference with $\kappa_d = q_{m_d}$ and $\kappa_0 = q_{m_d} + k_p - 1$. 🅃

Proof. We prove this theorem by explicitly constructing the required control law. Let us first construct the special coordinate basis of Σ_P. It follows from Theorem 2.4.1 that there exist nonsingular state, output and input transformations Γ_s, Γ_o and Γ_i, which will take Σ_P into the standard format of the special coordinate basis, i.e.,

$$x = \Gamma_s \tilde{x}, \quad h = \Gamma_o \tilde{h}, \quad u = \Gamma_i \tilde{u}, \quad r = \Gamma_o \tilde{r}, \tag{13.3.4}$$

$$\tilde{x} = \begin{pmatrix} x_a^- \\ x_c \\ x_d \end{pmatrix}, \quad \tilde{h} = \begin{pmatrix} h_0 \\ h_d \end{pmatrix}, \quad \tilde{u} = \begin{pmatrix} u_0 \\ u_d \\ u_c \end{pmatrix}, \quad \tilde{r} = \begin{pmatrix} r_0 \\ r_d \end{pmatrix}, \tag{13.3.5}$$

$$x_d = \begin{pmatrix} x_1 \\ \vdots \\ x_{m_d} \end{pmatrix}, \quad x_i = \begin{pmatrix} x_{i1} \\ \vdots \\ x_{iq_i} \end{pmatrix}, \quad h_d = \begin{pmatrix} h_1 \\ \vdots \\ h_{m_d} \end{pmatrix}, \tag{13.3.6}$$

$$r_d = \begin{pmatrix} r_1 \\ \vdots \\ r_{m_d} \end{pmatrix}, \quad u_d = \begin{pmatrix} u_1 \\ \vdots \\ u_{m_d} \end{pmatrix}, \tag{13.3.7}$$

and

$$\delta(x_a^-) = A_{aa}^- x_a^- + B_{0a}^- h_0 + L_{ad}^- h_d, \tag{13.3.8}$$

$$\delta(x_c) = A_{cc} x_c + B_{0c} h_0 + L_{cd} h_d + B_c E_{ca}^- x_a^- + B_c u_c, \tag{13.3.9}$$

$$h_0 = C_{2,0a}^- x_a^- + C_{2,0c} x_c + C_{2,0d} x_d + u_0, \quad u_0 \in \mathbb{R}^{m_0}, \tag{13.3.10}$$

and for each $i = 1, \cdots, m_d$, $x_i \in \mathbb{R}^{q_i}$ and

$$\delta(x_i) = A_{q_i} x_i + L_{i0} h_0 + L_{id} h_d + B_{q_i} \left[u_i + E_{ia}^- x_a^- + E_{ic} x_c + \sum_{j=1}^{m_d} E_{ij} x_j \right], \tag{13.3.11}$$

$$h_i = C_{q_i} x_i = x_{i1}, \quad h_d = C_d x_d, \tag{13.3.12}$$

where $\delta(\star) = \star(k+1)$, the triple $(A_{q_i}, B_{q_i}, C_{q_i})$ has the special structure as given in (2.4.16). It follows from Theorem 2.4.1 that L_{id} has the following special format,

$$L_{id} = [\, L_{i1} \quad L_{i2} \quad \cdots \quad L_{ii-1} \quad 0 \quad \cdots \quad 0\,],$$

with its last row always being identically zero. Next, we partition L_{i0} and L_{id}, $i = 1, \cdots, m_d$, as follows:

$$L_{i0} = \begin{bmatrix} L_{i0,1} \\ \vdots \\ L_{i0,q_i} \end{bmatrix}, \quad L_{id} = \begin{bmatrix} L_{id,1} \\ \vdots \\ L_{id,q_i} \end{bmatrix}, \tag{13.3.13}$$

and define a new controlled output,

$$\tilde{h}_{\rm n}(k) = \begin{bmatrix} h_0(k) \\ h_1(k+q_1) - \sum_{j=1}^{q_1} \Big[L_{10,j} \ \ L_{1d,j} \Big] \tilde{h}(k+q_1-j) \\ \vdots \\ h_{m_d}(k+q_{m_d}) - \sum_{j=1}^{q_{m_d}} \Big[L_{m_d0,j} \ \ L_{m_dd,j} \Big] \tilde{h}(k+q_{m_d}-j) \end{bmatrix}. \tag{13.3.14}$$

Then, it is straightforward to verify that $\tilde{y}_{\rm n}$ can be expressed as

$$\tilde{h}_{\rm n}(k) = \check{C}_2 \tilde{x}(k) + \check{D}_2 \tilde{u}(k), \tag{13.3.15}$$

with

$$\check{C}_2 = \begin{bmatrix} C_{2,0a}^- & C_{2,0c} & C_{2,0d} \\ E_{da}^- & E_{dc} & E_{dd} \end{bmatrix} \quad \text{and} \quad \check{D}_2 = \begin{bmatrix} I_{m_0} & 0 & 0 \\ 0 & I_{m_d} & 0 \end{bmatrix}, \tag{13.3.16}$$

where

$$E_{da}^- = \begin{bmatrix} E_{1a}^- \\ \vdots \\ E_{m_da}^- \end{bmatrix}, \quad E_{dc} = \begin{bmatrix} E_{1c} \\ \vdots \\ E_{m_dc} \end{bmatrix}, \tag{13.3.17}$$

and

$$E_{dd} = \begin{bmatrix} E_{11} & \cdots & E_{1m_d} \\ \vdots & \ddots & \vdots \\ E_{m_d1} & \cdots & E_{m_dm_d} \end{bmatrix}. \tag{13.3.18}$$

Let $\tilde{A} = \Gamma_s^{-1} A \Gamma_s$ and $\tilde{B} = \Gamma_s^{-1} B \Gamma_i$, and let $\check{\Sigma}_{\rm P}$ characterized by $(\tilde{A}, \tilde{B}, \check{C}_2, \check{D}_2)$. It is simple to show that the auxiliary system $\check{\Sigma}_{\rm P}$ is right invertible and of minimum phase with no infinite zeros.

We first assume that $C_1 = I$ and follow Steps 13.S.1 to 13.S.3 of the previous section to obtain a state feedback control law

$$\tilde{u}(k) = \tilde{F}\tilde{x}(k) + \tilde{H}\tilde{r}_{\rm n}(k), \tag{13.3.19}$$

where

$$\tilde{r}_{\rm n}(k) = \begin{bmatrix} r_0(k) \\ r_1(k+q_1) - \sum_{j=1}^{q_1} \begin{bmatrix} L_{10,j} & L_{1d,j} \end{bmatrix} \tilde{r}(k+q_1-j) \\ \vdots \\ r_{m_d}(k+q_{m_d}) - \sum_{j=1}^{q_{m_d}} \begin{bmatrix} L_{m_d0,j} & L_{m_dd,j} \end{bmatrix} \tilde{r}(k+q_{m_d}-j) \end{bmatrix}, \tag{13.3.20}$$

which has the following properties: i) $\tilde{A}+\tilde{B}\tilde{F}$ is asymptotically stable, and ii) the resulting $\tilde{h}_{\rm n}(k) = \tilde{r}_{\rm n}(k)$. This implies that the actual controlled output h is capable of precisely tracking the given reference $r(k)$ after q_{m_d} steps. Rewriting (13.3.19), we obtain

$$\begin{aligned} u(k) = \Gamma_i \tilde{u}(k) &= \Gamma_i \left[\tilde{F}\tilde{x}(k) + \tilde{H}\tilde{r}_{\rm n}(k)\right] \\ &= \Gamma_i \left[\tilde{F}\tilde{x}(k) + \tilde{H}L_0\tilde{r}(k) + \tilde{H}L_1\tilde{r}(k+1) + \cdots + \tilde{H}L_{m_d}\tilde{r}(k+q_{m_d})\right] \end{aligned} \tag{13.3.21}$$

for some appropriate matrices L_0, L_1, $\cdots$, L_{m_d}. Let $F = \Gamma_i \tilde{F} \Gamma_s^{-1}$, and

$$H_j = \Gamma_i \tilde{H} L_j \Gamma_o^{-1}, \tag{13.3.22}$$

for $j = 0, 1, \cdots, m_d$. We have

$$u(k) = Fx(k) + H_0 r(k) + H_1 r(k+1) + \cdots + H_{m_d} r(k+q_{m_d}). \tag{13.3.23}$$

Next, we proceed to construct a reduced order measurement feedback controller. We follow Steps 13.R.1 to 13.R.3 of the previous section to obtain matrices B_1, B_2, F_1, F_2 and gain matrices $A_{\rm cmp}$, $B_{\rm cmp}$, $C_{\rm cmp}$ and $D_{\rm cmp}$ as given in (13.2.32). Noting that the pair (A, C_1) is observable and (A', C_1') has a controllability index $\{k_1, \cdots, k_p\}$, it is simple to show that $(A_{\rm R}, C_{\rm R})$ is also observable and the controllability index of $(A_{\rm R}', C_{\rm R}')$ is given by $\{k_1-1, \cdots, k_p-1\}$. It follows from Theorem 2.3.1 of Chapter 2 that there exists a gain matrix $K_{\rm R}$ such that $A_{\rm R} + K_{\rm R} C_{\rm R}$ has all its eigenvalues at the origin and

$$(A_{\rm R} + K_{\rm R} C_{\rm R})^{k_p - 2} \equiv 0. \tag{13.3.24}$$

We thus choose such a $K_{\rm R}$ in constructing gain matrices $A_{\rm cmp}$, $B_{\rm cmp}$, $C_{\rm cmp}$ and $D_{\rm cmp}$. The reduced order measurement feedback law is then given by

$$\begin{cases} v(k+1) = A_{\rm cmp} v(k) + B_{\rm cmp} y(k) + \sum_{j=0}^{m_d} G_j r(k+j), \\ u(k) \quad = C_{\rm cmp} v(k) + D_{\rm cmp} y(k) + \sum_{j=0}^{m_d} H_j r(k+j), \end{cases} \tag{13.3.25}$$

where $G_j = (B_2 + K_{\rm R} B_1) H_j$, $i = 0, 1, \cdots, m_d$.

Let $x_s(k)=x_2(k)-v(k)+K_{\rm R}x_1(k)$. It is straightforward to verify that the closed-loop system comprising the given system (13.3.1) and the reduced order measurement feedback control law of (13.3.25) can be rewritten as follows,

$$x_s(k+1)=(A_{\rm R}+K_{\rm R}C_{\rm R})x_s(k), \tag{13.3.26}$$

$$x(k+1)=(A+BF)x(k)-BF_2x_s(k)+\sum_{j=0}^{m_d}BH_jr(k+j), \tag{13.3.27}$$

$$h(k)=(C_2+D_2F)x(k)-D_2F_2x_s(k)+\sum_{j=0}^{m_d}D_2H_jr(k+j). \tag{13.3.28}$$

Thus, it is simple to see that the closed-loop system is asymptotically stable as $A+BF$ and $A_{\rm R}+K_{\rm R}C_{\rm R}$ have eigenvalues inside the unit circle. Clearly, for any initial condition, (13.3.26) implies that $x_s(k)=0$ for all $k\geq k_{\rm p}-1$. Hence, for $k\geq k_{\rm p}-1$, (13.3.27) and (13.3.28) reduce to

$$x(k+1)=(A+BF)x(k)+\sum_{j=0}^{m_d}BH_jr(k+j), \tag{13.3.29}$$

$$h(k)=(C_2+D_2F)x(k)+\sum_{j=0}^{m_d}D_2H_jr(k+j), \tag{13.3.30}$$

which are precisely the same as the closed-loop dynamics under the state feedback law. If we treat $x(k_{\rm p}-1)$ as a new initial condition to (13.3.29) and (13.3.30), it will take another q_{m_d} steps for h to precisely track the reference r. Thus, we have $h(k)=r(k)$ for all $k\geq q_{m_d}+k_{\rm p}-1$. Hence, the $(\kappa_{\rm d},\kappa_0)$ almost perfect tracking problem is solved by the control law (13.3.2) with $\kappa_{\rm d}=q_{m_d}$ and $\kappa_0=q_{m_d}+k_{\rm p}-1$. ⊠

The following remarks are in order.

Remark 13.3.1. Consider the given plant (13.3.1) which has all properties as stated in Theorem 13.3.1. Then, the $(\kappa_{\rm d},\kappa_0)$ almost perfect tracking problem is solvable by a full order measurement feedback controller of the form (13.3.2) with $\kappa_{\rm d}=q_{m_d}$ and $\kappa_0=q_{m_d}+k_{\rm p}$. Following the similar lines of reasoning as in the proof of Theorem 13.3.1, one can show that the control law given below is the required solution,

$$\begin{cases} v(k+1)=A_{\rm cmp}v(k)+\sum_{j=0}^{m_d}BH_jr(k+j)-Ky(k), \\ u(k)\;=\;F\,v(k)\;+\sum_{j=0}^{m_d}H_jr(k+j), \end{cases} \tag{13.3.31}$$

where $A_{\rm cmp} = A + BF + KC_1$ with K being chosen such that all the eigenvalues of $A + KC_1$ are at the origin and $(A + KC_1)^{k_{\rm P}-1} = 0$. ▣

Remark 13.3.2. For simplicity, we consider $\Sigma_{\rm P}$ being a single output system, i.e., $\ell = 1$, with a relative degree q_1. Clearly, if the reference $r(k+d)$ is unknown for all $d > 0$, then the full order measurement feedback controller (13.3.31) with $r(k+\star)$ being replaced by $r(k)$ will be capable of tracking the reference with a delay of q_1 steps after $q_1 + k_{\rm P}$ initial steps. Similarly, under the same situation, the reduced order measurement feedback controller (13.3.25) with $r(k+\star)$ being replaced by $r(k)$ will track the reference with a delay of q_1 steps after $q_1 + k_{\rm P} - 1$ initial steps. ▣

We illustrate the main results of this section in the following example.

Example 13.3.1. Consider a given discrete-time system of (13.3.1) with

$$A = \begin{bmatrix} 1 & 1 & 0 \\ 1 & 0 & 1 \\ 1 & 1 & 1 \end{bmatrix}, \quad B = \begin{bmatrix} 1 & 1 \\ 0 & 0 \\ 1 & 2 \end{bmatrix}, \quad x(0) = x_0, \tag{13.3.32}$$

$$C_2 = \begin{bmatrix} 2 & 1 & 0 \\ 1 & 2 & 0 \end{bmatrix}, \quad D_2 = \begin{bmatrix} 0 & 0 \\ 0 & 0 \end{bmatrix}. \tag{13.3.33}$$

It is simple to show that $\Sigma_{\rm P}$ is invertible without any invariant zeros and with two infinite zeros of orders 1 and 2, respectively. Let the reference $r(k)$ be given as,

$$r(k) = \begin{bmatrix} r_1(k) \\ r_2(k) \end{bmatrix} = \begin{bmatrix} k/4 \\ \cos(k\pi/5) \end{bmatrix}. \tag{13.3.34}$$

A. State Feedback Case. We first assume that the measurement output $y = x$, i.e., $C_1 = I$. Then, following the proof of Theorem 13.3.1 and using the software reported in [14], we obtain the following nonsingular state, input and output transformations,

$$\Gamma_s = \begin{bmatrix} 0.5 & -0.59628 & 0 \\ 0 & 0.74536 & 0 \\ -0.5 & 0.59628 & 0.74536 \end{bmatrix}, \tag{13.3.35}$$

$$\Gamma_i = \begin{bmatrix} 1.5 & -0.74536 \\ -1.0 & 0.74536 \end{bmatrix}, \quad \Gamma_o = \begin{bmatrix} 1.0 & -0.44721 \\ 0.5 & 0.89443 \end{bmatrix}, \tag{13.3.36}$$

which take $\Sigma_{\rm P}$ into the following special coordinate basis form,

$$\tilde{A} = \Gamma_s^{-1} A \Gamma_s = \begin{bmatrix} 1 & 0.29814 & 1.19257 \\ 0 & 0 & 1 \\ 0.67082 & 1.2 & 1 \end{bmatrix}, \quad \tilde{B} = \Gamma_s^{-1} B \Gamma_i = \begin{bmatrix} 1 & 0 \\ 0 & 0 \\ 0 & 1 \end{bmatrix}, \tag{13.3.37}$$

and

$$\tilde{C}_2 = \Gamma_o^{-1} C_2 \Gamma_s = \begin{bmatrix} 1 & 0 & 0 \\ 0 & 1 & 0 \end{bmatrix}, \quad \tilde{D}_2 = \Gamma_o^{-1} D_2 \Gamma_i = \begin{bmatrix} 0 & 0 \\ 0 & 0 \end{bmatrix}. \tag{13.3.38}$$

We then obtain

$$\breve{C}_2 = \begin{bmatrix} 1 & 0.29814 & 1.19257 \\ 0.67082 & 1.2 & 1 \end{bmatrix}, \quad \breve{D}_2 = \begin{bmatrix} 1 & 0 \\ 0 & 1 \end{bmatrix}, \tag{13.3.39}$$

$$\tilde{F} = \begin{bmatrix} -1 & -0.29814 & -1.19257 \\ -0.67082 & -1.2 & -1 \end{bmatrix}, \quad \tilde{H} = \begin{bmatrix} 1 & 0 \\ 0 & 1 \end{bmatrix}, \tag{13.3.40}$$

and finally, we obtain the following internally stabilizing state feedback control law,

$$u(k) = Fx(k) + H_1 r(k+1) + H_2 r(k+2), \tag{13.3.41}$$

with

$$F = \begin{bmatrix} -3.4 & -1 & -1.4 \\ 1.6 & 0 & 0.6 \end{bmatrix}, \tag{13.3.42}$$

$$H_1 = \begin{bmatrix} 1.2 & 0.6 \\ -0.8 & -0.4 \end{bmatrix} \quad \text{and} \quad H_2 = \frac{1}{3} \begin{bmatrix} 1 & -2 \\ -1 & 2 \end{bmatrix}. \tag{13.3.43}$$

Simulation results in Figure 13.3.1 clearly show that the resulting controlled output $h(k)$ precisely tracks the reference $r(k)$ after two steps, i.e., the control law of (13.3.41) achieves an (κ_d, κ_0) almost perfect tracking performance with $\kappa_d = 2$ and $\kappa_0 = 2$. The simulation was done with an initial condition,

$$x_0 = \begin{pmatrix} -1.5 \\ 1.2 \\ 0.5 \end{pmatrix}. \tag{13.3.44}$$

Next, we deal with measurement feedback cases. Let

$$C_1 = [1 \quad 0 \quad 0], \tag{13.3.45}$$

which does not satisfy the condition $\mathrm{Ker}\,(C_2) \supset \mathrm{Ker}\,(C_1)$, as required in the RPT problem. However, for almost perfect tracking, we only need (A, C_1) to be observable. It is ready to verify that (A, C_1) is indeed observable and the controllability index of (A', C_1') is given by $\mathcal{C} = \{3\}$. Hence, $k_p = 3$.

B.1. Full Order Measurement Feedback Case. Following Remark 13.3.1, we obtain an internally stabilizing full order measurement feedback control law,

$$\begin{cases} v(k+1) = A_{\mathrm{cmp}}\, v(k) + B_{\mathrm{cmp}}\, y(k) + G_1\, r(k+1) + G_2\, r(k+2), \\ u(k) \;\; = C_{\mathrm{cmp}}\, v(k) + D_{\mathrm{cmp}}\, y(k) + H_1\, r(k+1) + H_2\, r(k+2), \end{cases} \tag{13.3.46}$$

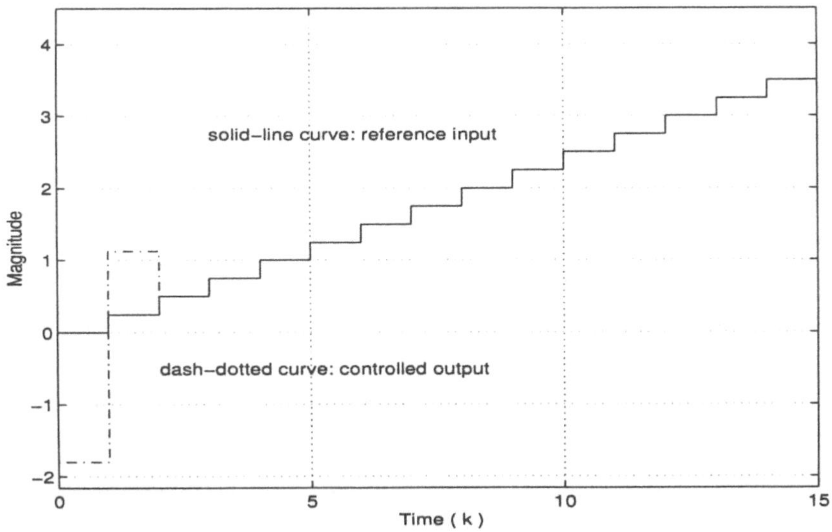

(a) Controlled output h_1 and reference r_1.

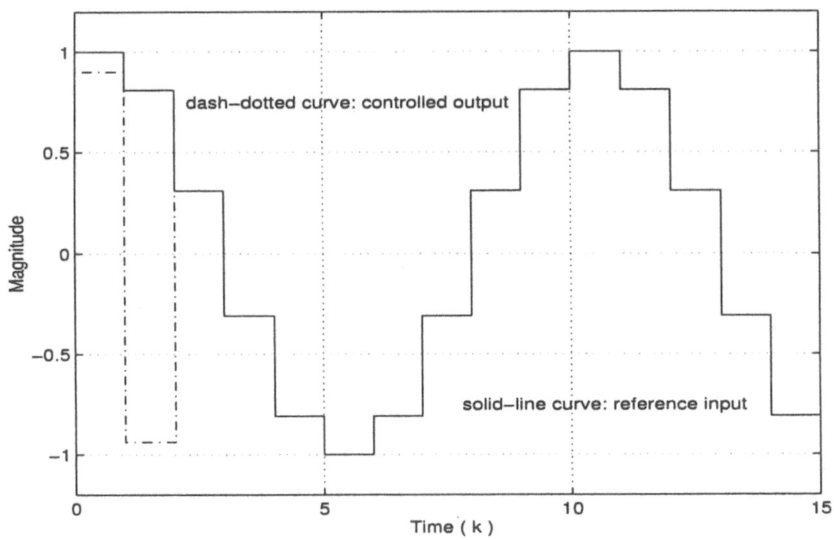

(b) Controlled output h_2 and reference r_2.

Figure 13.3.1: Closed-loop response under state feedback.

where

$$A_{\rm cmp} = \begin{bmatrix} -2.8 & 0 & -0.8 \\ -2.0 & 0 & -1.0 \\ -3.2 & 0 & 0.8 \end{bmatrix}, \quad B_{\rm cmp} = -K = \begin{bmatrix} 2 \\ 3 \\ 4 \end{bmatrix}, \tag{13.3.47}$$

$$C_{\rm cmp} = F, \quad D_{\rm cmp} = 0, \quad G_1 = BH_1, \quad G_2 = BH_2, \tag{13.3.48}$$

and where the observer gain matrix K is chosen such that all the eigenvalues of $A + KC_1$ are placed at the origin. Simulation results shown in Figure 13.3.2 were obtained from the closed-loop system comprising the given plant and the above full order measurement feedback law with an initial condition,

$$x_0 = \begin{pmatrix} -0.1 \\ 0.1 \\ -0.1 \end{pmatrix}. \tag{13.3.49}$$

It clearly shows that the (κ_d, κ_0) almost perfect tracking problem is solved with $\kappa_d = 2$ and $\kappa_0 = 5$, i.e., the closed-loop system is capable of precisely tracking the reference after five steps.

B.2. Reduced Order Measurement Feedback Case. Following the proof of Theorem 13.3.1, we obtain an internally stabilizing reduced order measurement feedback control law,

$$\begin{cases} v(k+1) = A_{\rm cmp}\, v(k) + B_{\rm cmp}\, y(k) + G_1\, r(k+1) + G_2\, r(k+2), \\ u(k) \;\;= C_{\rm cmp}\, v(k) + D_{\rm cmp}\, y(k) + H_1\, r(k+1) + H_2\, r(k+2), \end{cases} \tag{13.3.50}$$

with

$$A_{\rm cmp} = \begin{bmatrix} 0 & 1.8 \\ 0 & 2.4 \end{bmatrix}, \quad B_{\rm cmp} = \begin{bmatrix} 5.4 \\ 7.2 \end{bmatrix}, \quad G_1 = \begin{bmatrix} -0.4 & -0.2 \\ -1.2 & -0.6 \end{bmatrix}, \tag{13.3.51}$$

$$C_{\rm cmp} = \begin{bmatrix} -1 & -1.4 \\ 0 & 0.6 \end{bmatrix}, \quad D_{\rm cmp} = \begin{bmatrix} -7.2 \\ 2.8 \end{bmatrix}, \quad G_2 = \frac{1}{3}\begin{bmatrix} 0 & 0 \\ -1 & 2 \end{bmatrix}. \tag{13.3.52}$$

Simulation results shown in Figure 13.3.3 were obtained from the closed-loop system comprising the given plant and the above reduced order measurement feedback law with the same initial condition as in the full order measurement feedback case. Again, it is clear that the (κ_d, κ_0) almost perfect tracking problem is solved with $\kappa_d = 2$ and $\kappa_0 = 4$. The closed-loop system is capable of precisely tracking the reference after four steps. ▣

Finally, we conclude this chapter by the following example, which illustrates the situation when $r(k + d)$ is unknown for all $d > 0$.

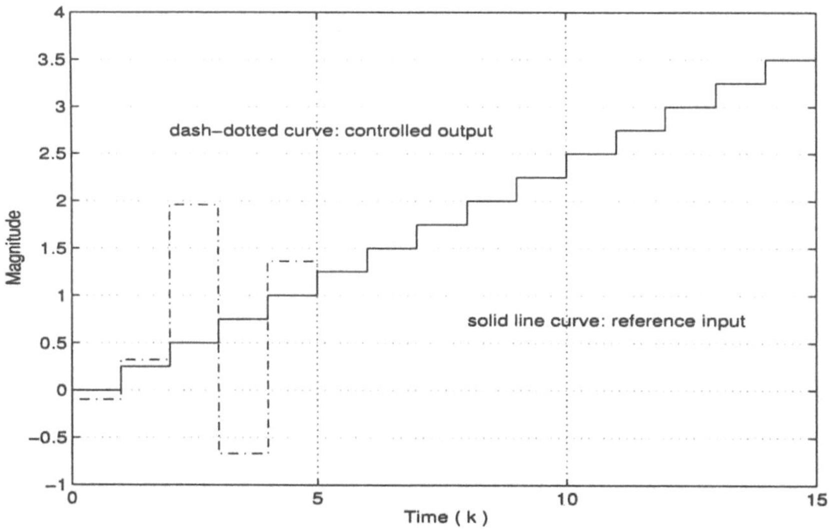

(a) Controlled output h_1 and reference r_1.

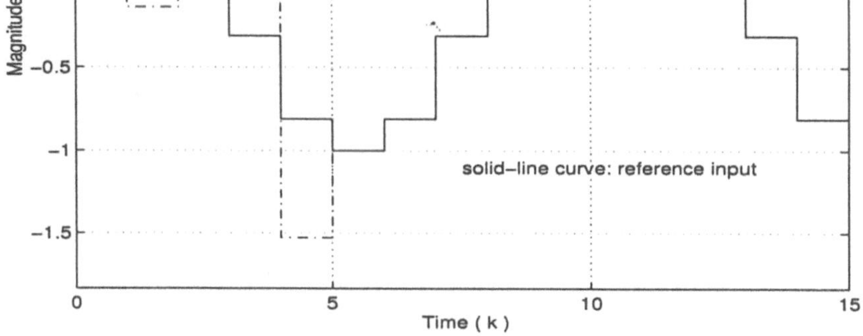

(b) Controlled output h_2 and reference r_2.

Figure 13.3.2: Closed-loop response under full order control law.

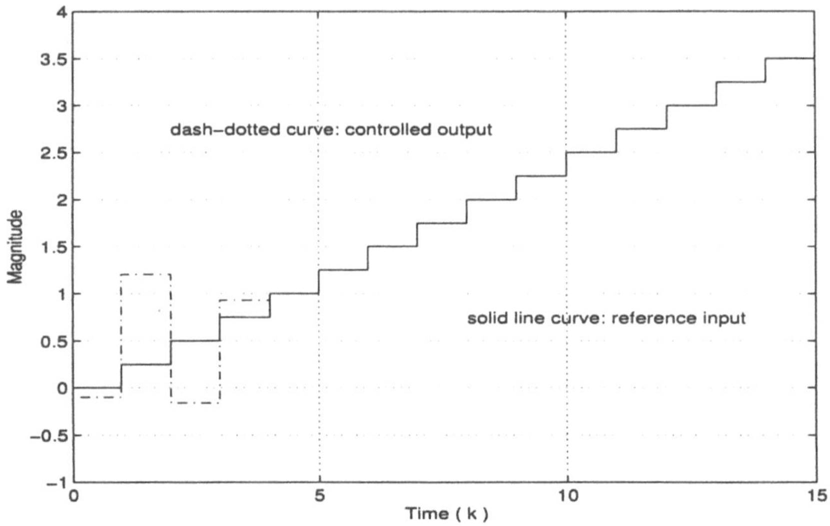

(a) Controlled output h_1 and reference r_1.

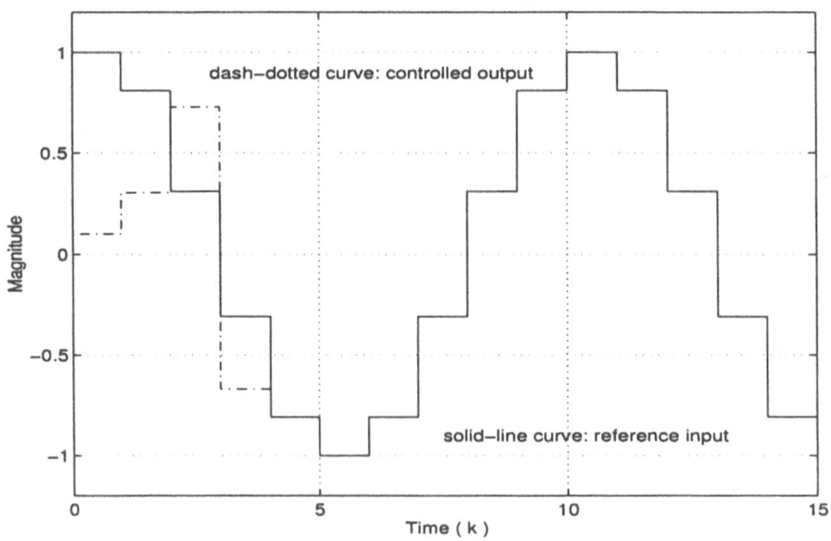

(b) Controlled output h_2 and reference r_2.

Figure 13.3.3: Closed-loop response under reduced order control law.

Example 13.3.2. Consider a discrete-time system characterized by (13.1.1) with

$$A = \begin{bmatrix} 1 & 1 & 1 \\ 3 & 2 & 0 \\ 2 & 0 & 1 \end{bmatrix}, \quad B = \begin{bmatrix} 0 & 0 \\ 0 & 1 \\ 1 & -1 \end{bmatrix}, \quad x_0 = \begin{pmatrix} 0.0 \\ -0.5 \\ 0.4 \end{pmatrix}, \tag{13.3.53}$$

and

$$C_1 = [1 \ \ 0 \ \ 0], \quad C_2 = [1 \ \ 1 \ \ 1], \quad D_2 = [0 \ \ 0]. \tag{13.3.54}$$

Let the reference be

$$r(k) = \frac{k}{4} \cos\left(\frac{k\pi}{5}\right). \tag{13.3.55}$$

Note that Σ_P is right invertible and of minimum phase with one invariant zero at the origin and one infinite zero of order 1, i.e., the relative degree of Σ_P is equal to 1. Also, (A, C_1) is completely observable and the controllability index of (A', C_1') is given by $\{3\}$, i.e., $k_p = 3$. Following the result in Remark 13.3.2, we obtain the following full order measurement feedback controller,

$$\begin{cases} v(k+1) = \begin{bmatrix} -3 & 1 & 1 \\ -11.5 & -0.5 & -0.5 \\ -1.5 & -0.5 & -0.5 \end{bmatrix} v(k) + \begin{bmatrix} 4 \\ 11 \\ 1 \end{bmatrix} y(k) + \begin{bmatrix} 0 \\ 0 \\ 1 \end{bmatrix} r(k), \\ u(k) \ \ = \begin{bmatrix} -6 & -3 & -2 \\ -3.5 & -2.5 & -0.5 \end{bmatrix} v(k) + \begin{bmatrix} 1 \\ 0 \end{bmatrix} r(k), \end{cases} \tag{13.3.56}$$

and reduced order measurement feedback controller,

$$\begin{cases} v(k+1) = \begin{bmatrix} -4.5 & -4.5 \\ 0.5 & 0.5 \end{bmatrix} v(k) + \begin{bmatrix} -18 \\ 2 \end{bmatrix} y(k) + \begin{bmatrix} 0 \\ 1 \end{bmatrix} r(k), \\ u(k) \ \ = \begin{bmatrix} -3 & -2 \\ -2.5 & -0.5 \end{bmatrix} v(k) + \begin{bmatrix} -16 \\ -13 \end{bmatrix} y(k) + \begin{bmatrix} 1 \\ 0 \end{bmatrix} r(k). \end{cases} \tag{13.3.57}$$

Simulation results shown in Figures 13.3.4 and 13.3.5 clearly confirm that the full and reduced order control laws are capable of tracking the reference with a delay of one step after 4 and 3 initial steps, respectively. ▣

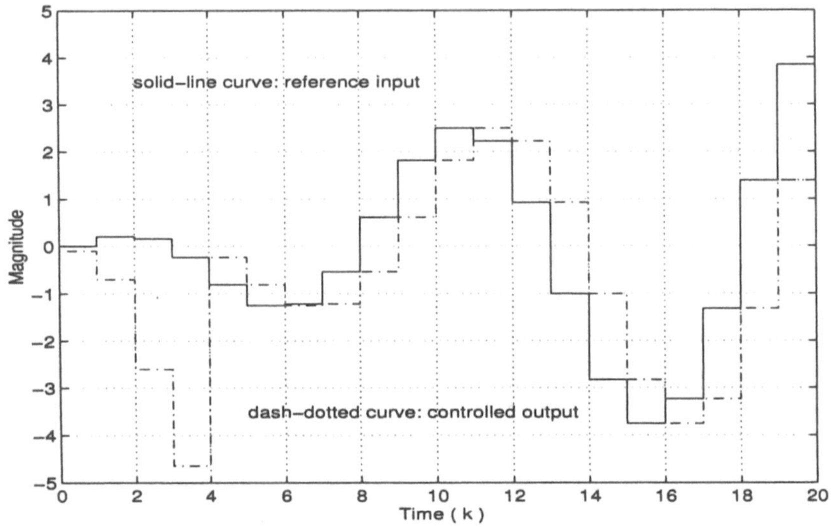

Figure 13.3.4: Response of h and r under full order control law.

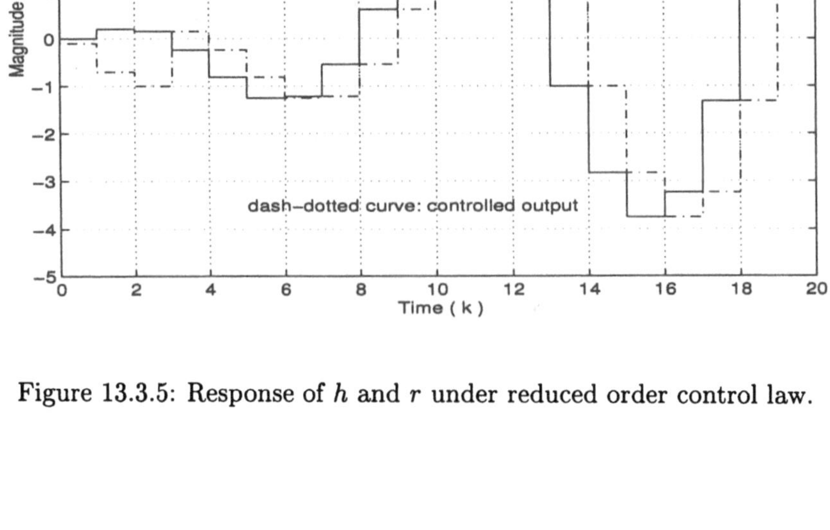

Figure 13.3.5: Response of h and r under reduced order control law.

Chapter 14

Design of a Hard Disk Drive Servo System

14.1. Introduction

HARD DISK DRIVES (HDDs) provide important data-storage medium for computers and other data-processing systems. In most hard disk drives, rotating disks coated with a thin magnetic layer or recording medium are written with data, which are arranged in concentric circles or tracks. Data are read or written with a read/write (R/W) head, which consists of a small horseshoe-shaped electromagnet. Figure 14.1.1 shows a simple illustration of a typical hard disk servo system. The two main functions of the R/W head positioning servomechanism in disk drives are track seeking and track following. Track seeking moves the R/W head from the present track to a specified destination track in minimum time using a bounded control effort. Track following maintains the head as close as possible to the destination track center while information is being read from or written to the disk. Track density is the reciprocal of the track width. It is suggested that on a disk surface, tracks should be written as closely spaced as possible so that we can maximize the usage of the disk surface. This means an increase in the track density, which subsequently means a more stringent requirement on the allowable variations of the position of the heads from the true track center.

The prevalent trend in hard disk design is towards smaller hard disks with increasingly larger capacities. This implies that the track width has to be smaller leading to lower error tolerance in the positioning of the head. The controller for track following has to achieve tighter regulation in the control of the servomechanism. Current hard disk drives use a combination of classical

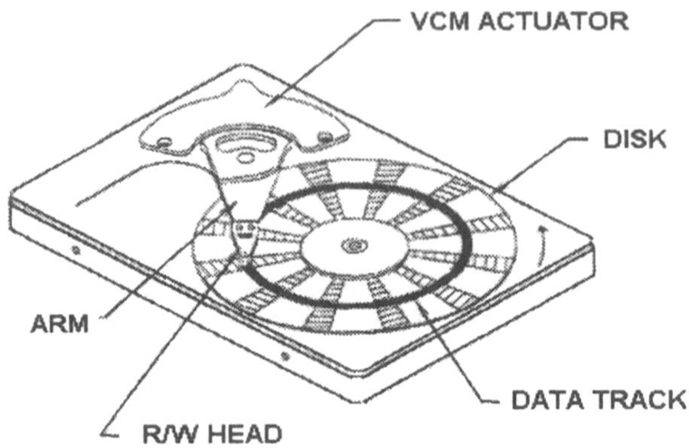

Figure 14.1.1: A hard disk drive with a VCM actuator servo system.

control techniques, such as lead-lag compensators, PI compensators, and notch filters. These classical methods can no longer meet the demand for hard disk drives of higher performance. So many control approaches have been tried, such as LQG and/or LTR approach (see e.g., [61] and [133]), and adaptive control (see e.g., [92]) and so on. Although much work has been done to date, more studies need to be conducted to use more control methods to achieve better performance of the hard disk drives.

The purpose of this chapter is to use the result of the robust and perfect tracking (RPT) control method of Chapter 9 to carry out a design of a hard disk drive servo system. We will first obtain a model of the VCM actuator and then cast the overall servo system design into an RPT design framework. A first order dynamic measurement feedback controller is then designed to achieve robust and perfect tracking for any step reference. Our controller is theoretically capable of making the L_p-norm of the resulting tracking error with $1 \leq p < \infty$ arbitrarily small in faces of external disturbances and initial conditions. Some trade-offs are then made in order for the RPT controller to be implementable using the existing hardware setup and to meet physical constraints such as sampling rates and the limit of control of the system. The implementation results of the RPT controller are compared with those of a PID controller. The results show that our servo system is simpler and yet has faster seeking times, lower overshoot and higher accuracy. The results of this chapter were reported earlier by Goh *et al.* [59].

14.2. Modeling of the VCM Actuator

In this section, we present the modeling of the VCM actuator, which is well known in the research community of the hard disk drive servo systems to have a characteristic of a double integrator cast with some high frequency resonance, which can reduce the system stability if neglected. There are some bias forces in the hard disk drive system which will cause steady state error in tracking performance. Moreover, there are also some nonlinearities in the system at low frequencies, which are primarily due to the pivot and bearing frictions. All these factors should be taken into consideration when considering the design of a controller for the VCM. For the purpose of developing a model, we have to compromise between accuracy and simplicity. In this section, a relatively simplified model of the VCM is identified and presented.

We will utilize the frequency response identification method (see e.g., [50]) to model our actuator. Such a method is applicable to minimum phase processes. We expect from the properties of the physical system that the VCM actuator should be of minimum phase. The detailed procedure proceeds as follows: We first assume that the transfer function of a minimum phase plant is given by

$$G(s) = \frac{N(s)}{D(s)} = \frac{b_0 + b_1 s + b_2 s^2 + \cdots + b_m s^m}{1 + a_1 s + a_2 s^2 + \cdots + a_n s^n}, \tag{14.2.1}$$

for some appropriate coefficients a_k, $k = 1, 2, \cdots, n$, and b_k, $k = 0, 1, \cdots, m$, with $n \geq m$. These parameters are to be identified. Then, its corresponding frequency response is given by

$$G(j\omega) = \frac{\alpha(\omega) + j\omega\beta(\omega)}{\sigma(\omega) + j\omega\tau(\omega)} = \frac{N(j\omega)}{D(j\omega)}, \tag{14.2.2}$$

where

$$\left.\begin{aligned} \alpha(\omega_i) &= b_0 - b_2\omega_i^2 + b_4\omega_i^4 - \cdots \\ \beta(\omega_i) &= b_1 - b_3\omega_i^2 + b_5\omega_i^4 - \cdots \\ \sigma(\omega_i) &= 1 - a_2\omega_i^2 + a_4\omega_i^4 - \cdots \\ \tau(\omega_i) &= a_1 - a_3\omega_i^2 + a_5\omega_i^4 - \cdots \end{aligned}\right\} \tag{14.2.3}$$

Let $R(\omega)$ and $I(\omega)$ be the real and imaginary part of the measured frequency response of the actuator system. The frequency response error between the model and the actual measurement data is given by

$$\mathcal{E}(j\omega) = [R(\omega) + jI(\omega)] - \frac{N(j\omega)}{D(j\omega)}. \tag{14.2.4}$$

Thus, the parameters of the system can be obtained by minimizing the following index,

$$J = \sum_{i=1}^{L} |\mathcal{E}(j\omega_i)|^2, \tag{14.2.5}$$

where L is the total number of points of the measured data. Unfortunately, this is a nonlinear optimization problem, and it is difficult to be solved. We then follow the results of [50] to modify the error norm as,

$$J = \sum_{i=1}^{L} |D(j\omega_i)\mathcal{E}(j\omega_i)|^2. \tag{14.2.6}$$

The original problem now becomes a linear optimization problem. Using (14.2.2) and (14.2.4), we can rewrite (14.2.6) as follows,

$$J = \sum_{i=1}^{L} \left\{ [X(\omega_i)]^2 + [Y(\omega_i)]^2 \right\}, \tag{14.2.7}$$

where

$$X(\omega_i) = \sigma(\omega_i)R(\omega_i) - \omega_i\tau(\omega_i)I(\omega_i) - \alpha(\omega_i), \tag{14.2.8}$$

and

$$Y(\omega_i) = \omega_i\tau(\omega_i)R(\omega_i) + \sigma(\omega_i)I(\omega_i) - \omega_i\beta(\omega_i). \tag{14.2.9}$$

Therefore, J can be minimized by finding $\hat{b}_0, \hat{b}_1, \cdots, \hat{b}_m$ and $\hat{a}_1, \hat{a}_2, \cdots, \hat{a}_n$ such that,

$$\left.\begin{array}{l}
\left.\dfrac{\partial J}{\partial b_0}\right|_{b_0=\hat{b}_0} = \left.\displaystyle\sum_{i=1}^{L} \left\{ 2X(\omega_i)(-1) \right\}\right|_{b_0=\hat{b}_0} = 0 \\
\left.\dfrac{\partial J}{\partial b_1}\right|_{b_1=\hat{b}_1} = \left.\displaystyle\sum_{i=1}^{L} \left\{ 2Y(\omega_i)(-\omega_i) \right\}\right|_{b_1=\hat{b}_1} = 0 \\
\qquad\vdots \\
\left.\dfrac{\partial J}{\partial a_1}\right|_{a_1=\hat{a}_1} = \left.\displaystyle\sum_{i=1}^{L} \left\{ 2X(\omega_i)[-\omega_i I(\omega_i)] + 2Y(\omega_i)[\omega_i R(\omega_i)] \right\}\right|_{a_1=\hat{a}_1} = 0 \\
\left.\dfrac{\partial J}{\partial a_2}\right|_{a_2=\hat{a}_2} = \left.\displaystyle\sum_{i=1}^{L} \left\{ 2X(\omega_i)[-\omega_i^2 R(\omega_i)] - 2Y(\omega_i)[\omega_i^2 I(\omega_i)] \right\}\right|_{a_2=\hat{a}_2} = 0 \\
\qquad\vdots
\end{array}\right\} \tag{14.2.10}$$

Rearranging the above equations, we obtain

$$\begin{pmatrix} \boldsymbol{A}_{11} & \boldsymbol{A}_{12} \\ \boldsymbol{A}_{21} & \boldsymbol{A}_{22} \end{pmatrix} \begin{pmatrix} \boldsymbol{b} \\ \boldsymbol{a} \end{pmatrix} = \begin{pmatrix} \boldsymbol{B}_1 \\ \boldsymbol{B}_2 \end{pmatrix}, \tag{14.2.11}$$

where

$$A_{11} = \begin{bmatrix} V_0 & 0 & -V_2 & 0 & V_4 & \cdots \\ 0 & V_2 & 0 & -V_4 & 0 & \cdots \\ V_2 & 0 & -V_4 & 0 & V_6 & \cdots \\ 0 & V_4 & 0 & -V_6 & 0 & \cdots \\ V_4 & 0 & -V_6 & 0 & V_8 & \cdots \\ \vdots & \vdots & \vdots & \vdots & \vdots & \ddots \end{bmatrix}, \qquad b = \begin{bmatrix} \hat{b}_0 \\ \hat{b}_1 \\ \hat{b}_2 \\ \hat{b}_3 \\ \hat{b}_4 \\ \vdots \end{bmatrix}, \tag{14.2.12}$$

$$A_{12} = \begin{bmatrix} T_1 & S_2 & -T_3 & -S_4 & T_5 & \cdots \\ -S_2 & T_3 & S_4 & -T_5 & -S_6 & \cdots \\ T_3 & S_4 & -T_5 & -S_6 & T_7 & \cdots \\ -S_4 & T_5 & S_6 & -T_7 & -S_8 & \cdots \\ T_5 & S_6 & -T_7 & -S_8 & T_9 & \cdots \\ \vdots & \vdots & \vdots & \vdots & \vdots & \ddots \end{bmatrix}, \qquad a = \begin{bmatrix} \hat{a}_1 \\ \hat{a}_2 \\ \hat{a}_3 \\ \hat{a}_4 \\ \hat{a}_5 \\ \vdots \end{bmatrix}, \tag{14.2.13}$$

$$A_{21} = \begin{bmatrix} T_1 & -S_2 & -T_3 & S_4 & T_5 & \cdots \\ S_2 & T_3 & -S_4 & -T_5 & S_6 & \cdots \\ T_3 & -S_4 & -T_5 & S_6 & T_7 & \cdots \\ S_4 & T_5 & -S_6 & -T_7 & S_8 & \cdots \\ T_5 & -S_6 & -T_7 & S_8 & T_9 & \cdots \\ \vdots & \vdots & \vdots & \vdots & \vdots & \ddots \end{bmatrix}, \qquad B_1 = \begin{bmatrix} S_0 \\ T_1 \\ S_2 \\ T_3 \\ S_4 \\ \vdots \end{bmatrix}, \tag{14.2.14}$$

$$A_{22} = \begin{bmatrix} U_2 & 0 & -U_4 & 0 & U_6 & \cdots \\ 0 & U_4 & 0 & -U_6 & 0 & \cdots \\ U_4 & 0 & -U_6 & 0 & U_8 & \cdots \\ 0 & U_6 & 0 & -U_8 & 0 & \cdots \\ U_6 & 0 & -U_8 & 0 & U_{10} & \cdots \\ \vdots & \vdots & \vdots & \vdots & \vdots & \ddots \end{bmatrix}, \qquad B_2 = \begin{bmatrix} 0 \\ U_2 \\ 0 \\ U_4 \\ 0 \\ \vdots \end{bmatrix}, \tag{14.2.15}$$

and where

$$V_k = \sum_{i=0}^{L} \omega_i^k, \qquad S_k = \sum_{i=0}^{L} \omega_i^k R(\omega_i), \tag{14.2.16}$$

$$T_k = \sum_{i=0}^{L} \omega_i^k I(\omega_i), \qquad U_k = \sum_{i=0}^{L} \omega_i^k [R^2(\omega_i) + I^2(\omega_i)]. \tag{14.2.17}$$

The desired parameters of the corresponding transfer function model can be obtained by solving the above equations.

The dynamics of an ideal VCM actuator can be formulated as a second order state space model as follows,

$$\begin{pmatrix} \dot{y} \\ \dot{v} \end{pmatrix} = \begin{pmatrix} 0 & k_y \\ 0 & 0 \end{pmatrix} \begin{pmatrix} y \\ v \end{pmatrix} + \begin{pmatrix} 0 \\ k_v \end{pmatrix} u, \tag{14.2.18}$$

where u is the actuator input (in volts), y and v are the position (in tracks) and the velocity of the R/W head, k_y is the position measurement gain and

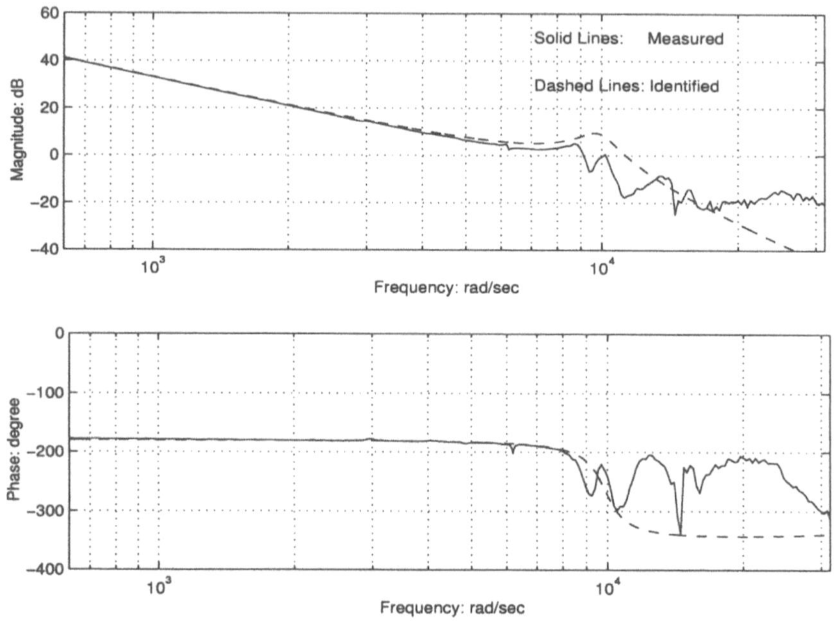

Figure 14.2.1: Frequency responses of the actual and identified VCM models.

$k_v = k_t/m$, with k_t being the current-force conversion coefficient and m being the mass of the VCM actuator. Thus, the transfer function of an ideal VCM model appears to be a double integrator, i.e.,

$$G_{v1}(s) = \frac{k_v k_y}{s^2}. \tag{14.2.19}$$

However, if we consider also the high frequency resonance modes, a more realistic model for the VCM actuator should be

$$G_v(s) = \frac{k_v k_y}{s^2} \frac{k_d s + \omega_n^2}{s^2 + 2\xi\omega_n s + \omega_n^2}. \tag{14.2.20}$$

Using the algorithm presented above and the measured data from the actual system (see Figure 14.2.1), we obtain a fourth order model for the actuator,

$$G_v(s) = \frac{4.3817 \times 10^{10} s + 4.3247 \times 10^{15}}{s^2(s^2 + 1.5962 \times 10^3 s + 9.7631 \times 10^7)}. \tag{14.2.21}$$

Figure 14.2.1 shows that the frequency response of the identified model matches the measured data very well for the frequency range from 0 to 10^4 rad/sec, which far exceeds the working range of the VCM actuator.

14.3. Servo System Design and Simulation Results

We now present the servo system design for the actuator identified in the previous section. Basically, almost all commercially available hard disk drive servo systems up-to-date are designed using conventional PID approach. For drives with a single VCM actuator, designers would encounter problems if they wish to push up the tracking following speed. Usually, there will be some huge peak overshoot in step response. Thus, in practice, one would have to make trade-offs between the track following speed and overshoot by selecting appropriate PID controller gains. We formulate our servo system design as a robust and perfect tracking (RPT) control problem. Such an approach will enable the designer to design a very low order control law, and moreover, the resulting closed-loop system will have fast track following speed and low overshoot as well as strong robustness.

We will design a servo system that meets the following design specifications:

1. The control input should not exceed ±2 volts due to physical constraints on the actual VCM actuator.

2. The overshoot and undershoot of the step response should be kept less than 5% as the R/W head can start to read or write within ±5% of the target.

3. The 5% settling time in the step response should be less than 2 milli seconds (to beat the PID controller).

4. Sampling frequency in implementing the actual controller is 4 kHz, which is the sampling frequency currently used in most commercial disk drives.

From experience that we gained in designing PID controllers, we know that it is quite safe to ignore the resonance models of the VCM actuator if we are focusing on tracking performance. Thus, we will consider only a second order model for the VCM actuator at this stage. We will then put the resonance modes back when we are to evaluate the performance of the overall design. Thus, in our design, we will first use the following simplified model of the VCM actuator,

$$\dot{x} = Ax + Bu = \begin{bmatrix} 0 & 1 \\ 0 & 0 \end{bmatrix} x + \begin{bmatrix} 0 \\ 44296000 \end{bmatrix} u, \tag{14.3.1}$$

and

$$y = C_1 x = [1 \quad 0]\, x. \tag{14.3.2}$$

Next, we define the output to be controlled as,

$$h = y = C_2 x + D_2 u = [\,1 \quad 0\,]\,x. \tag{14.3.3}$$

Let the reference $r(t)$ be a step function with a magnitude α, i.e., $r(t) = \alpha \cdot 1(t)$, where $1(t)$ is a unit step function. Then, we have

$$\dot{r}(t) = \alpha \cdot \delta(t), \tag{14.3.4}$$

where $\delta(t)$ is a unit impulse function. Following the results of Chapter 9, we obtain a corresponding auxiliary system,

$$\left\{\begin{array}{rcl} \dot{\boldsymbol{x}} &=& \begin{bmatrix} 0 & 0 & 0 \\ 0 & 0 & 1 \\ 0 & 0 & 0 \end{bmatrix} \boldsymbol{x} + \begin{bmatrix} 0 \\ 0 \\ 44296000 \end{bmatrix} u + \begin{bmatrix} 1 \\ 0 \\ 0 \end{bmatrix} w \\ \boldsymbol{y} &=& \begin{bmatrix} 1 & 0 & 0 \\ 0 & 1 & 0 \end{bmatrix} \boldsymbol{x} \qquad\qquad\qquad\quad + \begin{bmatrix} 0 \\ 0 \end{bmatrix} w \\ e &=& [\,-1 \quad 1 \quad 0\,]\, \boldsymbol{x} + \quad 0 \quad u \end{array}\right. \tag{14.3.5}$$

where

$$\boldsymbol{x} = \begin{pmatrix} r \\ x \end{pmatrix}, \quad w = \alpha \cdot \delta(t), \quad \boldsymbol{y} = \begin{pmatrix} r \\ y \end{pmatrix}, \quad e = h - r. \tag{14.3.6}$$

It is simple to see that (A, B, C_2, D_2) is invertible and free of invariant zeros, and $\mathrm{Ker}\,(C_1) = \mathrm{Ker}\,(C_2)$. Hence, it follows from the result of Chapter 9 that the robust and perfect tracking (RPT) performance is achievable. Following the results of Chapter 9, one can show that there exists a family of measurement feedback control laws, parameterized by a tuning parameter ε, such that when it is applied to the given VCM actuator,

1. The resulting closed-loop system is asymptotically stable for sufficiently small ε; and

2. For any given initial condition x_0i and any $p \in [1, \infty)$, the l_p-norm of the resulting tracking error, e, has the property $\|e\|_p \to 0$, as $\varepsilon \to 0$.

Following the construction algorithm for the reduced order RPT controller in Chapter 9, we obtained a parameterized first order measurement feedback control law of the form

$$\left\{\begin{array}{l} \dot{v} = A_{\mathrm{RC}}(\varepsilon)\; v + B_{\mathrm{RC}}(\varepsilon) \begin{pmatrix} r \\ y \end{pmatrix}, \\ u = C_{\mathrm{RC}}(\varepsilon)\; v + D_{\mathrm{RC}}(\varepsilon) \begin{pmatrix} r \\ y \end{pmatrix}, \end{array}\right. \tag{14.3.7}$$

with

$$\left.\begin{aligned} A_{\rm RC}(\varepsilon) &= -7800/\varepsilon \\ B_{\rm RC}(\varepsilon) &= \frac{1}{\varepsilon^2}\,[\,1.62\times 10^6 \quad -4.842\times 10^7\,] \\ C_{\rm RC}(\varepsilon) &= -4.063572\times 10^{-5}/\varepsilon \\ D_{\rm RC}(\varepsilon) &= \frac{1}{\varepsilon^2}\,[\,0.036572 \quad -0.280386\,]. \end{aligned}\right\} \tag{14.3.8}$$

Results in Figure 14.3.1 are obtained using a MATLAB package. They clearly show that the RPT problem is solved as we tune the tuning parameter ε to be smaller and smaller. Unfortunately, due to the constraints of the physical system, i.e., the limits in control inputs and sampling rates, as well as resonance modes, it is impossible to implement a controller that will track the reference with zero time. We would thus have to make some compromises in the track following speed because of these limitations. After several trials, we found that the controller parameters of (14.3.8) with $\varepsilon = 0.9$ would give us a satisfactory performance. We then discretize it using a bilinear transformation with a sampling frequency of 4 kHz. Note that it was shown in Chapter 3 that the bilinear transformation does not introduce additional nonminimum phase invariant zeros and it preserves the invertibility structure of the system. The discretized controller is given by

$$\left\{\begin{aligned} x_v(k+1) &= -0.04\, x_v(k) + 15178.933\, r(k) - 453681.43\, y(k), \\ u(k) &= -3.4267\times 10^{-7}\, x_v(k) + 0.03973\, r(k) - 0.18421\, y(k). \end{aligned}\right.$$

Figure 14.3.2 shows that the step response of the overall system comprising the fourth order model of the VCM actuator (we now put the resonance modes back into the VCM actuator model) and the discretized RPT controller, meets the design specifications. In actual hard disk drive manufacturing, the resonant frequency ω_n of the VCM actuator, see (14.2.21), for the same batch of drives might vary from one to the other. A common practice in the disk drive industry is to add some notch filters in the servo system to attenuate these resonant peaks as much as possible. Surprisingly, our RPT controller is capable of withstanding the variation of resonance frequencies as well. Figure 14.3.3 shows the step responses of the closed-loop systems of our RPT controller and the VCM model with two different resonant frequencies: one is 1.125 kHz, which is $\beta = 75\%$ of the nominal value, and the other is 2.25 kHz, which is $\beta = 150\%$ of the nominal resonant frequency. The results show that the RPT controller is very robust with respect to the change of resonant frequency in the actuator.

Although we do not consider the effects of run-out disturbances in our problem formulation, it turns out that our simple first order controller is capable

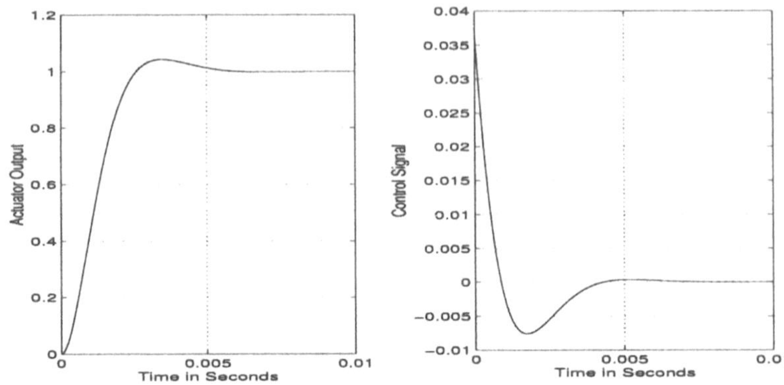

(a). $\varepsilon = 1$.

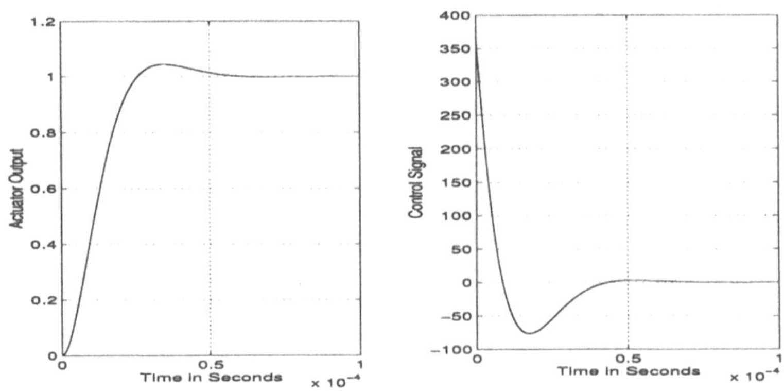

(b). $\varepsilon = 0.01$.

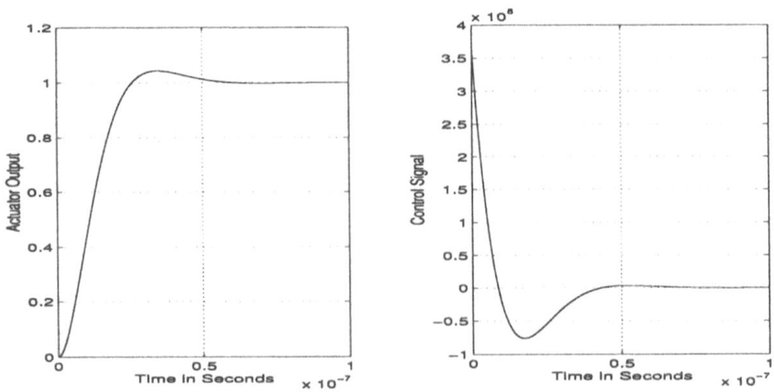

(c). $\varepsilon = 10^{-5}$.

Figure 14.3.1: Responses of the closed-loop systems with RPT controller.

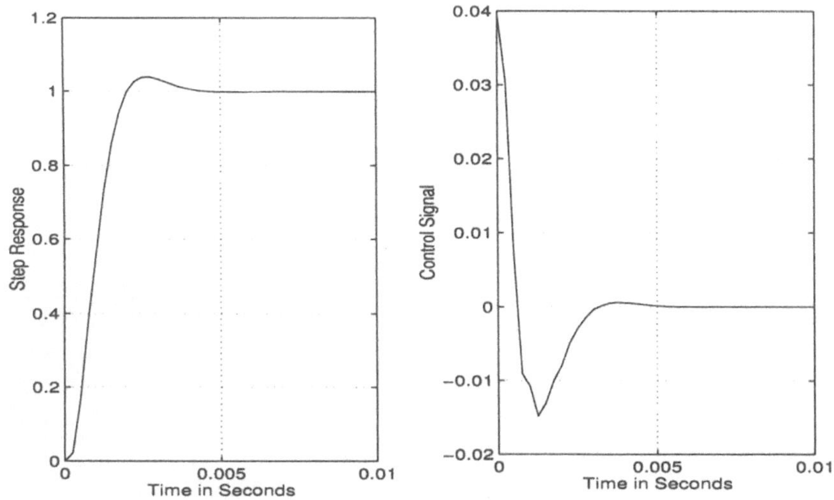

Figure 14.3.2: Closed-loop step response with discretized RPT controller.

of rejecting the first few modes of the run-out disturbances, which are mainly due to the imperfectness of the data tracks and the spindle motor speeds, and commonly have frequencies at the multiples of about 55 Hz. We simulate these run-out effects by injecting a sinusoidal signal into the measurement output, i.e., the new measurement output is the sum of the actuator output and the run-out disturbance. Figure 14.3.4 shows the simulation result of the output response of the overall servo system comprising the fourth order model of the VCM actuator model and the discretized RPT controller with a fictitious run-out disturbance injection $\tilde{w}(t) = 0.5 + 0.1\cos(110\pi t) + 0.05\sin(220\pi t)$ and a zero reference $r(t)$. The result shows that the effects of such a disturbance to the overall response are minimal. A more comprehensive test on run-out disturbances, i.e., the position error signal (PES) test on the actual system will be presented in the next section.

14.4. Implementation Results

In this section, we present the actual implementation results of our design and their comparison with those of a PID controller. Two major tests are presented: one is the track following of the closed-loop systems and the other is the position error signal (PES) test, which is considered to be a major factor in design hard disk drive servo systems. Our controller was implemented on an open hard

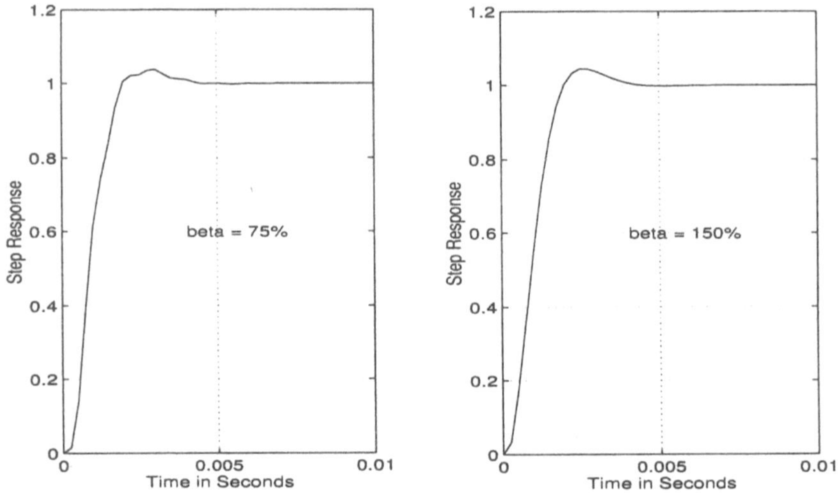

Figure 14.3.3: Closed-loop step responses with different resonant frequencies.

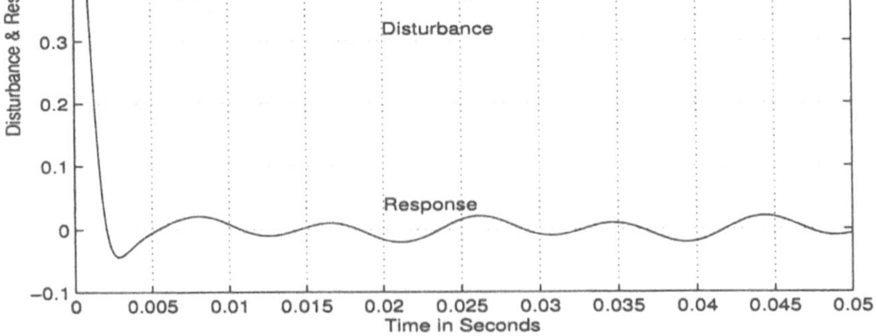

Figure 14.3.4: Closed-loop output response due to a run-out disturbance.

disk drive with a TMS320 digital signal processor (DSP) and a sampling rate of 4 kHz. Closed-loop actuation tests were performed using a Laser Doppler Vibrometer (LDV) to measure the R/W head position. The resolution used for LDV was 1μm/Volt. This displacement output is then fed into the DSP, which would then generate the necessary control signal to the VCM actuator. A digital signal analyzer (DSA) was used to assist in obtaining the frequency response of the overall control system. It can inject a swept sinusoidal reference signal, then read the output displacement from the LDV and calculate the frequency Bode plot using this information. Altogether, two sets of experiment were performed, one using the RPT controller and the other using a PID controller reported in Goh [58].

A. Track Following Test

The solid-line curve in Figure 14.4.1 shows the experimental step response of the RPT controller. In this figure, the response of the RPT controller is shown together with that of a PID controller of Goh [58] as a comparison. Note that the actual response of the closed-loop system with the RPT controller is slightly faster and its overshoot is slightly larger (about 7%) compared to the simulation results given in the previous section. The 5% settling time is about 1.6 milli seconds, which surely meets the design specifications. Figure 14.4.2 shows the experimental closed-loop Bode plot. It shows that the system has a closed loop bandwidth of about 500 Hz. At the roll-off frequency, there is no discernible resonance peak.

The dotted-line curve in Figure 14.4.1 shows the step response of the PID controller of Goh [58] (again using a 4kHz sampling rate). The PID controller had a usual structure and was tuned such that it could have fast time response. It is given by

$$u = \frac{0.13z^2 - 0.23z + 0.10}{z^2 - 1.25z + 0.25}(r - y). \tag{14.4.1}$$

Unfortunately, the overshoot of the controller is rather high, about 50% and this is a result of trading improved settling time at the expense of higher overshoot. To achieve a settling time of 4-5 mini seconds, it is necessary to tune the PID controller such that the overshoot is significant. Figure 14.4.3 shows the experimental closed-loop Bode plot of the PID controller. The closed loop bandwidth of this servo system is also about 500 Hz, with a slight peak of about 7dB at the roll-off frequency. This resonance peak would result in additional tracking errors close to the bandwidth frequency.

We believe that the shortcoming of the PID control is mainly due to its structure, i.e., it only feeds in the error signal, $y - r$, instead of feeding in both

y and r independently. We trust that the same problem might be present in other control methods if the only signal fed is $y - r$. The PID control structure might be simple as most of researchers and engineers have claimed. However, our RPT controller is even simpler, i.e., the RPT controller is of the first order and the PID controller is of the second order. But, we have fully utilized all available information associated with the actual system.

Unfortunately, we could not compare our results with those of other methods mentioned in the introduction. Most of references we found in the open literature contained only simulation results in this regard. Some of implementation results we found were, however, very different in nature. For example, Hanselmann and Engelke of [61] reported an implementation result of a disk drive servo system design using the LQG approach with a sampling frequency of 34 kHz. The overall step response of [61] with a higher order LQG controller and higher sampling frequency is worse than that of ours.

B. Position Error Signal Test

The disturbances in a real hard disk drive are usually considered as a lumped disturbance at the plant output, also known as run-outs. Repeatable run-outs (RRO) and non-repeatable run-outs (NRRO) are the major sources of track following errors. RRO is caused by the rotation of the spindle motor and consists of frequencies that are multiples of the spindle frequency. NRRO can be perceived as coming from three main sources: vibration shocks, mechanical disturbance and electrical noise. Static force due to flex cable bias, pivot-bearing friction and windage are all components of the vibration shock disturbance. Mechanical disturbances include spindle motor variations, disk flutter and slider vibrations. Electrical noises include quantization errors, media noise, servo demodulator noise and power amplifier noise. NRRO are usually random and unpredictable by nature, unlike repeatable run-outs. They are also of a lower magnitude (see e.g., [55]). A perfect servo system of hard disk drives should reject both the RRO and NRRO.

In our experiment, we have simplified the system somewhat by removing many sources of disturbances, especially that of the spinning magnetic disk. Therefore, we have to actually add the run-outs and other disturbances into the system manually. Based on previous experiments, we know that the run-outs in real disk drives is mainly composed of the RRO, which is basically sinusoidal with a frequency of about 55 Hz, equivalent to the spin rate of the spindle motor. By manually adding this “noise” to the output while keeping the reference signal to zero, we can then read off the subsequent position signal as the expected

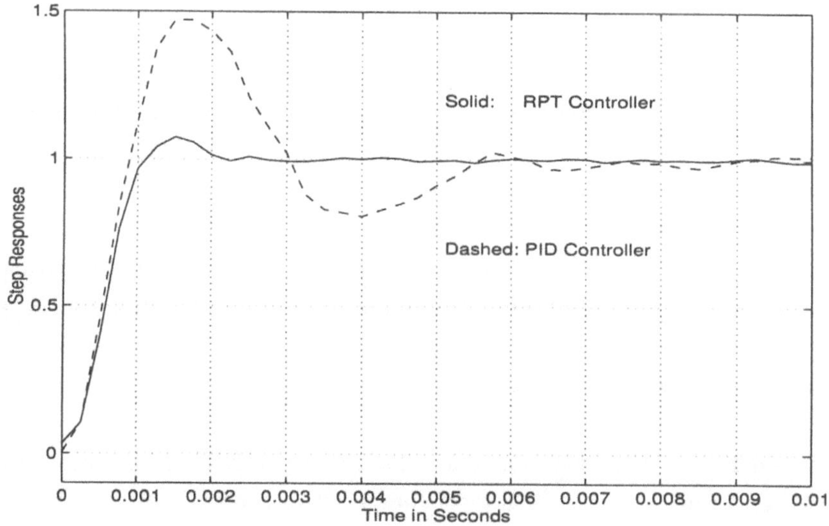

(a). Output responses.

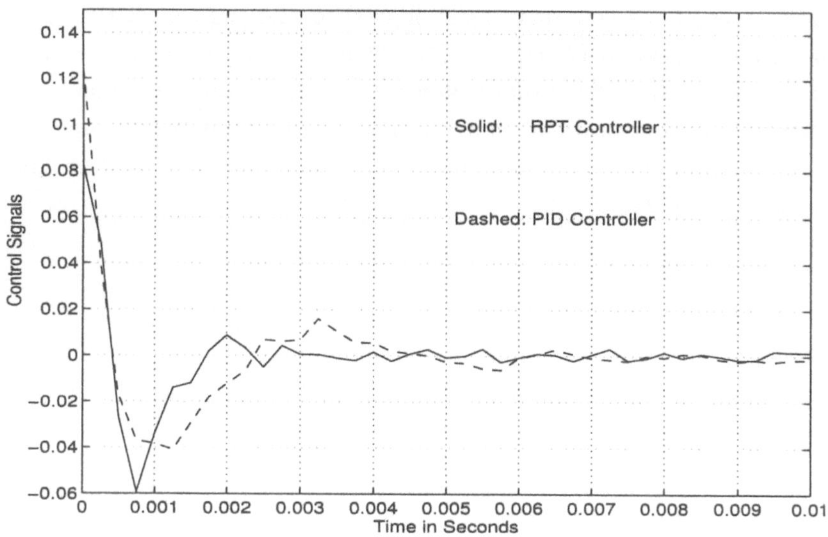

(b). Control signals.

Figure 14.4.1: Implementation result: Step responses with RPT & PID control.

PES in the presence of run-outs. In disk drive applications, the variations of the R/W head from the center of track during track following, which can be directly read off as the position error signal (PES), is very important. Track following servo systems have to ensure that the PES is kept to a minimum. Having deviations that are above the tolerance of the disk drive would result in too many read or write errors, making the disk drive unusable. A suitable measure is the standard deviation of the readings, σ. A useful guideline is to make the 3σ value less than 5% of the track width, which is about 0.1μm for a track density of 10-15 kTPI (kilo tracks per inch).

Figures 14.4.4 and 14.4.5 show the tracking errors of the robust and perfect tracking controller and PID controller respectively, under the disturbance of the run-outs. The 3σ value is about 0.095μm for the RPT controller, and about 0.175μm for the PID controller. Again, the RPT controller does better than the PID one in the PES test.

In conclusion, the RPT controller has a much better performance in track following as well as in the PES tests compared to those of the PID controller. The RPT controller utilized is first order. This is one order lower in comparison with the PID controller and would allow for quicker execution of the DSP codes during implementation. This would be an important consideration when the sampling rate of the disk drive servo is pushed higher to meet the increasing demands on the servo performance. The current results can be further improved if we used a better VCM actuator and arm assembly, with a higher resonance frequency. The control input limit has not been reached and theoretically, we should be able to tune the controller to achieve even faster settling time and higher servo bandwidth.

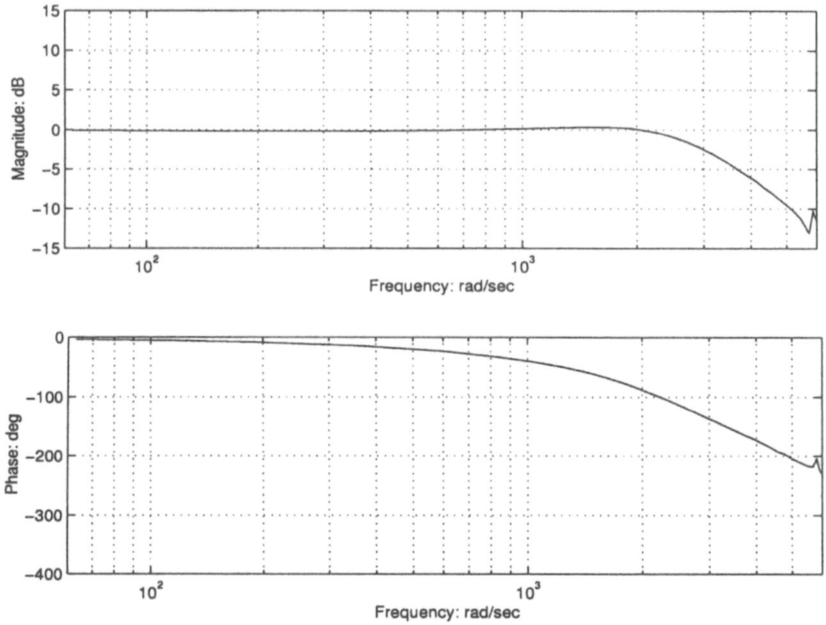

Figure 14.4.2: Implementation result: Closed-loop frequency response (RPT).

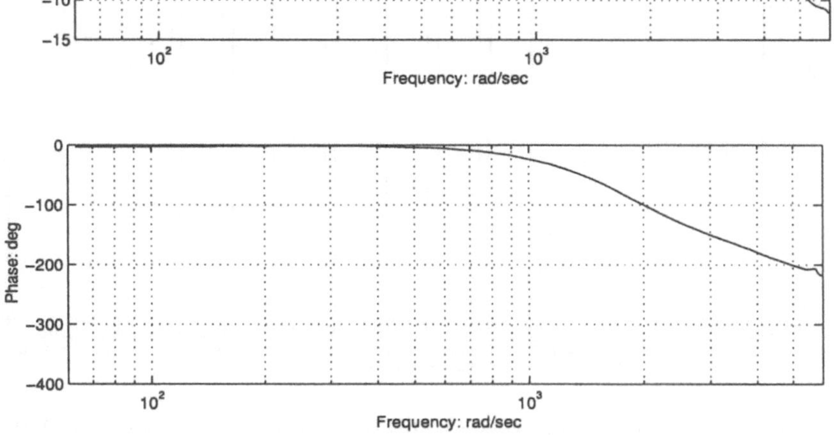

Figure 14.4.3: Implementation result: Closed-loop frequency response (PID).

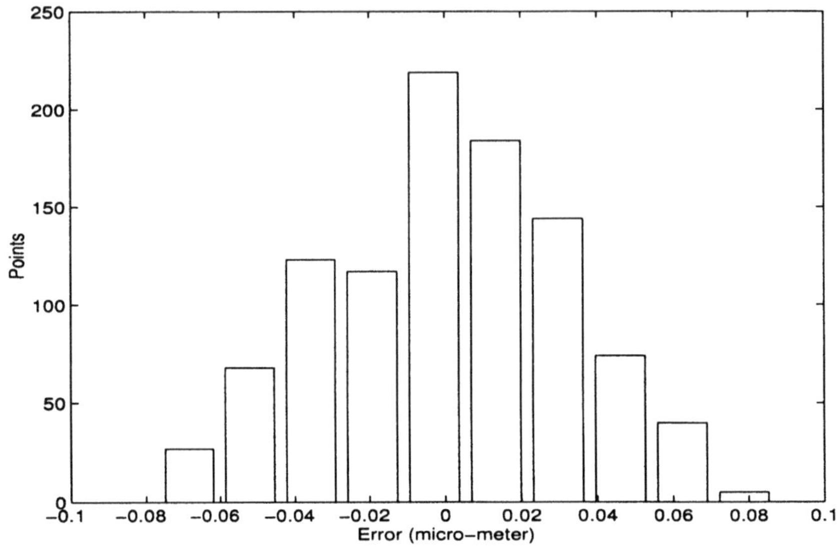

Figure 14.4.4: Implementation result: Histogram of the PES test (RPT).

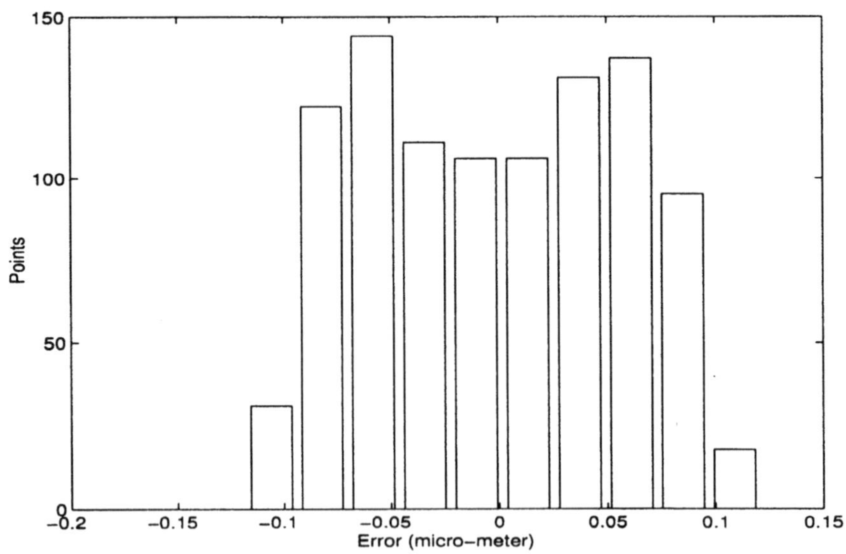

Figure 14.4.5: Implementation result: Histogram of the PES test (PID).

Chapter 15

Design of a Piezoelectric Actuator System

15.1. Introduction

WE PRESENT IN this chapter a case study on a piezoelectric bimorph actuator control system design using an H_∞ optimization approach. This work was originally reported in Chen *et al.* [23].

Piezoelectricity is a fundamental process in electromechanical energy conversion. It relates electric polarization to mechanical stress/strain in piezoelectric materials. Under the direct piezoelectric effect, an electric charge can be observed when the materials are deformed. The converse or the reciprocal piezoelectric effect is when the application of an electric field can cause mechanical stress/strain in the piezo materials. There are numerous piezoelectric materials available today including PZT (Lead Zirconate Titanate), PLZT (Lanthanum modified Lead Zirconate Titanate), and PVDF (Piezoelectric Polymeric Polyvinylidene Fluoride) to name a few (see Low and Guo [88]).

Piezoelectric structures are widely used in applications that require electrical to mechanical energy conversion coupled with size limitations, precision, and speed of operation. Typical examples are micro-sensors, micro-positioners, speakers, medical diagnostics, shutters and impact print hammers. In most applications, bimorph or stack piezoelectric structures are used because of the relatively high stress/strain to input electric field ratio (see Low and Guo [88]).

The present work is motivated by the possibility of applying piezoelectric micro-actuators in magnetic recording. The exponential growth of area densities seen in magnetic disk drives means that data tracks and data bits are being placed at closer proximity than ever before. The 25,000 TPI (tracks-per-inch) track densities envisaged at the turn of the century mean that the positioning of the read/write (R/W) heads could only tolerate at most 1 to 2 micro-inch error in track following. The closed loop positioning servo will also be required to have a bandwidth in excess of 1 to 2 kHz to be able to maintain this accuracy at the high spindle speeds required for channel data transfer rates, which will be in excess of 200 Mbits/s. Such a performance is clearly out of reach with the present voice coil motor (VCM) actuators used in disk drive access systems (see Chapter 14 for more information on hard disk drive servo systems with VCM actuators).

A dual actuator was successfully demonstrated by Tsuchiura *et al.* of Hitachi [130]. In [130], a fine positioner based on a piezoelectric structure was mounted at the end of a primary VCM stage to form the dual actuator. The higher bandwidth of the fine positioner allowed the R/W heads to be accurately positioned. There have been other instances where electromagnetic (see Miu and Tai [93]) and electrostatic (see Fan *et al.* [52]) micro-actuators have been used for fine positioning of R/W heads.

The focus of this chapter is to concentrate on the control issues involved in dealing with the nonlinear hysteresis behavior displayed by most piezoelectric actuators. More specifically, we consider a robust controller design for a piezoelectric bimorph actuator as depicted in Figure 15.1.1. A scaled up model of this piezoelectric actuator, which is targeted for use in the secondary stage of a future dual actuator for magnetic recording, was actually built and modeled by Low and Guo [88]. It has two pairs of bimorph beams which are subjected to bipolar excitation. The dynamics of the actuator were identified in [88] as a second order linear model coupled with a hysteresis. The linear model is given by

$$m\ddot{x}_1 + b\dot{x}_1 + kx_1 = k(du - z), \tag{15.1.1}$$

where m, b, k and d are the tangent mass, damping, stiffness and effective piezoelectric coefficients, while u is the input voltage that generates excitation forces to the actuator system. The variable x_1 is the displacement of the actuator and is also the only measurement we can have in this system. It should be noted that the working range of the displacement of this actuator is within $\pm 1\mu$m. The variable z is from the hysteretic nonlinear dynamics [88] and is governed by

$$\dot{z} = \alpha d\dot{u} - \beta|\dot{u}|z - \gamma\dot{u}|z|, \tag{15.1.2}$$

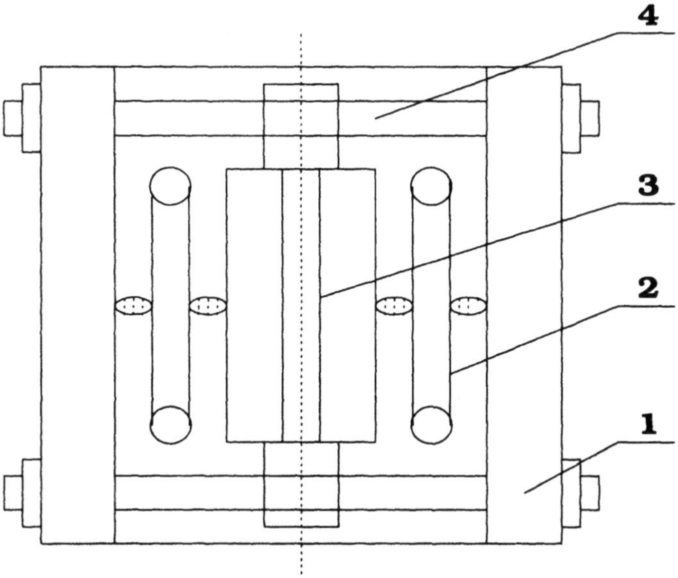

1-base; 2-piezoelectric bimorph beams; 3-moving plate; and 4-guides

Figure 15.1.1: Structure of the piezoelectric bimorph actuator.

where α, β and γ are some constants that control the shapes of the hysteresis. For the actuator system that we are considering in this paper, the above coefficients are identified as follows:

$$\left.\begin{aligned} m &= 0.01595 \text{ kg}, \\ b &= 1.169 \text{ Ns/m}, \\ k &= 4385 \text{ N/m}, \\ d &= 8.209 \times 10^{-7} \text{ m/V}, \\ \alpha &= 0.4297, \\ \beta &= 0.03438, \\ \gamma &= -0.002865. \end{aligned}\right\} \tag{15.1.3}$$

For a more detailed description of this piezoelectric actuator system and the identifications of the above parameters, we refer interested readers to the work of Low and Guo [88]. Our goal in this chapter is to design a robust controller, as in Figure 15.1.2, that meets the following design specifications:

1. The steady state tracking errors of the displacement should be less than 1% for any input reference signals that have frequencies ranging from 0 to 30 Hz, as the actuator is to be used to track certain color noise type of signals in disk drive systems.

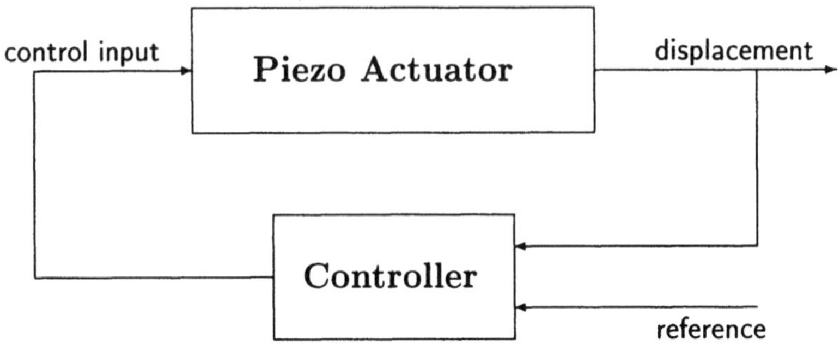

Figure 15.1.2: Piezoelectric bimorph actuator plant with controller.

2. The 1% settling time should be as fast as possible (we are able to achieve a 1% settling time of less than 0.003 seconds in our design).

3. The control input signal $u(t)$ should not exceed 112.5 volts because of the physical limitations of the piezoelectric materials.

Our approach is as follows: we will first use the stochastic equivalent linearization method proposed in Chang [10] to obtain a linearized model for the nonlinear hysteretic dynamics. Then we reformulate our design into an H_∞ almost disturbance decoupling problem in which the disturbance inputs are the reference input and the error between the hysteretic dynamics and that of its linearized model, while the controlled output is simply the double integration of the tracking error. Thus, our task becomes to design a controller such that when it is applied to the piezoelectric actuator, the overall system is asymptotically stable, and the controlled output, which corresponds to the tacking error, is as small as possible and decays as fast as possible.

The outline of this chapter is as follows: In Section 15.2, a first order linearized model is obtained for the nonlinear hysteresis using the stochastic equivalent linearization method. A simulation result is also given to show the match between the nonlinear and linearized models. In Section 15.3, we formulate our controller design into a standard almost disturbance decoupling problem by properly defining the disturbance input and the controlled output. Two integrators are augmented into the original plant to enhance the performance of the overall system. Then a robust controller that is explicitly parameterized by a certain tuning parameter and that solves the proposed almost disturbance decoupling problem, is carried out using a so-called asymptotic time-scale and eigenstructure assignment technique. In Section 15.4, we present the final controller and simulation results of our overall control system using MATLAB SIMULINK. We also obtain an explicit relationship between the peak values of the control signal and the tuning parameter of the controller, as well as an explicit linear relationship of the maximum trackable frequency, i.e, the corresponding tracking error can be settled to 1%, *vs* the tuning parameter of the controller. The simulation results of this section clearly show that all the design specifications are met and the overall performance is very satisfactory.

15.2. Linearization of the Nonlinear Hysteretic Dynamics

We will proceed to linearize the nonlinear hysteretic dynamics of (15.1.2) in this section. As pointed out in Chang [10], there are basically three methods available in the literature to linearize the hysteretic type of nonlinear systems. These are i) the Fokker-Planck equation approach (see for example Caughey [8]), ii) the perturbation techniques (see for example Crandall [43] and Lyon [90]) and iii) the stochastic linearization approach. All of them have certain advantages and limitations. However, the stochastic linearization technique has the widest range of applications compared to the other methods. This method is based on the concept of replacing the nonlinear system with an "equivalent" linear system in such a way that the "difference" between these two systems is minimized in a certain sense. The technique was initiated by Booton [6]. In this chapter, we will just follow the stochastic linearization method given in Chang [10] to obtain a linear model of the following form

$$\dot{z} = k_1 \dot{u} + k_2 z, \tag{15.2.1}$$

for the hysteretic dynamics of (15.1.2), where k_1 and k_2 are the linearization coefficients and are to be determined. The procedure is quite straightforward and proceeds as follows: First we introduce a so-called "difference" function e between $\dot{z}$ of (15.1.2) and $\dot{z}$ of (15.2.1),

$$e(k_1, k_2) = \alpha d\dot{u} - \beta|\dot{u}|z - \gamma\dot{u}|z| - (k_1\dot{u} + k_2 z). \tag{15.2.2}$$

Then minimizing $\boldsymbol{E}[e^2]$, where $\boldsymbol{E}$ is the expectation operator, with respect to k_1 and k_2, we obtain

$$\frac{\partial \boldsymbol{E}[e^2]}{\partial k_1} = \frac{\partial \boldsymbol{E}[e^2]}{\partial k_2} = 0, \tag{15.2.3}$$

from which the stochastic linearization coefficients k_1 and k_2 are determined. It turns out that if h and $\dot{u}$ are of zero means and jointly Gaussian, then k_1 and

k_2 can be easily obtained. Let us assume that h and $\dot{u}$ have a joint probability density function

$$f_{\dot{u}z}(\dot{u},z) = \frac{1}{2\pi\sigma_{\dot{u}}\sigma_z\sqrt{1-\rho_{\dot{u}z}^2}} \exp\left\{-\frac{\sigma_{\dot{u}}^2 z^2 - 2\sigma_{\dot{u}}\sigma_z\rho_{\dot{u}z}\dot{u}z + \sigma_z^2\dot{u}^2}{2\sigma_{\dot{u}}^2\sigma_z^2(1-\rho_{\dot{u}z}^2)}\right\}, \tag{15.2.4}$$

where $\rho_{\dot{u}z}$ is the normalized covariance of $\dot{u}$ and z, and $\sigma_{\dot{u}}$ and σ_z are the standard deviation of $\dot{u}$ and z, respectively. Then the linearization coefficients k_1 and k_2 can be expressed as follows:

$$k_1 = \alpha d - \beta c_1 - \gamma c_2, \tag{15.2.5}$$

and

$$k_2 = -\beta c_3 - \gamma c_4, \tag{15.2.6}$$

where c_1, c_2, c_3 and c_4 are given by

$$c_1 = 0.79788456\sigma_z \cos\left[\tan^{-1}\left(\frac{\sqrt{1-\rho_{\dot{u}z}^2}}{\rho_{\dot{u}z}}\right)\right], \tag{15.2.7}$$

$$c_2 = 0.79788456\sigma_z, \qquad c_4 = 0.79788456\rho_{\dot{u}z}\sigma_{\dot{u}}, \tag{15.2.8}$$

and

$$c_3 = 0.79788456\sigma_{\dot{u}}\left\{1-\rho_{\dot{u}z}^2 + \rho_{\dot{u}z}\cos\left[\tan^{-1}\left(\frac{\sqrt{1-\rho_{\dot{u}z}^2}}{\rho_{\dot{u}z}}\right)\right]\right\}. \tag{15.2.9}$$

After a few iterations, we found that a sinusoidal excitation $\dot{u}$ with frequencies ranging from 0 to 100 Hz (the expected working frequency range) and peak magnitude of 50 volts, which has a standard deviation of $\sigma_{\dot{u}} = 35$, would yield a suitable linearized model for (15.1.2). For this excitation, we obtain $\sigma_z = 5 \times 10^{-7}$, $\rho_{\dot{u}z} = 5 \times 10^{-3}$

$$c_1 = 1.9947 \times 10^{-9}, \qquad c_2 = 3.9894 \times 10^{-7}, \tag{15.2.10}$$

$$c_3 = 27.9260, \qquad c_4 = 0.1396, \tag{15.2.11}$$

and

$$k_1 = 3.5382 \times 10^{-7}, \qquad k_2 = -0.9597. \tag{15.2.12}$$

The stochastic linearization model of the given nonlinear hysteretic dynamics of (15.1.2) is then given by

$$\dot{\hat{z}} = k_1\dot{u} + k_2\hat{z} = 3.5382 \times 10^{-7}\dot{u} - 0.9597\hat{z}. \tag{15.2.13}$$

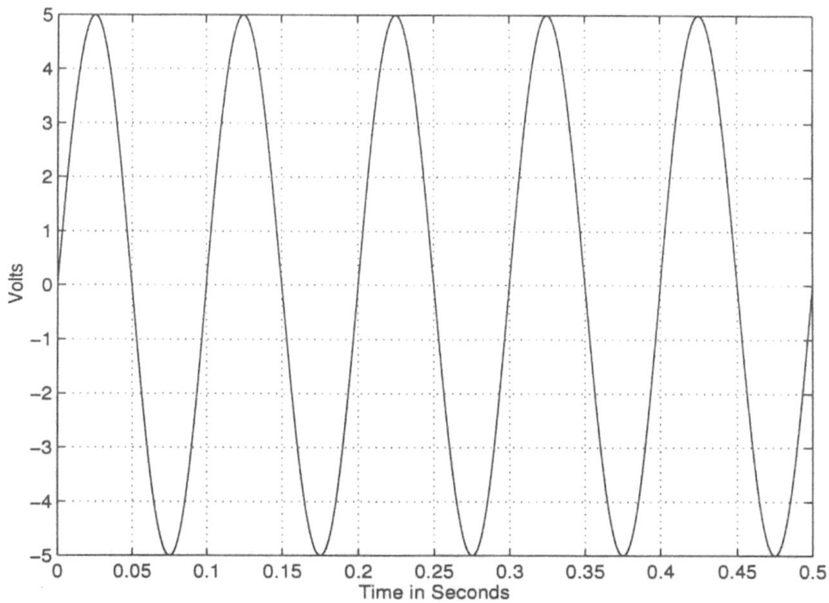

(a) The control input signal, u

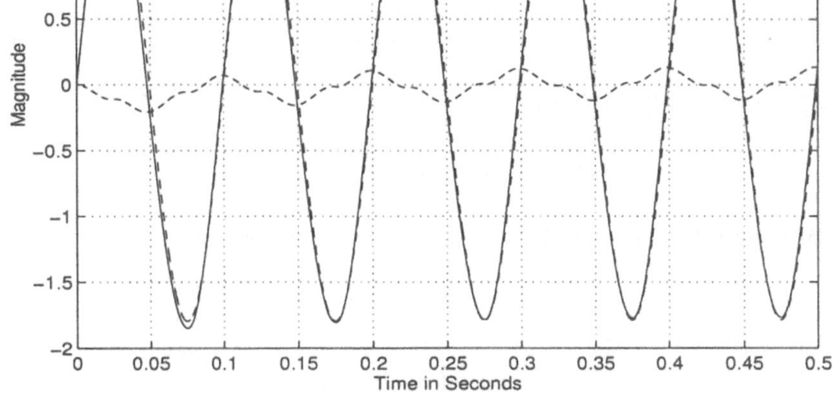

(b) z (solid), $\hat{z}$ (dashed), and e_z (dash-dotted)

Figure 15.2.1: Responses of hysteresis and its linearized model to a sine input.

For future use, let us define the linearization error as

$$e_z = z - \hat{z}. \tag{15.2.14}$$

Figure 15.2.1 shows the open-loop simulation results of the nonlinear hysteresis and its linearized model, as well as their error for a typical sine wave input signal u. The results are quite satisfactory. Here we should note that because of the nature of our approach in controller design later in the next section, the variation of the linearized model within a certain range, which might result in larger linearization error, e_z, will not much affect the overall performance of the closed-loop system. We will formulate e_z as a disturbance input and our controller will automatically reject it from the output response.

15.3. Formulation of the Problem as an H_∞-ADDPMS

This section is the heart of this chapter. We will first formulate our control system design for the piezoelectric bimorph actuator into a standard H_∞ almost disturbance decoupling problem, and then apply the results of Chapter 8 to check the solvability of the proposed problem. Finally, we will utilize the results in Chapter 8 to find an internally stabilizing controller that solves the proposed almost disturbance decoupling problem. Of course, most importantly, the resulting closed-loop system and its responses should meet all the design specifications as listed in Section 15.1. To do this, we will have to convert the dynamic model of (15.1.1) with the linearized model of the hysteresis into a state space form. Let us first define a new state variable

$$v = \hat{z} - k_1 u. \tag{15.3.1}$$

Then from (15.2.13), we have

$$\dot{v} = \dot{\hat{z}} - k_1 \dot{u} = k_2 \hat{z} = k_2 v + k_1 k_2 u. \tag{15.3.2}$$

Substituting (15.2.14) and (15.3.1) into (15.1.1), we obtain

$$\ddot{x}_1 + \frac{b}{m}\dot{x}_1 + \frac{k}{m}x_1 + \frac{k}{m}v = \frac{k(d-k_1)}{m}u - \frac{k}{m}e_z. \tag{15.3.3}$$

The overall controller structure of our approach is then depicted in Figure 15.3.1. Note that in Figure 15.3.1 we have augmented two integrators after e, the tracking error between the displacement x_1 and the reference input signal r. We have observed a very interesting property of this problem, i.e., the more integrators that we augment after the tracking error e, the smaller the tracking

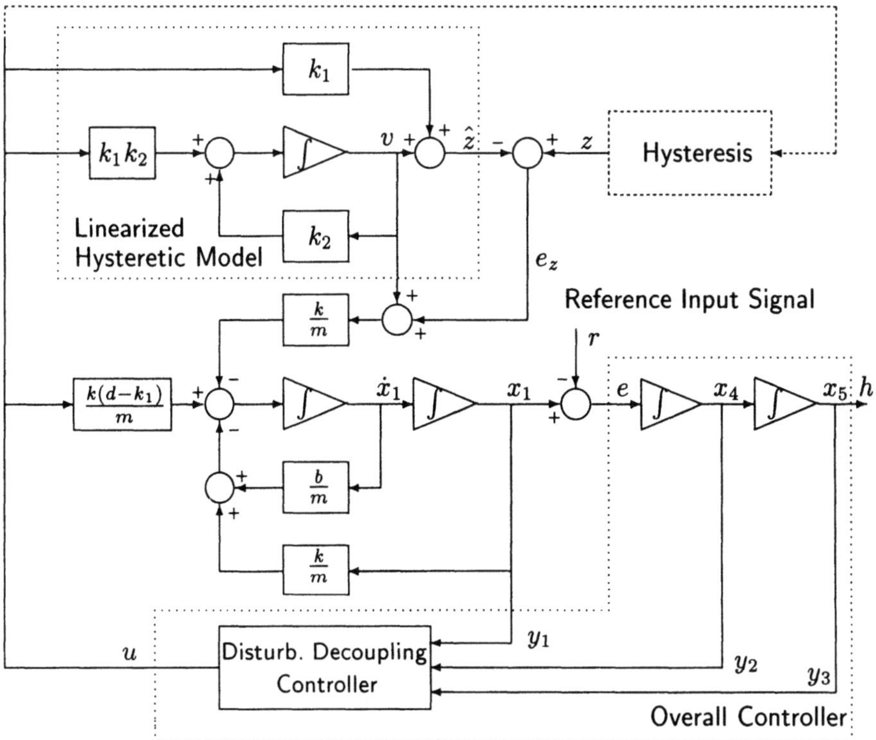

Figure 15.3.1: Augmented linearized model with controller.

error we can achieve for the same level of control input u. Because our control input u is limited to the range from -112.5 to 112.5 volts, it turns out that two integrators are needed in order to meet all the design specifications. It is clear to see that the augmented system has an order of five. Next, let us define the state of the augmented system as

$$x = (\,x_1 \quad \dot{x}_1 \quad v \quad x_4 \quad x_5\,)', \tag{15.3.4}$$

and the measurement output

$$y = \begin{pmatrix} y_1 \\ y_2 \\ y_3 \end{pmatrix} = \begin{pmatrix} x_1 \\ x_4 \\ x_5 \end{pmatrix}, \tag{15.3.5}$$

i.e., the original measurement of displacement x_1 plus two augmented states. The auxiliary disturbance input is

$$w = \begin{pmatrix} e_z \\ r \end{pmatrix}, \tag{15.3.6}$$

and the output to be controlled, h, is simply the double integration of the tracking error. The state space model of the overall augmented system is then given by

$$\Sigma : \begin{cases} \dot{x} = A\ x + B\ u + E\ w, \\ y = C_1\ x + D_1\ w, \\ h = C_2\ x + D_2\ u\ , \end{cases} \tag{15.3.7}$$

with

$$A = \begin{bmatrix} 0 & 1 & 0 & 0 & 0 \\ -k/m & -b/m & -k/m & 0 & 0 \\ 0 & 0 & k_2 & 0 & 0 \\ 1 & 0 & 0 & 0 & 0 \\ 0 & 0 & 0 & 1 & 0 \end{bmatrix}$$

$$= \begin{bmatrix} 0 & 1 & 0 & 0 & 0 \\ -274921.63 & -73.2915 & -274921.63 & 0 & 0 \\ 0 & 0 & -0.9597 & 0 & 0 \\ 1 & 0 & 0 & 0 & 0 \\ 0 & 0 & 0 & 1 & 0 \end{bmatrix}, \tag{15.3.8}$$

$$B = \begin{bmatrix} 0 \\ k(d-k_1)/m \\ k_1 k_2 \\ 0 \\ 0 \end{bmatrix} = \begin{bmatrix} 0 \\ 0.12841 \\ -3.39561 \times 10^{-7} \\ 0 \\ 0 \end{bmatrix}, \tag{15.3.9}$$

$$E = \begin{bmatrix} 0 & 0 \\ -k/m & 0 \\ 0 & 0 \\ 0 & -1 \\ 0 & 0 \end{bmatrix} = \begin{bmatrix} 0 & 0 \\ -274921.63 & 0 \\ 0 & 0 \\ 0 & -1 \\ 0 & 0 \end{bmatrix}, \tag{15.3.10}$$

$$C_1 = \begin{bmatrix} 1 & 0 & 0 & 0 & 0 \\ 0 & 0 & 0 & 1 & 0 \\ 0 & 0 & 0 & 0 & 1 \end{bmatrix}, \quad D_1 = \begin{bmatrix} 0 & 0 \\ 0 & 0 \\ 0 & 0 \end{bmatrix}, \tag{15.3.11}$$

and

$$C_2 = [0 \quad 0 \quad 0 \quad 0 \quad 1], \quad D_2 = 0. \tag{15.3.12}$$

For the problem that we are considering here, it is simple to verify that the system Σ of (15.3.7) has the following properties:

1. The subsystem (A, B, C_2, D_2) is invertible and of minimum phase with one invariant zero at -1.6867. It also has one infinite zero of order 4.

2. The subsystem (A, E, C_1, D_1) is left invertible and of minimum phase with one invariant zero at -0.9597 and two infinite zeros of orders 1 and 2, respectively.

Then it follows from Theorem 8.2.1 or Theorem 8.2.1 that the H_∞-ADDPMS for (15.3.7) is solvable. In fact, one can design either a full order observer based controller or a reduced order observer based controller to solve this problem. For the full order observer based controller, the order of the disturbance decoupling controller (see Figure 15.3.1) will be 5 and the order of the final overall controller (again see Figure 15.3.1) will be 7 (the disturbance decoupling controller plus two integrators). On the other hand, if we use a reduced order observer in the disturbance decoupling controller, the total order of the resulting final overall controller will be reduced to 4. From the practical point of view, the latter is much more desirable than the former. Thus, in what follows we will only focus on the controller design based on a reduced order observer. We can separate our controller design into two steps:

1. In the first step, we assume that all five states of Σ in (15.3.7) are available and then design a static and parameterized state feedback control law,

$$u = F(\varepsilon)x, \tag{15.3.13}$$

such that it solves the almost disturbance decoupling problem for the state feedback case, i.e., $y = x$, by adjusting the tuning parameter ε to an appropriate value.

2. In the second step, we design a reduced order observer based controller. It has a parameterized reduced order observer gain matrix $K_2(\varepsilon)$ that can be tuned to recover the performance achieved by the state feedback control law in the first step.

We will use the asymptotic time-scale and eigenstructure assignment (ATEA) design method of Chapter 8 to construct both the state feedback law and the reduced order observer gain. We would like to note that in principle, one can also apply the ARE (algebraic Riccati equation) based H_∞ optimization technique (see for example Zhou and Khargonekar [141]) to solve this problem. However, because the numerical conditions of our system Σ are very bad, we are unable to obtain any satisfactory solution from the ARE approach. We cannot get any meaningful solution for the associated H_∞-CARE in MATLAB. In this sense and at least for this problem, the ATEA method is much more powerful than the ARE one. The software realization of the ATEA algorithm can be found in the Linear Systems and Control Toolbox developed by Chen [14]. The following is a closed form solution of the static state feedback parameterized gain matrix $F(\varepsilon)$ obtained using the ATEA method.

$$F(\varepsilon) = \Big[(2.1410 \times 10^6 - 62.3004/\varepsilon^2) \quad (570.7619 - 31.1502/\varepsilon) \\ 2.1410 \times 10^6 \quad -62.3004/\varepsilon^3 \quad -31.1502/\varepsilon^4 \Big], \tag{15.3.14}$$

where ε is the tuning parameter that can be adjusted to achieve almost disturbance decoupling. It can be verified that the closed-loop system matrix, $A + BF(\varepsilon)$ is asymptotically stable for all $0 < \varepsilon < \infty$ and the closed-loop transfer function from the disturbance w to the controlled output h, $T_{hw}(\varepsilon, s)$, satisfying

$$\|T_{hw}(\varepsilon, s)\|_\infty = \|[C_2 + D_2F(\varepsilon)][sI - A - BF(\varepsilon)]^{-1}E\|_\infty \to 0, \tag{15.3.15}$$

as $\varepsilon \to 0$.

The next step is to design a reduced order observer based controller that will recover the performance of the above state feedback control law. First, let us perform the following nonsingular (permutation) state transformation to the system Σ of (15.3.7),

$$x = T\tilde{x}, \tag{15.3.16}$$

where

$$T = \begin{bmatrix} 1 & 0 & 0 & 0 & 0 \\ 0 & 0 & 0 & 1 & 0 \\ 0 & 0 & 0 & 0 & 1 \\ 0 & 1 & 0 & 0 & 0 \\ 0 & 0 & 1 & 0 & 0 \end{bmatrix}, \tag{15.3.17}$$

such that the transformed measurement matrix has the form of

$$C_1T = \begin{bmatrix} 1 & 0 & 0 & 0 & 0 \\ 0 & 1 & 0 & 0 & 0 \\ 0 & 0 & 1 & 0 & 0 \end{bmatrix} = [\, I_3 \quad 0\,]. \tag{15.3.18}$$

Clearly, the first three states of the transformed system, or x_1, x_4 and x_5 of the original system Σ in (15.3.7), need not be estimated as they are already available from the measurement output. Let us now partition the transformed system as follows:

$$T^{-1}AT = \left[\begin{array}{c|c} A_{11} & A_{12} \\ \hline A_{21} & A_{22} \end{array}\right] = \left[\begin{array}{ccc|cc} 0 & 0 & 0 & 1 & 0 \\ 1 & 0 & 0 & 0 & 0 \\ 0 & 1 & 0 & 0 & 0 \\ \hline -274921.63 & 0 & 0 & -73.2915 & -274921.63 \\ 0 & 0 & 0 & 0 & -0.9597 \end{array}\right], \tag{15.3.19}$$

$$T^{-1}B = \left[\begin{array}{c} B_1 \\ \hline B_2 \end{array}\right] = \left[\begin{array}{c} 0 \\ 0 \\ 0 \\ \hline 0.12841 \\ -3.39561\times 10^{-7} \end{array}\right], \tag{15.3.20}$$

$$T^{-1}E = \left[\begin{array}{c} E_1 \\ \hline E_2 \end{array}\right] = \left[\begin{array}{cc} 0 & 0 \\ 0 & -1 \\ 0 & 0 \\ \hline -274921.63 & 0 \\ 0 & 0 \end{array}\right]. \tag{15.3.21}$$

Also, we partition

$$F(\varepsilon)T = \left[\begin{array}{c|c} F_1(\varepsilon) & F_2(\varepsilon) \end{array}\right] \tag{15.3.22}$$

$$= \Big[(2.1410\times 10^6 - 62.3004/\varepsilon^2) \quad -62.3004/\varepsilon^3 \quad -31.1502/\varepsilon^4 \;\Big|$$
$$(570.7619 - 31.1502/\varepsilon) \quad 2.1410\times 10^6 \Big]. \tag{15.3.23}$$

Then the reduced order observer based controller (see Chapter 8) is given in the form of

$$\Sigma_{\rm cmp} : \begin{cases} \dot{v} = A_{\rm cmp}(\varepsilon)\, v + B_{\rm cmp}(\varepsilon)\, y, \\ u = C_{\rm cmp}(\varepsilon)\, v + D_{\rm cmp}(\varepsilon)\, y, \end{cases} \tag{15.3.24}$$

with

$$A_{\rm cmp}(\varepsilon) = A_{22} + K_2(\varepsilon)A_{12} + B_2F_2(\varepsilon) + K_2(\varepsilon)B_1F_2(\varepsilon), \tag{15.3.25}$$

$$B_{\rm cmp}(\varepsilon) = A_{21} + K_2(\varepsilon)A_{11} - \big[A_{22} + K_2(\varepsilon)A_{12}\big]K_2(\varepsilon) + \big[B_2 + K_2(\varepsilon)B_1\big]\big[F_1(\varepsilon) - F_2(\varepsilon)K_2(\varepsilon)\big], \quad (15.3.26)$$

$$C_{\rm cmp}(\varepsilon) = F_2(\varepsilon), \quad (15.3.27)$$

$$D_{\rm cmp}(\varepsilon) = F_1(\varepsilon) - F_2(\varepsilon)K_2(\varepsilon), \quad (15.3.28)$$

where $K_2(\varepsilon)$ is the parameterized reduced order observer gain matrix and is to be designed such that $A_{22} + K_2(\varepsilon)A_{12}$ is asymptotically stable for sufficiently small ε and also

$$\|[sI - A_{22} - K_2(\varepsilon)A_{12}]^{-1}[E_2 + K_2(\varepsilon)E_1]\|_\infty \to 0, \quad (15.3.29)$$

as $\varepsilon \to 0$. Again, using the software package of Chen [14], we obtained the following parameterized reduced order observer gain matrix

$$K_2(\varepsilon) = \begin{bmatrix} 73.2915 - 1/\varepsilon & 0 & 0 \\ 0 & 0 & 0 \end{bmatrix}. \quad (15.3.30)$$

Then the explicitly parameterized matrices of the state space model of the reduced order observer based controller are given by

$$A_{\rm cmp}(\varepsilon) = \begin{bmatrix} 73.2915 - 4/\varepsilon - 1/\varepsilon & 0 \\ -1.9381\times10^{-4} + 1.0577\times10^{-5}/\varepsilon & -1.6867 \end{bmatrix},$$

$$C_{\rm cmp}(\varepsilon) = [\,570.7619 - 31.1502/\varepsilon \quad 2140967\,],$$

$$D_{\rm cmp}(\varepsilon) = [\,2099135.4 + 2853.81/\varepsilon - 93.45/\varepsilon^2 \quad -62.3/\varepsilon^3 \quad -31.1502/\varepsilon^4\,],$$

$$B_{\rm cmp}(\varepsilon) = \begin{bmatrix} \psi_1 & -8/\varepsilon^3 & -4/\varepsilon^4 \\ \psi_2 & 2.1155\times10^{-5}/\varepsilon^3 & 1.0577\times10^{-5}/\varepsilon^4 \end{bmatrix},$$

where

$$\psi_1 = -5731.6533 - 13/\varepsilon^2 + 439.7492/\varepsilon, \quad (15.3.31)$$

and

$$\psi_2 = -0.7128 + 3.1732\times10^{-5}/\varepsilon^2 - 9.6904\times10^{-4}/\varepsilon. \quad (15.3.32)$$

The overall closed loop system comprising the system Σ of (15.3.7) and the above controller would be asymptotically stable as long as $\varepsilon \in (0,\infty)$. In fact, the closed loop poles are exactly located at -1.6867, two pairs at $-1/\varepsilon \pm j1/\varepsilon$, -0.9597 and $-1/\varepsilon$. The plots of the maximum singular values of the closed loop transfer function matrix from the disturbance w to the controlled output h, namely $T_{hw}(\varepsilon, s)$, for several values of ε, i.e., $\varepsilon = 1/100$, $\varepsilon = 1/400$ and $\varepsilon = 1/3000$, in Figure 15.3.2 show that as ε becomes smaller and smaller, the

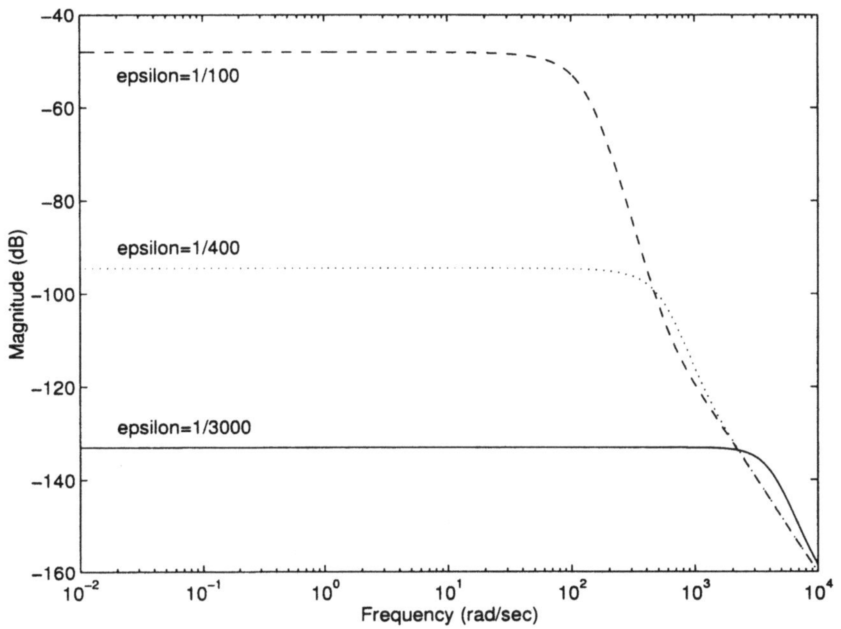

Figure 15.3.2: Max. singular values of closed loop transfer function $T_{hw}(\varepsilon, s)$.

H_∞ norms of $T_{hw}(\varepsilon, s)$ are also smaller and smaller. Hence, almost disturbance decoupling is indeed achieved. These are the properties of our control system in the frequency domain. In the next section, we will address its time domain properties, which are of course much more important as all the design specifications are in the time domain.

15.4. Final Controller and Simulation Results

In this section, we will put our design of the previous section into a final controller as depicted in Figure 15.1.2. It is simple to derive the state space model of the final overall controller by observing its interconnection with the disturbance decoupling controller $\Sigma_{\rm cmp}$ of (15.3.24) (see Figure 15.3.1). We will also present simulation results of the responses of the overall design to several different types of reference input signals. They clearly show that all the design specifications are successfully achieved. Furthermore, because our controller is explicitly parameterized by a tuning parameter, it is very easy to adjust to meet other design specifications without going through it all over again from the beginning. This will also be discussed next.

As mentioned earlier, the final overall controller of our design will be of the order of 4, of which two are from the disturbance decoupling controller and two from the augmented integrators. It has two inputs: one is the displacement x_1 and the other is the reference signal r. It is straightforward to verify that the state space model of the final overall controller is given by

$$\Sigma_{\rm oc}(\varepsilon): \quad \begin{cases} \dot{v} = A_{\rm oc}(\varepsilon)\ v + B_{\rm oc}(\varepsilon)\ x_1 + G_{\rm oc}\ r, \\ u = C_{\rm oc}(\varepsilon)\ v + D_{\rm oc}(\varepsilon)\ x_1\ , \end{cases} \tag{15.4.1}$$

where $A_{\rm oc}(\varepsilon)$ is given by

$$\begin{bmatrix} 73.2915 - 5/\varepsilon & 0 & -8/\varepsilon^3 & -4/\varepsilon^4 \\ -0.0002 + 1.0577\times10^{-5}/\varepsilon & -1.6867 & 2.1155\times10^{-5}/\varepsilon^3 & 1.0577\times10^{-5}/\varepsilon^4 \\ 0 & 0 & 0 & 0 \\ 0 & 0 & 1 & 0 \end{bmatrix},$$

$$G_{\rm oc} = \begin{bmatrix} 0 \\ 0 \\ -1 \\ 0 \end{bmatrix}, \qquad B_{\rm oc}(\varepsilon) = \begin{bmatrix} \psi_1 \\ \psi_2 \\ 1 \\ 0 \end{bmatrix},$$

with ψ_1 and ψ_2 given by (15.3.31) and (15.3.32), respectively,

$$C_{\rm oc}(\varepsilon) = [\,570.7619 - 31.1502/\varepsilon \quad 2140967 \quad -62.3004/\varepsilon^3 \quad -31.1502/\varepsilon^4\,],$$

and

$$D_{\rm oc}(\varepsilon) = 2099135.4 - 93.4506/\varepsilon^2 + 2853.8095/\varepsilon.$$

There are some very interesting and very useful properties of the above parameterized controller. After repeatedly simulating the overall design, we found that the maximum peak values of the control signal u are independent of the frequencies of the reference signals. They are only dependent on the initial error between displacement, x_1, and the reference, r. The larger the initial error is, the bigger the peak that occurs in u. Because the working range of our actuator is within $\pm 1\mu$m, we will assume that the largest magnitude of the initial error in any situation should not be larger that 1μm. This assumption is reasonable as we can always reset our displacement, x_1, to 0 before the system is to track any reference and hence the magnitude of initial tracking error can never be larger than 1μm. Let us consider the worst case, i.e., the magnitude of the initial error is 1μm. Then interestingly, we are able to obtain a clear relationship between the tuning parameter $1/\varepsilon$ and the maximum peak of u. The result is plotted in Figure 15.4.1. We also found that the tracking error is independent of initial errors. It only depends on the frequencies of the references, i.e., the larger the frequency that the reference signal r has, the larger the tracking error

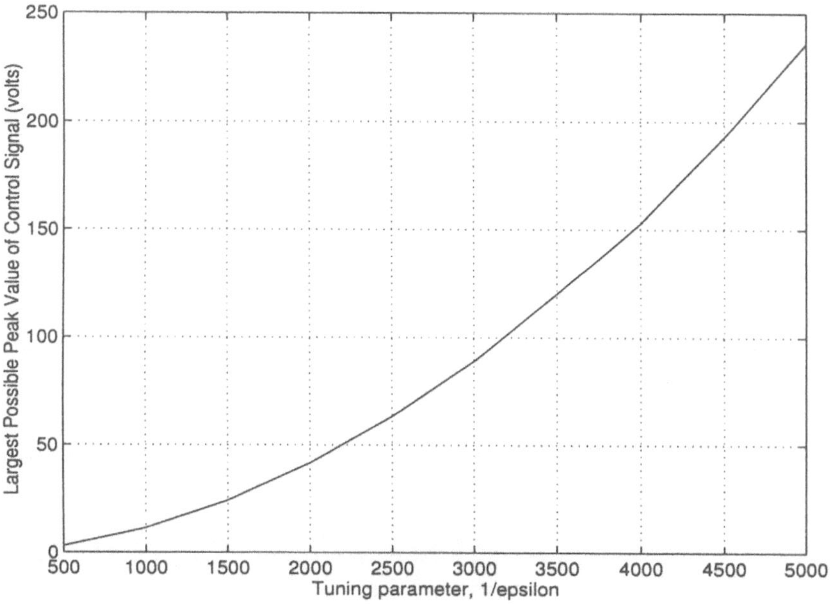

Figure 15.4.1: Parameter $1/\varepsilon$ *vs* max. peaks of u in worst initial errors.

that occurs. Again, we can obtain a simple and linear relationship between the tuning parameter ε and the maximum frequency that a reference signal can have such that the corresponding tracking error is no larger than 1%, which is one of our main design specifications. The result is plotted in Figure 15.4.2.

Clearly, from Figure 15.4.1, we know that due to the constraints on the control input, i.e., it must be kept within ± 112.5 volts, we have to select our controller with $\varepsilon > 1/3370$. From Figure 15.4.2, we know that in order to meet the first design specification, i.e., the steady state tracking errors should be less than 1% for reference inputs that have frequencies up to 30 Hz, we have to choose our controller with $\varepsilon < 1/2680$. Hence, the final controller as given in (15.4.1) to (15.4) will meet all the design goals for our piezoelectric actuator system. i.e., (15.1.1) and (15.1.2), for all $\varepsilon \in (1/3370, 1/2680)$. Let us choose $\varepsilon = 1/3000$. We obtain the overall controller as in the form of (15.4.1) with

$$A_{\rm oc} = \begin{bmatrix} -14926.7085 & 0 & -2.16 \times 10^{11} & 3.24 \times 10^{14} \\ 0.0315 & -1.6867 & 5.7118 \times 10^{5} & 8.5677 \times 10^{8} \\ 0 & 0 & 0 & 0 \\ 0 & 0 & 1 & 0 \end{bmatrix}, \qquad (15.4.2)$$

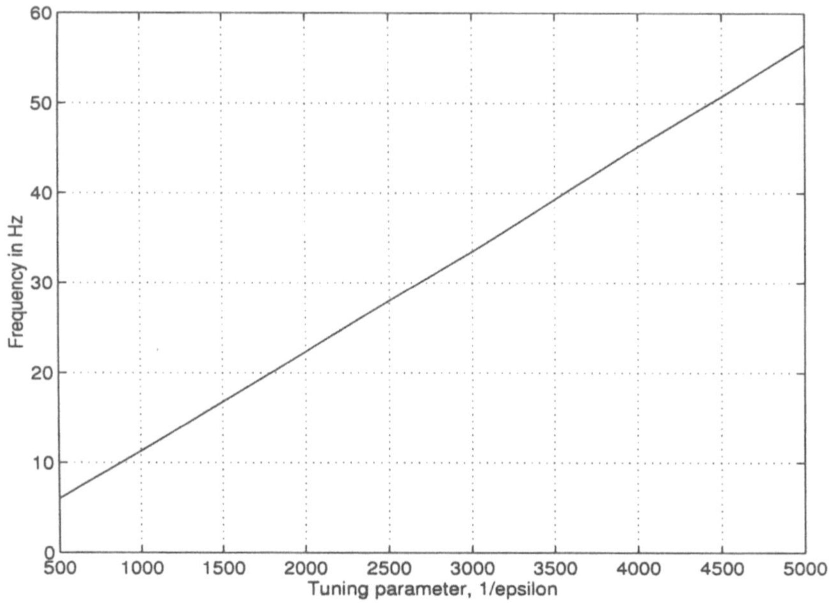

Figure 15.4.2: Parameter $1/\varepsilon$ *vs* max. frequency of r that has 1% tracking error.

$$B_{\rm oc} = \begin{bmatrix} -1.1569 \times 10^{8} \\ 281.9699 \\ 1 \\ 0 \end{bmatrix}, \quad G_{\rm oc} = \begin{bmatrix} 0 \\ 0 \\ -1 \\ 0 \end{bmatrix}, \tag{15.4.3}$$

$$C_{\rm oc} = [\,-92879.9041 \quad 2140967 \quad -1.6821 \times 10^{12} \quad -2.5232 \times 10^{15}\,], \tag{15.4.4}$$

and

$$D_{\rm oc} = -8.3040 \times 10^{8}. \tag{15.4.5}$$

The simulation results presented in the following are done using the MATLAB SIMULINK package, which is widely available everywhere these days. The SIMULINK simulation block diagram for the overall piezoelectric bimorph actuator system is given in Figure 15.4.3. Two different reference inputs are simulated using the Runge-Kutta 5 method in SIMULINK with a minimum step size of 10 micro-seconds and a maximum step size of 100 micro-seconds as well as a tolerance of 10^{-5}. These references are: 1) a cosine signal with a frequency of 30 Hz and peak magnitude of 1 μm, and 2) a sine signal with a frequency of 34 Hz and peak magnitude of 1 μm. The results for the cosine signal are given in Figures 15.4.4 to 15.4.6. In Figure 15.4.4, the solid-line curve is x_1 and the dash-dotted curve is the reference. The tracking error and the control signal corresponding to this reference are given in Figures 15.4.5 and

15.4.6, respectively. Similarly, Figures 15.4.7 to 15.4.9 are the results corresponding to the sine signal. All these results show that our design goals are fully achieved. To be more specific, the tracking error for a 30 Hz cosine wave reference is about 0.8%, which is better than the specification, and the worst peak magnitude of the control signal is less than 90 volts, which is of course less than the saturated level, i.e., 112.5 volts. Furthermore, the 1% tracking error settling times for both cases are less than 0.003 seconds.

Because the piezoelectric actuator is designed to be operated in a small neighborhood of its equilibrium point, the stability properties of the overall closed loop system of the nonlinear piezoelectric bimorph actuator should be similar to those of its linearized model. This fact can also be verified from simulations. In fact, the performance of the actual closed loop system is even better than that of its linear counterpart.

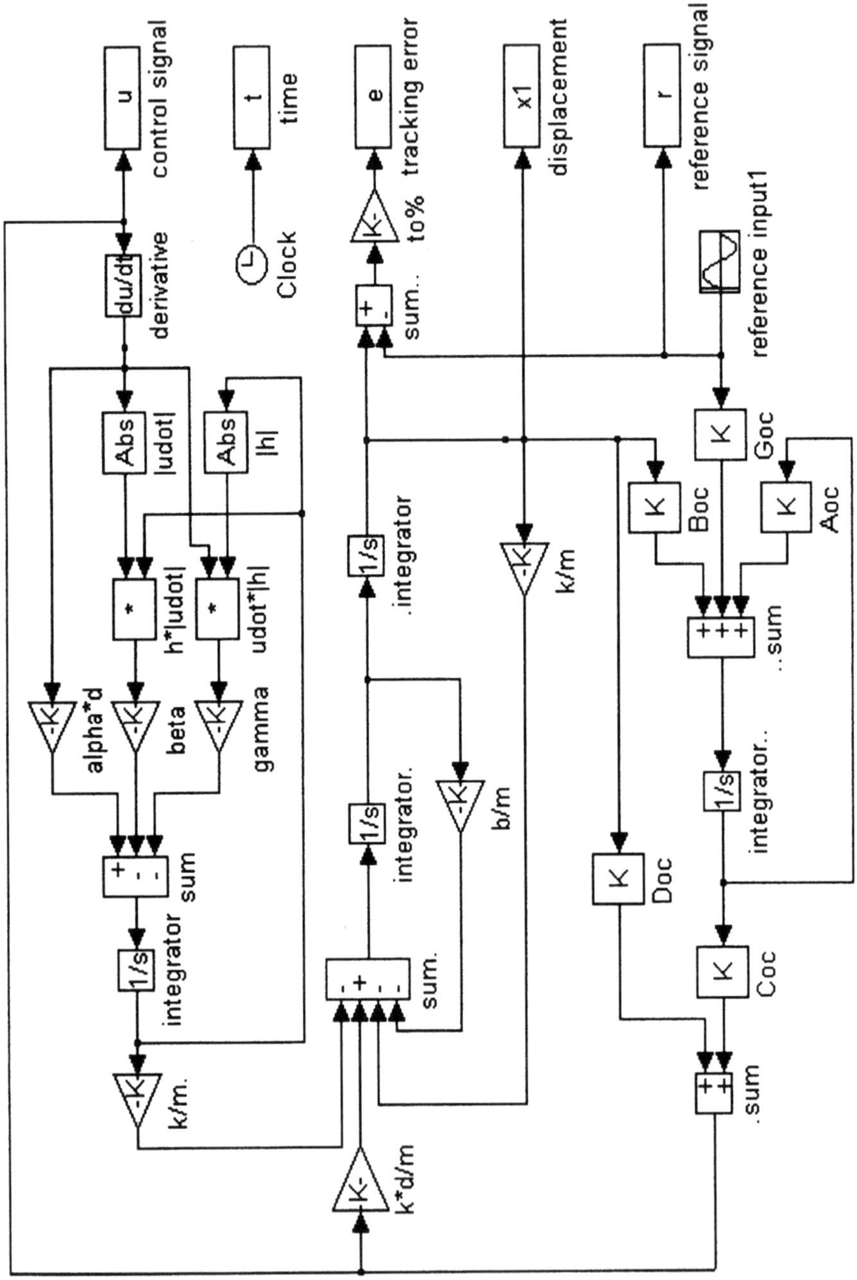

Figure 15.4.3: Simulation block diagram for the overall actuator control system.

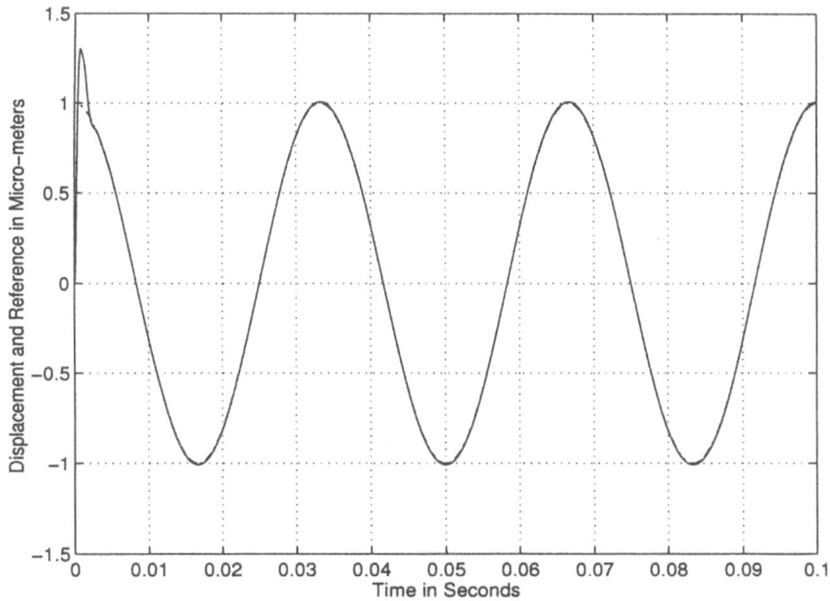

Figure 15.4.4: Responses of the displacement and the 30 Hz cosine reference.

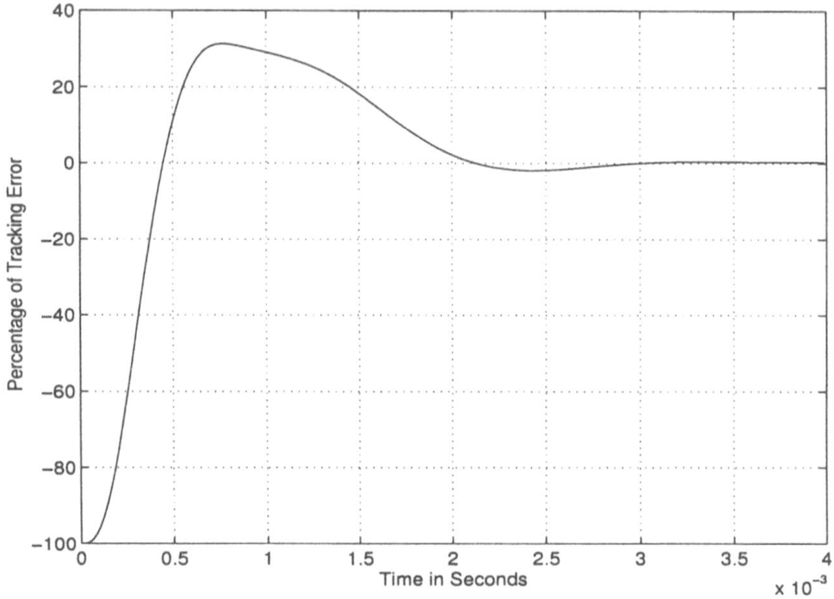

(a) Tracking error from 0 to 0.004 seconds

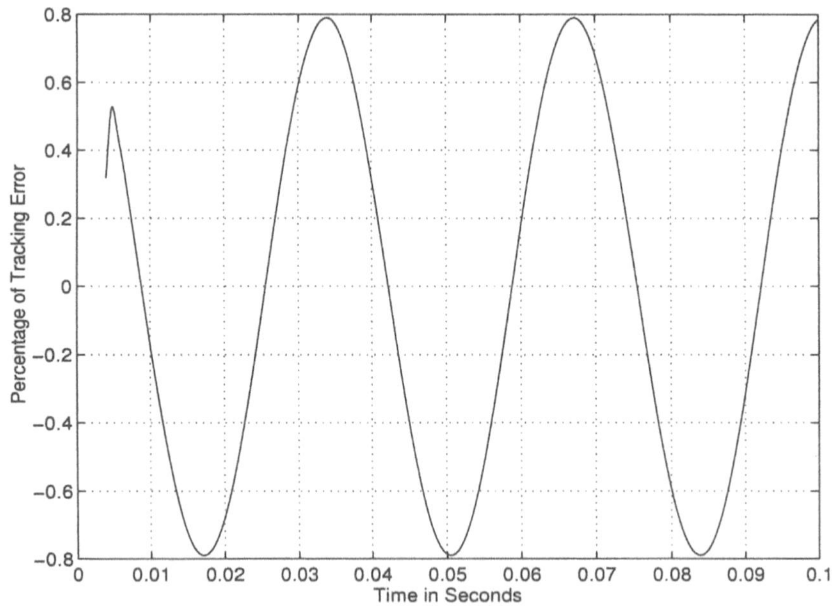

(b) Tracking error from 0.004 to 0.1 seconds

Figure 15.4.5: Tracking error for the 30 Hz cosine reference.

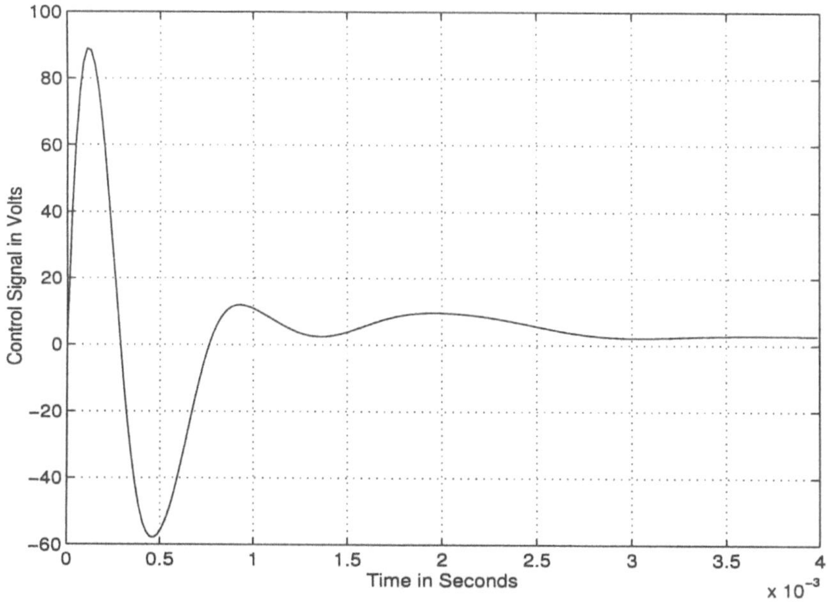

(a) Control signal from 0 to 0.004 seconds

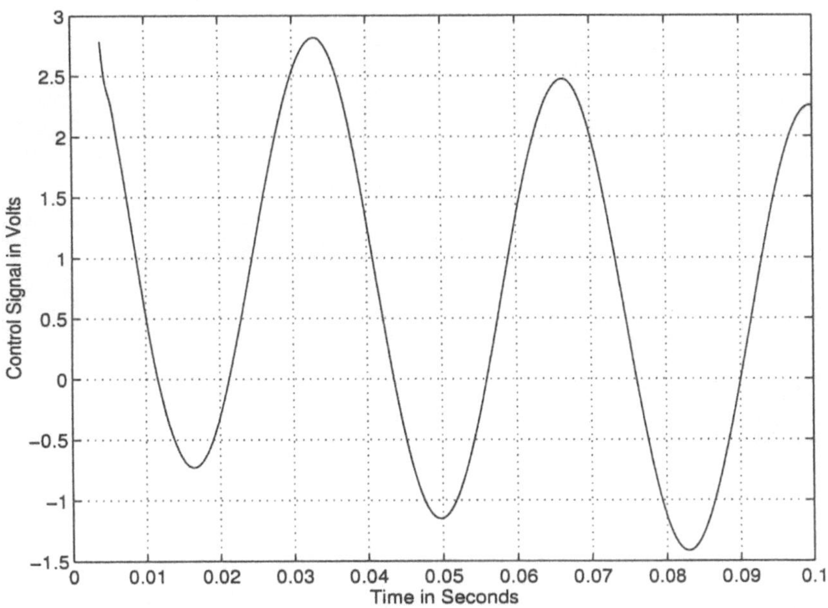

(b) Control signal from 0.004 to 0.1 seconds

Figure 15.4.6: Control signal for the 30 Hz cosine reference.

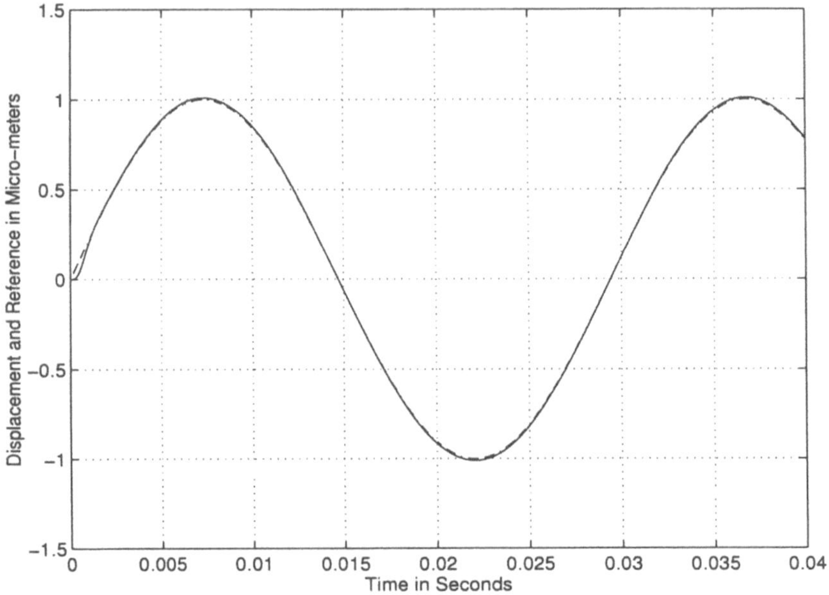

Figure 15.4.7: Responses of the displacement and the 34 Hz sine reference.

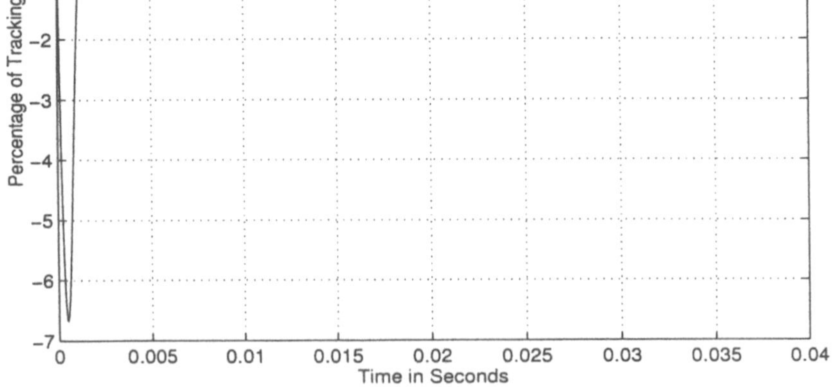

Figure 15.4.8: Tracking error for the 34 Hz sine reference.

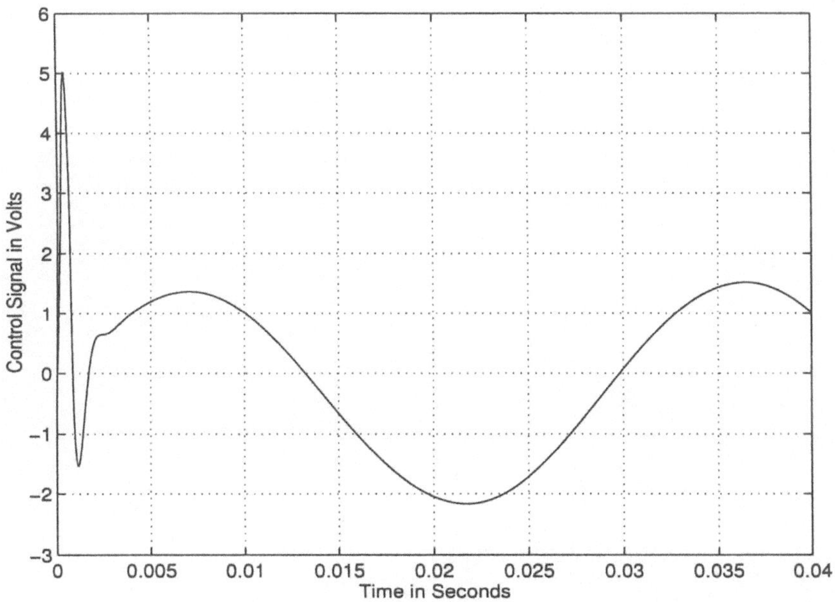

Figure 15.4.9: Control signal for the 34 Hz sine reference.

Chapter 16

Design of a Gyro-stabilized Mirror Targeting System

16.1. Introduction

ELECTRO-OPTICAL (E-O) SENSORS that are mounted on vehicles such as aircraft, helicopter and tanks are subjected to vibrations introduced by these platforms. These vibrations cause the line-of-sight (LOS) of the E-O sensors to shift, resulting in serious degradation of the image quality (see for example [5]). This problem is even more pronounced in systems with high magnification property. One way of overcoming it is to use free gyro stabilization. A gyroscope or gyro is basically an axially symmetrical mass rotating at a high constant speed. With the magnitude of the angular inertia and the speed of rotation both kept constant, the momentum generated is also fixed. Bearing in mind that the momentum is a vector quantity, this implies that the directional orientation is maintained. Therefore, in the absence of large external forces, a gyro is capable of maintaining the orientation of its spin axis in the inertia space. By choosing an appropriate high value for the speed of rotation, the vibrational torque produced by the platforms can be made insignificant as compared to the momentum generated. The LOS can thus be stabilized by simply designing a system such that the LOS and the gyro spin axis are parallel in space. However, a spinning gyro has another property known as precession. This means that if a torque is applied to one axis, it will, contrary to the intuitions of mechanics, rotate in the direction of another axis [102]. Thus, to enable for changes in the space orientation of the LOS, a gyro with at least two degrees of freedom is needed. This property also poses a problem in controlling the LOS because movement about one axis will cause a coupled movement in the other.

Therefore a controller has to be designed to provide the correct slewing (i.e., the application of a calculated torque to cause a desired precession).

In this chapter, we consider a multivariable servomechanism gyro-stabilized mirror system. More specifically, it is a two-input-and-two-output system. The control of this multiple-input-multiple-output system is not a simple problem solved by using conventional PID controllers, because there exist cross-coupling interactions between the dynamics of the two axes. In addition, the controller has to maintain stable operation even when there are changes in the system dynamics. Over the years, many researchers have worked on this system, and the control methodologies studied include adaptive with feedforward paradigm (see e.g., [76]), neural network control (see e.g., [56]) and fuzzy logic (see e.g., [75]). Unfortunately, the controllers obtained using these techniques, except the one of [56], are in general too complicated to be implemented in the real system. Here we tackle this problem using a robust and perfect tracking (RPT) approach to design a simple and low order controller such that the overall closed-loop system would have fast tracking and good robustness performance. The work of this chapter was originally reported in a recent work of Siew, Chen and Lee [121]. The outline of this chapter is as follows: In Section 16.2, the mechanical setup of the free gyro-stabilized mirror system as well as its dynamical equations are given. This is followed by Section 16.3 where we formulate our controller design into a robust and perfect tracking control problem. A technique based on the so-called asymptotic time-scale and eigenstructure assignment (ATEA) of Chapter 9 is then used to solve the proposed problem. Section 16.4 presents the simulation and implementation studies of our overall design. The results of both studies clearly show that all the design specifications are met and the overall performance is very satisfactory.

16.2. The Free Gyro-stabilized Mirror System

This section aims to give a brief overview of the hardware used in the whole free gyro-stabilized mirror system. The whole system consists of four main parts: a) a gyro mirror; b) a system interface assembly; c) a data acquisition board; and d) a personal computer. The overall hardware setup was pictured in Figure 16.2.1. In what follows, we give some brief descriptions of these four hardware parts.

A. The Gyro Mirror

The most crucial part of the free gyro-stabilized mirror system is naturally the gyro-mirror itself. Figure 16.2.2 is a schematic diagram of the gyro mirror. It

Figure 16.2.1: A gyro-stabilized mirror system.

consists of the following essential components: i) a flywheel and its spin motor; ii) gimbals that provide two degrees of freedom to the flywheel and two torque motors for slewing purposes; and iii) a mirror that is geared to the gimbals through a 2 : 1 reduction drive mechanism.

Because no rigid body can spin forever, a piece of a pancake spin motor (flywheel) is used as the gyroscope (gyro). By adjusting the input torque, the flywheel can be made to spin at a high constant velocity about its *spin axis* (Axis 3 in Figure 16.2.2). The flywheel is mounted on an inner gimbal so that it can rotate freely up and down. This axis of rotation is called the *pitch axis* and corresponds to Axis 2 in Figure 16.2.2. The inner gimbal is in turn mounted on an outer gimbal, which provides another axis of freedom (the *yaw axis* or Axis 1) which moves left and right. Note that with these three axes being orthogonal to each other, the system line-of-sight (LOS) can be made parallel to Axis 3 by aligning the mirror axis to the pitch axis.

A torque motor is attached to each of the inner and outer gimbals. These torque motors move the gyro either in the yaw or in the pitch direction, and are thus named the yaw and the pitch motors, respectively. Providing appropriate torque through these motors causes the system to precess relative to the inertia space to achieve some desired line-of-sight (LOS). Removing these input torques stabilizes the LOS in its new position. The angular positions about which the yaw and the pitch axes are defined as θ_1 and θ_2, respectively. θ_1 and

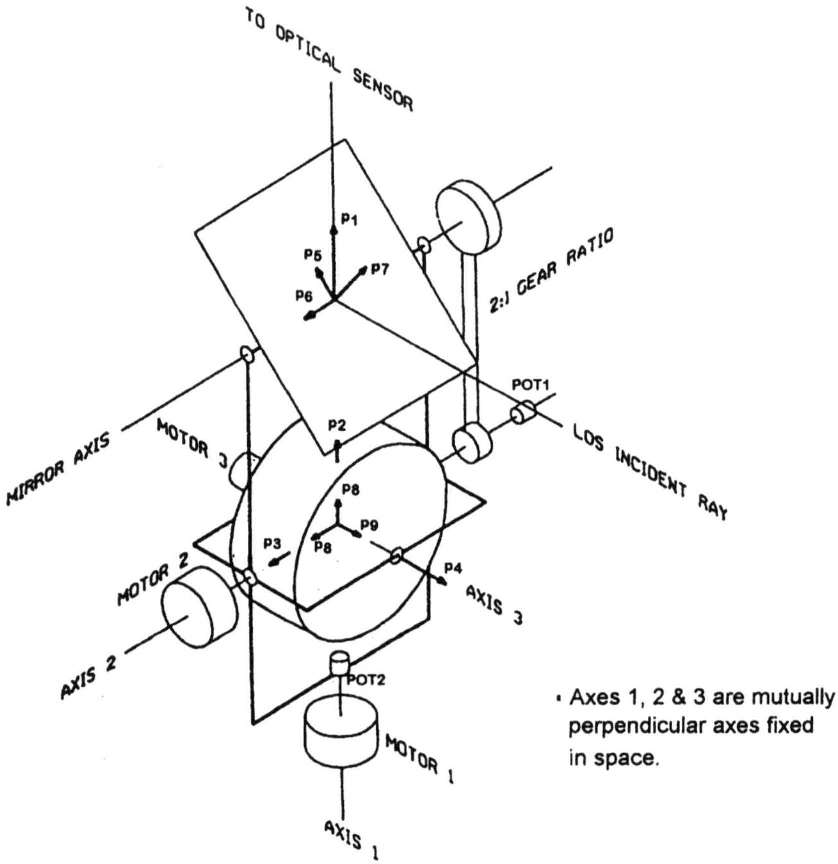

Motor 1 & Motor 2	-	Torque Motor
Motor 3	-	Spin Motor
POT1 & POT2	-	Potentiometers

Figure 16.2.2: Schematic diagram of the gyroscope mirror.

θ_2 can be measured through potentiometers mounted on the inner and outer gimbals. There are however, no velocity sensor to sense $\dot{\theta}_1$ and $\dot{\theta}_2$. Due to physical constrains, the workspace for the gyro-stabilized mirror is limited to $-50° \leq \theta_1 \leq 50°$ and $-30° \leq \theta_2 \leq 30°$. Also, the maximum torques for both yaw and pitch motors are physically limited to a range from -0.5Nm to 0.5Nm.

In this particular system, a mirror is used in place of the actual electro-optical (E-O) sensors. The advantage of doing this is that the E-O sensors will not form an integral part of the system. Therefore any E-O sensor can be used without affecting the system dynamics. The mirror is connected to the flywheel-gimbal structure via a 2 : 1 reduction drive. This 2 : 1 reduction drive is required because when the mirror is tilted by an angle α, the reflected LOS is rotated by 2α.

The dynamical equations of the gyro mirror were developed by applying the well-known Lagrange's motion equation [96]:

$$M_1(\theta)\ddot{\theta}_1 + H_1(\theta,\dot{\theta}) + G_1(\theta,\dot{\theta},\dot{\theta}_3) = u_1, \tag{16.2.1}$$

$$M_2(\theta)\ddot{\theta}_2 + H_2(\theta,\dot{\theta}) + G_2(\theta,\dot{\theta},\dot{\theta}_3) = u_2, \tag{16.2.2}$$

where $\theta = (\theta_1, \theta_2)'$; u_1 and u_2 are the actuator torques for the yaw and the pitch axes; $\dot{\theta}_3$ is the spin velocity of the flywheel. The parameters in equations (16.2.1)-(16.2.2) are defined as follows:

$$M_1 = \bar{a}+\bar{d}+(\bar{b}-\bar{d}+\bar{\ell})\cos^2\theta_2+\frac{1}{2}(\bar{e}+\bar{g})+\frac{1}{2}(\bar{e}-\bar{g})\sin\theta_2, \tag{16.2.3}$$

$$H_1 = -(\bar{b}-\bar{d}+\bar{\ell})\dot{\theta}_1\dot{\theta}_2\sin 2\theta_2+\frac{1}{2}(\bar{e}-\bar{g})\dot{\theta}_1\dot{\theta}_2\cos\theta_2+\bar{k}\dot{\theta}_1\dot{\theta}_2\sin\theta_2\cos\theta_2, \tag{16.2.4}$$

$$G_1 = \bar{k}\dot{\theta}_2\dot{\theta}_3\cos\theta_2, \tag{16.2.5}$$

$$M_2 = \bar{c}+\frac{\bar{f}}{4}+\bar{\ell}, \tag{16.2.6}$$

$$H_2 = \frac{1}{2}(\bar{b}-\bar{d}+\bar{\ell})\dot{\theta}_1^2\sin 2\theta_2-\frac{1}{4}(\bar{e}-\bar{g})\dot{\theta}_1^2\cos\theta_2-\bar{k}\dot{\theta}_1^2\sin\theta_2\cos\theta_2, \tag{16.2.7}$$

$$G_2 = -\bar{k}\dot{\theta}_1\dot{\theta}_3\cos\theta_2, \tag{16.2.8}$$

where $\bar{a}$, $\bar{b}$, $\bar{c}$, $\bar{d}$, $\bar{e}$, $\bar{f}$, $\bar{g}$, $\bar{\ell}$ and $\bar{k}$ are all physical constants representing the various moment of the inertia of the system. These constants were identified earlier by [96] and [75], and took on the following values:

$$\bar{a} = 0.004, \quad \bar{b} = 0.00128, \quad \bar{c} = 0.00098, \quad \bar{d} = 0.02, \tag{16.2.9}$$

$$\bar{e} = 0.0049, \quad \bar{f} = 0.0025, \quad \bar{g} = 0.00125, \quad \bar{\ell} = 0.0032, \quad \bar{k} = 0.0025. \tag{16.2.10}$$

The above parameters all have units of kg·m^2. As can be seen from the above equations, the system is highly nonlinear and there exist cross-coupling terms between the yaw and the pitch axes.

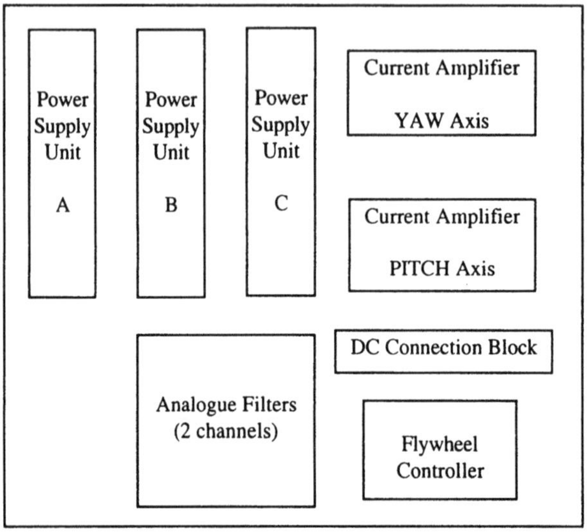

Figure 16.2.3: System interface assembly layout.

B. The System Interface Assembly

The torque motors and position sensors on the gyro-mirror have to be connected to a data acquisition board on the personal computer. This is accomplished via the system interface assembly. Figure 16.2.3 shows the layout of the components assembled in this platform.

1. POWER SUPPLIES. The power supply Units A and B are of single 28V DC regulated type. They are connected in series to give a -24V - 0V - $+24$V DC supply. This combined power unit supplies all the currents required by the torque motors, the position sensors and the analogue filters. Unit C is rated 24V DC, which is used to drive the flywheel controller.
2. FLYWHEEL CONTROLLER. This is a dedicated driver unit (model MCH20-20-002CL). It provides adjustable speed control to the spin motor via a potentiometer. The spin velocity ranges from 0 up to around 5000rpm.
3. CURRENT AMPLIFIERS. There are two current amplifiers, one for the yaw motor and the other for the pitch motor. The inputs of the amplifiers are connected directly to the D/A outputs of the AD/DA card, and their outputs are connected to the torque motors. They are built using a power operational amplifier with the outputs ranging from -25V to $+25$V. These outputs will produce the corresponding motor torques ranging respectively from -0.5Nm to 0.5Nm.

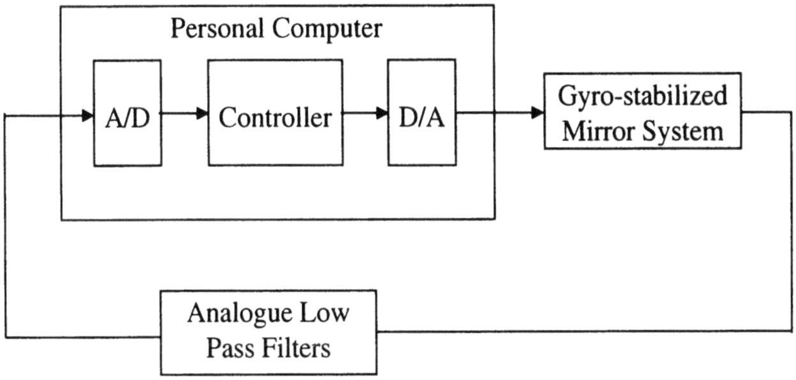

Figure 16.2.4: Block diagram of experimental setup.

4. ANALOGUE FILTERS. The position signals from the potentiometers are first passed through these filters before being connected to the A/D inputs of the AD/DA card. They are low pass filters with cutoff frequency at 19Hz so as to reject high frequency noises.

C. The Data Acquisition Board

The analog-digital and digital-analog (AD/DA) card used in our implementation has two analog input channels and two analog output channels. The analog inputs are the filtered position signals of the yaw and the pitch axes while the analog outputs are the torques to control the motors. The signals in all channels range from −10V to +10V DC, with a 12 bit accuracy.

D. The Personal Computer

The controller is implemented on a personal computer via an AD/DA card mounted within. The block diagram of the experimental setup is given in Figure 16.2.4. The personal computer configuration is an IBM PC compatible with an Intel Pentium 75 Processor and an MS-DOS 6.0 Operating System.

16.3. Controller Design Using the RPT Approach

In this section, we formulate our controller design for the free gyro-stabilized mirror system as a robust and perfect tracking (RPT) control problem and then use the design method of Chapter 9 to carry out the design of the controller. Our goal is to design a simple and low order controller as structured in Figure 16.3.1 such that the overall system will meet the following specifications:

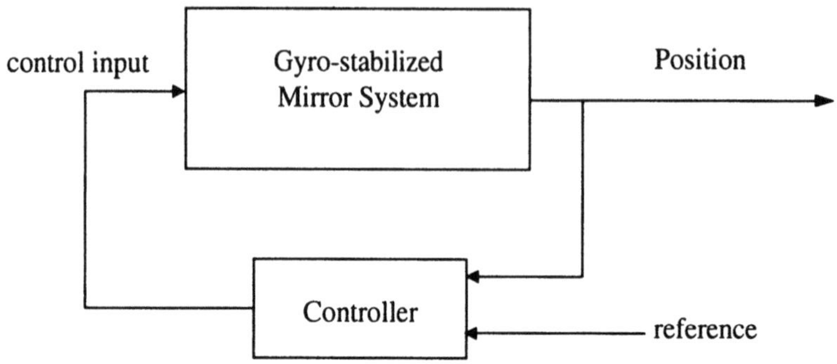

Figure 16.3.1: Structure of control system for gyro-stabilized mirror system.

1. The closed-loop system should have fast tracking in both the yaw and the pitch axes for step input commands with small or no overshoot;

2. The cross-coupling interactions between the yaw and the pitch axes should be minimized; and

3. The overall system should be robust to external disturbances and changes in system parameters.

As will be seen shortly, our controller is very simple and has low order. Thus, it can easily be implemented using low speed personal computers and A/D and D/A cards.

First of all, we need to linearize the dynamical model given in equations (16.2.1)-(16.2.2) and cast it into the standard state space form. The linearized state space model is given as follows:

$$\dot{x}_g = A_g x_g + B_g u + E_g w_g, \tag{16.3.1}$$

where

$$x_g = \begin{pmatrix} \theta_1 \\ \dot{\theta}_1 \\ \theta_2 \\ \dot{\theta}_2 \end{pmatrix} \quad \text{and} \quad u = \begin{pmatrix} u_1 \\ u_2 \end{pmatrix}, \tag{16.3.2}$$

and $w_g \in L_2$ is the viscous damping coefficients for the system, which can be regarded as disturbances. The matrices A_g, B_g and E_g are given by

$$A_g = \begin{bmatrix} 0 & 1 & 0 & 0 \\ 0 & 0 & 0 & -\bar{k}\dot{\theta}_3/N_1 \\ 0 & 0 & 0 & 1 \\ 0 & \bar{k}\dot{\theta}_3/N_2 & 0 & 0 \end{bmatrix}, \tag{16.3.3}$$

and

$$B_g = \begin{bmatrix} 0 & 0 \\ 1/N_1 & 0 \\ 0 & 0 \\ 0 & 1/N_2 \end{bmatrix}, \quad E_g = \begin{bmatrix} 0 & 0 \\ -1 & 0 \\ 0 & 0 \\ 0 & -1 \end{bmatrix}, \tag{16.3.4}$$

where

$$N_1 = \bar{a} + \bar{b} + \frac{\bar{e} + \bar{g}}{2} + \bar{\ell}, \quad N_2 = \bar{c} + \frac{\bar{f}}{4} + \bar{\ell}. \tag{16.3.5}$$

The measurement output of the free gyro-stabilized mirror system is

$$\theta = \begin{pmatrix} \theta_1 \\ \theta_2 \end{pmatrix}. \tag{16.3.6}$$

Since we are interested in the changes in the orientation of the LOS, we focus only on the case where the command input $r(t)$ is a step function. To be more specific, we consider

$$r(t) = \begin{bmatrix} r_1(t) \\ r_2(t) \end{bmatrix} = \begin{bmatrix} \psi_1 \\ \psi_2 \end{bmatrix} 1(t) = \Psi \cdot 1(t), \tag{16.3.7}$$

where $1(t)$ is the unit step function, and ψ_1, ψ_2 are some constants. Then, we have

$$\dot{r}(t) = \begin{bmatrix} \dot{r}_1(t) \\ \dot{r}_2(t) \end{bmatrix} = \begin{bmatrix} \psi_1 \\ \psi_2 \end{bmatrix} \delta(t) = \Psi \cdot \delta(t), \tag{16.3.8}$$

where $\delta(t)$ is the unit impulse function. Let us define a controlled output h as the difference between the actual output θ and the command input r, i.e.,

$$e = \theta - r = \begin{pmatrix} \theta_1 - r_1 \\ \theta_2 - r_2 \end{pmatrix}. \tag{16.3.9}$$

Obviously, e is simply the tracking error. Finally, we obtain the following auxiliary system,

$$\Sigma \ : \ \begin{cases} \dot{\boldsymbol{x}} = \boldsymbol{A} \ \boldsymbol{x} + \boldsymbol{B} \ \boldsymbol{u} + \boldsymbol{E} \ \boldsymbol{w}, \\ \boldsymbol{y} = \boldsymbol{C}_1 \ \boldsymbol{x} \qquad\qquad + \boldsymbol{D}_1 \ \boldsymbol{w}, \\ e = \boldsymbol{C}_2 \ \boldsymbol{x} + \boldsymbol{D}_2 \ \boldsymbol{u}\ , \end{cases} \tag{16.3.10}$$

with

$$\boldsymbol{x} = \begin{pmatrix} r \\ x_g \end{pmatrix} = \begin{pmatrix} r_1 \\ r_2 \\ \theta_1 \\ \dot{\theta}_1 \\ \theta_2 \\ \dot{\theta}_2 \end{pmatrix}, \quad u = \begin{pmatrix} u_1 \\ u_2 \end{pmatrix}, \quad \boldsymbol{w} = \begin{pmatrix} w_g \\ \delta(t) \end{pmatrix}, \quad \boldsymbol{y} = \begin{pmatrix} r_1 \\ r_2 \\ \theta_1 \\ \theta_2 \end{pmatrix}, \tag{16.3.11}$$

$$\boldsymbol{A} = \begin{bmatrix} 0 & 0 \\ 0 & A_g \end{bmatrix}, \quad \boldsymbol{B} = \begin{bmatrix} 0 \\ B_g \end{bmatrix}, \quad \boldsymbol{E} = \begin{bmatrix} 0 & \Psi \\ E_g & 0 \end{bmatrix}, \tag{16.3.12}$$

and

$$\boldsymbol{C}_1 = \begin{bmatrix} I_2 & 0 \\ 0 & C_1 \end{bmatrix} = \begin{bmatrix} 1 & 0 & 0 & 0 & 0 & 0 \\ 0 & 1 & 0 & 0 & 0 & 0 \\ 0 & 0 & 1 & 0 & 0 & 0 \\ 0 & 0 & 0 & 0 & 1 & 0 \end{bmatrix}, \quad \boldsymbol{D}_1 = \begin{bmatrix} 0 \\ D_1 \end{bmatrix} = 0, \tag{16.3.13}$$

and

$$\boldsymbol{C}_2 = [\,-I_2 \quad C_2\,] = \begin{bmatrix} -1 & 0 & 1 & 0 & 0 & 0 \\ 0 & -1 & 0 & 0 & 1 & 0 \end{bmatrix}, \quad \boldsymbol{D}_2 = D_2 = 0. \tag{16.3.14}$$

It is simple to show now that i) the subsystem (A_g, B_g, C_2, D_2) is invertible with two infinite zeros of order 2, and ii) $\mathrm{Ker}\,(C_2) = \mathrm{Ker}\,(C_1)$. It follows from the results of Chapter 9, one can show that a robust and perfect tracking performance can be achieved for such a system, i.e., there exists a family of measurement feedback control law of the form,

$$\Sigma_{\mathrm{cmp}} : \begin{cases} \dot{v} = A_{\mathrm{cmp}}(\varepsilon)\; v + B_{\mathrm{cmp}}(\varepsilon)\; \boldsymbol{y}, \\ u = C_{\mathrm{cmp}}(\varepsilon)\; v + D_{\mathrm{cmp}}(\varepsilon)\; \boldsymbol{y}, \end{cases} \tag{16.3.15}$$

such that when it is applied to the gyro-stabilized mirror targeting system,

1. The resulting closed-loop system is asymptotically stable for sufficiently small ε; and

2. The resulting tracking error e which is of course depended on ε has the property, $\|e\|_2 \to 0$, as $\varepsilon \to 0$, for any initial condition x_0 and any disturbance $w \in L_2$.

Thus, in principle, $\theta(t)$ is capable of tracking the command $r(t)$ perfectly with no overshoot and with no time. Of course, the price one needs to pay for this kind of excellent performance is that the control input must be unlimited, i.e., using infinite gains. This is not possible in the real world. As mentioned earlier, the control inputs u_1 and u_2 of our problem are actually bounded from -0.5Nm to 0.5Nm. Therefore, a trade-off is needed.

Using the result of Chapter 9, one can either design a full order observer based controller or a reduced order observer based controller to solve the above problem. For the full order observer based controller, the order of the controller will be 6. On the other hand, a reduced order observer based controller will have an order of 2 since we only need to reconstruct the velocity states. Therefore from the practical point of view, a reduced order observer based controller is more desirable. We separate our controller design into the following two steps:

1. In the first step, we assume that all six states of Σ in (16.3.10) are available and then design a static state feedback control law,

$$u = \boldsymbol{F}\boldsymbol{x}, \tag{16.3.16}$$

such that the closed-loop system has desired properties.

2. In the second step, we design a reduced order observer based controller. It has a reduced order observer gain matrix $K_{\rm R}$ that can recover the performance achieved by the state feedback control law in the first step.

Using the m-function `atea.m` of the toolbox [14] and after a few iterations, we obtained the following state feedback gain:

$$\boldsymbol{F} = -\begin{bmatrix} -2.3732 & -1.4264 & 2.3732 & 1.0271 & 1.4264 & 0.0000 \\ 1.4264 & -2.3732 & -1.4264 & 0.0000 & 2.3732 & 1.0113 \end{bmatrix}. \tag{16.3.17}$$

Simulation result showed that the performance of the closed-loop system with the above state feedback law is quite satisfactory. Next, we follow the algorithm given in Chapter 9 and obtain a second order measurement feedback law of the form (16.3.15) with

$$A_{\rm cmp} = \begin{bmatrix} -174.3280 & -74.2370 \\ 106.2743 & -332.7939 \end{bmatrix}, \tag{16.3.18}$$

$$B_{\rm cmp} = \begin{bmatrix} 1.0269 & 0.6172 & -83.3798 & -64.5160 \\ -1.4843 & 2.4695 & 11.5772 & -194.7265 \end{bmatrix}, \tag{16.3.19}$$

$$C_{\rm cmp} = \begin{bmatrix} -205.4112 & 0 \\ 0 & -202.2678 \end{bmatrix}, \tag{16.3.20}$$

$$D_{\rm cmp} = \begin{bmatrix} 2.3732 & 1.4264 & -90.1288 & -23.2207 \\ -1.4264 & 2.3732 & -20.0343 & -126.0777 \end{bmatrix}. \tag{16.3.21}$$

16.4. Simulation and Implementation Results

In order to implement our controller designed in the previous section using our hardware setup, we need to discretize it. The performance of this discretized controller is then evaluated using MATLAB SIMULINK. Finally, it is applied to the actual free gyro-stabilized mirror system. Using the well-known bilinear transformation (see also Chapter 3) with a sampling time of 4ms, we obtained the following discretized controller,

$$\Sigma_{\rm dcmp} : \begin{cases} v(k+1) = A_{\rm dcmp}\, v(k) + B_{\rm dcmp}\, y(k), \\ u(k) \;\;\; = C_{\rm dcmp}\, v(k) + D_{\rm dcmp}\, y(k), \end{cases} \tag{16.4.1}$$

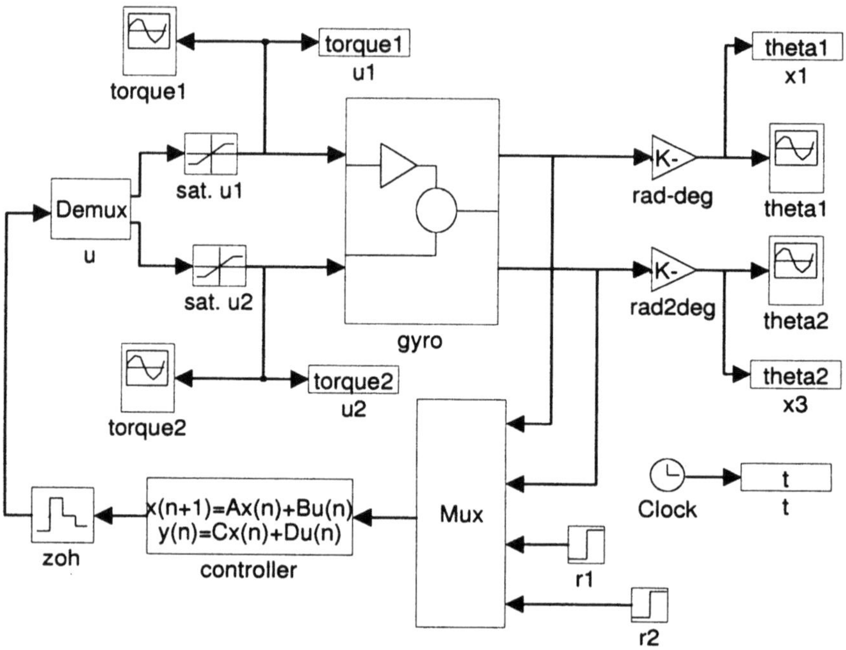

Figure 16.4.1: Simulation block patched up in SIMULINK.

where

$$A_{\text{dcmp}} = \begin{bmatrix} 0.4624 & -0.1304 \\ 0.1866 & 0.1841 \end{bmatrix}, \tag{16.4.2}$$

$$B_{\text{dcmp}} = \begin{bmatrix} 0.8476 & 0.2904 & -61.7225 & -34.4820 \\ -0.7830 & 1.5197 & -0.9257 & -121.3119 \end{bmatrix}, \tag{16.4.3}$$

$$C_{\text{dcmp}} = \begin{bmatrix} -0.6008 & 0.0536 \\ -0.0755 & -0.4790 \end{bmatrix}, \tag{16.4.4}$$

$$D_{\text{dcmp}} = \begin{bmatrix} 2.0249 & 1.3072 & -64.7719 & -9.0547 \\ -1.1097 & 1.7584 & -19.6598 & -77.0027 \end{bmatrix}. \tag{16.4.5}$$

The SIMULINK simulation block diagram for the free gyro-stabilized mirror system is given in Figure 16.4.1 (note that the inputs to the controller in the simulation block diagram are reordered, i.e., r_1 and θ_1, and r_2 and θ_2 are swapped, respectively). In order to achieve more accurate results, the nonlinear model given in equations (16.2.1)-(16.2.2) is used in the *gyro* block. Simulations are carried out using the *Runge-Kutta 5* method with both minimum and maximum step sizes set to be the same as the sampling period, i.e., 4ms. To account for the limitations in the torque motors, a *saturation* block is added to each of them. The limits are set to be ± 0.5Nm. Throughout the simulations, the gyro spin velocity is set to be 2500rpm.

The gyro is first commanded to move simultaneously to (yaw, pitch) = $(5^\circ, -5^\circ)$. On the fifth seconds, it is moved from this new position to $(20^\circ, -20^\circ)$. A horizontal span is then carried out, i.e., the gyro is moved horizontally from 20° to -5° while keeping the pitch position at -20°. This is followed by a vertical span; this time the yaw position is fixed at -5° while the pitch position is changed from -20° to 5°. Finally, it is pushed to its extreme position $(-50^\circ, 30^\circ)$ before returning back to its zero position. The gyro response as well as the torque input to each axis are plotted in Figures 16.4.2-16.4.3.

The various set-points in the above tests are chosen such that from one position to another, the displacement ranges from as small as 5° up to 45°. This is to verify that our controller works well within the whole workspace although it is designed based on a linearized model. The simultaneous movement is to test whether our controller is capable of achieving perfect tracking in both axes while the spans are conducted to investigate how well does our controller 'decouple' the gyro-stabilized mirror system. As can be seen from the responses in Figure 16.4.2, the gyro is able to reach all commanded positions without steady state errors. Furthermore, none of the responses exhibits any overshoot. The settling time from its extreme position back to the zero position is about 3.5 seconds. The maximum coupled movement in θ_1 caused by moving θ_2 is around 0.15°. The maximum coupled movement in θ_2 caused by moving θ_1 is about 0.5°. A check with Figure 16.4.3 shows that all these are accomplished with the torques kept within the constraint of ± 0.5Nm. Thus we conclude that our controller designed in the previous section is very satisfactory.

Next, we implement this controller on the actual free gyro-stabilized mirror system via a computer (see Figure 16.2.1) and perform the whole test once again. The results obtained are shown in Figures 16.4.4-16.4.5.

Comparing Figures 16.4.2-16.4.3 with Figures 16.4.4-16.4.5, we note that the general waveforms are the same. However, there exist steady state errors in both axes. Furthermore, the real system takes a slightly longer time before settling at its set-point. For example, it now takes about 5 seconds instead of 3.5 seconds to move from its extreme position back to zero. The coupled interaction caused by movement in the other axis is also larger than our simulation results (1.6° in the yaw axis and 0.55° in the pitch axis). The performance of the controller during the implementation is clearly not as good as in the simulation. The reason is due to the imperfection of the hardware system.

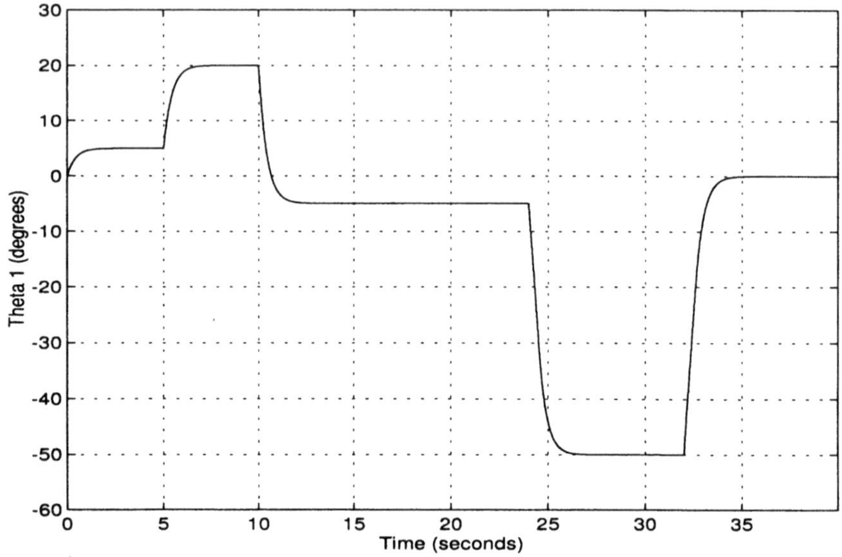

(a) Response for θ_1.

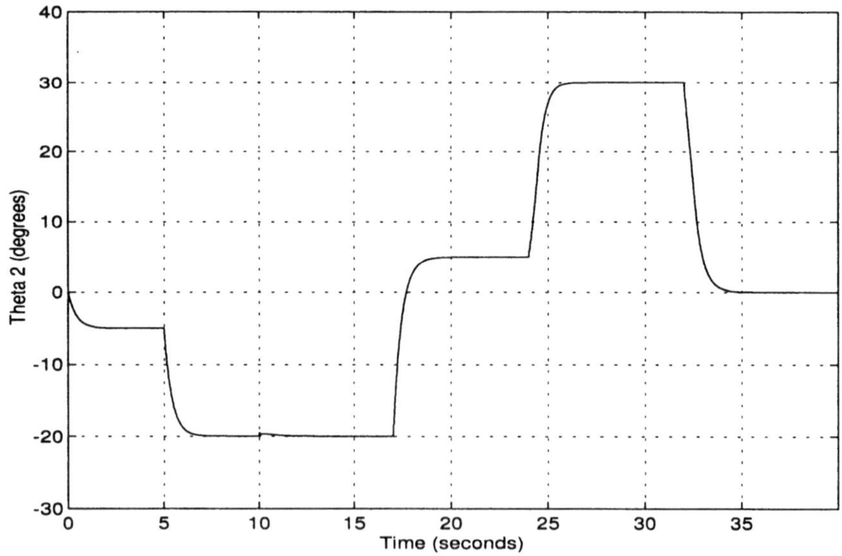

(b) Response for θ_2.

Figure 16.4.2: Simulation result: Responses of θ_1 and θ_2.

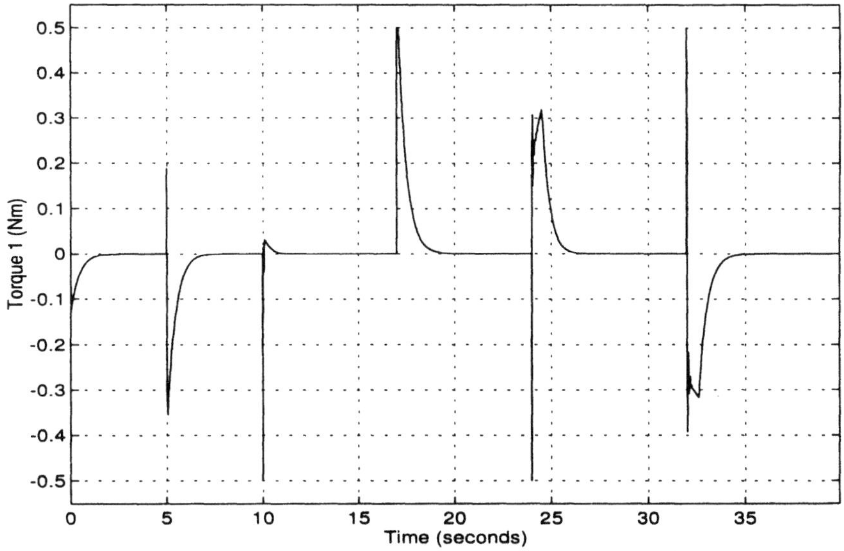

(a) Control input, u_1.

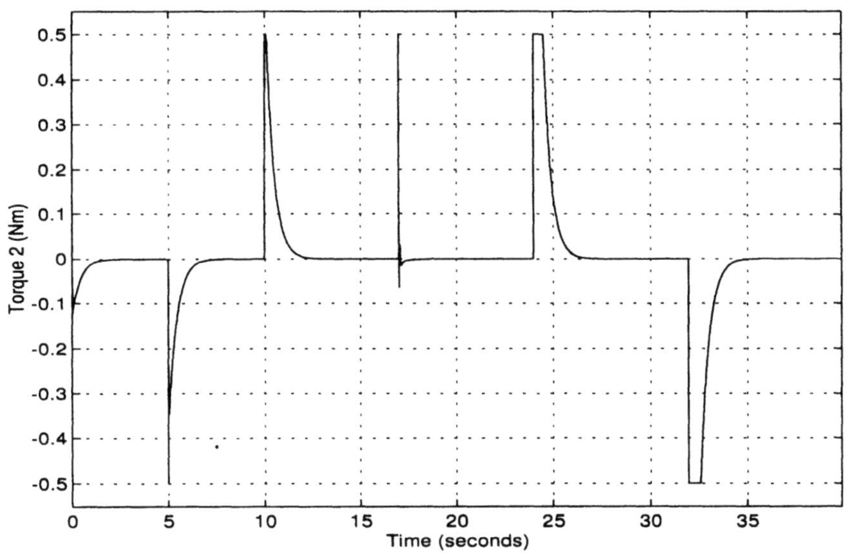

(b) Control input, u_2.

Figure 16.4.3: Simulation result: Control inputs u_1 and u_2.

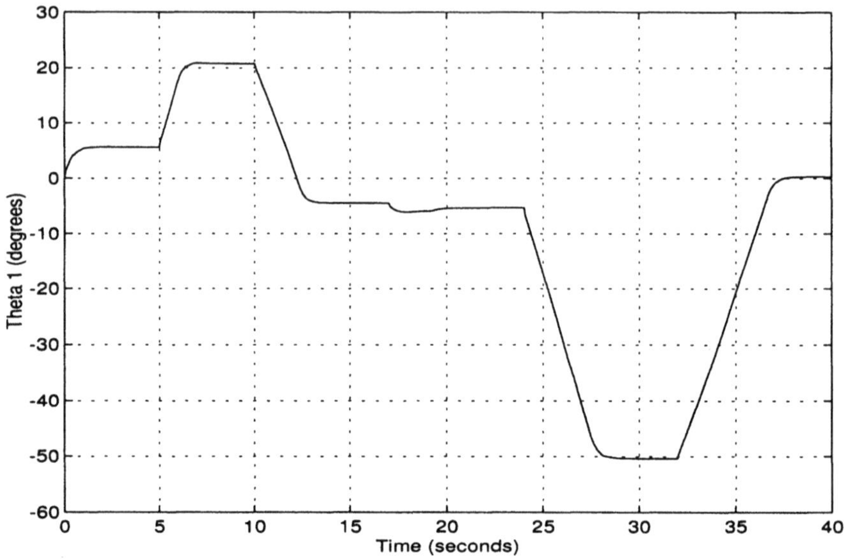

(a) Response for θ_1.

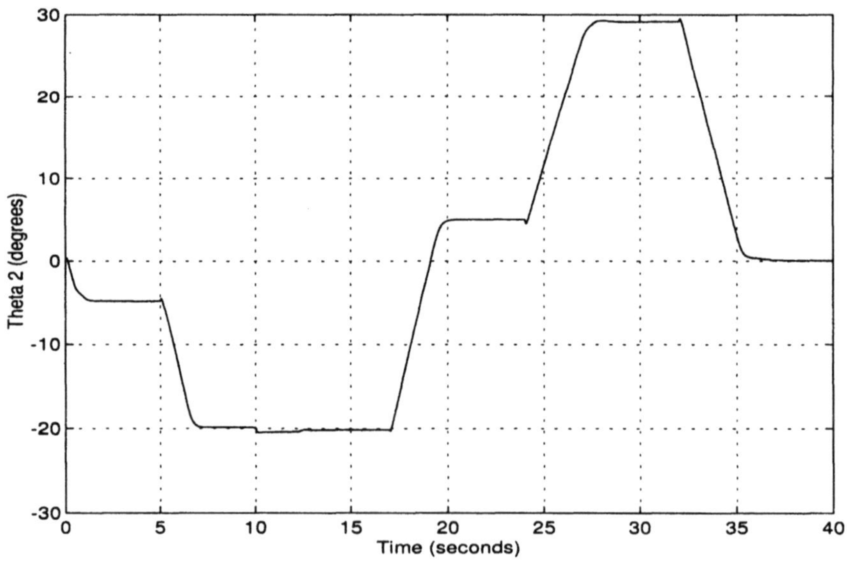

(b) Response for θ_2.

Figure 16.4.4: Implementation result: Responses of θ_1 and θ_2.

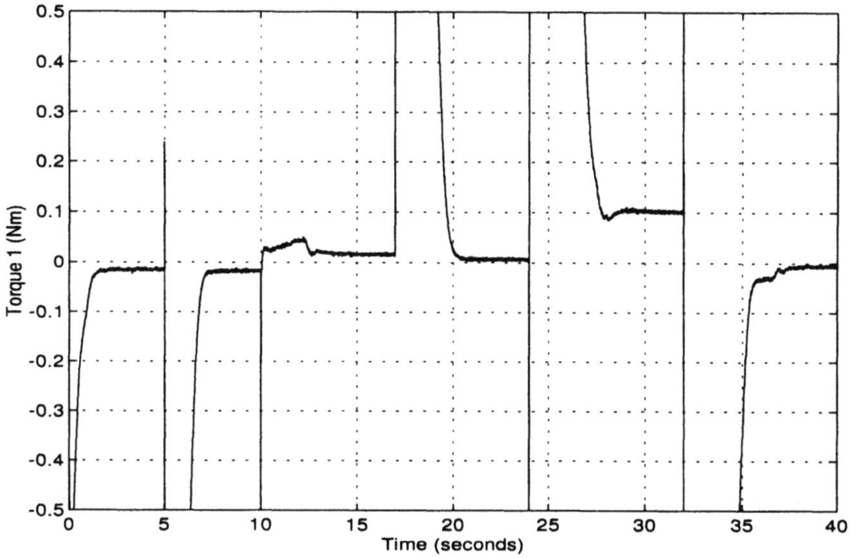

(a) Control input, u_1.

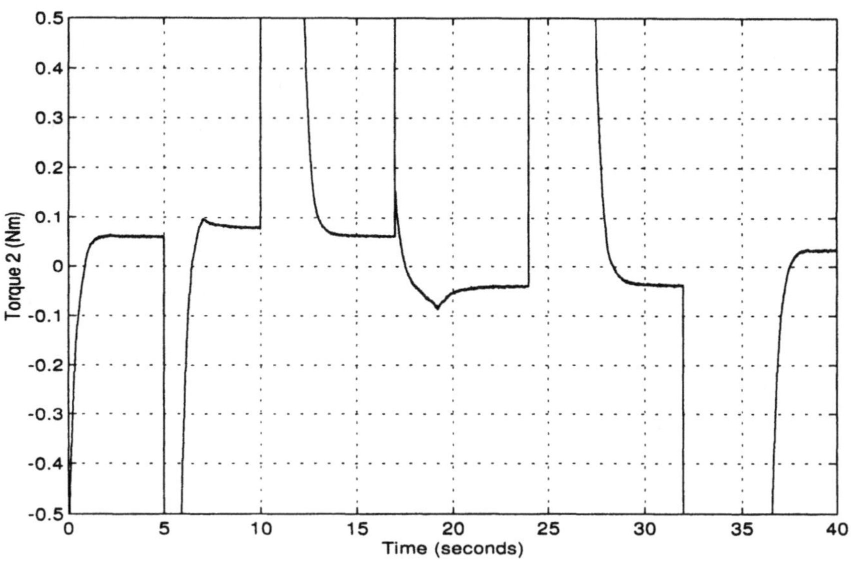

(b) Control input, u_2.

Figure 16.4.5: Implementation result: Control inputs u_1 and u_2.

The biggest defect that the system has may be the dead zones of the torque motors. Studying Figure 16.4.5, we observe that although the torques are still nonzero, the positions have already reached their steady states. This can only happen if the torque motors are working within their dead zones. In fact, after running a few tests, we find that the dead zone in the pitch motor is more pronounced and it does not remain constant throughout operation. According to one past documentation (see e.g., [75]), *the dead zone is related to the mechanical vibration on the gyro-mirror. In situations when the gyro-mirror vibrates, the vibrations cause the system to 'loosen up' and result in a small dead zone; at other times when the gyro-mirror is stabilized and spinning smoothly, a large dead zone exists.* This behavior makes the dead zone compensation extremely difficult. Nevertheless through trial and error, we observe that the magnitude of the dead zone compensation seems to be related to the set-points in the following way:

$$u_{os1} = \alpha_1 r_1 + \beta_1 r_2, \tag{16.4.6}$$

and

$$u_{os2} = \alpha_2 r_1 + \beta_2 r_2, \tag{16.4.7}$$

where u_{os1} and u_{os2} are the values to be added to u_1 and u_2, respectively. Various sets of (r_1, r_2) are used to tune α_1, α_2, β_1 and β_2 so as to obtain suitable offsets to be added to the control inputs such that the dead zone effects can be minimized. Figures 16.4.6-16.4.7 are the results we obtain from our controller with a dead zone compensation whose parameters are chosen as follows:

$$\alpha_1 = -0.001125, \quad \beta_1 = -0.000125, \quad \alpha_2 = -0.0049875, \quad \beta_2 = -0.00059375. \tag{16.4.8}$$

With these results, we once again show that our controller is able to perform fast tracking without overshoot in both axes and minimize the coupled effect (0.8° in the yaw axis and 0.5° in the pitch axis).

In order to test the robustness of this controller, we send a command to move the gyro simultaneously in the yaw (+20°) and pitch (−20°) direction. Then we purposely introduce some disturbance (through knocking on the gimbals) to the system. As shown in Figure 16.4.8, our controller is robust to this external disturbance.

During implementation, the gyro spin velocity is controlled via a potentiometer. Hence it is very difficult to set an exact speed of rotation. To make things worse, the gyro will vary its spinning velocity by itself. Since the free gyro-stabilized mirror system dynamics are dependent on its spin velocity (see equations (16.2.1)-(16.2.2)), the system dynamics is changed too. Furthermore, the physical constants $\bar{a}$, $\bar{b}$, $\bar{c}$, $\bar{d}$, $\bar{e}$, $\bar{f}$, $\bar{g}$, $\bar{\ell}$ and $\bar{k}$ were obtained from experiments conducted on the free gyro-stabilized mirror system a few years back. Over these years, the free gyro-stabilized mirror system has broken down and has been serviced for many times. Thus, these values may not be accurate anymore. Yet in view of these model uncertainties, the performance of our controller remains very satisfactory.

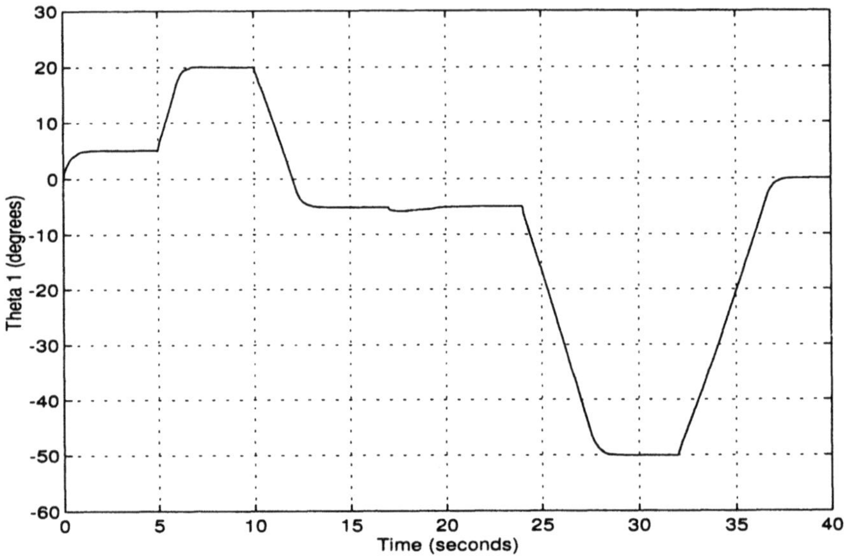

(a) Response for θ_1.

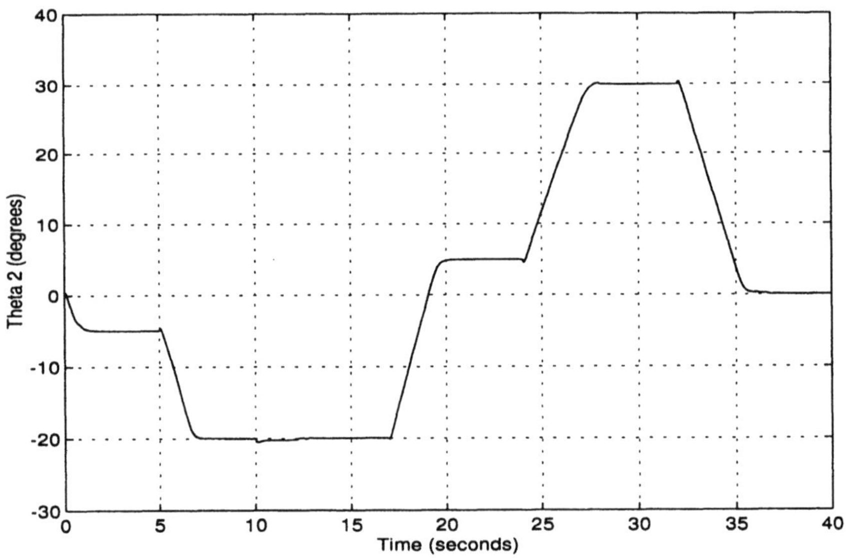

(b) Response for θ_2.

Figure 16.4.6: Implementation with dead zone compensation: θ_1 and θ_2.

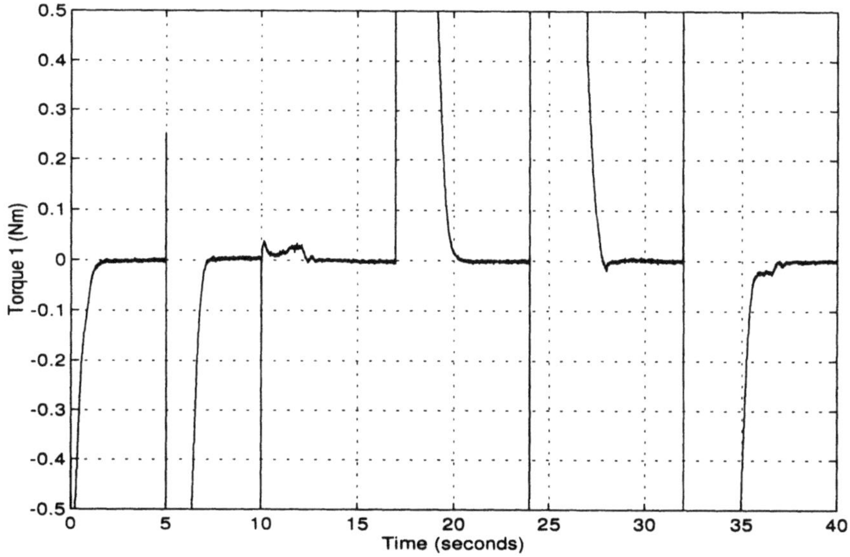

(a) Control input, u_1.

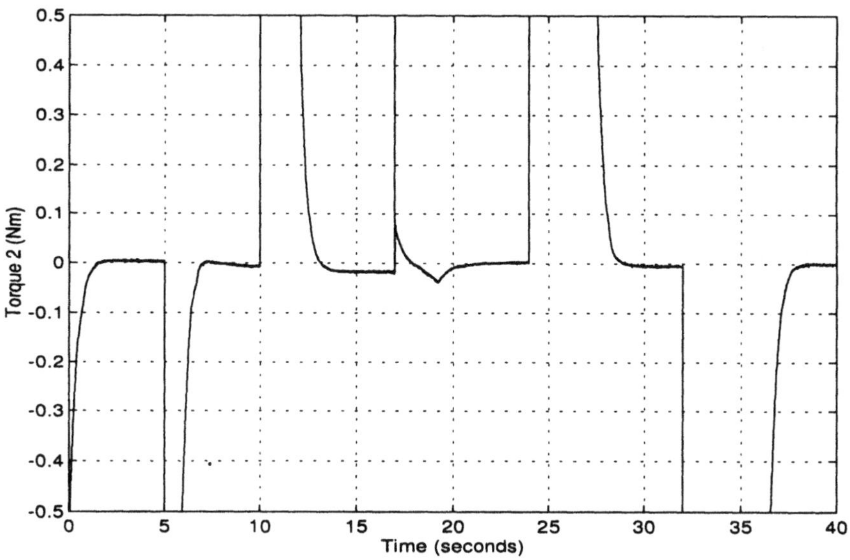

(b) Control input, u_2.

Figure 16.4.7: Implementation with dead zone compensation: u_1 and u_2.

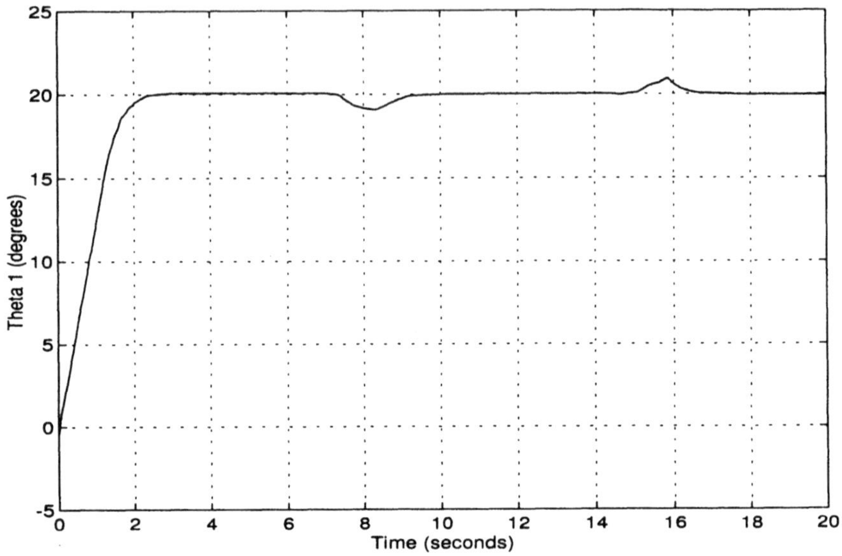

(a) Response for θ_1.

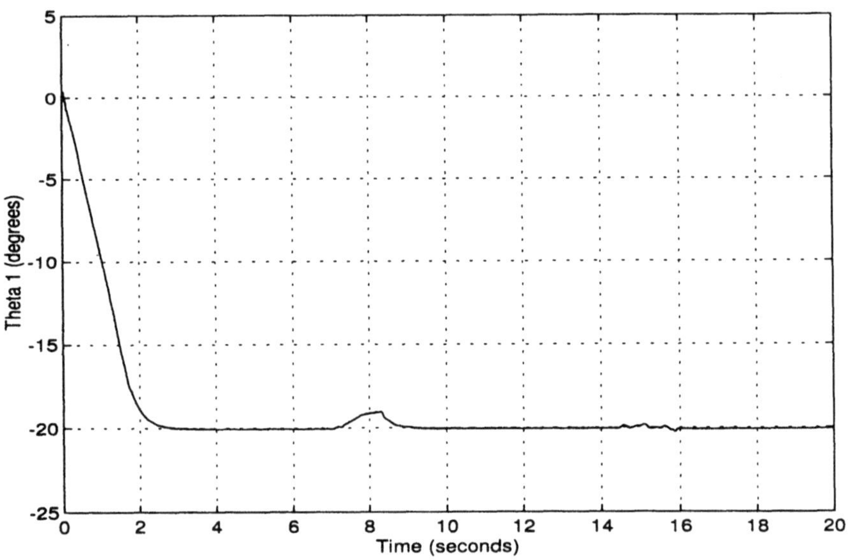

(b) Response for θ_2.

Figure 16.4.8: Robustness test for the final system.

Bibliography

[1] K. J. Aström, P. Hagander and J. Sternby, "Zeros of sampled systems," *Automatica*, Vol. 20, pp. 21–38, 1984.

[2] K. J. Aström and B. Wittenmark, *Computer Controlled Systems: Theory and Design*, Prentice Hall, Englewood Cliffs, 1984.

[3] S. Barnett, *Matrices in Control Theory*, Robert E. Krieger Publishing Company, Malabar, Florida, 1984.

[4] T. Başar and P. Bernhard, *H_∞ Optimal Control and Related Minimax Design Problems: A Dynamic Game Approach*, Second Edition, Birkhäuser, Boston, Massachusetts, 1995.

[5] W. J. Bigley and V. J. Rizzo, "Wideband linear quadratic control of a gyro-stabilized electro-optical sight system," *IEEE Control Systems Magazine*, Vol. 7, pp. 20-24, 1987.

[6] R. C. Booton, Jr., "Nonlinear control systems with random inputs," *IRE Transactions on Circuit Theory*, CT-1, pp. 9-18, 1954.

[7] P. Brunovsky, "A classification of linear controllable systems," *Kybernetika (Praha)*, Vol. 3, pp. 173-187, 1970.

[8] T. K. Caughey, "Derivation and application of the Fokker-Planck equation to discrete nonlinear dynamic systems subjected to white random excitation," *Journal of Acoustical Society of America*, Vol. 35, pp. 1683-1692, 1963.

[9] S. W. Chan, G. C. Goodwin and K. S. Sin, "Convergence properties of the Riccati difference equation in optimal filtering of nonstabilizable systems," *IEEE Transactions on Automatic Control*, Vol. 29, pp. 110-118, 1984.

[10] T. P. Chang, *Seismic Response Analysis of Nonlinear Structures Using the Stochastic Equivalent Linearization Technique*, Ph.D. Dissertation, Columbia University, 1985.

[11] B. M. Chen, *Software Manual for the Special Coordinate Basis of Multivariable Linear Systems*, Washington State University Technical Report Number: ECE 0094, Pullman, Washington, 1988.

[12] B. M. Chen, *Theory of Loop Transfer Recovery for Multivariable Linear Systems*, Ph.D. Dissertation, Washington State University, 1991.

[13] B. M. Chen, "A simple algorithm for the stable/unstable decomposition of a linear discrete-time system," *International Journal of Control*, Vol. 61, pp. 255-260, 1995.

[14] B. M. Chen, *Linear Systems and Control Toolbox*, Technical Report, Department of Electrical Engineering, National University of Singapore, Singapore, 1997.

[15] B. M. Chen, "Exact computation of infimum for a class of continuous-time H_∞ optimal control problem with a nonzero direct feedthrough term from the disturbance input to the controlled output," *Systems & Control Letters*, Vol. 32, pp. 99-109, 1997.

[16] B. M. Chen, "Solvability conditions for the disturbance decoupling problems with static measurement feedback," *International Journal of Control*, Vol. 68, pp. 51-60, 1997.

[17] B. M. Chen, "On properties of the special coordinate basis of linear systems," *International Journal of Control*, Vol. 71, pp. 981-1003, 1998.

[18] B. M. Chen, *H_∞ Control and Its Applications*, Lecture Notes in Control and Information Sciences, Vol. 235, Springer, London, 1998.

[19] B. M. Chen, "Direct computation of infimum in discrete-time H_∞-optimization using measurement feedback," *Systems & Control Letters*, Vol 35, pp. 269-278, 1998.

[20] B. M. Chen and Y.-L. Chen, "Loop transfer recovery design via new observer based and CSS architecture based controllers," *International Journal of Robust & Nonlinear Control*, Vol. 5, pp. 649-669, 1995.

[21] B. M. Chen, Y. Guo and Z. L. Lin, "Non-iterative computation of infimum in discrete-time H_∞-optimization and solvability conditions for the discrete-time disturbance decoupling problem," *International Journal of Control*, Vol. 65, pp. 433-454, 1996.

[22] B. M. Chen, J. He and Y.-L. Chen, "Explicit solvability conditions for general discrete-time H_∞ almost disturbance decoupling problem with internal stability," *International Journal of Systems Science*, Vol. 30, pp. 105-115, 1999.

[23] B. M. Chen, T. H. Lee, C. C. Hang, Y. Guo and S. Weerasooriya, "An H_∞ almost disturbance decoupling robust controller design for a piezoelectric bimorph actuator with hysteresis," *IEEE Transactions on Control Systems Technology*, Vol 7, pp. 160-174, 1999.

[24] B. M. Chen, Z. Lin and C. C. Hang, "Design for general H_∞ almost disturbance decoupling problem with measurement feedback and internal stability — An eigenstructure assignment approach," *International Journal of Control*, Vol. 71, pp. 653-685, 1998.

[25] B. M. Chen, Z. Lin and K. Liu, "On the problem of robust and perfect tracking of discrete-time systems," Submitted for publication.

[26] B. M. Chen, K. Liu and Z. Lin, "Solvability conditions and solutions to perfect regulation problem under measurement output feedback," Submitted for publication.

[27] B. M. Chen, I. M. Y. Mareels, Y. F. Zheng and C. Zhang, "Solutions to disturbance decoupling problem with constant measurement feedback for linear systems," *Proceedings of the 38th IEEE Conference on Control and Decision*, Phoenix, Arizona, pp. 4062-4067, 1999.

[28] B. M. Chen and A. Saberi, "Non-iterative computation of infimum in H_∞-optimization for plants with invariant zeros on the $j\omega$ axis," *IEE Proceedings–Part D: Control Theory and Applications*, Vol. 140, pp. 298-304, 1993.

[29] B. M. Chen, A. Saberi, S. Bingulac and P. Sannuti, "Loop transfer recovery for non-strictly proper plants," *Control–Theory and Advanced Technology*, Vol. 6, pp. 573–594, 1990.

[30] B. M. Chen, A. Saberi and U. Ly, "Exact computation of the infimum in H_∞-optimization via state feedback," *Control–Theory and Advanced Technology*, Vol. 8, pp. 17-35, 1992.

[31] B. M. Chen, A. Saberi and U. Ly, "Exact computation of the infimum in H_∞-optimization via output feedback," *IEEE Transactions on Automatic Control*, Vol. 37, pp. 70-78, 1992.

[32] B. M. Chen, A. Saberi and U. Ly, "A non-iterative method for computing the infimum in H_∞-optimization," *International Journal of Control*, Vol. 56, pp. 1399-1418, 1992.

[33] B. M. Chen, A. Saberi and U. Ly, "Closed loop transfer recovery with observer based controllers – Part 1: Analysis," *Control and Dynamic Systems*, Vol. 51, Part 2, pp. 247-293, 1992.

[34] B. M. Chen, A. Saberi and U. Ly, "Closed loop transfer recovery with observer based controllers – Part 2: Design," *Control and Dynamic Systems*, Vol. 51, Part 2, pp. 295-348, 1992.

[35] B. M. Chen, A. Saberi and P. Sannuti, "Explicit expressions for cascade factorization of general nonminimum phase systems," *IEEE Transactions on Automatic Control*, Vol. 37, pp. 358-363, 1992.

[36] B. M. Chen, A. Saberi and P. Sannuti, "On blocking zeros and strong stabilizability of linear multivariable systems," *Automatica*, Vol. 28, pp. 1051-1055, 1992.

[37] B. M. Chen, A. Saberi, P. Sannuti and Y. Shamash, "Construction and parameterization of all static and dynamic H_2-optimal state feedback solutions, optimal fixed modes and fixed decoupling zeros," *IEEE Transactions on Automatic Control*, Vol. 38, pp. 248-261, 1993.

[38] B. M. Chen, A. Saberi and Y. Shamash, "A non-recursive method for solving the general discrete time algebraic Riccati equation related to the H_∞ control problem," *International Journal of Robust and Nonlinear Control*, Vol. 4, pp. 503-519, 1994.

[39] B. M. Chen, A. Saberi, Y. Shamash and P. Sannuti, "Construction and parameterization of all static and dynamic H_2-optimal state feedback solutions for discrete time systems," *Automatica*, Vol. 30, pp. 1617-1624, 1994.

[40] B. M. Chen and S. R. Weller, "Mappings of the finite and infinite zero structures and invertibility structures of general linear multivariable systems under the bilinear transformation," *Automatica*, Vol. 34, pp. 111-124, 1998.

[41] B. M. Chen and D. Z. Zheng, "Simultaneous finite and infinite zero assignments of linear systems," *Automatica*, Vol. 31, pp. 643-648, 1995.

[42] C. Commault and J. M. Dion, "Structure at infinity of linear multivariable systems: A geometric approach," *IEEE Transactions on Automatic Control*, Vol. AC-27, pp. 693-696, 1982.

[43] S. T. Crandall, "Perturbation techniques for random vibration of nonlinear systems," *Journal of Acoustical Society of America*, Vol. 35, pp. 1700-1705, 1963.

[44] E. J. Davison and S. G. Chow, "Perfect control in linear time-invariant multivariable systems: The control inequality principle," *Control System Design by Pole-Zero Assignment*, pp. 1-15, Academic Press, New York, 1977.

[45] E. J. Davison and B. M. Scherzinger, "Perfect control of the robust servomechanism problem," *IEEE Transactions on Automatic Control*, Vol. 32, pp. 689-702, 1987.

[46] C. A. Desoer and M. Vidyasagar, *Feedback Systems: Input-Output Properties*, Academic Press, New York, 1975.

[47] J. C. Doyle, *Lecture Notes in Advances in Multivariable Control*, ONR-Honeywell Workshop, 1984

[48] J. C. Doyle and K. Glover, "State-space formulae for all stabilizing controllers that satisfy an H_∞-norm bound and relations to risk sensitivity," *Systems & Control Letters*, Vol. 11, pp. 167-172, 1988.

[49] J. Doyle, K. Glover, P. P. Khargonekar and B. A. Francis, "State space solutions to standard H_2 and H_∞ control problems," *IEEE Transactions on Automatic Control*, Vol. 34, pp. 831-847, 1989.

[50] P. Eykhoff, *System Identification - Parameter and State Estimation*, John Wiley, New York, 1981.

[51] C. Fama and K. Matthews, *Linear Algebra IIH*, Lecture Notes MP274, Department of Mathematics, The University of Queensland, 1991.

[52] L. S. Fan, H. H. Ottesen, T. C. Reiley and R. W. Wood, "Magnetic recording head positioning at very high track densities using a microactuator based, two stage servo system," *IEEE Transaction on Industrial Electronics*, pp. 222-233, 1995.

[53] B. A. Francis, "The optimal linear quadratic time-invariant regulator with cheap control," *IEEE Transactions on Automatic Control*, Vol. 24, pp. 616-621, 1979.

[54] B. A. Francis, *A Course in H_∞ Control Theory*, Lecture Notes in Control and Information Sciences, Vol. 88, Springer, Berlin, 1987.

[55] G. F. Franklin, J. D. Powell and M. L. Workman, *Digital Control of Dynamic Systems*, Addison-Wesley, Reading, Massachusetts, 1990.

[56] S. S. Ge, T. H. Lee and C. J. Harris, *Adaptive Neural Network Control of Robotic Manipulators*, World Scientific, Singapore, 1998.

[57] K. Glover, "All optimal Hankel-norm approximations of linear multivariable systems and their $\mathcal{L}_\infty$ error bounds," *International Journal of Control*, Vol. 39, pp. 1115-1193, 1984.

[58] T. B. Goh, *Development of a Dual Actuator Controller in Hard Disk Drives*, Master of Engineering Thesis, Department of Electrical Engineering, National University of Singapore, 1999.

[59] T. B. Goh, Z. Li, B. M. Chen, T. H. Lee and T. Huang, "Design and implementation of a hard disk drive servo system using robust and perfect tracking approach," *Proceedings of the 38th IEEE Conference on Control and Decision*, Phoenix, Arizona, pp. 5247-5252, 1999.

[60] J. W. Grizzle and M. H. Shor, "Sampling, infinite zeros and decoupling of linear systems," *Automatica*, Vol. 24, pp. 387–396, 1988.

[61] H. Hanselmann and A. Engelke, "LQG-control of a highly resonant disk drive head positioning actuator," *IEEE Transactions on Industrial Electronics*, Vol. 35, pp. 100-104, 1988.

[62] D. Hinrichsen and A. J. Pritchard, "Real and complex stability radii: A survey," *Proceedings of an International Workshop*, Bremen, Germany, pp. 119-162, 1989.

[63] P. A. Iglesias and K. Glover, "State space approach to discrete time H_∞ control," *International Journal of Control*, Vol. 54, pp. 1031-1073, 1991.

[64] T. Kailath, *Linear Systems*, Prentice Hall, Englewood Cliffs, 1980.

[65] P. Khargonekar, I. R. Petersen and M. A. Rotea, "H_∞-optimal control with state feedback," *IEEE Transactions on Automatic Control*, Vol. AC-33, pp. 786-788, 1988.

[66] H. Kimura, "A new approach to the perfect regulation and the bounded peaking in linear multivariable control systems," *IEEE Transactions on Automatic Control*, Vol. 26, pp. 253-270, 1981.

[67] H. Kimura, *Chain-Scattering Approach to H^∞-Control*, Birkhäuser, Boston, Massachusetts, 1997.

[68] V. Kučera, "The discrete Riccati equation of optimal control, *Kybernetika*, Vol. 8, pp. 430-447, 1972.

[69] H. Kwakernaak, "A polynomial approach to minimax frequency domain optimization of multivariable feedback systems," *International Journal of Control*, Vol. 41, pp. 117-156, 1986.

[70] H. Kwakernaak and R. Sivan, "The maximally achievable accuracy of linear optimal regulators and linear optimal filters," *IEEE Transactions on Automatic Control*, Vol. 17, pp. 79-86, 1972.

[71] H. Kwakernaak and R. Sivan, *Linear Optimal Control Systems*, John Wiley, New York, 1972.

[72] P. Lancaster, A. C. M. Ran and L. Rodman, "State space approach to discrete time H_∞ control," *International Journal of Control*, Vol. 44, pp. 777-802, 1986.

[73] A. J. Laub, "A Schur method for solving algebraic Riccati equations," *IEEE Transactions on Automatic Control*, Vol. 24, pp. 913-921, 1979.

[74] D. A. Lawrence and W. J. Rugh, "Almost perfect tracking for linear systems," *Systems & Control Letters*, Vol. 17, pp. 281-289, 1991.

[75] M. W. Lee, *An Investigation in Fuzzy Logic*, Bachelor of Engineering Thesis, Department of Electrical Engineering, National University of Singapore, 1995.

[76] T. H. Lee, E. K. Koh and M. K. Loh, "Stable adaptive control of multivariable servomechanisms, with application to a passive line-of-sight stabilization system," *IEEE Transactions on Industrial Electronics*, Vol. 43, pp. 98-105, 1996.

[77] D. J. N. Limebeer and B. D. O. Anderson, "An interpolation theory approach to H_∞ controller degree bounds," *Linear Algebra and its Applications*, Vol. 98, pp. 347-386, 1988.

[78] D. J. N. Limebeer, M. Green and D. Walker, "Discrete time H_∞ control," *Proceedings of IEEE Conference on Decision and Control*, Tampa, FL, pp. 392-396, 1989.

[79] Z. Lin, *The Implementation of Special Coordinate Basis for Linear Multivariable Systems in MATLAB*, Washington State University Technical Report Number ECE0100, Pullman, Washington, 1989.

[80] Z. Lin, *Global and Semi-global Control Problems for Linear Systems Subject to Input Saturation and Minimum-Phase Input-Output Linearizable Systems,* Ph.D. Dissertation, Washington State University, Pullman, Washington, 1994.

[81] Z. Lin, "Almost disturbance decoupling with global asymptotic stability for nonlinear systems with disturbance affected unstable zero dynamics," *Systems & Control Letters*, Vol. 33, pp. 163-169, 1998.

[82] Z. Lin, *Low Gain Feedback*, Lecture Notes in Control and Information Sciences, Vol. 240, Springer, London, 1999.

[83] Z. Lin, X. Bao and B. M. Chen, "Further results on almost disturbance decoupling with global asymptotic stability for nonlinear systems," *Automatica*, Vol. 35, pp. 709-717, 1999.

[84] Z. Lin and B. M. Chen, "Solutions to general H_∞ almost disturbance decoupling problem with measurement feedback and internal stability for discrete-time systems," *Automatica*, in press.

[85] Z. Lin, B. M. Chen, A. Saberi and Y. Shamash, "Input-output factorization of discrete-time transfer matrices," *IEEE Transactions on Circuits and Systems - I: Fundamental Theory and Applications*, Vol. 43, pp. 941-945, 1996.

[86] Z. Lin, A. Saberi, P. Sannuti and Y. Shamash, "Perfect regulation of linear multivariable systems – A low-and-high-gain design," *Proceedings of the Workshop on Advances on Control and Its Applications, Lecture Notes in Control and Information Sciences*, Vol. 208, pp. 172-193, 1996.

[87] K. Liu, B. M. Chen and Z. Lin, "On the problem of robust and perfect tracking for linear systems with external disturbances," Submitted for publication.

[88] T. S. Low and W. Guo, "Modeling of a three-layer piezoelectric bimorph beam with hysteresis," *Journal of Microelectromechanical Systems*, Vol. 4, pp. 230-237, 1995.

[89] D. G. Luenberger, "Canonical forms for linear multivariable systems," *IEEE Transactions on Automatic Control*, Vol. 12, pp. 290-293, 1967.

[90] R. H. Lyon, "Response of a nonlinear string to random excitation," *Journal of Acoustical Society of America*, Vol. 32, pp. 953-960, 1960.

[91] A. G. J. MacFarlane and N. Karcanias, "Poles and zeros of linear multivariable systems: A survey of the algebraic, geometric and complex variable theory," *International Journal of Control*, Vol. 24, pp. 33-74, 1976.

[92] J. McCormick and R. Horowitz, "A direct adaptive control scheme for disk file servos," *Proceedings of the 1993 American Control Conference*, pp. 346-351, San Francisco, California, 1993.

[93] D. K. Miu and Y. C. Tai, "Silicon micromachined SCALED technology," *IEEE Transaction on Industrial Electronics*, pp. 234-239, 1995.

[94] A. S. Morse, "Structural invariants of linear multivariable systems," *SIAM Journal on Control*, Vol. 11, pp. 446-465, 1973.

[95] P. Moylan, "Stable inversion of linear systems," *IEEE Transactions on Automatic Control*, Vol. 22, pp. 74-78, 1977.

[96] W. K. Ng, *Design Considerations of a Gyro-stabilized Mirror System*, Bachelor of Engineering Thesis, Department of Electrical Engineering, National University of Singapore, 1986.

[97] D. H. Owens, "Invariant zeros of multivariable systems: A geometric analysis," *International Journal of Control*, Vol. 28, pp. 187-198, 1978.

[98] H. K. Ozcetin, A. Saberi and P. Sannuti, "Design for H_∞ almost disturbance decoupling problem with internal stability via state or measurement feedback — Singular perturbation approach," *International Journal of Control*, Vol. 55, pp. 901-944, 1993.

[99] H. K. Ozcetin, A. Saberi and Y. Shamash, "H_∞-almost disturbance decoupling for non-strictly proper systems — A singular perturbation approach," *Control–Theory & Advanced Technology*, Vol. 9, pp. 203-245, 1993.

[100] G. P. Papavassilopoulos and M. G. Safonov, "Robust control design via game theoretic methods," *Proceedings of the 28th Conference on Decision and Control*, Tampa, Florida, pp. 382-387, 1989.

[101] T. Pappas, A. J. Laub and N. R. Sandell, Jr., "On the numerical solution of the discrete-time algebraic Riccati equation," *IEEE Transactions on Automatic Control*, Vol. 25, pp. 631-641, 1980.

[102] J. Perry, *Spinning Tops and Gyroscopic Motion*, Dover Publications, New York, 1957.

[103] I. R. Petersen, "Disturbance attenuation and H_∞-optimization: A design method based on the algebraic Riccati equation," *IEEE Transactions on Automatic Control*, Vol. AC-32, pp. 427-429, 1987.

[104] A. C. Pugh and P. A. Ratcliffe, "On the zeros and poles of a rational matrix," *International Journal of Control*, Vol. 30, pp. 213-227, 1979.

[105] A. C. M. Ran and R. Vreugdenhill, "Existence and comparison theorems for algebraic Riccati equations for continuous- and discrete-time systems," *Linear Algebra and its Applications*, Vol. 99, pp. 63-83, 1988.

[106] T. J. Richardson and R. H. Kwong, "On positive definite solutions to the algebraic Riccati equation," *Systems & Control Letters*, Vol. 7, pp. 99-104, 1986.

[107] H. H. Rosenbrock, *State-space and Multivariable Theory*, John-Wiley, New York, 1970.

[108] A. Saberi, B. M. Chen and Z. L. Lin, "Closed-form solutions to a class of H_∞-optimization problem," *International Journal of Control*, Vol. 60, pp. 41-70, 1994.

[109] A. Saberi, B. M. Chen and P. Sannuti, "Theory of LTR for nonminimum phase systems, recoverable target loops, recovery in a subspace – Part 1: Analysis and Part 2: Design," *International Journal of Control*, Vol. 53, pp. 1067-1160, 1991.

[110] A. Saberi, B. M. Chen and P. Sannuti, *Loop Transfer Recovery: Analysis and Design*, Springer, London, 1993.

[111] A. Saberi and P. Sannuti, "Squaring down of non-strictly proper systems," *International Journal of Control*, Vol. 51, pp. 621-629, 1990.

[112] A. Saberi and P. Sannuti, "Time-scale structure assignment in linear multivariable systems using high-gain feedback," *International Journal of Control*, Vol. 49, pp. 2191-2213, 1989.

[113] A. Saberi and P. Sannuti, "Observer design for loop transfer recovery and for uncertain dynamical systems," *IEEE Transactions on Automatic Control*, Vol. 35, pp. 878-897, 1990.

[114] A. Saberi, P. Sannuti and B. M. Chen, H_2 *Optimal Control*, Prentice Hall, London, 1995.

[115] M. Sampei, T. Mita and M. Nakamichi, "An algebraic approach to H_∞-output feedback control problem," *Systems & Control Letters*, Vol. 14, pp. 13-24, 1990.

[116] P. Sannuti and A. Saberi, "A special coordinate basis of multivariable linear systems – Finite and infinite zero structure, squaring down and decoupling," *International Journal of Control*, Vol. 45, pp. 1655-1704, 1987.

[117] C. Scherer, "H_∞-control by state feedback and fast algorithm for the computation of optimal H_∞ norms," *IEEE Transactions on Automatic Control*, Vol. 35, pp. 1090-1099, 1990.

[118] C. Scherer, "H_∞-control by state-feedback for plants with zeros on the imaginary axis," *SIAM Journal on Control and Optimization*, Vol. 30, pp. 123-142, 1992.

[119] C. Scherer, "H_∞-optimization without assumptions on finite or infinite zeros," *SIAM Journal on Control and Optimization*, Vol. 30, pp. 143-166, 1992.

[120] B. M. Scherzinger and E. J. Davison, "The optimal LQ regulator with cheap control for not strictly proper systems," *Optimal Control Applications & Methods*, Vol. 6, 291-303, 1985.

[121] B. C. Siew, B. M. Chen and T. H. Lee, "Design and implementation of a robust controller for a Free Gyro-stabilized Mirror System," *Journal of Dynamic Systems, Measurement, and Control*, Vol. 121, pp. 550-556, 1999.

[122] L. M. Silverman, "Inversion of multivariable linear systems," *IEEE Transactions on Automatic Control*, Vol. 14, pp. 270-276, 1969.

[123] L. M. Silverman, "Discrete Riccati equations: Alternative algorithms, asymptotic properties, and system theory interpretations," *Control and Dynamical Systems*, Vol. 12, pp. 313-386, 1976.

[124] A. A. Stoorvogel, *The H_∞ Control Problem: A State Space Approach*, Prentice Hall, Englewood Cliffs, 1992.

[125] A. A. Stoorvogel, A. Saberi and B. M. Chen, "The discrete-time H_∞ control problem with measurement feedback," *International Journal of Robust and Nonlinear Control*, Vol. 4, pp. 457-479, 1994.

[126] A. A. Stoorvogel, A. Saberi and B. M. Chen, "A reduced order observer based controller design for H_∞-optimization," *IEEE Transactions on Automatic Control*, Vol. 39, pp. 355-360, 1994.

[127] A. A. Stoorvogel and H. L. Trentelman, "The quadratic matrix inequality in singular H_∞-control with state feedback," *SIAM Journal on Control and Optimization*, Vol. 28, pp. 1190-1208, 1990.

[128] A. A. Stoorvogel and J. W. van der Woude, "The disturbance decoupling problem with measurement feedback and stability for systems with direct feedthrough matrices," *Systems & Control Letters*, Vol. 17, pp. 217-226, 1991.

[129] G. Tadmor, "Worst-case design in the time domain: The maximum principle and the standard H_∞ problem," *Mathematics of Control, Signals and Systems*, Vol. 3, pp. 301-324, 1990.

[130] K. M. Tsuchiura, H. H. Tsukuba, H. O. Toride and T. Takahashi, "Disk system with sub-actuators for fine head displacement," *US Patent No: 5189578*, 1993.

[131] G. Verghese, *Infinite Frequency Behavior in Generalized Dynamical Systems*, Ph.D. Dissertation, Stanford University, 1978.

[132] M. Vidyasagar, *Control System Synthesis: A Factorization Approach*, MIT Press, Cambridge, Massachusetts, 1985.

[133] S. Weerasooriya and D. T. Phan, "Discrete-time LQG/LTR design and modeling of a disk drive actuator tracking servo system," *IEEE Transactions on Industrial Electronics*, Vol. 42, pp. 240-247, 1995.

[134] S. Weiland and J. C. Willems, "Almost disturbance decoupling with internal stability," *IEEE Transactions on Automatic Control*, Vol. 34, pp. 277-286, 1989.

[135] H. Wielandt, "On the eigenvalues of $A + B$ and AB," *Journal of Research of the National Bureau of Standards–B. Mathematical Science*, Vol. 77B, pp. 61-63, 1973.

[136] J. C. Willems, "Almost invariant subspaces: An approach to high gain feedback design — Part I: Almost controlled invariant subspaces," *IEEE Transactions on Automatic Control*, Vol. 26, pp. 235-252, 1981.

[137] J. C. Willems, "Almost invariant subspaces: An approach to high gain feedback design — Part II: Almost conditionally invariant subspaces," *IEEE Transactions on Automatic Control*, Vol. 27, pp. 1071-1085, 1982.

[138] W. M. Wonham, *Linear Multivariable Control: A Geometric Approach*, Springer, New York, 1979.

[139] G. Zames, "Feedback and optimal sensitivity: Model reference transformations, multiplicative seminorms, and approximate inverses," *IEEE Transactions on Automatic Control*, Vol. 26, pp. 301-320, 1981.

[140] K. Zhou, J. Doyle and K. Glover, *Robust & Optimal Control*, Prentice Hall, New York, 1996.

[141] K. Zhou and P. Khargonekar, "An algebraic Riccati equation approach to H_∞-optimization," *Systems & Control Letters*, Vol. 11, pp. 85-91, 1988.

Index

MIX
Papier aus verantwortungsvollen Quellen
Paper from responsible sources
FSC® C105338
FSC
www.fsc.org

If you have any concerns about our products,
you can contact us on
ProductSafety@springernature.com

In case Publisher is established outside the EU,
the EU authorized representative is:
Springer Nature Customer Service Center GmbH
Europaplatz 3, 69115 Heidelberg, Germany

Printed by Libri Plureos GmbH
in Hamburg, Germany